Plant Essential Oils

Bhanu Prakash •
Nawal Kishore Dubey •
Jackline Freitas Brilhante de São José
Editors

Plant Essential Oils

From Traditional to Modern-day Application

Springer

Editors
Bhanu Prakash
Department of Botany
Banaras Hindu University
Varanasi, India

Nawal Kishore Dubey
Department of Botany
Banaras Hindu University
Varanasi, India

Jackline Freitas Brilhante de São José
Department of Integrated Health Education
Federal University of Espírito Santo
Vicotria, Brazil

ISBN 978-981-99-4369-2 ISBN 978-981-99-4370-8 (eBook)
https://doi.org/10.1007/978-981-99-4370-8

This Springer imprint is published by the registered company Springer Nature Singapore Pte Ltd.
The registered company address is: 152 Beach Road, #21-01/04 Gateway East, Singapore 189721, Singapore

Paper in this product is recyclable.

Preface

Plant essential oils and their volatile ingredients possess remarkable antimicrobial, antioxidant, and health benefit effects; hence, they could be used as a preferred alternative to synthetic preservative agents. Since antiquity, essential oils have been used for preservation as well as for the treatment of human illness, especially in biodiversity-rich countries as traditional practices. In general, plant essential oils are a complex mixture of bioactive compounds synthesized in different plant parts such as bark, stem, leaves, flower, and root. These compounds were developed to protect the plant from microbial infection, repel insect pests, and aid in pollination. In the past few decades, plant volatiles have been considered as a golden liquid with a multipurpose role, such as fragrances, flavors, antimicrobial properties, and therapeutic agents. Many of them have been placed in the Generally Recognized As Safe (GRAS) category by the US-FDA. Therefore, food scientists and researchers are inclined towards the elucidation of the preservative and therapeutic role of plant volatiles and to address the existing lacuna for their commercial application as rational preservatives and therapeutic products at low cost that could be used in a sustainable manner.

The chapters in the book provide an overview of the role of plant essential oils as natural preservatives and therapeutic agents. In addition, a brief overview of various information regarding the different sources of essential oils, such as higher plants, herbs, and spices, and the current research and technological advances related to extraction methods, chemical profile characterization, and industrial application in the agri-food sectors have been highlighted. Chapter 1 provides a brief overview of the traditional use of essential oils of medicinal and aromatic plants in India. In addition, it also highlights the biological activities of essential oils and their compounds and potential application in different sectors. Chapter 2 briefly explores the various extraction methods and chemotypic standardization techniques with special reference to phytochemical genomics. Chapters 3 and 4 explore the antioxidant and antibacterial properties of essential oils with detailed molecular mechanisms of action. Chapters 5 and 6 discuss the efficacy of essential oils against fruit spoilage fungi and mycotoxins contamination in food products including carcinogenic aflatoxin B_1. Chapter 7 explores the efficacy of essential oils against major insect pests of stored food commodities such as pulses, cereal grains, and their shelved products. Chapter 8 discusses the pharma-kinetic facet of plant essential oils and their potential uses in aromatherapy. Chapters 9 and 10 discuss the role of microbial consortia, biofertilizer, ozone, and edaphic factor on essential oils biosynthesis and their respective chemical constituents, different storage conditions on stability, chemical constituents, and biological properties of essential oils. Chapters 11 and 12 deal with information on recent molecular-assisted engineering of essential oils-based compounds using biotechnology, combinatorial approaches, and also dose optimization methods using different combinations based on mathematical modeling. Chapter 13 explores the role of bioinformatics and data acquirement tools to enhance the industrial application of phytochemicals in food sciences. Chapter 14 provides a brief account of different nanocarrier agents used as carrier agents for essential oils to protect against the loss of volatility, negative impacts of the intense aroma of essential oils on applicable food, and to enhance stability, targeted delivery, and the potential application in food packaging to extend the shelf-life of packed food commodities.

Varanasi, India — Bhanu Prakash
Varanasi, India — Nawal Kishore Dubey
Vitória, Brazil — Jackline Freitas Brilhante de São José

Acknowledgments

We would like to express our sincere gratitude to all the personnel associated during the project, especially the anonymous reviewers for their valuable suggestion and the authors of chapters for their admirable contributions. Dr. Bhanu Prakash acknowledges the financial support to the laboratory under the Institute of Eminence (IoE; 6031) scheme, Banaras Hindu University, Varanasi, India. Dr. Jackline Freitas Brilhante de São José gratefully acknowledges Federal University of Espírito Santo (Brazil), the Foundation for the Support of Research and Innovation of Espírito Santo, Brazil (Fundação de Amparo a Pesquisa do Espírito Santo-FAPES), and the National Council for Scientific and Technological Development (Conselho Nacional de Desenvolvimento Científico e Tecnológico - CNPq; Brazil) that support the research projects and scholarships. Thanks are also due to members of our families, and research team of our laboratory for their great support. We are greatly indebted to Springer Publisher, and staff members, especially Aakanksha Tyagi and Nandhini for their consistent professional support throughout the process for publication of the book as scheduled.

Contents

Editors and Contributors

About the Editors

Bhanu Prakash, Ph.D., FLS, MNASc is an Assistant Professor at the Department of Botany, Banaras Hindu University (BHU), Varanasi, India. He received his Ph.D. from the Department of Botany, Banaras Hindu University (BHU), Varanasi, India, and later served as Scientist-C at CSIR-Central Food Technological Research Institute, Mysore, Karnataka, India. He has over ten years of research experience in post-harvest management of stored food grains against microbial and insect pest deterioration. He has published 70 research papers, and 15 book chapters, and edited three books for the international publisher Elsevier with above 4850 citations and an h-index of 36 (Google scholar), and 2773 citations and an h-index of 30 by Scopus. He has one national patent granted by Indian Patent Office.

Dr. Prakash is the recipient of several prestigious awards, coveted honors, and fellowships such as Fellow of Linnean Society of London (FLS) and Associateship by National Academy of Agricultural Sciences, New Delhi, India. He received the prestigious NASI-Young Scientist Platinum Jubilee Award (2017) by the National Academy of Sciences, Allahabad, Uttar Pradesh, India; Young Scientist Award (2017) by the Association of Food Scientists & Technologists (Mysore, Karnataka, India); Most Productive Research Awards-2019 and 2020 by the Institute of Science, Banaras Hindu University, Varanasi, India; and Young Scientist and Early Carrier Research Award (ECR) by the Department of Science and Technology (DST), Government of India. Dr. Prakash is also listed among the world's top 2% scientists in the food science category (year 2020, and 2021) reported by Stanford University, USA. He is the reviewer of several nationally and internationally reputed journals. He also served as Guest Editor of the *International Journal of Food Science and Technology* and Associate Editor of *Frontiers in Plant Science* (Section Bioinformatics).

N. K. Dubey received his Ph.D. degree in Botany from Gorakhpur University, Uttar Pradesh, India. He has more than thirty-eight years of research experience in the area of essential oils-based botanical pesticides at the Department of Botany, Banaras Hindu University, Varanasi, India. He has explored the potential application of essential oils of aromatic plants against aflatoxin, storage fungi, and insect pest. He is the author of over 360 publications and 14 books published by reputed national and international journals and publishers. He has four national patents granted by Indian patent office based on the bioactive formulation of plant essential oils. He is also listed among the World's top 2 % list of scientists (Category: Food Science) in 2020, 2021, and 2022 by Stanford University, USA. He is also selected among the top 1% highly cited researcher by Web of Science 2022. He has been conferred with serval prestigious awards such as Fellow, National Academy of Science, Allahabad, India, and the National Academy of Agricultural Sciences, India, Indian Phytopathological Society, and Indian Mycological Society. He also served as an editorial board member and subject expert reviewer of national and international journals.

Jackline Freitas Brilhante de São José is a nutritionist with a master's degree in Agricultural Microbiology, specializing in Food Microbiology, and a Ph.D. in Food Science and Technology, all obtained from the Federal University of Viçosa in Viçosa, Brazil. After completing her Ph.D., she joined the Federal University of Espírito Santo, Vitória, Brazil, where she works as a professor and researcher. She is an adjunct professor in the Department of Integrated Health Education, Health Sciences Center—Federal University of Espírito Santo, Vitória, Espírito Santo, Brazil. Her research is related to food microbiology, food safety, decontamination of fruits and vegetables, ultrasound in food processing and food technology. She has more than ten years of research experience in food microbiology and food science and technology. She organized the book *Microbiology and Food Hygiene: Theory and Practice* published by Rubio Publisher in Brazil, and she has published chapters in national and international books. She has published 68 research papers, 27 book chapters (national and international), and has 1997 citations and an h-index of 19 by Google Scholar and 812 citations and an of h-index 12 by Scopus. She is a reviewer of scientific papers in several national and international journals. Furthermore, she has guided nine graduate students at the master's level and 30 undergraduate students in scientific research.

List of Contributors

Ayda Alavian Sari Agriculture Sciences and Natural Resources University, Mazandaran, Iran

Tarsila Rodrigues Arruda Department of Food Technology, Federal University of Viçosa, Viçosa, Minas Gerais, Brazil

Anindita Barua National Institute for Plant Biotechnology, New Delhi, India

Patrícia Campos Bernardes Department of Food Engineering, Center of Agrarian Sciences and Engineering, Federal University of Espírito Santo, Alegre, Espírito Santo, Brazil

Aleksandra Bogdanovic Faculty of Technology and Metallurgy, University of Belgrade, Belgrade, Serbia

Bruno Ricardo de Castro Leite Júnior Department of Food Technology, Federal University of Viçosa, Viçosa, Minas Gerais, Brazil

Somenath Das Department of Botany, Burdwan Raj College, Purba Bardhaman, West Bengal, India

Manueli Monciozo Domingos Postgraduate Program in Nutrition and Health, Federal University of Espírito Santo, Vitória, Espírito Santo, Brazil

Nilda de Fátima Ferreira Soares Department of Food Technology, Federal University of Viçosa, Viçosa, Minas Gerais, Brazil

Sara Jarske Geringer Department of Integrated Education in Health, Federal University of Espírito Santo, Vitória, Espírito Santo, Brazil

Food Tecnology Laboratory, Federal University of Espírito Santo, Vitória, Espírito Santo, Brazil

Alessandra Peres Guimarães Postgraduate Program in Nutrition and Health, Federal University of Espírito Santo, Vitória, Espírito Santo, Brazil

Atul Kumar Jaiswal Department of Computer Science and Engineering, School of Engineering Sciences and Technology, Jamia Hamdard University, New Delhi, India

Akash Kedia Department of Botany, Govt. General Degree College Mangalkote, Purba Bardhaman, West Bengal, India

Esmail Khosropour Department of Forestry and Forest Economics, University of Tehran, Tehran, Iran

Niraj Kohar Centre of Advanced Study in Botany, Institute of Science, Banaras Hindu University, Varanasi, India

Horesh Kumar Department of Physics, Institute of Science, Banaras Hindu University, Varanasi, India

Rhaiza Marcia Lopes Leal Department of Integrated Education in Health, Federal University of Espírito Santo, Vitória, Espírito Santo, Brazil

Food Tecnology Laboratory, Federal University of Espírito Santo, Vitória, Espírito Santo, Brazil

Bárbara Morandi Lepaus Postgraduate Program in Nutrition and Health, Federal University of Espírito Santo, Vitória, Espírito Santo, Brazil

Clara Suprani Marques Federal Institute of Minas Gerais, Campus Bambuí, Bambuí, Minas Gerais, Brazil

Akash Maurya Department of Botany, Shri Murli Manohar PG College, Ballia, Uttar Pradesh, India

Allan Robledo Fialho e Moraes Agricultural Science Institute, Federal University of Viçosa, Rio Paranaiba, Minas Gerais, Brazil

Taíla Veloso de Oliveira Department of Food Technology, Federal University of Viçosa, Viçosa, Minas Gerais, Brazil

Parshant Laboratory of Bio-resource Technology, Department of Botany, Indira Gandhi National Tribal University, Amarkantak, Madhya Pradesh, India

Patrícia Fontes Pinheiro Department of Chemistry, Federal University of Viçosa, Viçosa, Minas Gerais, Brazil

Bhanu Prakash Centre of Advanced Study in Botany, Institute of Science, Banaras Hindu University, Varanasi, Uttar Pradesh, India

Surya Pratap Department of Physics, Institute of Science, Banaras Hindu University, Varanasi, India

Department of Physics, Harish Chandra Post Graduate College, Varanasi, India

Tanya Singh Raghuvanshi Centre of Advanced Study in Botany, Institute of Science, Banaras Hindu University, Varanasi, India

Jackline Freitas Brilhante de São José Department of Integrated Education in Health, Federal University of Espírito Santo, Vitória, Espírito Santo, Brazil

Ravindra Shukla Laboratory of Bio-resource Technology, Department of Botany, Indira Gandhi National Tribal University, Amarkantak, Madhya Pradesh, India

Bijendra Kumar Singh Department of Botany, Feroze Gandhi College, Raebareli, Uttar Pradesh, India

Prem Pratap Singh Centre of Advanced Study in Botany, Institute of Science, Banaras Hindu University, Varanasi, Uttar Pradesh, India

Ritu Singh National Institute of Plant Genome Research (NIPGR), Aruna Asaf Ali Marg, New Delhi, India

Dejan Skala Faculty of Technology and Metallurgy, University of Belgrade, Belgrade, Serbia

Anjana Tiwari Laboratory of Bio-resource Technology, Department of Botany, Indira Gandhi National Tribal University, Amarkantak, Madhya Pradesh, India

Daniel Sgrancio Uliana Department of Integrated Education in Health, Federal University of Espírito Santo, Vitória, Espírito Santo, Brazil

Food Tecnology Laboratory, Federal University of Espírito Santo, Espírito Santo, Brazil

Bárbara Santos Valiati Postgraduate Program in Nutrition and Health, Federal University of Espírito Santo, Vitória, Espírito Santo, Brazil

Vandana National Institute for Plant Biotechnology, New Delhi, India

Banaras Hindu University, Varanasi, Uttar Pradesh, India

Weria Weisany Department of Agriculture and Food Science, Science and Research Branch, Islamic Azad University, Tehran, Iran

Essential Oils: From Traditional to Modern-Day Applications with Special Reference to Medicinal and Aromatic Plants in India

1

Tanya Singh Raghuvanshi, Prem Pratap Singh, Niraj Kohar, and Bhanu Prakash

Abstract

Essential oils (EOs) are complex mixtures of volatile secondary metabolites extracted from different parts of aromatic plants such as leaves, flowers, fruits, and seeds. They are also referred to as aromatic oils. In general, EOs are cocktail of different low-molecular-weight compounds, such as flavonoids, terpenoids, terpenes, and phenylpropanoids. They represent a diverse class of stereochemistry that results in a range of medicinal properties, viz., antimicrobial, antioxidant, anti-inflammatory, and antiviral effects. Additionally, Eos are biodegradable and hardly affect non-targeted species, which can be beneficial in delaying the development of resistance. EOs have been used for a long time in various areas such as food, medicine, cosmetics, and aromatherapy. However, due to certain limitations associated with them such as high volatility, intense aroma, and chemotypic variation, etc., they were replaced by chemical alternatives that were more efficient and better in terms of bioactivity. However, in view of green consumerism and the increased negative concerns (non-biodegradability and the adverse effects of their by-products on the environment and human health) associated with the indiscriminate use of synthetic chemicals, industries are looking toward green chemicals as a preferred alternative to synthetic ones.

Keywords

Essential oils · Medicinal and aromatic plants · Industrial application in food and pharma · Technological advancement

T. S. Raghuvanshi · P. P. Singh · N. Kohar · B. Prakash (✉)
Centre of Advanced Study in Botany, Institute of Science, Banaras Hindu University, Varanasi, India
e-mail: bprakash@bhu.ac.in

B. Prakash et al. (eds.), *Plant Essential Oils*,
https://doi.org/10.1007/978-981-99-4370-8_1

1.1 Introduction

Since antiquity, medicinal and aromatic plants have been used in traditional medicine systems against a range of microbes due to their inherent biological properties such as antimicrobial, antiviral, antioxidant, and anti-inflammatory. Plants synthesize a range of natural compounds often known as plant metabolites during their life cycle. In general, plant metabolites can be divided into two categories: primary metabolites, which are involved in processes that directly affect growth and development, and secondary metabolites, which carry out a variety of crucial functions, including defense and interactions with the environment (Pott et al. 2019; Kanani et al. 2021; Önder et al. 2022). Synthesis of secondary metabolites shows much more variation at the species, tissue, and cellular levels, as well as at various developmental stages, whereas primary metabolites are generally conserved. Additionally, it is impacted by the availability of nutrients, stress, interactions with microbes, and the environment (Wink 2010). The production of secondary metabolites in plants is however linked to different intermediates of primary metabolites. Since secondary metabolic pathways show interaction with various primary metabolic pathways, it is possible that during plant development and defense, enzymatic activities directed against primary metabolites result in the production of novel compounds that are then gradually transformed into distinctive metabolites (Weng 2014).

Among plant products, essential oils (EOs) produced by aromatic plants is a complex blend of volatile secondary metabolites which are produced in a variety of plant organs to offer defense against external threats such as UV radiation, herbivores, insects, diseases, a variety of stress and also changing environmental conditions (El Asbahani et al. 2015). In general, EOs are cocktail of different low-molecular-weight compounds, such as flavonoids, isoflavones, terpenoids, terpenes, phenylpropanoids, phenolic acids, and aldehydes, which are complex and are often lipophilic and hardly water-soluble (Seow et al. 2014; Donsì and Ferrari 2016). The literature review revealed a plethora of information about the biological activities of EOs such as antibacterial, antioxidant, anti-inflammatory, and antiviral effects (Prakash et al. 2015). Angiosperms plant families such as Asteraceae, Apiaceae, Geraniaceae, Lamiaceae, Lauraceae, Myrtaceae, Pinaceae, Verbenaceae, and Zingiberaceae are the richest source of essential oils (Fig. 1.1).

However, due to the inherent drawbacks of EOs such as intense aroma, impurities, chemotypic variation, high volatility, high reactivity, poor water solubility, and availability of raw materials their industrial application is not well explored so far. The recent science and technological advancement such as extraction methods, boosting the production of EOs with elicitors, separation, characterization, and nanotechnology could address the aforesaid drawback to enhance its industrial application in various fields such as food, agriculture, nutraceuticals, and pharmaceuticals as a considered alternative to the synthetic chemicals.

This chapter deals with the traditional use of essential oils, the drawbacks associated with them, modern applications, and recent advancements in science and technology to address the challenges associated with their industrial application, sustainable uses, and future prospects.

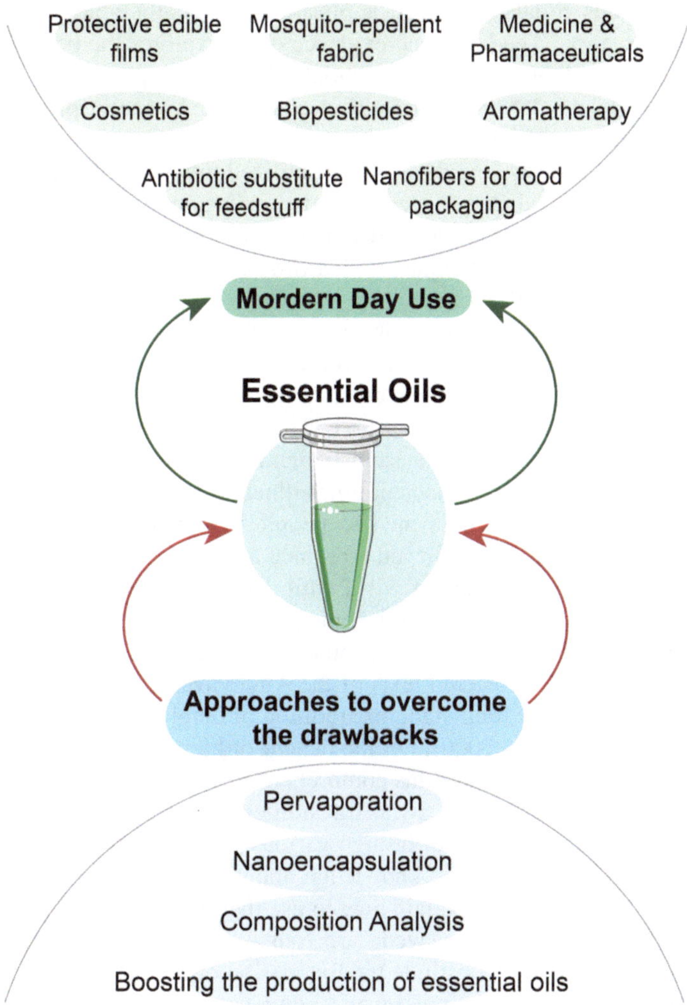

Fig. 1.1 Schematic representation of the industrial application of EOs and approaches to overcome the existing limitations

1.2 Traditional Use of Aromatic Plants and Essential Oils

Plants are one of the most important parts of human life covering a wide area of usage. Although one cannot imagine the existence of human life without plants as they provide us a favorable green environment in which we live, the oxygen that we breathe, and act as the major source of food. With the growing imagination, need, and innovations, from ancient to contemporary human beings have derived a plethora of other ways by which we can benefit from plants and their products. Aromatic plants and their essential oils have been utilized for thousands of years in traditional medicine, religious rituals, cosmetic applications, aromatherapies, food preparation, and preservation. The historical evidence often explores the beneficial effects of herbs, aromatic plants, and their essential oils in treating a variety of ailments (Zollman and Vickers 1999; Giacometti et al. 2018). Aromatic plants and herbs were the first pharmacological substances used to treat illnesses or other aberrant circumstances in antiquity, and they are still utilized today in folk or ethnomedical treatments (Christaki et al. 2012). Herbal medicines are defined by the World Health Organization (WHO) as plant-derived materials or products with therapeutic or other benefits for humans that contain either raw or processed substances from one or more plants (WHO 2001).

Nearly all ancient civilizations show evidence that they were aware of the therapeutic potential of plants. Indigenous peoples used aromatic plants and herbs to treat both physical and mental illnesses in earlier civilizations. The aromatic plant extracts and essential oils have been used from ancient civilization of Egyptian, Chinese, Native American, Tibetan, and Indian Ayurvedic systems as the core of their treatment process. During ancient times illness was thought to have a spiritual origin, caused by evil, and happened to people who have committed some sin (Kankara et al. 2015). It was believed that once a person gets an illness there is a significantly low chance of their survival until and unless their body is capable of healing on its own. During that time the emergence of herbal medicine was contradictory to their belief yet significantly ground-breaking.

The manuscripts of the sacred texts of the Atharvaveda, Rigveda, and Sushruta Samhita, demonstrate the long-standing tradition of employing herbs and spices as medicine in India (Sumner 2000). For the benefit of humanity, it is thought that the universe's creator gave the sages this comprehensive understanding of healing. The knowledge of traditional remedies was transmitted from the sages to the disciples, who then passed it on orally and through various documents to the people. In the form of poetry called "Shlokas," knowledge about the medicinal virtues of herbs was compiled and utilized to explain how to use therapeutic plants. The Yajurveda, Rigveda, Samaveda, and Atharvaveda are four renowned knowledge collections (Vedas) that are thought to constitute the foundation of the Indian traditional medicine system. The Rigveda, the most well known of the four Vedas, contains 1028 shlokas and names 67 medicinally important plants and their uses in the traditional Indian system of medicine. Similarly, a large number of medicinally beneficial plants are described in the Atharvaveda and Yajurveda. Holistic medicine is still used today, not just in India but also in Western countries. Traditional medicine is being used in primary healthcare in the form of herbal medicine in a large part of the world. With recent developments and findings, the use of plant bioactive in modern medicine has also gained a prominent interest.

In Ayurvedic medicine, the fragrant basil plant (*Ocimum* spp.), also known as Tulsi in India and Nepal, is frequently used to treat a variety of respiratory, gastrointestinal, kidney, blood, skin, and other conditions (Singletary 2018). In the manufacture of numerous drug therapies around the world, turmeric, another significant herb that includes the polyphenolic component curcumin, is still used (Henrotin et al. 2013). Peppermint, *Mentha piperita* L., is one of the most well known and frequently utilized species in the world in the culinary, industrial, and medical areas. *M. piperita's* value is mostly attributable to its polyphenols and essential oils (EOs) (Gholamipourfard et al. 2021). Rosmarinic acid and various flavonoids (eriocitrin, luteolin, and hesperidin), menthol, and menthone are the components of the essential oils. Peppermint exhibits notable antibacterial and antiviral properties in vitro conditions, along with substantial antioxidant and anticancer properties with moderate antiallergenic potential. Studies show that certain compounds have a calming effect on gastrointestinal tissue, analgesic and anesthetic effects on the nervous systems. It also has immunomodulating and chemopreventive potential (McKay and Blumberg 2006).

Essential oils isolated from *Cymbopogon citratus* contain bioactives such as citral, genariol, α-oxobisabolene, and myrcene. The leaves have long been consumed as tea or decoction throughout Asia, South America, and Africa. The plant leaves have antispasmodic, analgesic, antipyretic, tranquilizer, anti-hermetic, and diuretic capabilities and also have anti-inflammatory, antiseptic, anti-dyspeptic, and anti-fever actions. It is used in many household products including perfume, regional soaps, cleansers, aroma candles, and other insect repellents. In some parts of Asia and several African nations, it has been utilized as a snake and reptile repellent (Oladeji et al. 2019). Based on the texts of the ancient Indian literature and increasing demand of natural medicinal systems across the globe, Government Of India (GOI) established "The Ministry of Ayush" on November 9 t, 2014, with the goal of revitalizing the rich knowledge of Indian traditional medical systems and ensuring their successful growth and dissemination which was earlier governed by Department of Indian System of Medicine and Homoeopathy (ISM&H), established in 1995. In November 2003, it was renamed to become the Department of Ayurveda, Yoga, and Naturopathy, Unani, Siddha, and Homoeopathy (Ayush), with a concentration on education and research in these fields. Table 1.1 represents the major bioactive compounds of traditionally used aromatic plants/essential oils and their biological activities.

1.3 Modern-Day Use in Different Sectors

With the advancement in science and technology, the challenges associated with the use of essential oils in different fields are being tackled. Nanotechnology is one of the most promising fields of research, especially in pharmaceuticals, cosmetics, and food science. With the increasing indiscriminate application of synthetic chemicals in medicine and agriculture and their harmful effects on health and the environment, there is an increase in demand for sustainable alternatives. Thus, during the past few decades tremendous progress in nanotechnology and the use of green chemicals have been observed and thus interlinking of these areas can be a potential boon to medicine, agriculture, and the environment. Some modern-day applications of essential oils have been discussed further in this chapter.

Table 1.1 Plant essential oils and their biological activities

Sr. No.	Plants	Essential oil (major component)	Sector/Use	References
1.	*Acorus calamus*	β-Asarone	Strong antifungal, antiaflatoxigenic, and antioxidant activity	Shukla et al. (2013)
2.	*Bunium persicum*	Cuminaldehyde, sabinene and γ-terpinene	Antifungal, anti-aflatoxin B1, and radical scavenging activity	Yadav et al. (2020)
3.	*Capsicum annuum*	Capsaicin	Coloring and flavoring of foods. Fruits used as anti-hemorrhoidal, antiseptic, anti-rheumatic. Used against lung cancer, breast cancer, gastric and prostrate human cancer cell lines Flavoring and fragrance agent in cosmetic products	Ralte et al. (2021); Baenas et al. (2019)
4.	*Carum carvi*	Carvone	Antimicrobial, antidiabetic, gastrointestinal problem, CNS activity, and immunomodulatory activity Flavoring foods and beverages Food preservative against *Erwinia, Agrobacteriu, Staphylococcus aureus, Escherichia coli*	Joshi and Soulimani (2020); Borozan et al. (2022)
5.	*Callistemon lanceolatus*	Eucalyptol	Potent antifungal and toxin-inhibitory actions against food-borne *aspergillus flavus* and aflatoxin AFB1	Singh et al. (2020)
6.	*Chrysopogon zizanioides*	Vetivone and Khusimol	Regulatory activity of skin Fragrances in perfumes and soaps Aromatherapy Use to treat notably flu, colic, nausea, and pleurisy	Burger et al. (2017)
7.	*Cinnamomum cassia*	Cinnamaldehyde	Used as preservatives of food against microbes Antimutagenic, antimicrobial, immunomodulatory, antidiabetic, and antitumorigrnic Insect and animal repellant Used in skin whitening	de Almeida et al. (2023); Chang et al. (2013)
8.	*Citrus lemon*	Limonene	Antimicrobial, anti-carcinogenic and anti-cardiovascular Skin whitening, scalp disorder, skin aging and wrinkles Used as food preservatives against bacteria, fungus, and insects As a flavoring agent in foods	Rasool et al. (2022); Bora et al. (2020)

(continued)

Table 1.1 (continued)

Sr. No.	Plants	Essential oil (major component)	Sector/Use	References
9.	*Curcuma longa*	Curcumin	Anti-inflammatory, antiarthritic, anticancer, and antiseptic Reducing inflammation, indigestion, antioxidant Bacteriostatic nature against food-borne bacteria like *mycobacterium smegmatis*	Asif and Mohd (2019); Mishra et al. (2018)
10.	*Cuminum cyminum*	Cymene, γ-terpinene, cuminaldehyde and (−)-β-pinene.	Potent fumigant, repellent, oviposition deterrent, ovicidal, larvicidal, and pupaecidal activity against pests	Kedia et al. (2015)
11.	*Eucalyptus globules*	1,8-cineole (eucalyptol)	Insecticide and insect repellent Boost immunity against measles, flu, cold, and chicken pox Used to treat respiratory related problems Muscle pain, joint pain, and aches Used as a flavoring and fragrance agent in food and cosmetics	Ali et al. (2015); Almas et al. (2021)
12.	*Foeniculum vulgare*	Anethole	Significant antifungal activity against food-borne molds, viz., *aspergillus Niger, aspergillus candidus, aspergillus fumigatus, aspergillus sydowi, aspergillus terreus, fusarium verticillioides, Penicillum italicum, Alternaria alternata, Curvularia lunata, Cladosporium cladosporioides, and aspergillus flavus* Potent antioxidant	Kumar et al. (2020)
13.	*Helichrysum italicum*	Neryl acetate and g-curcumene	Used to treat respiratory tract infection Used to stimulate blood circulation in the skin and to reduce fine lines and wrinkle in the skin	Sarkic and Stappen (2018)
14.	*Lavandula angustifolia*	Linalool and trans-beta-ocimene	Mental disorder, sedative and hypnotic activities, and treatment of insomnia Food additive due to its pleasant flavor and odor Effective against food-borne pathogen *salmonella, Escherichia*	Xu et al. (2023); Erland and Mahmoud (2016)
15.	*Mentha piperita*	Menthol and menthanone	Hair growth and anti-dandruff Fragrance in soaps and cosmetics Flavoring agent in toothpaste and mouthwash	Guzmán and Lucia (2021)

(continued)

Table 1.1 (continued)

Sr. No.	Plants	Essential oil (major component)	Sector/Use	References
16.	*Ocimum basilicum*	Linalool, geranial, and neral	Antimicrobial, antifungal, antioxidant, mosquito repellent, antidiabetic and immunomodulatory Flavoring and fragrance agents in the food and cosmetic industry Insecticidal and larvicidal	Kholiya et al. (2022); Kathirvel Poonkodi (2016)
17.	*Origanum vulgare*	Carvacrol, thymol, and γ-terpinene	*Flavoring and fragrance agents in the food and cosmetic industry* *Antioxidant and biocide in the food and cosmetic industry* *Used as Antiacne, and antiaging* *Used as an antimicrobial, antioxidant, anti-inflammatory, antiproliferative, vasoprotective, and antidiabetic*	Asensio et al. (2015); Bora et al. (2022)
18.	*Oscimum sanctum*	Caryophyllene and eugenol	Flavoring in foods Fragrance in perfumes Bronchial asthma, chronic fever, cold, cough, malaria, dysentery, convulsions, diabetes, diarrhea, arthritis, emetic syndrome skin disease, insect bite, hepatic cardiovascular and immunological disorder, antiallergic, antifungal, and antibacterial Potential preservative for fermented dairy product against *Arthrobacter globiformis*, *Bacillus cereus*, *Escherichia coli*, *pseudomonas* spp.	Iqbal et al. (2020); Anand et al. (2016)
19.	*Rosmarinus officinalis*	Eucalyptol and pinene	Aromatherapy Anti-dandruff and hair growth Fragrance in soaps Treatment of androgenic alopecia	Sarkic and Stappen (2018)
20.	*Salvia sclarea*	Linalool and linalyl acetate	Regulate menstrual periods Used for womb and uterus-associated problems Ease tension and muscle cramps Used to remove acne and wrinkle and to control sebum production	Ali et al. (2015)
21.	*Syzygium aromaticum*	Eugenol	Insecticide Antitumor, antiviral, antinociceptive, and anticancer Food preservative against molds and bacteria (*aspergillus* spp., *Escherichia coli*)	Isman (2020); Haro-González et al. (2021)

(continued)

Table 1.1 (continued)

Sr. No.	Plants	Essential oil (major component)	Sector/Use	References
22.	*Thymus vulgaris*	Thymol	Used as preservatives of food against microbes Used for skin issues like oily skin, sciatica, acne, dermatitis, and bug bites Used in intestinal treatment against hookworms, ascarids, and gram-positive and negative bacteria Improve liver functioning and act as an appetite stimulant	de Almeida et al. (2023); Prasanth Reddy et al. (2014)
23.	*Zingiber officinale*	Zingerone, gingerols, and zingiberebe Verbenol	Antidiabetic, antiulcer, anti-gastrointestinal, antimicrobial Wrinkles, dark spots on the skin, aging Anti-dermatophyte against *Trichophyton rubrum* and *Microsporum gypseum* Antifungal efficacy against *aspergillus flavus* and potent inhibitor of aflatoxin AFB_1 production	Rasool et al. (2022); Mahboubi (2019); Singh et al. (2021)

1.3.1 Protective Edible Films

Edible films are being used for a long time in one way or another. Wax coatings were first used for fruits and vegetables, and now active edible films with selectivity for mass transfer, light barrier, and antibacterial activity are now available. Over the past few years, the market for edible films and coatings has grown remarkably (Gennadios and Weller 1990). This increase is being accompanied by an increase in knowledge on edible films and coatings that have been collected through research, product development effort, advancements in material science, and processing technologies. To increase food safety and shelf life, more focus has recently been placed on creating and testing films with antimicrobial capabilities. Essential oils exhibit strong antibacterial activity against a variety of strains of Gram-positive and Gram-negative bacteria as well as a wide range of fungi due to their high amounts of plant bio-actives (Elsabee et al. 2016). They not only increase the shelf life but also protect from microbial growth and accumulation of their by-product which can be dangerous for human health. Active biomolecules like chitosan, and its derivatives, which are biodegradable, biocompatible, nontoxic, and antibacterial agents with good film characteristics, play a vital role in the food packaging industry. With all of this, it is being widely used in the food packaging sector (Muzzarelli et al. 2000).

1.3.2 EOs-Loaded Nanofibers for Food Packaging

Traditional food packaging consists of items constructed from a wide range of materials and is designed to shield food from harmful external influences on quality while it is being delivered to the consumer. However, throughout the transition from packing to consumption, microbial and oxidative deterioration can take place. Additionally, there is a demand for food with a longer shelf life due to the rise in food retail industry. One of the greatest options that go beyond simply shielding meals from external influences is active food packaging, which has been heavily emphasized in recent years. These are one of the best options to increase the protection and palatability of food for a longer duration without the addition of any harmful chemicals. Typically, active packages are created by encapsulating a polymer with a plant-based bioactive ingredient. Essential oils stand out among the active ingredients due to their antibacterial and antioxidant characteristics (Raut and Karuppayil 2014). By using a variety of techniques, including molding, extrusion, compression, and thermal forming, bioactive-added natural or synthetic polymer matrices can be processed into thin films that can be employed as food packaging (Chen et al. 2021; Wu et al. 2019).

1.3.3 Mosquito-Repellent Fabric

Mosquitoes pose the great threat to public health because they can spread a wide variety of diseases. With the increasing risk of dengue and malaria and their high fatality, especially during monsoon, there is a need for management. Effective insect repellents have certain advantages for public health (Bernhard et al. 2003). The textile-based strategy is thought to be significant among all mosquito prevention techniques because after food and shelter, textiles are considered to be the third most important component for human survival. Humans use a variety of chemical and fumigation methods for protection against mosquitoes which can be harmful to their health. On the other hand, the number of these issues can be solved by using mosquito-repellent textiles which will not have the risk associated with them. By offering the essential characteristics of repelling mosquitoes, particularly in tropical locations, mosquito-repellent textiles contribute to the advancement of the textile industry. They serve as a textile barrier that prevents adult mosquito bites and deters insects from flying nearby. Fabrics used in homes, tents, curtains, uniforms for the military, and other items can all be made to repel mosquitoes. The repellent qualities of essential oils can be used as fumigants in the vapor phase to repel the mosquitoes. Researchers have employed the microencapsulation approach to retain essential oils and achieve controlled release in order to stabilize and control the volatility of the active substances (Muhoza et al. 2022; Tariq et al. 2022).

1.3.4 Cosmetics

Cosmetics include a wide variety of items for both internal and external use. They are typically used to improve the appearance of the body part and skin to which they are applied or to give a pleasant scent. The use of cosmetic items has greatly increased for the following reasons: increase in people's awareness of beauty; to prevent skin aging; to remain healthy; and to increase life expectancy. The market has been dominated by more sophisticated products (serums, toners, moisturizers, etc.) with properties like antiaging, anti-wrinkle, skin brightening, skin whitening, anti-acne, anti-blemish, etc., as opposed to simple soaps and creams. However, most cosmetics contain harsh chemicals that, over time, are bad for the skin and health, thus to reduce toxicity, the industries are trying to include natural or herbal ingredients in cosmetic formulations. As a result, essential oils and other herbal items with strong antioxidant effects are extensively utilized today because they have many useful features for humans, including as antioxidants, anti-hairfall, anti-dandruff, antibiotics, anti-inflammatory agents, wound-healing aids, anti-acne, and many more (Jennings and Parks 2020). Tea tree essential oil is one of the prominent anti-acne active used these days; you can find a variety of moisturizers that claim to fight acne have tea tree EO as one of the major ingredients. Peppermint EO is one of the prominent solutions against dandruff and is widely used in anti-dandruff shampoo. Rosemary EOs have significant activity to prevent hair loss and premature graying of hair. Limonene having antioxidant activity is used in many cosmetics for skin aging prevention. Limonene oil and olive oil have been utilized for the scalp disorder (Kishimoto and Kashiwagi 2022). An abundance of antioxidant chemicals including zingiberene, cinnamaldehyde, and limonene can be found in the essential oils derived from cinnamon, ginger, and lemon leaves. These substances have the capacity to shield cells and tissue from UV radiation. As a result, lotions or ointments containing these essential oils might make the claim that they shield skin from UV rays and prevent oxidation (Bertuzzi et al. 2013; Wei and Shibamoto 2007). The term "nano-cosmeceuticals" refers to a cosmetic formulation that uses nanotechnology as a means of delivering bioactive ingredients to the skin (Hougeir and Kircik 2012; Kaul et al. 2018). In order to create anti-wrinkle, moisturizing, and skin-whitening products, and hair-repairing shampoos, conditioners, and serums, nanomaterials are frequently used. This method makes it possible to create small nanoparticles of cosmetic components that are readily absorbed by the skin, easily repair the damage, and increase product efficacies.

1.3.5 Biopesticides

Biopesticides are natural, biologically occurring compounds that are used to control various agricultural pests infesting plants in forests, gardens, farmlands, etc. (Kumar et al. 2021). The world needs to adopt new and improved agricultural methods and techniques for high sustainability and production if it is to survive the challenges offered by population growth and climate change. A variety of strategies can be employed to increase agricultural productivity, such as increasing crop output by using sufficient organic pesticides and manures, or reducing yield losses brought on by harsh environmental circumstances like abiotic and biotic stresses (Walker 1983; Gharde et al. 2018).

Essential oils and plant-based extracts have been used extensively to manage insect pests and demonstrate a variety of insecticidal actions. Due to the complexity of their chemical makeup, it is highly challenging for pests to develop resistance to such biopesticides. These can serve as insect repellents, oviposition deterrence, and antifeedants depending on how they work. They may also interfere with the respiration process of the pests or make it difficult to identify the host plant. By having ovicidal and larvicidal actions, these pesticides can prevent oviposition and reduce adult emergence. Even commercial versions of some insecticides based on essential oils are available such as *Cupressus sempervirens* EO that are adulticidal and exhibit fumigant toxicity against *Aedes aegypt* and *Thrips palmi. Hyptis spicigera* shows fumigant toxicity, and is repulsive and insecticidal against *Callosobruchus maculatus*. *Cymbopogon nardus* shows knockdown and mortality against *Musca domestica* (Khursheed et al. 2022a, b).

1.3.6 Antibiotic Substitute for Feedstuff

In the past century, as the world's population has continued to increase rapidly, the need for meat and meat products has expedited the transition from family-owned and downsized feeding practices to industrialized and intensive animal feeding of livestock (Singh et al. 2019). As a result, the poultry and cattle sectors now place a high premium on issues related to epidemic prevention, animal health, and nutritional intake (Zhang et al. 2020). Due to their strong antibacterial and anti-inflammatory properties, antibiotics have frequently been the first choice as an animal-feeding supplement in animal husbandry in order to prevent and treat infectious diseases that could potentially be induced by high population density feeding (Han et al. 2020; Rizal et al. 2018). Essential oils (EOs) generated from plants are depicted as promising antibiotic replacements utilized in animal feeds, as well as their present and projected uses in the poultry and cattle industries. Numerous studies have shown that the phytochemicals included in EOs have many functionalities and are less prone to cause bacterial resistance. On intestinal inflammation, intestinal flora, immunity, digestion, and growth performances, the positive benefits of EOs feed supplementation have already been thoroughly studied (Pan et al. 2022).

1.3.7 Aromatherapy

Aromatherapy, which refers to the use of essential oils in a variety of methods, includes oil massage, oral delivery, and inhalation through the nose (Johnson et al. 2017). EOs, such as lavender oil, rose oil, tea tree oil, rosemary oil, and peppermint oil, are frequently utilized for their aromatic fragrance. It has been found significant that while utilizing products containing EOs, these oils have particular effects on the central nervous system (CNS), such as enhancing intellect, influencing sleep, enhancing cognition and memory, acting as an anti-anxiety and anti-depressant, sedating, etc. (Soares et al. 2022). The use of essential oils in aromatherapy has an effect on calming the nervous system and thus can serve as a potential co-treatment along with medicine and therapy in mental health-related diseases (Liang et al. 2022).

1.3.8 Medicine and Pharmaceuticals

Plant bioactive, especially essential oils are widely used in medicines and pharmaceuticals due to their inherent biological properties such as antimicrobial, anti-inflammatory, antioxidant, antidiabetic, antirheumatic, antitumor, and anticancer.

A significant number of antioxidants are found in the essential oil-based blended mixture. For example, the essential oil-based mixtures of ginger, cinnamon, and lemon leaves contain significant amounts of cinnamaldehyde, eugenols, zingiberene, and limonene, which are well-known for their antioxidant properties. Many industries produce functional foods and nutraceuticals based on aromatic plants and essential oils for therapeutic purposes. They also enhance blood circulation, lower the level of harmful LDL cholesterol, and lower the risk of heart disease. These oils are also excellent at lowering blood pressure, which lowers the risk of cardiovascular disease (Rasool et al. 2022). EOs can control irritable bowel syndrome by suppressing pro-inflammatory cytokine expression and increasing anti-inflammatory cytokine levels. Zingiberene and cinnamaldehyde are plant antioxidants that can repair oxidation-related cell damage. These substances restore cell damage, prevent additional cellular destruction, act as a soother, and lessen the severity of migraine headaches (Chhikara et al. 2018; Helli et al. 2021; Rasool et al. 2022). Some important plant-based compounds and their mechanism of action are mentioned in Table 1.2.

1.4 Challenges Associated with the Use of Essential Oils

Although EOs have a number of uses and benefits, they also have a few drawbacks which limit their industrial application in agri-food and pharma industries. Some of the major drawbacks associated with EOs are their intense aroma, presence of impurities, complex mixture of various metabolites, high volatility, high reactivity, poor water solubility, and economic use. Sometimes a property that can serve as a boon to one field can be a bane to another. Taking the example of its intense aroma, if we consider the perfumery and aroma industries, this property is one of the reasons for its use while if we consider the food industry, intense aroma can interfere with the organoleptic properties of food. Therefore, the area in which the EOs are being used determines the potential for which we are using them. Apart from the mentioned drawbacks the separation of the complex mixture was also one of the major challenges associated with their use, especially in pharmaceuticals. To

Table 1.2 Plant-based compounds and their use in pharmaceuticals

S.N.	Plant	Compounds	Uses	Mechanism	Reference
1	*Acacia arabica*	Arabin	Diabetes	Initiates release of insulin	Mishra et al. (2018)
2	*Allium sativum*	Allicin	Diabetes	Prevents the inactivation of insulin in the liver	Sudhakar et al. (2021)
3	*Anisakis simplex*	Geraniol	Colon cancer	It induces membrane depolarization and interferes with ionic canals and signaling pathways	Bakkali et al. (2008)
4	*Artemisia annua*	Artemisin	Malaria	Inhibits the sarco-endoplasmic reticulum Ca^{2+} ATPase (SERCA)	
5	*Atractylodes lancea*	Beta-eudesmol	Angiogenesis	Blocks the ERK signaling pathway	Bakkali et al. (2008)
6	*Berberis vugaris*	Beberine	Diabetes	Reduces insulin resistance and improves insulin sensitivity which activates AMpk-mediated GLUT-4 gene expression, which increases glucose uptake	Sudhakar et al. (2021)
7	*Cannabis sativa*	Phytocannabinoids	Cardiovascular diseases	Modulation of autonomic outflow in both the central and peripheral nervous systems as well as direct effects on the myocardium and vasculature	Dimmito et al. (2021)
8	*Carum carvi*	Carvone	Diabetes	Inhibition of hepatic glucose production and/or stimulation of glucose utilization by peripheral tissues	Agrahari and Singh (2014)
9	*Catharanthus roseus*	Catharanthine, vinblastine, and vincristine	Cancer	Arrests dividing cells in metaphase by binding tubulin and preventing its polymerization into microtubules	Chandran et al. (2020)
10	*Cephalotaxus harrintonia*	Homoharringtonine	Cancer	It inhibits chain elongation during translation by suppressing the substrate binding to the receptor site on the 60S ribosome unit	Efferth et al. 2007

S.N.	Plant	Compounds	Uses	Mechanism	Reference
11	*Coriandrum sativum*	Linalool	Hypoglycemic	Increases pancreatic beta-cell activity and increases the release of insulin	Rahman et al. (2022)
12	*Croton lecheri*	Crofelemer (oligomeric proanthocyanidin)	Diarrhea	It blocks chloride ion secretion via cystic fibrosis transmembrane conductance regulator channel, normalizes water flow in the gut	Saklani and Kutty (2008)
13	*Curcuma longa*	Curcumin	Colon cancer	It modifies the expression of DNA methyltransferases and histone deacetylases, thereby inhibiting the growth of HT-29 colon cancer cells	Esmeeta et al. (2022)
14	*Dendrobium officinale*	Gigantol	Antibiotic and anticancer	Inhibits the growth of breast cancer cells and enhances the down regulation of p13k/Akt/mTOR signaling via causing DDP-induced apoptosis	Shukla et al. (2022)
15	*Emblica ribes*	Embelin	Anticancer	Binds to the BIR3 domain of XIAP, preventing the association of XIAP and caspase-9 resulting in the suppression of cell growth, proliferation, and migration of various types of cancer cells	Sharma et al. (2022)
16	*Ephemeranthalon chophylla*	Denbinobin	Breast cancer, spleen enlargement, tumor metastasis	Inhibiting the Src-mediated signaling pathways, it stimulates the tumor suppressor gene, upregulates various downstream effectors, and causes DNA damage, responsible for apoptosis	Shukla et al. (2022)
17	*Erythroxylum coca*	Cocaine	Analgesic	It binds to the dopamine transporter, blocking the removal of dopamine from the synapse	Anand et al. (2019)
18	*Eucalyptus*	1,8-cineole	Herpes simplex virus	Direct inactivation of the virus by binding with the viral proteins involved in the adsorption and penetration of host cell	Mieres-Castro et al. (2021)

(continued)

Table 1.2 (continued)

S.N.	Plant	Compounds	Uses	Mechanism	Reference
19	*Gardenia jasminoides*	Genipin	Headache, inflammation, fever, and hepatic disorder	Inhibits the expression of inducible nitric oxide synthase (iNOS) and Nitric oxide (NO) production in lipopolysaccharide	Salminen et al. (2008)
20	*Glycyrrhiza uralensis*	Glycyrrhizic acid	Hepatoprotective	It reduces inflammation and apoptosis by suppressing THF-alpha and caspase-3	Foghis et al. (2023)
21	*Origanum vulgare*	Carvacrol and thymol	Cancer	It induces cell cycle arrest or apoptosis	Nieto (2017)
22	*Otanthus maritimus*	Sasamine	Toothache, asthmatic bronchitis and urinary bladder inflammation	Inhibits the NO production stimulated by LPS in macrophages	Dimmito et al. (2021)
23	*Papever somniferum*	Morphine and codeine	Analgesic	It binds to the opioid receptors in the brain, thereby blocking pain signals from the nervous system	Anand et al. (2019)
24	*Polygonum hydropiper*	Caryophylene	Alzheimer and dementia	It inhibits acetylcholinesterase (AChE) and butyrylcholinesterase (BChE) enzymes involved in the degradation of the essential neurotransmitter acetylcholine	Ayaz et al. (2017)
25	*Schisandra chinensis*	Schisandrin, bifendate, and bicyclol	Hepatoprotective and anti-hepatitis B virus	It prohibited hepatitis B virus replication in chronic hepatitis B patients	Yuan et al. (2016)
26	*Swietenia macrophylla*	Limonoids	Dengue virus	It inhibits the growth of viruses by inhibiting their activity	Daud et al. (2022)
27	*Syzygium claviflorum*	Bevirimat {3-O-(3′,3′-dimethyl succinyl) betulinic acid}	Antiretroviral	Action, blocking a late step in the processing of the group-specific antigen (gag) protein and resulting virus particles are structurally defective and stop spreading infection	Saklani and Kutty (2008)
28	*Taxus brevifolia*	Taxol	Cancer	It stabilizes and prevents microtubule depolymerization leading to cell cycle arrest at the G2/M phase and cell death	Yuan et al. (2016)

S.N.	Plant	Compounds	Uses	Mechanism	Reference
29	*Thymus vulgaris*	Thymol	Leishmaniasis	Due to its lipophilic nature, it can easily be incorporated into cell membranes and disrupt the phospholipid bilayer	Kowalczyk et al. (2020)
30	*Zingiber officale*	α-Zingiberene	Cancer	It causes nucleosomal DNA fragmentation, increases the percentage of sub-diploid cells, apoptosis, activates the caspases in SiHa cells	Mahboubi (2019)

prevent contamination of the extract, the decomposition of significant metabolites, or the formation of compounds by the interaction of metabolites as a result of extraction conditions or solvent impurities, it is crucial to minimize interference from compounds that may coextract with the target compounds during the extraction of plant materials.

1.5 Approaches to Overcome the Drawbacks

With the recent advances in science and technology, considerable progress has been made to explore the industrial application of plant bioactive in modern days. Thus, addressing the inherent challenges associated with essential oils could provide a sustainable alternative to the synthetic chemicals that often impose negative effects on human health and the environment. Some approaches to overcome the drawbacks associated with the modern use of EOs are enlisted below.

1.5.1 Boosting the Production of Essential Oils

With the increasing use of essential oils, there is a parallelly increasing demand for aromatic plants which however is not much economically feasible. Large biomass of plants is required for the production of comparatively lesser amounts of essential oils and thus techniques to enhance the content of essential oils in the plants can be a beneficial advancement in this field.

Numerous biotic and abiotic factors largely affect the formation of EOs in plants, which in turn affect the yield and bioactive components of the oils. Different physiological, ecological, and defense uses are made of the EOs. The microbes have the ability to release chemicals known as elicitors that interact with plant cell receptors. According to the research, the elicitors and their actions act as signals to trigger different biological reactions in plants, such as the formation of secondary metabolites and thus essential oils (Chamkhi et al. 2021). These elicitors can be a microbe, disarmed pathogen, nutritional aberration, or environmental stress. These techniques can be used to alter the production of a secondary metabolite which can vary in both quantity and composition.

1.5.2 Composition Analysis

The most popular method for separating and chemically characterizing EOs is gas chromatography electron impact ionization mass spectrometry (GC-EI-MS) (Rubiolo et al. 2010). In order to help with metabolite identification, GC-EI-MS generates representative and repeatable chromatograms of volatile metabolites along with the accompanying mass spectra for each separated component peak. EOs are intricate samples that could have up to several hundred different components in them. For instance, using a thorough two-dimensional gas chromatography-mass spectrometry technology (GC GC-MS), the composition of vetiver EO has been found to contain almost 500 chemicals (Cordero et al. 2015). With the estimation and knowledge of the composition, one can proceed to purify or separate the individual components and thus can use it in a different field, especially in pharmaceuticals.

1.5.3 Pervaporation

The pervaporation technique, which is still relatively new, offers considerable promise for the purification of combinations that produce azeotropes or thermolabile chemicals, which break down at high temperatures. The key element of this process is the presence of a particular membrane that is selective to one or more components and leads to the formation of a permeate where the concentration of the substance that is more related to the membrane occurs (Cheng et al. 2017; Ong et al. 2016).

In case of separation of essential oil via pervaporation it is important to pay attention to the relative polarities of the constituents because those that are more non-polar (like hydrocarbon monoterpenes and sesquiterpenes) tend to permeate organophilic membranes more readily, whereas those that are more polar (like oxygenated monoterpenes and sesquiterpenes) are absorbed to a lesser extent.

The membrane is an essential part of the pervaporation process, and together with other characteristics, membrane-substance selectivity is one of the most important characteristics of pervaporation process. Although there are "classes" of pervaporation membranes, such as organophilic, hydrophilic, and separation between organics, the current challenge is focused on membrane development for specific separations, whose separation by classical methods (distillation, liquid-liquid separation, crystallization, etc.) is difficult (Jyoti et al. 2015; Van der Bruggen and Luis 2015).

1.5.4 Nanoencapsulation

EOs can be used in various fields including agri-food, medicine, and cosmetics in a variety of ways that can include either directly or indirectly. However, despite the excellent efficiency of essential oils, most of them possess poor physical properties such as low water solubility and low stability toward environmental factors including heat, moisture, and oxygen (Lertsutthiwong et al. 2009). Some of them possess intense aromas which can interfere with the properties of the system in which they are being used. To tackle the problems, protect the active component, and enhance the stability of the EOs the technique of nanoencapsulation serves the best (Ibrahim et al. 2022).

Encapsulation is a technology that allows for the safe and regulated release of solids, liquids, and gases contained in small capsules. This method guards against oxidation throughout the manufacturing and storage processes, preventing the development of undesirable aromas and odors as well as the loss of nutritional and metabolic qualities (Sotelo-Boyás et al. 2017). Nanoencapsulation can be used to extend the shelf life of EOs and result in controlled release delivery systems that can be used for a longer duration of time (Vahedikia et al. 2019). Several nanoencapsulation techniques have been identified, which include the use of various nanoencapsulation materials such as Alginate (AL), chitosan (CS), Zea protein-based, alfalfa protein-based nanoparticles which are naturally occurring polymers, and are frequently employed in the biomedical and pharmaceutical industries in the form of nanoparticles, capsules, and emulsions. The boosting effects on antimicrobial efficacy of essential oils of aromatic plants have been observed in the past few decades (Prakash et al. 2018; Natrajan et al. 2015; Assadpour and Mahdi Jafari 2019).

1.6 Conclusion and Future Prospects

Advancements in science and technology have explicitly made human life easier and technically sound but as it is said there are two sides to every story. The boon of science and technology is also accompanied by the bane. With the indiscriminate use of synthetic chemicals in pharma and Agri-food sector, several complications associated with them have been witnessed in the past few decades. These can be linked to their harmful impacts on the environment, biodiversity, and human beings, thus there is an increasing demand to address these issues with sustainable alternatives. The use of renewable resources in a sustainable manner can play an important role in the management of these alarming issues. Considering the above situation, use of plant-based chemicals has gained significant interest in the recent past because of their natural, safe, eco-friendly, cost-effective, renewable, and biodegradable nature. Essential oils are one of them. Essential oils are being used since antiquity in aromatherapies, food, agriculture, and traditional medicine due to their inherent biological properties such as antimicrobial, antiviral, antioxidant, and anti-inflammatory but was slowly replaced by synthetic chemical alternatives due to some of their drawbacks like intense aroma, hydrophobic nature, dearth of availability, etc. In the past few decades, the concept of green economy has emerged as one of the best ways to tackle the problems associated with synthetic chemicals and thus the increased demands for plant-based chemicals have been witnessed. With the advancements in science and technology, the drawbacks associated with the essential oils-based green chemicals are currently being rectified which could enhance their application in agri-food and pharma industries.

References

Agrahari P, Singh DK (2014) A review on the pharmacological aspects of Carum carvi. J Biol Earth Sci 4(1):1–13

Ali B, Al-Wabel NA, Shams S, Ahamad A, Khan SA, Anwar F (2015) Essential oils used in aromatherapy: a systemic review. Asian Pac J Trop Biomed 5(8):601–611

Almas I, Innocent E, Machumi F, Kisinza W (2021) Chemical composition of essential oils from Eucalyptus globulus and Eucalyptus maculata grown in Tanzania. Scientific African 12:e00758

Anand S, Grover CR, Beniwal A (2016) Evaluation of Ocimum sanctum essential oil as potential preservative for fermented dairy products. J Pure Appl Microbiol 10(4):2763–2771

Anand U, Jacobo-Herrera N, Altemimi A, Lakhssassi N (2019) A comprehensive review on medicinal plants as antimicrobial therapeutics: potential avenues of biocompatible drug discovery. Meta 9(11):258

Asensio CM, Grosso NR, Juliani HR (2015) Quality characters, chemical composition and biological activities of oregano (Origanum spp.) essential oils from central and southern Argentina. Ind Crop Prod 63:203–213

Asif M, Mohd I (2019) Prospects of medicinal plants derived nutraceuticals: a re-emerging new era of medicine and health aid. Prog Chem Biochem Res 2(4):150–169

Assadpour E, Mahdi Jafari S (2019) A systematic review on nanoencapsulation of food bioactive ingredients and nutraceuticals by various nanocarriers. Crit Rev Food Sci Nutr 59(19):3129–3151

Ayaz M, Sadiq A, Junaid M, Ullah F, Subhan F, Ahmed J (2017) Neuroprotective and anti-aging potentials of essential oils from aromatic and medicinal plants. Front Aging Neurosci 9:168

Baenas N, Belović M, Ilic N, Moreno DA, García-Viguera C (2019) Industrial use of pepper (Capsicum annum L.) derived products: technological benefits and biological advantages. Food Chem 274:872–885

Bakkali F, Averbeck S, Averbeck D, Idaomar M (2008) Biological effects of essential oils–a review. Food Chem Toxicol 46(2):446–475

Bernhard L, Bernhard P, Magnussen P (2003) Management of Patients with Lymphoedema Caused by Filariasis in North-eastern Tanzania: Alternative approaches. Physiotherapy 89(12):743–749

Bertuzzi G, Tirillini B, Angelini P, Venanzoni R (2013) Antioxidative action of Citrus limonum essential oil on skin. Eur J Med Plants 3:1–9

Bora H, Kamle M, Mahato DK, Tiwari P, Kumar P (2020) Citrus essential oils (CEOs) and their applications in food: an overview. Plan Theory 9(3):357

Bora L, Avram S, Pavel IZ, Muntean D, Liga S, Buda V et al (2022) An up-to-date review regarding cutaneous benefits of Origanum vulgare L. essential oil. Antibiotics 11(5):549

Borozan AB, Popescu S, Bordean D, Dobrei A (2022) Carum carvi–ASPECIES of food interest with antimicrobial properties. A short review. Curr Trend Nat Sci 11(21):315–321

Burger P, Landreau A, Watson M, Janci L, Cassisa V, Kempf M et al (2017) Vetiver essential oil in cosmetics: what is new? Medicines 4(2):41

Chamkhi I, Benali T, Aanniz T, El Menyiy N, Guaouguaou FE, El Omari N et al (2021) Plant-microbial interaction: the mechanism and the application of microbial elicitor induced secondary metabolites biosynthesis in medicinal plants. Plant Physiol Biochem 167:269–295

Chandran H, Meena M, Barupal T, Sharma K (2020) Plant tissue culture as a perpetual source for production of industrially important bioactive compounds. Biotechnol Rep 26:e00450

Chang CT, Chang WL, Hsu JC, Shih Y, Chou ST (2013) Chemical composition and tyrosinase inhibitory activity of Cinnamomum cassia essential oil. Bot Stud 54(1):1–7

Chen HZ, Zhang M, Rao Z (2021) Development of cellulose nanofibrils reinforced polyvinyl alcohol films incorporated with alizarin for intelligent food packaging. Int J Food Sci Technol 56(9):4248–4257

Cheng X, Pan F, Wang M, Li W, Song Y, Liu G et al (2017) Hybrid membranes for pervaporation separations. J Membr Sci 541:329–346

Chhikara N, Kour R, Jaglan S, Gupta P, Gat Y, Panghal A (2018) Citrus medica: nutritional, phytochemical composition and health benefits–a review. Food Funct 9(4):1978–1992

Christaki E, Bonos E, Giannenas I, Florou-Paneri P (2012) Aromatic plants as a source of bioactive compounds. Agriculture 2(3):228–243

Cordero C, Rubiolo P, Cobelli L, Stani G, Miliazza A, Giardina M, Bicchi C et al (2015) Potential of the reversed-inject differential flow modulator for comprehensive two-dimensional gas chromatography in the quantitative profiling and fingerprinting of essential oils of different complexity. J Chromatogr A 1417:79–95

Daud NM, Putra NR, Jamaludin R, Norodin NSM, Sarkawi NS, Hamzah MHS, Salleh LM et al (2022) Valorisation of plant seed as natural bioactive compounds by various extraction methods: a review. Trends Food Sci Technol 119:201–214
de Almeida JM, Crippa BL, de Souza VVMA, Alonso VPP, Júnior EDMS, Picone CSF, Silva NCC et al (2023) Antimicrobial action of oregano, thyme, clove, cinnamon and black pepper essential oils free and encapsulated against foodborne pathogens. Food Control 144:109356
Dimmito MP, Stefanucci A, Della Valle A, Scioli G, Cichelli A, Mollica A (2021) An overview on plants cannabinoids endorsed with cardiovascular effects. Biomed Pharmacother 142:111963
Donsì F, Ferrari G (2016) Essential oil nanoemulsions as antimicrobial agents in food. J Biotechnol 233:106–120
Efferth T, Li PC, Konkimalla VSB, Kaina B (2007) From traditional Chinese medicine to rational cancer therapy. Trends Mol Med 13(8):353–361
El Asbahani A, Miladi K, Badri W, Sala M, Addi EA, Casabianca H et al (2015) Essential oils: from extraction to encapsulation. Int J Pharm 483(1–2):220–243
Elsabee MZ, Morsi RE, Fathy M (2016) Chitosan-oregano essential oil blends use as antimicrobial packaging material. In: Antimicrobial food packaging. Academic Press, pp 539–551
Erland LA, Mahmoud SS (2016) Lavender (Lavandula angustifolia) oils. Essential oils in food preservation, flavor and safety, 501–508
Esmeeta A, Adhikary S, Dharshnaa V, Swarnamughi P, Maqsummiya ZU, Banerjee A, Duttaroy AK et al (2022) Plant-derived bioactive compounds in colon cancer treatment: an updated review. Biomed Pharmacother 153:113384
Foghis M, Bungau SG, Bungau AF, Vesa CM, Purza AL, Tarce AG, Radu AF et al (2023) Plants-based medicine implication in the evolution of chronic liver diseases. Biomed Pharmacother 158:114207
Gennadios A, Weller CL (1990) Edible films and coatings from wheat and corn proteins. Food Technol (Chicago) 44(10):63–69
Gharde Y, Singh PK, Dubey RP, Gupta PK (2018) Assessment of yield and economic losses in agriculture due to weeds in India. Crop Prot 107:12–18
Gholamipourfard K, Salehi M, Banchio E (2021) Mentha piperita phytochemicals in agriculture, food industry and medicine: features and applications. S Afr J Bot 141:183–195
Giacometti J, Kovačević DB, Putnik P, Gabrić D, Bilušić T, Krešić G et al (2018) Extraction of bioactive compounds and essential oils from mediterranean herbs by conventional and green innovative techniques: a review. Food Res Int 113:245–262
Guzmán E, Lucia A (2021) Essential oils and their individual components in cosmetic products. Cosmetics 8(4):114
Han T, Zhang Q, Liu N, Wang J, Li Y, Huang X, Qi K et al (2020) Changes in antibiotic resistance of Escherichia coli during the broiler feeding cycle. Poult Sci 99(12):6983–6989
Haro-González JN, Castillo-Herrera GA, Martínez-Velázquez M, Espinosa-Andrews H (2021) Clove essential oil (Syzygium aromaticum L. Myrtaceae): extraction, chemical composition, food applications, and essential bioactivity for human health. Molecules 26(21):6387
Helli B, Gerami H, Kavianpour M, Heybar H, Hosseini SK, Haghighian HK (2021) Curcumin nanomicelle improves lipid profile, stress oxidative factors and inflammatory markers in patients undergoing coronary elective angioplasty; a randomized clinical trial. Endocr Metab Immune Disord-Drug Targets (Formerly Current Drug Targets-Immune, Endocrine & Metabolic Disorders) 21(11):2090–2098
Henrotin Y, Priem F, Mobasheri A (2013) Curcumin: a new paradigm and therapeutic opportunity for the treatment of osteoarthritis: curcumin for osteoarthritis management. Springerplus 2:1–9
Hougeir FG, Kircik L (2012) A review of delivery systems in cosmetics. Dermatol Ther 25(3):234–237
Ibrahim SS, Abou-Elseoud WS, Elbehery HH, Hassan ML (2022) Chitosan-cellulose nanoencapsulation systems for enhancing the insecticidal activity of citronella essential oil against the cotton leafworm Spodoptera littoralis. Ind Crop Prod 184:115089
Iqbal Z, Akhtar M, Sabri MU, Altaf A (2020) Chemical composition of ocimum sanctum essential oil by GC-MS analysis. Nat Prod Chem Res 8:1–5

Isman MB (2020) Bioinsecticides based on plant essential oils: a short overview. Zeitschrift für Naturforschung C 75(7–8):179–182

Jennings MR, Parks RJ (2020) Curcumin as an antiviral agent. Viruses 12(11):1242

Johnson K, West T, Diana S, Todd J, Haynes B, Bernhardt J, Johnson R (2017) Use of aromatherapy to promote a therapeutic nurse environment. Intensive and Critical Care Nursing 40:18–25

Joshi RK, Soulimani R (2020) Ethno-medicinal and phytochemical potential of Carum carvi Linn. And Cuminum cyminum: a review. International journal of Pharmacognosy and life. Science 1(1):33–37

Jyoti G, Keshav A, Anandkumar J (2015) Review on pervaporation: theory, membrane performance, and application to intensification of esterification reaction. J Eng 2015

Kanani M, Chamani E, Shokouhian AA, Torabi-Giglou M (2021) Plant secondary metabolism and flower color changes in damask rose at different flowering development stages. Acta Physiol Plant 43(4):1–10

Kankara SS, Ibrahim MH, Mustafa M, Go R (2015) Ethnobotanical survey of medicinal plants used for traditional maternal healthcare in Katsina state, Nigeria. S Afr J Bot 97:165–175

Kaul S, Gulati N, Verma D, Mukherjee S, Nagaich U (2018) Role of nanotechnology in cosmeceuticals: a review of recent advances. J Pharm 2018

Kedia A, Prakash B, Mishra PK, Dwivedy AK, Dubey NK (2015) Biological activities of Cuminum cyminum seed oil and its major components against Callosobruchus chinensis and Sitophilus oryzae. J Asia Pac Entomol 18(3):383–388

Kholiya S, Punetha A, Venkatesha KT, Kumar D, Upadhyay RK, Padalia RC (2022) Essential oil yield and composition of Ocimum basilicum L. at different phenological stages, plant density and post-harvest drying methods. S Afr J Bot 151:919–925

Khursheed A, Rather MA, Jain V, Rasool S, Nazir R, Malik NA, Majid SA (2022a) Plant based natural products as potential ecofriendly and safer biopesticides: a comprehensive overview of their advantages over conventional pesticides, limitations and regulatory aspects. Microb Pathog 105854.tabolism:1–19

Khursheed A, Rather MA, Jain V, Rasool S, Nazir R, Malik NA, Majid SA (2022b) Plant based natural products as potential ecofriendly and safer biopesticides: a comprehensive overview of their advantages over conventional pesticides, limitations and regulatory aspects. Microb Pathog 105854:105854

Kishimoto N, Kashiwagi A (2022) Evaluation of the process for transferring limonene from lemon to olive oil. Food Sci Technol Res 28(2):113–118

Kowalczyk A, Przychodna M, Sopata S, Bodalska A, Fecka I (2020) Thymol and thyme essential oil—new insights into selected therapeutic applications. Molecules 25(18):4125

Kumar A, Singh PP, Prakash B (2020) Unravelling the antifungal and anti-aflatoxin B1 mechanism of chitosan nanocomposite incorporated with Foeniculum vulgare essential oil. Carbohydr Polym 236:116050

Kumar J, Ramlal A, Mallick D, Mishra V (2021) An overview of some biopesticides and their importance in plant protection for commercial acceptance. Plan Theory 10(6):1185

Lertsutthiwong P, Rojsitthisak P, Nimmannit U (2009) Preparation of turmeric oil-loaded chitosan-alginate biopolymeric nanocapsules. Mater Sci Eng C 29(3):856–860

Liang J, Zhang Y, Chi P, Liu H, Jing Z, Cao H, Kong D et al (2022) Essential oils: chemical constituents, Potential Neuropharmacological Effects and Aromatherapy-A Review. Pharmacol Res-Modern Chinese Med 6:100210

Mahboubi M (2019) Zingiber officinale Rosc. Essential oil, a review on its composition and bioactivity. Clinical Phytoscience 5(1):1–12

McKay DL, Blumberg JB (2006) A review of the bioactivity and potential health benefits of peppermint tea (Mentha piperita L.). Phytother Res: Int J Devoted Pharmacol Toxicol Eval Nat Prod Deriv 20(8):619–633

Mieres-Castro D, Ahmar S, Shabbir R, Mora-Poblete F (2021) Antiviral activities of eucalyptus essential oils: their effectiveness as therapeutic targets against human viruses. Pharmaceuticals 14(12):1210

Mishra R, Gupta AK, Kumar A, Lal RK, Saikia D, Chanotiya CS (2018) Genetic diversity, essential oil composition, and in vitro antioxidant and antimicrobial activity of Curcuma longa L. germplasm collections. J. Appl Res Med Aromat Plants 10:75–84

Muhoza B, Xia S, Wang X, Zhang X, Li Y, Zhang S (2022) Microencapsulation of essential oils by complex coacervation method: preparation, thermal stability, release properties and applications. Crit Rev Food Sci Nutr 62(5):1363–1382

Muzzarelli RAA, Frega N, Miliani M, Muzzarelli C, Cartolari M (2000) Interactions of chitin, chitosan, N-lauryl chitosan and N-dimethylaminopropyl chitosan with olive oil. Carbohydr Polym 43(3):263–268

Natrajan D, Srinivasan S, Sundar K, Ravindran A (2015) Formulation of essential oil-loaded chitosan–alginate nanocapsules. J Food Drug Anal 23(3):560–568

Nieto G (2017) Biological activities of three essential oils of the Lamiaceae family. Medicines 4(3):63

Oladeji OS, Adelowo FE, Ayodele DT, Odelade KA (2019) Phytochemistry and pharmacological activities of Cymbopogon citratus: a review. Scientific African 6:e00137

Önder S, Tonguç M, Erbaş S, Önder D, Mutlucan M (2022) Investigation of phenological, primary and secondary metabolites changes during flower developmental of Rosa damascena. Plant Physiol Biochem 192:20–34

Ong YK, Shi GM, Le NL, Tang YP, Zuo J, Nunes SP, Chung TS (2016) Recent membrane development for pervaporation processes. Prog Polym Sci 57:1–31

Pan J, Zhu Y, Abdel-Samie MA, Li C, Cui H, Lin L (2022) Biological properties of essential oil emphasized on the feasibility as antibiotic substitute in feedstuff. Grain Oil Sci Technol 6:10

Poonkodi KATHIRVEL (2016) Chemical composition of essential oil of Ocimum basilicum L.(basil) and its biological activities-an overview. J Crit Rev 3(3):56–62

Pott DM, Osorio S, Vallarino JG (2019) From central to specialized metabolism: an overview of some secondary compounds derived from the primary metabolism for their role in conferring nutritional and organoleptic characteristics to fruit. Front Plant Sci 835

Prakash B, Kujur A, Yadav A, Kumar A, Singh PP, Dubey NK (2018) Nanoencapsulation: an efficient technology to boost the antimicrobial potential of plant essential oils in food system. Food Control 89:1–11

Prakash B, Mishra PK, Kedia A, Dwivedy AK, Dubey NK (2015) Efficacy of some essential oil components as food preservatives against food contaminating molds, aflatoxin B 1 production and free radical generation. J Food Qual 38(4):231–239

Prasanth Reddy V, Ravi Vital K, Varsha PV, Satyam S (2014) Review on Thymus vulgaris traditional uses and pharmacological properties. Med Aromat Plants 3(164):2167–0412

Rahman MM, Dhar PS, Anika F, Ahmed L, Islam MR, Sultana NA, Rauf A et al (2022) Exploring the plant-derived bioactive substances as antidiabetic agent: an extensive review. Biomed Pharmacother 152:113217

Ralte L, Bhardwaj U, Singh YT (2021) Traditionally used edible Solanaceae plants of Mizoram, India have high antioxidant and antimicrobial potential for effective phytopharmaceutical and nutraceutical formulations. Heliyon 7(9):e07907

Rasool N, Saeed Z, Pervaiz M, Ali F, Younas U, Bashir R, Sikandar R et al (2022) Evaluation of essential oil extracted from ginger, cinnamon and lemon for therapeutic and biological activities. Biocatalysis and agricultural. Biotechnology 102470:102470

Raut JS, Karuppayil SM (2014) Bioprospecting of plant essential oils for medicinal uses. In: Environment and sustainable development, pp 59–76

Rizal GM, Gyeltshen J, Namgay K (2018) Evaluation of animal feeds for the presence of three important antibiotic classes in Bhutan. J Glob Antimicrob Resist 15:228–231

Rubiolo P, Sgorbini B, Liberto E, Cordero C, Bicchi C (2010) Essential oils and volatiles: sample preparation and analysis. A review. Flavour Fragr J 25(5):282–290

Saklani A, Kutty SK (2008) Plant-derived compounds in clinical trials. Drug Discov Today 13(3–4):161–171

Salminen A, Lehtonen M, Suuronen T, Kaarniranta K, Huuskonen J (2008) Terpenoids: natural inhibitors of NF-κB signaling with anti-inflammatory and anticancer potential. Cell Mol Life Sci 65(19):2979–2999

Sarkic A, Stappen I (2018) Essential oils and their single compounds in cosmetics—a critical review. Cosmetics 5(1):11

Seow YX, Yeo CR, Chung HL, Yuk HG (2014) Plant essential oils as active antimicrobial agents. Crit Rev Food Sci Nutr 54(5):625–644

Sharma V, Gautam DNS, Radu AF, Behl T, Bungau SG, Vesa CM (2022) Reviewing the traditional/modern uses, Phytochemistry, essential oils/extracts and pharmacology of Embelia ribes Burm. Antioxidants 11(7):1359

Shukla MK, Thakur A, Verma R, Lalhlenmawia H, Bhattacharyya S, Bisht D, Kumar D et al (2022) Unravelling the therapeutic potential of orchid plant against cancer. S Afr J Bot 150:69–79

Shukla R, Singh P, Prakash B, Dubey NK (2013) Efficacy of a corus Calamus L. essential oil as a safe plant-based antioxidant, a flatoxin B 1 suppressor and broad spectrum antimicrobial against food-infesting fungi. Int J Food Sci Technol 48(1):128–135

Singh PP, Jaiswal AK, Kumar A, Gupta V, Prakash B (2021) Untangling the multi-regime molecular mechanism of verbenol-chemotype Zingiber officinale essential oil against aspergillus flavus and aflatoxin B1. Sci Rep 11(1):6832

Singh PP, Kumar A, Prakash B (2020) Elucidation of antifungal toxicity of Callistemon lanceolatus essential oil encapsulated in chitosan nanogel against aspergillus flavus using biochemical and in-silico approaches. Food Additives & Contaminants: Part A 37(9):1520–1530

Singh R, Singh AP, Kumar S, Giri BS, Kim KH (2019) Antibiotic resistance in major rivers in the world: a systematic review on occurrence, emergence, and management strategies. J Clean Prod 234:1484–1505

Singletary KW (2018) Basil: a brief summary of potential health benefits. Nutr Today 53(2):92–97

Soares GA, Bhattacharya T, Chakrabarti T, Tagde P, Cavalu S (2022) Exploring pharmacological mechanisms of essential oils on the central nervous system. Plan Theory 11(1):21

Sotelo-Boyás ME, Correa-Pacheco ZN, Bautista-Baños S, Corona-Rangel ML (2017) Physicochemical characterization of chitosan nanoparticles and nanocapsules incorporated with lime essential oil and their antibacterial activity against food-borne pathogens. LWT 77:15–20

Sudhakar K, Mishra V, Hemani V, Verma A, Jain A, Jain S, Charyulu RN (2021) Reverse pharmacology of phytoconstituents of food and plant in the management of diabetes: current status and perspectives. Trends Food Sci Technol 110:594–610

Sumner J (2000) The natural history of medicinal plants. Timber Press

Tariq Z, Izhar F, Malik MH, Oneeb M, Anwar F, Abbas M, Khan A (2022) Development of functional textile via microencapsulation of peppermint oils: a novel approach in textile finishing. Res J Text Appar. (ahead-of-print)

Vahedikia N, Garavand F, Tajeddin B, Cacciotti I, Jafari SM, Omidi T, Zahedi Z (2019) Biodegradable zein film composites reinforced with chitosan nanoparticles and cinnamon essential oil: physical, mechanical, structural and antimicrobial attributes. Colloids Surf B: Biointerfaces 177:25–32

Van der Bruggen B, Luis P (2015) Pervaporation. In: Progress in filtration and separation. Academic Press, pp 101–154

Walker PT (1983) Crop losses: the need to quantify the effects of pests, diseases and weeds on agricultural production. Agric Ecosyst Environ 9(2):119–158

Wei A, Shibamoto T (2007) Antioxidant activities of essential oil mixtures toward skin lipid squalene oxidized by UV irradiation. Cutan Ocul Toxicol 26(3):227–233

Weng JK (2014) The evolutionary paths towards complexity: a metabolic perspective. New Phytol 201(4):1141–1149

Wink M (2010) Introduction: biochemistry, physiology and ecological functions of secondary metabolites. Annual plant reviews volume 40: biochemistry of plant secondary me McKay, D. L., & Blumberg, J. B. (2006). A review of the bioactivity and potential health benefits of chamomile tea (Matricaria recutita L.). Phytotherapy Res 20(7):519–530

World Health Organization (2001) Legal status of traditional medicine and complementary

Wu C, Zhu Y, Wu T, Wang L, Yuan YI, Chen J et al (2019) Enhanced functional properties of biopolymer film incorporated with curcurmin-loaded mesoporous silica nanoparticles for food packaging. Food Chem 288:139–145

Xu Y, Ma L, Liu F, Yao L, Wang W, Yang S, Han T (2023) Lavender essential oil fractions alleviate sleep disorders induced by the combination of anxiety and caffeine in mice. J Ethnopharmacol 302:115868

Yadav A, Kujur A, Kumar A, Singh PP, Gupta V, Prakash B (2020) Encapsulation of Bunium persicum essential oil using chitosan nanopolymer: preparation, characterization, antifungal assessment, and thermal stability. Int J Biol Macromol 142:172–180

Yuan H, Ma Q, Ye L, Piao G (2016) The traditional medicine and modern medicine from natural products. Molecules 21(5):559

Zhang S, Pang X, Yue Z, Zhou Y, Duan H, Shen W, Cheng Q et al (2020) Sulfonamides removed from simulated livestock and poultry breeding wastewater using an in-situ electro-Fenton process powered by photovoltaic energy. Chem Eng J 397:125466

Zollman C, Vickers A (1999) What is complementary medicine? BMJ 319(7211):693–696

Extraction and Chemotypic Standardization of Plant Essential Oils with Special Reference to Phytochemical Genomics

2

Vandana and Anindita Barua

Abstract

Essential oils are the mixture of volatile ingredients and have been utilized extensively since antiquity. Regarding its bioactivity as an antibacterial, antiviral, antioxidant, and antidiabetic, the major significance and mode of action of these naturally occurring compounds are reviewed. Further, its crucial role in the suppression of cancer and chemoprevention is also emphasized. The traditional extraction methods such as steam distillation, hydrodistillation, hydro diffusion, and solvent extraction, and advanced (non-conventional) extraction methods such as solvent-free microwave extraction, subcritical extraction liquid, and supercritical fluid extraction have been discussed. Modern extraction techniques are thought to have the most potential because they require less time to extract, use less energy and less solvent, and emit less carbon dioxide. The chemical profile of essential oils has been explored in relation to several modern characterization approaches.

Keywords

Essential oils · Antimicrobial activities · Antidiabetic · Chemoprevention · GC-MS · NMR · Chemotypes

Vandana (✉)
National Institute for Plant Biotechnology, New Delhi, India

Banaras Hindu University, Varanasi, Uttar Pradesh, India

A. Barua
National Institute for Plant Biotechnology, New Delhi, India

B. Prakash et al. (eds.), *Plant Essential Oils*,
https://doi.org/10.1007/978-981-99-4370-8_2

2.1 Introduction

Essential oils are liquid, aromatic, volatile, odiferous, natural, and complex blends of low-molecular-weight compounds. According to Hyldgaard et al. (2012), aerial parts of plants, which are typically made up of flowers, leaves, stems (chamomile, peppermint, lavender), seeds (nutmeg), fruits (anise), bark (cinnamon), as well as in the radix and rhizomes (curcuma and ginger), are the main sources of essential oils, and secretory organs like trichomes, glandular hairs (Lamiaceae and Asteraceae), secretory ducts (schizogenous in Myrtaceae and schizolysigenous in Rutaceae), cavities (Conifers), or undifferentiated cells (Lauraceae) are the storage sites. Due to the various ancient texts currently in use around the world, essential oils' high refraction index and rotatory power may be useful for their identification and quality control. However, these properties are inconsistent. Essential oils play a variety of roles in a particular plant, including pollination, defensive mechanisms, and irritation or repulsive effects. Additionally, there are various hypotheses regarding their potential involvement as antioxidants in that they donate hydrogen in oxidative processes, particularly when light is present. They may also be antibacterial and antifungal, shielding the plant from any pathogenic attack. The oil can be used as an antibacterial agent with a low risk of microbial resistance development due to the inclusion of several components (Veras et al. 2012).

These oils have garnered a lot of attention throughout history, and while many of their original applications have already been forgotten, it is widely believed that humans have been harvesting these substances from fragrant plants since prehistoric times. Strong aromas are an essential attribute of essential oils. Typically, steam distillation, hydrodistillation, or solvent extraction is used to extract them (Rivera Calo et al. 2015; Bakkali et al. 2008). The uses of essential oils for various tasks are numerous and include making perfumes and cosmetics in addition to using them in the kitchen to improve the flavor and health benefits of food. They are frequently employed as flavoring ingredients in food, beverages, cosmetics, fragrances, and pharmaceuticals. Due to the essential oil's potent therapeutic effects, either alone or in combination, it can be used as chemotherapy for both infectious and non-infectious diseases (Raut and Karuppayil 2014).

Essential oils were employed by the ancient Egyptians in medicine, perfumery, and in the practice of mummification, the art of embalming, and prepping remains for burial. The Vedas of ancient Asia defined the use of fragrances and aromatics for both therapeutic and ceremonial purposes. Uncountable civilizations have employed essential oils and scents for a variety of uses throughout history, including religious rituals, the creation of perfumes, and the treatment of infectious diseases. The enormous degree of variation in the chemical makeup of aromatic plants, however, poses a potentially serious issue for the perfume production sector. This has led to much investigation on variables other than pure genetic ones that may contribute to this variant. For instance, researchers have identified many races of the same species, such as Melaleuca bracteata, that are rich in various primary ingredients and generate various essential oils with methyl eugenol, methyl isoeugenol, and elemicin. Numerous chemotypes and chemical races have been identified in other aromatic plants, including sweet flag (Acorus calamus), wormwood (Artemisia absinthium), sweet basil (Ocimum basilicum), lemon balm (Melissa officinalis), thyme (Thymus vulgaris), camphorwood (Cinnamomum camphora), peppermint (Mentha piperita), and tansy (Tanacetum vulgare) (Evans 2009). It has also been found that other elements, such as the climate, rainfall, or geographic origin of the plant, can alter the chemical composition of essential oils.

There are several definitions of essential oils; however Sonwa (2000) gave the summary of Schilcher, Hegnauer, and Cohn-Riechter's which is likely the most accurate: Essential oils are substances or mixes of substances that are created in the cytoplasm and often exist as minute droplets between cells. They are both aromatic and volatile. According to the definition of a fragrant substance, it is a "chemically pure compound that is volatile under standard conditions and which, owing to its odour, can be beneficial to society." They are made up of "combinations of fragrant substances or mixtures of fragrant and odourless compounds." According to the International Organization for Standardization (ISO), essential oils are "products obtained from a natural raw material of plant origin, by steam distillation, by mechanical processes from the epicarp of citrus fruits, or by dry distillation, after separation of the aqueous phase—if any—by physical processes, " adding that "the essential oil can undergo physical treatments, which do not result in any significant change in its composition." Other organizations, societies, and businesses have developed their own standards for concept and excellence.

The molecules that make up essential oils are a complex mixture, typically containing more than 20 different, low-molecular-weight components in a wide range of quantities. Although diterpenes and phenylpropanoids can be present to varying degrees, terpenoids (monoterpenes [C10] and sesquiterpenes [C15]) are often the major constituents of essential oils. Terpenoids are organic hydrocarbons that a wide range of plants naturally make. Based on their five carbon (isoprene) units, terpenes are divided into many categories based on their structure and function. Hemiterpene (C5), diterpene (C20), triterpene (C30), and tetraterpene (C40) are also significant terpenes and make up around 90% of essential oils (Bakkali et al. 2008). Over 3000 essential oils have been described up to this point, of which about 10% are pertinent to the pharmaceutical, nutritional, or cosmetic industries. For their ability to cause cell death, a number of essential oils are the subject of intense research. A lot of work is put into examining the possible therapeutic effects of oils against a variety of illnesses, particularly those marked by excessive cell growth and proliferation like cancer or bacterial diseases (Freires et al. 2015; Russo et al. 2015). The enormous range of secondary metabolites that plants produce, many of which have cytotoxic or less appetizing characteristics to herbivores or predators, contribute to plant defense against both of these threats. Additionally, research into the use of essential oils as viable substitutes has been prompted by the rise in germs that are resistant to numerous antibiotics. Dorman and Deans (2000) and O'Bryan et al. (2015) found that the thymol, carvacrol, eugenol, cinnamaldehyde, and citral are examples of compounds that have phenolic structures or aldehydes that are highly effective against microbes. Instead of examining the processes behind their bioactivity, the majority of studies, however, were only concerned with examining the antibacterial properties of some of the main chemicals of essential oils.

2.2 Mechanism of Action of Essential Oils

Antibacterial, antifungal, and antiviral properties of essential oils and their constituents have been researched, but little is known about their mechanism. In addition to knowing the mechanism of action of the essential oil mixture, it is crucial to identify the active components of essential oils. In order to acquire the perfect and most active composition of an essential oil, it will be possible to choose the best and appropriate growing conditions, harvesting, and extraction methods. Additionally, once the active component is identified, synthetic analogues can be created and prepared. In terms of preparation reproducibility and increased economic viability, these synthetic components would be easier to manage. It has been hypothesized that when essential oils are breathed or applied to the skin, they interact with the lipid moieties of cell membranes via their lipophilic components, altering the functioning of calcium and potassium ion channels (Shah et al. 2011). The membranes are saturated when essential oils are applied in a specific dose. Depending on the composition and physicochemical characteristics of the components, they tend to interact with cell membranes in a way that affects the functioning of a variety of membrane molecular structures (Turina et al. 2006), such as transport systems, enzymes, ion channels (Huang et al. 2008), and receptors. There have been several research on the various physiological effects of essential oils on humans, such as their boosting, sedative, and antidepressant benefits. Similar to this, several research have concentrated on how fragrance elements impact mood, memory, and intelligence. In these situations, the scent elements are inhaled and interact with central nervous system (Aoshima and Hamamoto 1999) receptors to penetrate the blood-brain barrier. Generally, bioassays that tried to describe and explain the action of essential oils are carried out on mice, rats, and frogs. Instances include the investigation of their analgesic properties, the effect of peppermint oil on intestinal transport (Beesley et al. 1996), the research of essential oil absorption through the skin, and the influence of some essential oils on skeletal muscle fibres (Serpa et al. 1997). Numerous microorganisms have been used to illustrate the antibacterial activity of essential oils. Although many methods of action have been put forth, none is fully understood (Fisher and Phillips 2008; Holley and Patel 2005). The majority of these mechanisms merely explain how the chemical constituents in essential oils work, but antibacterial activity involves a variety of mechanisms that target harmful bacteria (Benchaar et al. 2008).

Terpenes have a strong affinity for lipids due to their hydrophobic nature, which is linked to their antimicrobial effect. Their lipophilic nature and the outer microbe structures are clearly linked to their antibacterial capabilities (Paduch et al. 2007). Due to the increased fluidity and permeability of the membrane, changes in the topology of the membrane proteins, and resulting disruptions in the respiration chain, monoterpenes are able to permeate internal cell membrane structures (Bajpai et al. 2012; Friedly et al. 2009). Essential oils contain phenolic chemicals that damage cell membranes, impairing cell functionality and finally leading to spillage of internal cell material (Trombetta et al. 2005). Helander et al. (1998) has reported that the phenolic compounds thymol and carvacrol have stopped the growth of Gram-negative bacteria by rupturing the outer cell membrane. Electron transport, ions, protein translocation, phosphorylation, and other enzyme-dependent activities are additional functions of the cell membrane. Cell death may occur as a result of the cytoplasmic membrane's disturbed permeability. Gram-positive bacteria have reportedly been shown to be more vulnerable to the antibacterial effects of essential oil components than Gram-negative bacteria. As expected, this is because the cell wall of Gram-negative bacteria is protected by an outer layer that prevents hydrophobic substances from penetrating the cell.

2.3 Chemistry of Essential Oils

An aromatic compound, or moiety, has a chemical arrangement that leads to the delocalization of electrons, creating higher molecular stability. This is true even though the term "aromatic" in current terminology describes the quality of giving off an aroma that is either pleasant or odious to the nose. As a result, the scent of essential oils may be a combination of aromatic and aliphatic components (Sadgrove and Jones 2015). Essential oils are insoluble in water but soluble in ether, alcohol, and volatile oils. With a few exceptions, they have a distinctive smell and a density that is lower than unity (cinnamon, clove, sassafras, and vetiver). Essential oils are extremely complex natural mixes that might have 20–60 different components at widely different concentrations, with two or three major components often present at comparatively elevated concentrations (20–70%) in comparison to other components that are present in low concentrations (Chouhan et al. 2017). Nevertheless, the flavor contribution of a single component is not just influenced by its concentration; it also depends on the particular odour threshold, which is influenced by structure and volatility. Therefore, even tiny components resulting from oxidation or degradation events may have a significant effect on the flavor if their aroma value is high enough (Grosch 2007). Lipophilic terpenoids, phenylpropanoids, or short-chain aliphatic hydrocarbon derivatives of low molecular weight can be classified as the majority of essential oil ingredients, with the first being the most prevalent and distinctive ones. Among these, allylic, mono-, bi-, or tricyclic mono- and sesquiterpenoids of various chemical classes, such as hydrocarbons, ketones, alcohols, oxides, aldehydes, phenols, or esters, make up a significant portion of essential oils (Turek and Stintzing 2013). Even one component removed from the mixture might alter the aroma. In general, oxygenated molecules, which are very odoriferous, are more valuable in terms of their contribution to the scent of the essential oil than monoterpene hydrocarbons (Kusuma and Mahfud 2016). Essential oils differ significantly in terms of their physical characteristics, such as density, refractive index, and rotatory power, in addition to their chemical characteristics. For instance, it has been noted that there are substantial differences between the essential oils of

Siphonochilus ethiopicus's various organs (limbo, foliar sheath, and rhizomes). These characteristics help characterize plant essential oils more accurately (Noudogbessi et al. 2012). The most significant terpenes and ones that are frequently employed in the fragrance industries include geraniol/nerol, linalool, citronellol, citronellal, and citral (Pavela 2015; Sell 2006). Terpenoids and phenylpropanoids are produced through various biosynthetic pathways and have various primary metabolic substrates from biosynthetic pathways. The mevalonate and mevalonate-independent (DXP) pathways are involved in terpenoids biosynthesis, whereas the shikimate pathway is where phenylpropanoids are produced (Dhifi et al. 2016).

Numerous asymmetric carbon atoms with optical activity can be found in one or more of the essential oils' constituent parts. These chiral chemicals have evolved through enzymatically regulated biosynthetic synthesis, and they are typically found in distinctive enantiomeric distributions. They also have a variety of applications in the food, drug, and fragrance sectors. Additionally, the chirality of the odorants affects their mode of action, suggesting that the antipodes may behave differently in comparison to other medications. Essential oil chemical composition can vary significantly due to physiological (plant organ, ontogenesis), environmental (soil conditions, climatic conditions), and genetic variables (Salgueiro et al. 2010). Geographical variation (Sanli and Karadogan 2017), plant parameters (such as species, cultivated or wild plants), harvest and postharvest/predistillation parameters (such as harvest season, pretreatment of biomass, and storage conditions of source plant materials), production parameters (such as mode of production, distillation parameters, commercial oil, or laboratory-produced), and other parameters (such as storage conditions) are additional factors that may affect the chemical composition of the oils.

2.4 Benefits of Essential Oils

We obtain a great variety of drugs from plants, such as morphine from Papaver somniferum, ashwagandha from Withania somnifera, ephedrine from Ephedra vulgaris, etc., therefore plants are the significant source of medicines (Prakash and Gupta 2005). Due to their essential oils, plants in the Apiaceae family have been known to have therapeutic effects since ancient times (Valente et al. 2013). According to a World Health Organization (WHO) survey conducted in 1993, these traditional medicines are used to treat 90% of patients in Bangladesh, 85% of patients in Burma, and 80% of patients in India (Prakash and Gupta 2005) (Fig. 2.1).

2.4.1 Antiviral and Antifungal Activity

Without being toxic, essential oils have antiviral properties. Multi lamella liposomes were made more effective against intracellular herpes simplex virus type 1 (HSV1) by adding Artimisia arboreseens essential oil. Due to the presence of citral

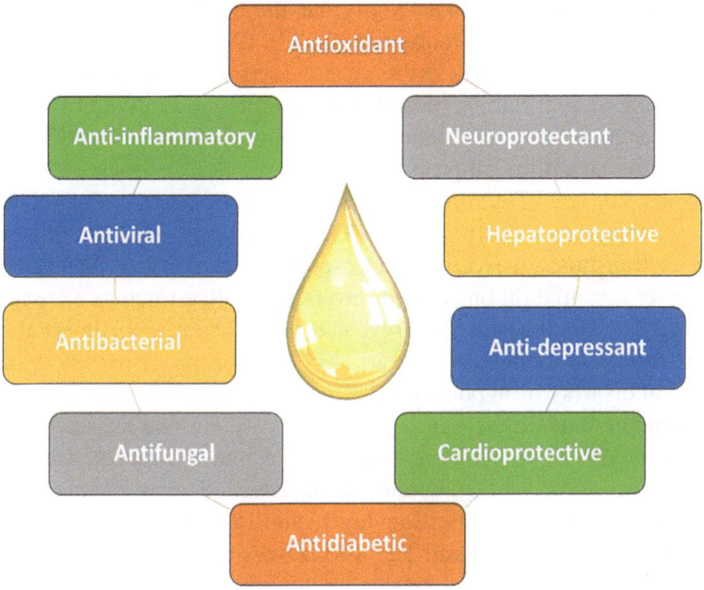

Fig. 2.1 Benefits of essential oils

and citronellal in Melissa officinalis L. essential oil, HSV-2 replication is also suppressed (Allahverdiyev et al. 2004), while HSV-1 replication can be suppressed through in vitro incubation with several essential oils. According to Allahverdiyev et al. (2004), the most significant HSV1 activity and replication inhibitor is found in lemon grass essential oil. In viral suspension assays against HSV-1, HSV-2, and acyclovir-resistant strains of HSV-1, peppermint (Mentha piperita) essential oil exhibits high levels of virucidal activity (Schuhmacher et al. 2003). HSV-1 and HSV-2 are both resistant to eucalyptus essential oil's antiviral effects (Schnitzler et al. 2001). De Logu et al. (2000) reported that Santolina insularis demonstrated antiviral efficacy against HSV-1 and HSV-2 in vitro and inhibited the transmission of the virus from effective cells to other healthy cells. Because of the synergism phenomenon, the complex combination of essential oils had more antiviral activity than individual components (Djilani and Dicko 2012). Additionally, fascinating novel therapeutic alternatives to synthetic medications exist in the form of natural compounds with antifungal characteristics. The fact that fungal drug-resistant strains are continuously evolving, much like bacteria, makes these items even more crucial (Bajwa and Kulshrestha 2013). Though few research have looked into the underlying mechanisms of action, many have shown how potent essential oils are against various fungal species (Da Cruz et al. 2013). With significant morphological changes in hyphae, the use of essential oils caused the cell wall's integrity to be lost and the plasma membrane to become more permeable. Candida albicans have been shown to be resistant to the antifungal effects of Lavandula multifida L. essential oil (Zuzarte et al. 2012). Propidium iodide staining, cytoplasmic membrane rupture, and suppression of filamentation are signs that apoptosis was induced. According to a different investigation by Ferreira et al. (2013), Curcuma longa L. was cytotoxic to Aspergillus flavus and reduced the synthesis of aflatoxin. When A. flavus was exposed to the essential oil, analysis using scanning electron microscopy revealed severe damage to hyphae membranes and conidiophores. Aspergillus niger was used to research the antifungal properties of essential oils extracted Thymus eriocalyx (Ronniger) Jalas and Thymus x-porlock. A. niger treated to essential oils showed irreversible damage to its cell wall, cell membrane, and other cellular organelles, according to transmission electron microscopy. The effects of Matricaria chamomilla L. flower essential oil on A. niger growth and ultrastructure were investigated (Tolouee et al. 2010). The findings revealed a clear rupture of internal organelles and cytoplasmic membranes, a separation of the plasma membrane from the cell wall, and total disarray of hyphal compartments. According to the authors, the fungal plasma membrane's increased cell permeability may be the cause of the morphological changes. Some investigations looked into the mechanisms underlying the pro-apoptotic effects of essential oils in addition to ultrastructural changes. A. flavus plasma membrane, fibrillar layer, and cytoplasm significantly changed when Cinnamomum jensenianum Hand-Mazz essential oil was used (Tian et al. 2012). With less mitochondrial cristae, mitochondria also experienced a wide-ranging disturbance of their internal structure. The amount of ergosterol in the plasma membrane was also significantly decreased as a result of the essential oil. As the primary sterol of the cell membrane and a unique chemical found in fungus, ergosterol is essential for maintaining the health and functionality of the fungal cell (Iwaki et al. 2008). Furthermore, it has been noted that Coriandrum sativum L. essential oil binds to membrane ergosterol, increasing ionic permeability and causing membrane damage that results in cell death in various Candida strains (Freires De Almeida et al. 2014). Likewise, the proteolytic activity of Candida albicans was reduced by C. sativum essential oil. Similar outcomes were noted when Mentha piperita L. essential oil was used to treat the Candida strains C. albicans, C. tropicalis, and C. glabrata (Samber et al. 2015). Cell membrane rupture, a sharp drop in ergosterol concentration, and morphological changes were all present in exposed cells. It has also been suggested that the cytotoxic effects of essential oils are moderated by the induction of reactive oxygen species (ROS) overproduction and oxidative stress. A decrease in ATPase activity, chromatin condensation, DNA fragmentation, phosphatidylserine exposure, cytochrome c release, and metacaspase activation were all observed as signs of apoptosis in a C. albicans strain that is a human pathogen (Chen et al. 2014). The production of mycotoxins, which can be hazardous to humans, is a further consideration. Aflatoxin are extremely dangerous fungal toxins, thus it is highly interesting to see how natural chemicals could inhibit the production of aflatoxins (Da Cruz et al. 2013). Zataria multiflora Boiss. essential oil was shown to inhibit A. parasiticus proliferation and aflatoxin accumulation (Yahyaraeyat et al. 2013).

2.4.2 Antibacterial and Antiparasitic Activity

Many harmful bacterial species, including Listeria monocytogenes, L. innocua, and Salmonella typhimurium, are susceptible to the antibacterial effects of essential oils. Some hazardous bacterial species, including E. coli, Salmonella choleraesuis, and Salmonella typhimurium, are inhibited by thyme and oregano essential oils, and this inhibition is directly related to the phenolic components of carvacrol and thymol. The ability of carvacrol to inhibit infections like Bacillus cereus is attributed to the presence of a phenolic hydroxyl group (Ultee et al. 1999). Against vegetative cells, alcohols have bactericidal rather than bacteriostatic action (Dorman and Deans 2000). The bacterial growth of Enterococcus Hiral is inhibited by Artemisia annua's essential oil. β-caryophyllene, camphor, β-selinene, germacrene D, artemisia ketone, and trans pinocarveol are among its constituents. Additionally, it has been claimed that using tea tree oil in formulations does not cause dermatological issues or alter the skin's natural protective microbiota. The active element for tea tree oil's antibacterial function is the large level of terpenen-4-ol, and in veterinary medicine, cream formulations containing 10% tea tree oil provide relief from canine localized acute and chronic dermatitis considerably more quickly than any commercial skin care product. Essential oils have also been demonstrated to have bactericidal effects against dental and oral pathogenic microorganisms, and as a result, they are added to rinses or mouthwashes for pre-procedural mouth control. Since essential oils can penetrate the plaque biofilm, where they destroy the pathogenic wall and block their enzymatic activity, mouthwashes containing these oils may be utilized to control plaque. Thymol, menthol, and eucalyptol, which are all found in Listerine and are derived from essential oils, are particularly effective against dental issues (Sharma et al. 2010). For the treatment of oropharyngeal candidiasis in AIDS, essential oil has also seen to be effective when gargled. Against specific pathogens of the respiratory tract, they can also be utilized as antibacterial agents. When used against Streptococcus pneumoniae, that is, penicillin susceptible as well as penicillin resistant, the oil of Achillea clavennae exhibited its highest level of activity. Against endoparasites and ectoparasites, plant essential oils can be utilized as a substitute. In this respect, Plasmodium falciparum and Leishmania donovani constitute protozoan parasites that are developing drug resistance, which raises the rates of mortality and morbidity (Singh 2006; Antony and Parija 2016). The development of pesticide resistance is another factor that makes the management of ectoparasites in veterinary care crucial (Ellse and Wall 2014). The possible anti-parasitic effects of essential oils against a variety of parasites, including protozoa, helminths, and arthropods, have been studied by several writers; however, the molecular mechanisms of action have received little attention. The antiparasitic activity of Lavandula angustifolia Mill. and Lavandula x intermedia essential oils was tested against the fish disease Hexamita inflata as well as the human protozoal infections Trichomonas vaginalis and Giardia duodenalis (Moon et al. 2006). In the in vitro test, both essential oils totally got rid of all three protozoa. Significant cytotoxic effects against intracellular amastigote form were observed in the antileishmanial activity of C. ambrosioides essential oil (Monzote et al. 2011, 2014). When tested against

Trypanosoma brucei and Plasmodium falciparum, the essential oils of four *Cymbopogon* species—C. citratus (DC) Stapf., C. giganteus Chiov., C. nardus (L.) Rendle, and C. schoenantus (L.) Spreng.—were cytotoxic (Kpoviessi et al. 2014). The essential oils from *Ocimum gratissimum L.* were found to have antitrypanosomal and antiplasmodial properties (Kpoviessi et al. 2014). Using the essential oil from the leaves of Artemisia annua L. against visceral leishmaniasis caused by Leishmania donovani, researchers looked into the molecular mechanisms underlying the antiparasitic actions of essential oils (Islamuddin et al. 2014). As evidenced by the externalization of phosphatidylserine, DNA fragmentation, dyskinetoplastidy, cell cycle arrest, loss of mitochondrial membrane potential, and ROS production in promastigotes, the cytotoxic action was caused by apoptosis. The antiparasitic effectiveness of essential oils of P. betle landrace Bangla Mahoba against intestinal Leishmaniasis was also tested in a different study (Misra et al. 2009). In L. donovani promastigotes and intracellular amastigotes, the investigated essential oils triggered apoptosis in conjunction with the production of ROS that targeted the mitochondria without exerting any harm on macrophages. In treated promastigote cells, transmission electron microscopy revealed abnormally shaped cells, swollen mitochondria, and autophagosomal structures. Additionally, reports of phosphatidylserine externalization, mitochondrial membrane potential loss, and cell-cycle arrest were made.

2.4.3 Antioxidant Activity

The oxidation of macromolecules such as proteins, amino acids, DNA, and others by free radicals and other reactive oxygen species leads to molecular changes associated with aging, arteriosclerosis, and cancer. "Oxidative stress" is caused by an imbalance between the creation of free radicals and their elimination by the antioxidant system in the human body (Abdollahi et al. 2004). Antioxidants must therefore be supplied externally in order to achieve a balance between free radicals and antioxidants. In the DPPH (2,2-Diphenyl-1pycrylhydrazyl) radical assay at room temperature, the essential oils of basil, cinnamon, clove, nutmeg, oregano, and thyme—natural sources of phenolic components—have demonstrated radical-scavenging and antioxidant capabilities (Tomaino et al. 2005). Thymus serpyllus and Thymus spathulifolius contain large concentrations of phenolic thymol and carvacrol, which are responsible for the antioxidant action (Sokmen et al. 2004). Thymol and carvacrol are once more responsible for the antioxidant activity of oregano (Origanum vulgare L.) essential oil, which is comparable to that of tocopherol and BHT (butylated hydroxyl toluene) (Kulisic et al. 2004). Additionally, it has been discovered that Agnus castus seeds essential oil is a superb DPPH radical scavenger (Asdadi et al. 2015). Some essential oils, such as Thymus caespititius, Thymus camphoratus, and Thymus mastichina, show antioxidant activity with high contents of linalool and 1,8-cineole, while thymol and carvacol are almost absent. The antioxidant activity cannot be solely attributed to the presence of phenolic group. Ketones, aldehydes, hydrocarbons, and ether also show free radical scavenging activity. The presence of α-terpinene, γ-terpinene, and α-terpinolene in tea tree oil makes it a natural antioxidant against BHT (Kim et al. 2004). Therefore, essential oils should be included as a component of daily supplements or additives to lower oxidative stress.

2.4.4 Antidiabetic Activity

Diabetes is a chronic disease in which the body either does not produce enough insulin or does not utilize it appropriately. Numerous studies were done to investigate the antidiabetic properties of essential oils. For example, rosemary essential oil inhibited insulin release and hyperglycemia in diabetic rabbits. Studies have shown that the lipophilic fraction of aromatic plants is not only responsible for this activity but has also been shown to improve insulin sensitivity in type II diabetics when administered orally in combination with other essential oils like cumin, cinnamon, oregano, fennel, myrtle, etc. (Al-Hader et al. 1994). In diabetic rats, Satureja khuzestanica essential oil significantly lowers fasting blood glucose levels. Eugenol and other significant essential oils from tulsi (Ocimum sanctum L.) operate as effective antidiabetic agents by lowering blood sugar levels, triglyceride, and cholesterol levels, as well as the activities of alkaline phosphatases, LDH, GPT, and GOT in blood serum (Prakash and Gupta 2005).

2.4.5 Cancer Chemoprevention

Typically, it is believed that substances that activate phase I or phase II drug metabolizing enzymes can guard against chemical deterioration during the beginning phase. Numerous dietary monoterpenes not only have anti-tumor properties but also stop the spread of cancer. Orange peel oil's D-limonene prevents chemically generated mouse mammary skin, liver, lung, and anterior stomach tumors from developing (Crowell 1999). The growth of lung and anterior stomach tumors brought on by chemicals is inhibited by caraway seed oil containing carvone. Additionally, menthol and carveol have chemoprevention effects against dimethylbenz[a]anthracene (DMBA)-induced cancer. The group of chemo preventive agents includes garlic essential oils that contain organosulfur elements (OSCs), which affect phase I and phase II drug detoxification enzymes (Milner 2001). When evaluated against various human cancer cell types, including melanoma, breast, ovarian, and blood lymphocytes, Tetraclinis articulate essential oil displayed the signature of apoptosis. Perillyl alcohol (POH) inhibits the growth of chemically generated liver cancer cells in rats at the early stages of the chemotherapy process (Crowell 1999). In National Cancer Institute (NCI)-sponsored Phase I, II, and III chemoprevention trials for prostate, breast, and colon cancers, POH was studied. As a glutathione-S-transferase class π (GSTP) inducer, citral is another significant component present in many essential oils, such as lemon grass oil. As a result, it is an extremely important chemo preventative drug for inflammatory carcinomas including skin cancer and colon cancer (Edris 2007). Most essential oils were initially investigated for their antioxidant and anti-inflammatory properties, as well as for their potential use in the treatment of inflammatory diseases. Essential oils may also have anticancer effects as a result of the direct relationship between the production of reactive oxygen species (ROS) and the oxidative and inflammatory states that can cause cancer. On the one hand, persistent inflammation is associated with an excess of ROS and can damage DNA, increasing the rate of mutation and the risk that cells will undergo an oncogenic transformation (Storz 2005). The ability of ROS, on the other hand, to affect redox-mediated signaling pathways, which can lead to the formation of tumors, is well recognized. The essential oils that might cause cancer cells to undergo apoptosis may be useful tools for dealing with cancer. The activation of the detoxification and DNA repair systems, the suppression of metastasis, and angiogenesis are additional processes that aid in the treatment of cancer in addition to the induction of apoptosis (Reed 2003; Gautam et al. 2014). Through several mechanisms, essential oils have anti-proliferative effects on a variety of cancer cell cultures. A well-defined type of programmed cell death that maintains homeostasis, apoptosis can be induced by either internal or external stimuli. Tests on tumor cell lines (MCF-7, PC-3, A-549, DLD-1, M4BEU, and CT-26) using the essential oil of Abies balsamea (L.) Mill. (balsam fir oil) revealed considerable cytotoxicity in each of these cell lines (Legault et al. 2003). In a dose- and time-dependent way, the use of the essential oil lowered cellular reduced glutathione (GSH) concentration and enhanced ROS generation. The essential oil of Melissa officinalis L. was also assessed on glioblastoma multiforme cells, and the results revealed a considerable induction of apoptosis as shown by DNA fragmentation and upregulation of caspase-3 (Queiroz et al. 2014). In HepG2 human hepatoma cells, the volatile extract from the dried pericarp of Zanthoxylum schinifolium Siebold & Zucc. likewise caused apoptosis and markedly enhanced ROS generation (Paik et al. 2005). Caspase-3 activity, however, showed no effect, indicating that the extract-induced death of hepatoma cells is caspase-3 independent. On the human epidermoid carcinoma cancer cell line A431 as well as on immortal HaCaT cells, the effects of the essential oil from rosewood Aniba rosaeodora Ducke were examined (Soeur et al. 2011). The therapy caused the mitochondrial membrane to depolarize and caspase-dependent cell death, which were both clearly harmful effects in both cell types. In a different test, YD-8 human oral squamous cell carcinoma cells were exposed to the essential oil from the leaf of Pinus densiflora Siebold & Zucc., which decreased their growth and survival while inducing apoptosis (Jo et al. 2012). The use of essential oils caused the production of reactive oxygen species (ROS), which then activated caspase-9 activity, cleaved the DNA repair enzyme poly(ADP-ribose) polymerase (PARP), down-regulated the anti-apoptotic protein B-cell lymphoma 2 (Bcl-2), and phosphorylated extracellular signal-regulated protein kinase (ERK)-1/2 and c-Jun N-terminal kinase (JNK)-1/2. The benefits of Pinus koraiensis essential oil against cancer, a study on the essential oil of Pinus koraiensis (EOPK) in HCT116 colorectal cancer cells was conducted by Siebold & Zucc. (Cho et al. 2014). The essential oil drastically decreased the colorectal cancer cells' ability to proliferate and migrate. It also lowered the expression of PAK1 (a family of serine/threonine p21-activated kinase), a hub for several oncogenic signaling pathways, which in turn decreased the phosphorylation of Akt and ERK. Essential targets for essential oils include mitogen-activated protein kinases (MAPKs). Artemisia capillaris Thunb. essential oil showed cytotoxicity in human oral epidermoid cancerous cells (Cha et al. 2009). The findings imply that the process of cell death involves activation of caspase, JNK/Bcl-2, and p38/nuclearfactor-kappa (NF-kappa) pathways. The cytotoxicity of Thymus vulgaris L. essential oil against the squamous cell cancer UMSCC1 cell line has been evaluated (Sertel et al. 2011). Human cervical cancer cells (HeLa, ME-180, and SiHa) were exposed to Cephalotaxus griffithii Hook. f. needle essential oil, and the results showed mitochondria-initiated apoptosis (Moirangthem et al. 2015). The essential oil improved the expression of caspases and PARP cleavage while increasing mitochondrial membrane depolarization.

2.4.6 Aromatherapy

Essential oils having an aroma, they are inhaled for psychological and physical health. They can alter behavior and affect brainwave activity. Essential oils' olfactory properties have both objective and subjective effects on cognitive performance and mood (Moss et al. 2003). For the treatment of epilepsy, smell inward breath of storax pill rejuvenating ointment and pre inward breath of Acorus gramineus rhizome (AGR) natural oil are utilized in Chinese people medication. The activity of Gamma-Aminobutyric acid (GABA) transaminase in the mouse brain is inhibited prior to inhalation of AGR essential oil. This results in an increase in the level of GABA and a decrease in the amount of glutamate in the brain, both of which have an effect on the length of time mice, induced by pentobarbital, spend sleeping (Koo et al. 2003). In addition, when mice were subjected to inhalation of essential oils and then artificially induced into overagitation by intraperitoneal application of caffeine, their motility decreased (Buchbauer et al. 1993). Because it interferes with GABA neurotransmission in the central nervous system, lavender oil has sedative and relaxant effects. On inhalation, dental patients experienced a reduction in anticipatory anxiety (Woronuk et al. 2011). Both gram-positive and gram-negative pathogenic bacteria were inhibited by cinnamon and clove oil during the vapor phase (Lopez et al. 2005). While rosemary oil improved the overall quality of memory but slowed its speed, lavender oil hampered working memory performance and slowed reaction times for memory and attention-based tasks (Moss et al. 2003). Lavender essential oil is responsible for the spasmolytic activity in the calcium chloride-induced contraction. 1,8-cineole is primarily responsible for its activity. As a result, its primary function is as a sedative (Gamez et al. 1990). Balm oil has been shown to be effective in the treatment of agitation in people with dementia, according to one study. Neroli oil proved to be very helpful for patients who had heart surgery; a foot massage had a long-lasting psychological effect, according to the study (Stevensen 1994). Aromatherapy was given to hypertensive women in a study, and the results showed that it had a significant impact on systolic blood pressure (SBP), diastolic blood pressure (DBP), and quality of sleep. This led the researchers to the conclusion that aromatherapy can greatly improve daily life quality.

2.4.7 Rectal and Vaginal Suppositories

Essential oil is used in a vaginal douche to treat candida albicans, a type of vaginal infection. Tea tree oil applied to a tampon is very effective for many infections in the vaginal area. Essential oils are used to treat gynecological or urinary conditions through the vaginal route because they are directly absorbed into the surrounding tissue. Additionally, urinary infections were treated with sandalwood oil. An ointment containing tea tree oil can be used in a double-blind study to treat hemorrhoids (Joksimovic et al. 2012).

2.4.8 Effective Pesticides

Additionally, mites and insects are neurotoxic when exposed to essential oil with particular constituents. Natural alternatives have been viewed as superior to conventional synthetic sources due to their lower risk to the environment and detrimental impact on humans and other living things. Clove oil (eugnol), thyme oil (thymol and carvacrol), mint oil (menthol, pulegone), lemon grass oil (citronellal, citral), cinnamon oil (cinnamaldehyde), rosemary oil (1,8-cineole), and oregano oil (carvacrol) are the essential oils used as pesticides and their constituents (Rai and Carpinella 2006). They use a variety of different mechanisms on insects to cause toxicity. Many essential oil constituents work by blocking these receptors in arthropods and many insects that use octopamine as a neurotransmitter and neuromodulator (Enan 2001, 2005). In insects, some essential oils may work by disrupting the cell membrane or by blocking the tracheal system. However, when compared to the synthetic pesticides that are currently in use, their efficacy is still low. Nevertheless, phenols outperform other monoterpenes as a pesticide.

2.4.9 Insect Repellents

Lemongrass oil and mint oil are combined to make insect repellent that repels mosquitoes. Citronella is its active ingredient (Fradin and Day 2002). Also, these repellents are used to get rid of cockroaches and keep them from coming back into human homes (Ngoh et al. 1998). The study reveals that pulegone is the most effective against adult lice, followed by terpinen-4-ol. Nerolidol and thymol, on the other hand, are more active in P. humanus eggs (Priestley et al. 2006).

2.5 Extraction of Plant Essential Oils

Using a variety of extraction techniques, essential oils can be extracted from a variety of plant parts. The botanical material that is used typically determines both the method that is utilized for extraction and the manufacturing process for essential oils. Another factor taken into consideration is the state and form of the material.

One of the most important factors in determining the quality of essential oil is the extraction method.

According to Reverchon and Senatore (1992), the most common method for extracting plant essential oils is steam distillation. According to Masango (2005), steam distillation extracts 93% of essential oils, while the remaining 7% can be further extracted in other ways.

The secretory glandules are specialized histological structures in which essential oils are biosynthesized, accumulated, and stored (Bouwmeester et al. 1995; Bruneton et al. 1987). According to Svoboda and Greenaway (2003), there are two kinds of secretory glandules: those that secrete from the plant through exogenous means and those that secrete from the plant through its own internal organs. In one or more plant organs, they are also concentrated in the cytoplasm of some secretory cells.

2.5.1 Tissue for External Secretion

The papillae of the skin are conical epidermal cells that secrete essences typically found in Rosa spp. flower petals. The secretory bristles, or glandular trichrome originate from epidermal cells. They are typical of the Lamiaceae family and serve as the site of essential oil biosynthesis and accumulation (Turner et al. 2000). A pocket between secretory cells and a common cuticle holds the produced essential oil.

2.5.2 Tissue for Internal Secretion

The plant itself contains this tissue. The secretory canals that separate them are tiny canals (Apiaceae) whose walls are made of seated secreting cells and sometimes span the entire length of the plant. The secretory, or schizogenous pocket is a spherical intercellular space that is filled by droplets of essential oils made by the cells that border it. Cells that secrete intracellularly are isolated with the ability to secrete and store EOs in their vacuoles. These cells, such as cinnamon cells, laurel leaves, and the calamus rhizome, die when the concentration of EOs reaches high levels.

Experts consider essential oils to be the source of chemical signals that enable plants to control and regulate their environment. Climate, soil composition, plant organ, age, and vegetative cycle stage may all influence the quality, quantity, and composition of the essential oil extracts (Masotti et al. 2003; Angioni et al. 2006).

2.5.3 Methods for Essential Oil Extraction

Several extraction techniques are used to obtain essential oils from plant raw materials (Wang and Weller 2006; Starmans and Nijhui 1996). There are two categories of these techniques—advanced/innovative methods in addition to conventional/classical methods. In recent decades, research into new technologies (ultrasound, microwave) has resulted in the development of novel, more effective extraction procedures (reduction of extraction time and energy consumption, increase of extraction yield, enhancement of essential oils' quality).

There are two standard methods for extracting essential oils:

2.5.3.1 Traditional/Conventional Approach

To extract essential oils from the plant matrix, these are standard procedures that use heating to distill water.

Hydrodistillation

According to Meyer-Warnod (1984), this is the oldest and most straightforward method for extracting essential oils. Avicenna, was the first to develop alembic extraction. He has obtained the rose's first pure essential oil. In the alembic, the plant material is directly submerged in the water and brought to a boil. The extraction apparatus has a heating source that is topped by an alembic-shaped vessel in which we could put plant material and water. In point of fact, the molecules of water and essential oils combine to form a heterogeneous mixture at atmospheric pressure and during the extraction process (heating). While the essential oil components' boiling point is extremely high, the water molecules' boiling point is closer to 100 °C. Water has the advantage of being immiscible with the majority of essential oils' molecules. As a result, essential oils can be easily separated from water by decanting after condensation. Additionally, it makes it possible to recover almost all of the essential oils in the vapor through the plate column.

Steam-Water Entrainment

One of the official ways to get essential oils is through this method. According to Masango (2005), it is a widely used method for extracting essential oils. The only difference between this and hydrodistillation is that the plant does not come into direct contact with the water.

Vapor-Hydrodistillation

Except for a system of perforated plates or grids that keep the plant suspended above the still base and out of direct contact with the water, extraction takes place within the alembic. Water vapors are injected into the plant matter to carry the volatile materials and complete the extraction. There are fewer artifacts. The extraction time is reduced, and polar molecules are lost.

Steam Distillation or Vapor Distillation

Although the production of vapors takes place outside of the distillation alembic, this method has the same principles and benefits as vapor-hydrodistillation (Masango 2005). After that, the steam can be superheated or saturated; the steam enters the lower portion of the extractor at slightly above atmospheric pressure and thus passes through.

Hydrodiffusion

This is an example of vapor distillation, in which vapors flow downward. It is also known as down hydrodiffusion.

Extraction with Organic Solvents

An organic solvent is used to macerate the plant material, by removing the solvent at a lower pressure; the extract is concentrated. Compared to hydrodistillation, this cold extraction method prevents chemical effects and alterations. In fact, when plants are submerged in boiling water during hydrodistillation, some fragrance components become soluble in water and the medium pH drops to 4–7 (sometimes less than 4 for some fruits). The original plant species' constituents undergo chemical modifications (hydrolysis, deprotonations, hydrations, and cyclizations) as well as the combined effects of heat and acid. Obtained essential oils differ significantly from the original essence, particularly when pH is low and boiling is prolonged. Contrarily, the foods and fragrances to which they are added are polluted by residues in organic solvent-derived extracts (Faborode and Favier 1996).

Cold Pressing

The most common method for extracting essential oils from citrus fruit zest is cold pressing. Oil sacs, also known as oil glands, break open during extraction and release volatile oils that are localized in the exterior of the mesocarp. By cold pressing, this oil is removed mechanically, resulting in a watery emulsion. Centrifugation is then used to recover the oil (Ferhat et al. 2007). In this instance, we get the citrus zest's vegetable essence, which is used as a flavoring ingredient or additive in the food and pharmaceutical industries (including cosmetics and some home care products).

2.5.3.2 Innovative Methods

The thermolability of essential oil components, which undergo chemical alterations (hydrolyse, isomerization, oxidation) at high applied temperatures, is one of the drawbacks of conventional methods. As a result, the extracted essential oils suffer greatly in quality, particularly when extracted for an extended period of time. It is critical that extraction techniques preserve the chemical composition and natural proportions of essential oils. Additionally, new extraction methods must cut down on extraction times, energy use, solvent consumption, and CO_2 emissions.

Supercritical Fluid Extraction (SCFE)

At well-defined conditions, fluids reach the supercritical state, critical temperature (Tc), and pressure (Pc). Then, fluids might exhibit properties that are very interesting: (i) a density that is close to that of liquids; (ii) a high diffusivity; and (iii) a low viscosity. Due to its numerous benefits, carbon dioxide is typically the solvent of choice for the extraction of essential oils: (i) it is simple to reach the critical point; (ii) insensitive to thermolabile plant essence molecules; (iii) it is chemically inert and nontoxic; (iv) it is not flammable; (v) it is readily available in high purity at a relatively low cost; v) it is simple to remove its traces from the extract by simple depression (Pourmortazavi and Hajimirsadeghi 2007), and (vii) its polarity is similar to pentane, making it suitable for lipophilic compound extraction. Several essential oils were extracted using SCFE (Braga et al. 2005; Carvalho et al. 2005; Khajeh et al. 2004; Aghel et al. 2004). In order to separate the obtained extract from the fluid, the extract is routed to one or more separators, where the CO_2 is gradually decompressed (thereby losing its solvent power).

Liquids for Subcritical Extraction (H_2 and CO_2)

The use of water in its subcritical state for the extraction of essential oils was demonstrated in some studies. When the temperature is lower than the critical temperature (Tc) but the pressure is higher than the critical pressure (Pc), or vice versa, the system enters the subcritical state. Water and CO_2 are the most commonly used fluids for EOs extraction at this point. The properties of obtained fluids are very interesting: low viscosity, a density that is comparable to that of liquids, and diffusivity that is comparable to that of liquids and gas. Compounds that are thermolabile and volatile are not lost or degraded in this way. The essential oils produce little residues with excellent efficiency and quality.

Extraction with Subcritical CO_2

When the temperature is between 31 and 55 °C and the pressure is between 0.5 and 7.4 MPa, CO_2 enters the subcritical state. According to Moyler (1993), CO_2 acts as a non-polar solvent under these conditions. The high temperatures and presence of water in steam distillation and entrainment by vapor cause degradations, which this method avoids. The flavors of the extracts made using this method are very similar to those of fresh vegetable raw materials. Additionally, subcritical CO_2 extracts have significantly higher quality than subcritical water extracts.

Ultrasound-Assisted Extraction of Essential Oils (UAE)

In 1950, laboratory-scale equipment was used to develop this method (Vinatoru 2001). When combined with other methods (hydrodistillation and solvent extraction), ultrasound speeds up the release of essential oils from plant material, allowing for selective and intensified extraction. The vegetable raw material is submerged in water or a solvent while simultaneously being affected by ultrasound. Many essential oils, particularly from seeds, have been extracted using this method (Assami et al. 2012; Sereshti et al. 2012). However, Chemat et al. (2006) state that it was developed specifically for the purpose of extracting specific molecules of therapeutic interest.

Two kinds of phenomena are involved in the extraction mechanism, once the cell walls are broken, diffusion moves through them and removes the contents of the cells (Vinatoru 2001). In point of fact, the plant stores essential oils in particular internal or external structures in the form of glands containing essential oil droplets. Sonication is capable of easily destroying their extremely thin skins (in the case of external structures). The milling degree of plant material has a significant impact on yield for internal ones. Increased cell exposure to ultrasonically induced cavitations is evident when plant material is reduced in size.

Microwave-Assisted Extraction (MAE)

The most prevalent frequency, 2450 MHz, which corresponds to a wavelength of 12.2 cm, is an electromagnetic wave with a frequency between 300 MHz and 30 GHz and a wavelength between 1 cm and 1 m. The use of microwave dielectric heating for the extraction of essential oils has received more attention. Starting with compressed air microwave distillation (CAMD) and vacuum microwave hydrodistillation (VMHD), innovation in microwave-assisted extraction (MAE) led to the creation of numerous variations, including microwave-assisted hydrodistillation (Golmakani and Rezae 2008), solvent-free microwave extraction (SFME), microwave-accelerated steam distillation (MASD) (Chemat et al. 2006), microwave steam distillation (Sahraoui et al. 2008), microwave hydrodiffusion and gravity (MHG) (Vian et al. 2008) as well as portable microwave-assisted extraction (PMAE). Microwave-assisted extraction, primarily created by Chemat and his associates (2006), quickly became one of the newest and most promising methods for extracting essential oils. It has simplified manipulation, reduced solvent consumption, and lower energy requirements in addition to offering high reproducibility in shorter times.

Solvent-Free Microwave Extraction (SFME)

Chemat and his associates created this strategy. It consists of microwave dry distillation of a fresh plant at atmospheric pressure without the addition of water or any organic solvent, based on the combination of microwave heating energy and dry distillation. Tissues swell and the glands and oleiferous receptacles burst when the in situ water content of plant material is selectively heated. Thus, essential oils are liberated, and they spontaneously evaporate through azeotropic distillation with the plant material's water. A pilot scale was suggested, and it was found to be practical for use in industry compared to a scale from the SFME Lab. This method was used to extract a large number of essential oils at a laboratory scale. While conventional hydrodistillation takes 2 h, this method can isolate and concentrate volatile compounds in just 30 min.

Microwave Hydrodiffusion and Gravity (MHG)

Chemat and his associates were the first to conceptualize and create MHG. Vian et al. (2008) came up with a method that used earth gravity and microwave heating of a reversed alembic at atmospheric pressure. In a reversed microwave reactor, no water or solvent is added to the plant material. Essential oils and plant water are released outside of the plant material as a result of the internal heating of the plant material's water, which disperses the plant cells and causes the glands and oleiferous receptacles to rupture (by heating microwave action). Extracts are driven from the microwave reactor's top to its bottom by gravity to the cooling system. Chemat et al. (2006) identified microwave hydrodiffusion and gravity (MHG) as an environmentally friendly, cost-effective strategy.

Microwave Steam Diffusion (MSDf) and Distillation (MSD) Processes

Sahraoui et al. (2008) have investigated the MSD and the extraction of dry lavender flowers and essential oils extracted from orange peel. This innovative method outperforms traditional steam distillation, providing important benefits such as noticeably reduced extraction time and better features. It also gives essential oils better sensory properties—better reproduction of the citrus essential oil's natural fresh fruit aroma without significantly altering the volatile oil composition. With the addition of a microwave coaxial antenna that is suitable for the extraction of 100 kg of fresh plant material, MSD could be exploited at a large industrial scale due to its performance (Flamini et al. 2007). We looked into the microwave steam diffusion (MSDf).

The Instantaneous Controlled Drop in Pressure

K. Allaf and his associates specifically looked into this method for extracting essential oils (Kristiawan et al. 2008; Berka-Zougali et al. 2012). Essential oils and antioxidants from vegetable matrices (Allaf et al. 2013), as well as both a pilot plan and a laboratory apparatus, were used to test this. For Lavandin essential oil, the instant controlled pressure drop (DIC) technology provides a better extraction yield, a shorter extraction time, and a significant reduction in energy and water consumption. Following the treatment type HTST (high temperature/high pressure-short term), DIC is primarily characterized by a sharp decrease in pressure to the vacuum. When compared to progressive autovaporization, the phenomenon of abrupt autovaporization permits the evaporation of a greater number of volatile molecules and a rapid drop in temperature. There are typically four stages to DIC treatment: (i) creating an initial void; (ii) utilizing a steam bath at a predetermined temperature and pressure; (iii) instantaneous surrender to the void; and (iv) cell processing is brought back to atmospheric pressure. The recovered plant material could then be dried at room temperature and stored for future extractions (Allaf et al. 2013).

Solvent Extraction

Solvent extraction has been used traditionally for flower materials that are fragile or delicate and cannot withstand the heat of steam distillation. For extraction, a variety of solvents, such as acetone, hexane, petroleum ether, methanol, or ethanol, can be used (Areias et al. 2000; Pizzale et al. 2002; Kosar et al. 2005). However, this method takes a lot of time, so the oils cost more than with other methods. Researchers used the solvent extraction method to extract *Thymus praecox* subsp. *skorpilii* var. *skorpilii* (TPS) essential oil with antioxidant activity. Thymol (40.31%) and o-cymene (13.66%) were found to be the primary components of TPS essential oil. Significant free-radical scavenging was accomplished by the water, ethanol, and methanol extracts. Total phenolics (6.211 mg gallic acid/g dry weight) and flavonoids (0.809 mg quercetin/g dry weight) are highest in the water extract. Additionally, the antioxidant activity of the water extract was higher than that of other extracts (hexane, dichloromethane, ethyl acetate, and methanol). However, due to inadequate removal, solvent residue may remain in the final product. According to Ferhat et al. (2007), this may affect the immune system, cause allergies, and be toxic.

According to Deng et al. (2005), conventional methods like steam distillation and solvent extraction have drawbacks like lengthy preparation times and a high concentration of organic solvents. As a result, the use of supercritical fluids as a substitute medium for essential oil extraction has been considered. Due to its low critical conditions, carbon dioxide (CO_2) is the most frequently utilized supercritical fluid (Hawthorne et al. 1993; Jimenez-Carmona et al. 1999; Senorans et al. 2000). The volatile compounds, like monoterpenes, can be extracted from the supercritical fluid extraction (SFE) effluent by more than 90%, which is equivalent to Hydrodistillation (HD) after 4 h. Some organic compounds that HD could not extract were retrieved by SFE (Hawthorne et al. 1993). According to Pereira and Meireles (2007), supercritical fluid extraction is more cost-effective than steam distillation. This is primarily due to the latter's lower yield and higher energy consumption.

Under dynamic conditions, the subcritical water, also known as pressurized hot water, has been introduced as an extractant. This type of water has a pressure that is high enough to keep water in the liquid state and at a temperature of between 100 and 374 °C. This method saves a lot of money on energy and plant material, yields a more valuable essential oil (with a higher concentration of oxygenated compounds and no significant terpenes), and is quicker (15 min compared to 3 h). When compared to that which was obtained by boiling the mixture for 2 h and then extracting it using the Soxhlet apparatus for 6 h, the yields were 40 to 60% lower.

2.6 Plant Essential Oil Chemotypes

Due to their bioactive potential, a wide range of chemical compounds found in plants have been investigated as alternatives to conventional insecticides (Silva et al. 2007; Rattan 2010). The essential oils of plants, which are mixtures of various compounds and typically have high volatility, have demonstrated lethal and sublethal effects on a variety of organisms. The complexity of essential oils and the potential for their components to have synergistic or additive effects are major contributors to their effectiveness in insect control. The rapid degradation (e.g., high volatility) of plant chemical compounds, which reduces environmental contamination and adverse effects on non-target organisms, has served as the primary justification for their use in pest management (Isman and Machial 2006).

Due to the fact that some of these insects result in significant financial losses, conventional pest control of stored grain is a global concern. In the case of stored grains, insecticide residues could be harmful to human health. As a result, the potential of plant essential oils as an alternative for controlling stored grain insects through a variety of exposure routes has been examined in a number of studies (Liu and Ho 1999; Bouda et al. 2001). The insects Sitophilus zeamais and Tribolium castaneum are regarded as widespread and significant pests due to their losses in stored grains. On S. zeamais and T. castaneum, the toxicity and repellency of essential oils from various *Lippia alba* genotypes (carvone chemotypes LA-13 and LA-57 and citral chemotypes LA-10 and LA-44) as well as their major monoterpenes, carvone and citral, were examined. To see if they could be used as infraspecific taxonomic characters, the essential oil compositions of 16 accessions of Ocimum basilicum belonging to various varieties were determined using fresh and freeze-dried leaves. Additionally, one O. citriodorum accession was investigated. There were 30 monoterpenoids, sesquiterpenoids, and phenylpropanoids found, with geranial and neral in O. x citriodorum and linalool, methyl chavicol, eugenol, methyl eugenol, and geraniol in O. basilicum accounting for more than 20% of the essential oil composition. In the O. basilicum accessions studied, five major essential oil profiles could be identified using a combination of the latter compounds. Within O. basilicum, essential oil patterns and varietal classification appeared to have little correlation. O. basilicum's use as a crop that produces essential oils, a culinary herb, a medicinal plant, and an insect-controlling agent all involve chemicals, so the infraspecific classification of this taxon should take chemical characteristics into account. The essential oil chemotypes in O. basilicum are proposed to be categorized using a method.

2.7 Plant Secondary Metabolites Characterization Techniques

2.7.1 Gas Chromatography

Gas chromatography-mass spectrometry (GC-MS) methods were used to examine bioactive metabolites, and after that, the methanolic extract's in vitro antibacterial and antifungal activity was assessed. Compound analysis using gas chromatography and mass spectrometry was done on an *E. coli* methanolic extract. The 23 identified compound peaks from the GC-MS chromatogram were displayed. Dodecanoic acid, 3-hydroxyl, was identified as the first set up peak. The 13-Tetradecynoic acid, methyl ester, was identified as the second peak (Al-Rubaye et al. 2017).

Eight intriguing matrices from an olive tree, Picudo cv., were examined using the robust LC-ESI/APCI-QTOF MS and GC-APCI-QTOF MS methodologies. The results offered a thorough coverage of their secondary metabolites (141 substances were identified in LC-MS and 58 in GC-MS), as well as trustworthy information about the distribution of their phytochemicals (Olmo-García et al. 2018).

2.7.2 Liquid Chromatography

The large variety of compounds that can be present, which vary in polarity and size (from simple phenolic acids to tannins), as well as the fact that many of these compounds are present in food products at low concentration levels, make the analysis of polyphenols in food samples relatively difficult. The extraction and treatment of samples, as well as their separation, determination, and identification, have all been hampered by the chemical diversity of polyphenols. The most efficient method for structural characterization and identification of both low and high molecular weight polyphenols in food samples is liquid chromatography combined with mass spectrometry (LC-MS) or tandem mass spectrometry (LC-MS/MS) (Lucci et al. 2017).

For additional investigation using chemometric techniques like principal component analysis (PCA) and partial least square discriminant analysis, both LC-MS and LC-HRMS (Liquid chromatography–high resolution mass spectrometry) produce data of remarkable quality (PLS-DA). Concentrations of relevant polyphenols (profiling technique) or instrumental signals made up of intensity counts as a function of m/z and retention time make up the data to be evaluated (fingerprinting approach). Additional data treatments have proven to be quite effective at making it easier to extract pertinent data on the descriptive and functional properties of food products that can subsequently be used for characterization, classification, and authentication.

2.7.3 Nuclear Magnetic Resonance (NMR)

Nuclear magnetic resonance (NMR) is frequently used to investigate secondary metabolites in unprocessed plant extracts. It is the most effective method for identifying unknown complex compounds' structural composition. However, before acquiring NMR spectra, substances must first be isolated and purified for classical NMR research. Over the past 20 years, physical coupling between liquid chromatography and NMR has been developed to shorten these time-consuming stages. Only in the past 10 years have routine LC-NMR applications been successfully used in practice.

2.8 Phytochemical Genomics: An Emerging Trend

Plants produce a wide range of phytochemicals, which are specialized or secondary metabolites and many of which are useful to people as medications and other health-improving elements. In contrast, some phytochemicals are toxic to people, hence measures must be taken to lower their concentration in diet. The study of the genetic basis of the biosynthesis and functionality of phytochemicals, notably based on sophisticated metabolomics, is known as phytochemical genomics. Metabolomics is frequently used in conjunction with Quantitative trait locus (QTL) analysis of inbred lines and wild variants to understand the genetic basis of phytochemical synthesis (Kliebenstein 2009; Keurentjes et al. 2006; Fernie and Schauer 2009; Schauer et al. 2006). A crucial aspect of phytochemical genomics is metabolomics (Saito and Matsuda 2010). To pinpoint the precise loci linked to these metabolic characteristics, relatively large numbers of samples need to be studied. The mechanisms behind the production and use of a great variety of plant metabolites, as well as their control and evolution, are currently the subject of intense research in the field of secondary plant metabolism. A decade ago, such incredible improvement would not have been anticipated, and it has just recently occurred (Yonekura-Sakakibara and Saito 2009). No one would have predicted genomics-based studies of hundreds of medicinal or exotic plant species before the development of phytochemical genomics technology; genomics was only used for model plants or important crops, whose genome sequences could only be revealed by sizable international research consortia. Yet, recently created mass DNA sequencers have made thorough genome and transcriptome sequencing feasible for regular phytochemical researchers who are looking into the synthesis of certain plant metabolites out of curiosity. One can create testable hypotheses regarding the activities of genes and metabolites by merging data sets from metabolomics and genomics/transcriptomics and understanding interactions between them (Higashi and Saito 2013; Yonekura-Sakakibara et al. 2013). The model plant Arabidopsis thaliana has a greater chemical diversity than previously thought, and an integrated investigation of the transcriptomes and metabolomes has largely helped to explain the relationships between genes and metabolites.

2.9 Conclusion

Essential oils are among the natural plant products that merit special consideration due to their widespread use in traditional therapeutic practices around the world. In order to increase recovery yields and ensure that the extraction processes do not damage the bioactive components of the essential oils, future studies should concentrate on extraction techniques that are both practical and affordable. In a variety model organisms, antibacterial, antioxidant, anti-inflammatory, and anticancer actions have been particularly proven. Conducting studies to produce higher oil yields is one of the most crucial factors to take into account for the use of essential oils in combating pathogenic bacteria resistant to antibiotics. Moreover, the extraction of essential oils from plant organs through distillation is today a very dependable and cost-effective process. Many preclinical research have examined the biological functions of essential oils in relation to their effectiveness, illuminating their mode of action and pharmacological targets in the process. To attain a high level of scientific evidence and determine the true efficacy and safety of plant products, more carefully planned clinical trials are required. Essential oils are generally thought to be safe, but because they are complex mixtures of substances, some of which are known to be skin sensitizers and allergens, they must be disclosed on cosmetics' labels, especially for customers with sensitive, allergic-prone skin or pre-existing skin disorders who may choose to undergo patch testing before using products containing them. Specificity and cytotoxicity must be considered for an effective preventive and therapeutic usage of essential oils or their active components. In addition, there is still a need to design and engineer novel techniques, in order to selectively address essential oil bioactive components to selected targets, which might accommodate the cytotoxicity feature. Considering their high molecular diversity and broad spectrum of activity, essential oils and their components would be among the finest agents to suggest for preventative medicine once cytotoxicity is under control.

References

Abdollahi M, Ranjbar A, Shadnia S, Nikfar S, Rezaie A (2004) Pesticides and oxidative stress: a review. Med Sci Monit 10:RA141–RA147

Aghel N, Yamini Y, Hadjiakhoondi A, Pourmortazavi SM (2004) Supercritical carbon dioxide extraction of *Mentha pulegium* L. essential oil. Talanta 62:407–411

Allaf T, Tomao V, Ruiz K, Chemat F (2013) Instant controlled pressure drop technology and ultrasound assisted extraction for sequential extraction of essential oil and antioxidants. Ultrason Sonochem 20:239–246

Allahverdiyev A, Durgran N, Ozguven M (2004) Antiviral activity of volatile oils of Melissa L. against herpes simplex virus type-2. Phytomedicine 11:657–661

Al-Rubaye AF, Hameed IH, Kadhim M (2017) Uses of gas chromatography-mass spectrometry (GC-MS) technique for analysis of bioactive natural compounds of some plants. Int J Toxicol Pharmacol Res 9:948–953

Angioni A, Barra A, Coroneo V, Dessi S, Cabras P (2006) Chemical composition, seasonal variability, and antifungal activity of Lavandula stoechas L. ssp. stoechas essential oils from stem/ leaves and flowers. J Agri Food Chem 54:4364–4370

Antony HA, Parija SC (2016) Antimalarial drug resistance: an overview. Trop Parasitol 6:30–41

Aoshima H, Hamamoto K (1999) Potentiation of GABAA receptors expressed in Xenopus oocytes by perfume and phytoncid. Biosci Biotechnol Biochem 63:743–748

Areias F, Valentão P, Andrade PB, Ferreres F, Seabra RM (2000) Flavonoids and phenolic acids of sage: influence of some agricultural factors. J Agri Food Chem 48:6081–6084

Asdadi A, Hamdouch A, Oukacha A, Moutaj R (2015) Study on chemical analysis, antioxidant and in vitro antifungal activities of essential oil from wild Vitex agnus-castus L. seeds growing in area of Argan tree of Morocco against clinical -strains of Candida responsible for nosocomial infections. J Mycol Med 25:118–127

Assami K, Pingret D, Chemat S, Meklati BY, Chemat F (2012) Ultrasound induced intensification and selective extraction of essential oil from Carum carvi L. seeds. Chem Eng Process Process Intensif 62:99–105

Bajpai VK, Baek KH, Kang SC (2012) Control of salmonella in food by using essential oils: a review. Food Res Int 45:722–734

Bajwa S, Kulshrestha A (2013) Fungal infections in intensive care unit: challenges in diagnosis and management. Ann Med Health Sci Res 3:238–244

Bakkali F, Averbeck S, Averbeck D, Idaomar M (2008) Biological effects of essential oils- a review. Food Chem Toxicol 46:446–475

Beesley A, Hardcastle J, Hardcastle PT, Taylor CJ (1996) Influence of peppermint oil on absorptive and secretory processes in rat small intestine. Gut 39:214–219

Benchaar C, Calsamiglia S, Chaves AV, Fraser GR, Colombatto D, McAllister TA (2008) A review of plant derived essential oils in ruminant nutrition and production. Anim Feed Sci Technol 145:209–228

Berka-Zougali B, Ferhat MA, Hassani A, Chemat F, Allaf KS (2012) Comparative study of essential oils extracted from *Algerian Myrtus* communis L. leaves using microwaves and hydrodistillation. Internat J Mol Sci 13:4673–4695

Bouda H, Tapondjou LA, Fontem DA, Gumedzoe MYD (2001) Effect of essential oils from leaves of Ageratum conyzoides, Lantana camara and Chromolaena odorata on the mortality of Sitophilus zeamais (Coleoptera, Curculionidae). J Stored Prod Res 37:103–109

Bouwmeester HJ, Davies JA, Smid HG, Welten RS (1995) Physiological limitations to carvone yield in caraway (Carum carvi L.). Indus Crop Prod 4:39–51

Braga ME, Ehlert PA, Ming LC, Meireles MA (2005) Supercritical fluid extraction from Lippia alba: global yields, kinetic data, and extract chemical composition. J Supercritical Fluids 34:149–156

Bruneton JN, Normand F, Balu-Maestro C, Kerboul P, Santini N, Thyss A, Schneider M (1987) Lymphomatous superficial lymph nodes: US detection. Radiology 165:233–235

Buchbauer G, Jirovetz L, Jäger W, Plank C, Dietrich H (1993) Fragrance compounds and essential oils with sedative effects upon inhalation. J Pharm Sci 82:660–664
Carvalho RN, Moura LS, Rosa PT, Meireles MA (2005) Supercritical fluid extraction from rosemary (Rosmarinus officinalis): kinetic data, extract's global yield, composition, and antioxidant activity. J Supercritical Fluids 35:197–204
Cha JD, Moon SE, Kim HY, Cha IH, Lee KY (2009) Essential oil of Artemisia capillaris induces apoptosis in KB cells via mitochondrial stress and caspase activation mediated by MAPK-stimulated signaling pathway. J Food Sci 74:T75–T81
Chemat F, Lucchesi ME, Smadja J, Favretto L, Colnaghi G, Visinoni F (2006) Microwave accelerated steam distillation of essential oil from lavender: a rapid, clean and environmentally friendly approach. Anal Chim Acta 555:157–160
Chen Y, Zeng H, Tian J, Ban X, Ma B, Wang Y (2014) Dill (Anethum graveolens L.) seed essential oil induces Candida albicans apoptosis in a metacaspase-dependent manner. Fungal Boil 118:394–401
Cho SM, Lee EO, Kim SH, Lee HJ (2014) Essential oil of Pinus koraiensis inhibits cell proliferation and migration via inhibition of p21-activated kinase 1 pathway in HCT116 colorectal cancer cells. BMC Complement Altern Med 14:275
Chouhan S, Sharma K, Guleria S (2017) Antimicrobial activity of some essential oils-present status and future perspectives. Medicines (Basel) 4:58
Crowell PL (1999) Prevention and therapy of cancer by dietary monoterpenes. J Nutr 129:775S–778S
Da Cruz CL, Pinto VF, Patriarca A (2013) Application of plant derived compounds to control fungal spoilage and mycotoxin production in foods. Int J Food Microbiol 166:1–14
De Logu A, Loy G, Pellerano ML, Bonsignore L, Schivo ML (2000) Inactivation of HSV-1 and HSV-2 and prevention of cell-to-cell virus spread by Santolina insularis essential oil. Antivir Res 48:177–185
Deng C, Yao N, Wang A, Zhang X (2005) Determination of essential oil in a traditional Chinese medicine, *Fructus amomi* by pressurized hot water extraction followed by liquid-phase microextraction and gas chromatography–mass spectrometry. Anal Chim Acta 536:237–244
Dhifi W, Bellili S, Jazi S, Bahloul N, Mnif W (2016) Essential oils' chemical characterization and investigation of some biological activities: a critical review. Medicines 3:25
Djilani A, Dicko A (2012) The therapeutic benefits of essential oil. In: Nutritional well-being and health. IntechOpen, pp 155–172
Dorman H, Deans S (2000) Antimicrobial agents from plants: antibacterial activity of plant volatile oils. J Appl Microbiol 88:308–316
Edris AE (2007) Pharmaceutical and therapeutic potentials of essential oils and their individual volatile constituents: a review. Phytother Res 21:308–323
Ellse L, Wall R (2014) The use of essential oils in veterinary ectoparasite control: a review. Med Vet Entomol 28:233–243
Enan E (2001) Insecticidal activity of essential oils: octopaminergic sites of action. Comparative biochemistry and physiology. Toxicology & Pharmacol:CBP 130:325–337
Enan EE (2005) Molecular and pharmacological analysis of an octopamine receptor from american cockroach and fruit fly in response to plant essential oils. Arch Insect Biochem Physiol 56:161–171
Evans WC (2009) Trease and Evans Pharmacognosy. Saunders–Elsevier, Edinburgh, pp 106–116
Faborode MO, Favier JF (1996) Identification and significance of oil-point in seed-oil expression. J Agri Engg Res 65:335–345
Ferhat MA, Meklati BY, Chemat F (2007) Comparison of different isolation methods of essential oil from citrus fruits: cold pressing, hydrodistillation and microwave 'dry'distillation. Flav Frag J 22:494–504
Fernie AR, Schauer N (2009) Metabolomics-assisted breeding: a viable option for crop improvement? Trends Genet 25:39–48

Ferreira FD, Mossini SAG, Ferreira FMD, Arrotéia CC, da Costa CL, Nakamura CV, Machinski Junior M (2013) The inhibitory effects of Curcuma longa L. essential oil and curcumin on aspergillus flavus link growth and morphology. Sci World J 343804

Fisher K, Phillips C (2008) Potential antimicrobial uses of essential oils in food: is citrus the answer? Trends Food Sci Technol 19:156–164

Flamini G, Tebano M, Cioni PL, Ceccarini L, Ricci AS, Longo I (2007) Comparison between the conventional method of extraction of essential oil of Laurus nobilis L. and a novel method which uses microwaves applied in situ, without resorting to an oven. J Chromatogr A 1143:36–40

Fradin MS, Day JF (2002) Comparative efficacy of insect repellents against mosquito bites. N Engl J Med 347:13–18

Freires De Almeida I, Murata RM, Furletti VF, Sartoratto A, de Alencar SM, Figueira GM, de Oliveira Rodrigues JA, MCT D, Rosalen PL (2014) Coriandrum sativum L. (coriander) essential oil: Antifungal activity and mode of action on Candida spp., and molecular targets affected in human whole-genome expression. PLoS One 9:e99086

Freires IA, Denny C, Benso B, de Alencar SM, Rosalen PL (2015) Antibacterial activity of essential oils and their isolated constituents against cariogenic bacteria: a systematic review. Molecules 20:7329–7358

Friedly EC, Crandall PG, Ricke SC, Roman M, O'Bryan C, Chalova VI (2009) In vitro antilisterial effect of citrus oil fractions in combination with organic acids. J Food Sci 74:M67–M72

Gamez MJ, Jimenez J, Navarro C, Zarzuelo A (1990) Study of the essential oil of Lavandula dentata L. Pharmazie 45:69–70

Gautam N, Mantha AK, Mittal S (2014) Essential oils and their constituents as anticancer agents: a mechanistic view. Biomed Res Int 154106:1

Golmakani MT, Rezae K (2008) Comparison of microwave-assisted hydrodistillation with the traditional hydrodistillation method in the extraction of essential oils from Thymus vulgaris L. Food Chem 109:925–930

Grosch W (2007) Gas chromatography-Olfactometry of aroma compounds. In: Flavours and Fragrances. Springer, Berlin Germany, pp 363–378

Hader AA, Hasan Z, Agel M (1994) Hyperglycemic and insulin release inhibitory effects of Rosmarinus officinalis. J Ethanopharmacol 43:217–221

Hawthorne SB, Rickkola ML, Screnius K, Holm Y, Hiltunen R, Hartonen K (1993) Comparison of hydrodistillation and supercritical fluid extraction for the determination of essential oils in aromatic plants. J Chromatogr A 634:297–308

Helander IM, Alakomi HL, Latva-Kala K, Mattila-Sandholm T, Pol L, Smid EJ (1998) Characterization of the action of selected essential oil components on gram negative bacteria. J Agric Food Chem 46:3590–3595

Higashi Y, Saito K (2013) Network analysis for gene discovery in plant-specialized metabolism. Plant Cell Environ 36:1597

Holley RA, Patel D (2005) Improvement in shelf-life and safety of perishable foods by plant essential oils and smoke antimicrobials. Food Microbiol 22:273–292

Huang L, Abuhamdah S, Howes MJ, Dixon CL, Elliot MS, Ballard C, Holmes C, Burns A, Perry EK, Francis PT, Lees G, Chazot PL (2008) Pharmacological profile of essential oils derived from Lavandula angustifolia and Melissa officinalis with anti-agitation properties: focus on ligand-gated channels. J Pharm Pharmaco 60:1515–1522

Hyldgaard M, Mygind T, Meyer RL (2012) Essential oils in food preservation: mode of action, synergies, and interactions with food matrix components. Front Microbiol 3:12

Islamuddin M, Chouhan G, Want MY, Tyagi M, Abdin MZ, Sahal D, Afrin F (2014) Leishmanicidal activities of Artemisia annua leaf essential oil against visceral Leishmaniasis. Front Microbial 5:626

Isman MB, Machial CM (2006) Pesticides based on plant essential oils: from traditional practice to commercialization. Adv Phytomed 3:29–44

Iwaki T, Iefuji H, Hiraga Y, Hosomi A, Morita T, Giga-Hama Y, Takegawa K (2008) Multiple functions of ergosterol in the fission yeast Schizosaccharomyces pombe. Microbiology 154:830–841

Jimenez-Carmona MM, Ubera JL, De Castro ML (1999) Comparison of continuous subcritical water extraction and hydrodistillation of marjoram essential oil. J Chromatogr A 855:625–632

Jo JR, Park JS, Park YK, Chae YZ, Lee GH, Park GY, Jang BC (2012) Pinus densiflora leaf essential oil induces apoptosis via ROS generation and activation of caspases in YD-8 human oral cancer cells. Int J Oncol 40:1238–1245

Joksimovic N, Spasovski G, Joksimovic V, Andreevski V, Zuccari C, Omini CF (2012) Efficacy and tolerability of hyaluronic acid, tea tree oil and methyl-sulfonyl-methane in a new gel medical device for treatment of haemorrhoids in a double-blind, placebo-controlled clinical trial. Updat Surg 64:195–201

Keurentjes JJ, Fu J, Vos de CH, Lommen A, Hall RD, Bino RJ et al (2006) The genetics of plant metabolism. Nat Genet 38:842–849

Khajeh M, Yamini Y, Sefidkon F, Bahramifar N (2004) Comparison of essential oil composition of *Carum copticum* obtained by supercritical carbon dioxide extraction and hydrodistillation methods. Food Chem 86:587–591

Kim H, Chen F, Wu C, Wang X, Chung H, Jin Z (2004) J Agric Food Chem 52:2849–2854

Kliebenstein D (2009) Advancing genetic theory and application by metabolic quantitative trait loci analysis. Plant Cell 21:1637–1646

Koo B, Park K, Ha J, Park J, Lim J, Lee D (2003) Inhibitory effects of the fragrance inhalation of essential oil from Acorus gramineus on central nervous system. Biol Pharm Bull 26:978–982

Kosar M, Tunalier Z, Özek T, Kürkcüogl M, Baser KHC (2005) A simple method to obtain essential oils from Salvia triloba L. and Laurus nobilis L. by using microwave-assisted hydrodistillation. Zeitschrift für Naturforschung C 60:501–504

Kpoviessi BGK, Kpoviessi SD, Ladekan EY, Gbaguidi F, Frédérich M, Moudachirou M, Quetin-Leclercq J, Accrombessi GC, Bero J (2014) In vitro antitrypanosomal and antiplasmodial activities of crude extracts and essential oils of Ocimum gratissimum Linn from Benin and influence of vegetative stage. J Ethnopharmacol 155:1417–1423

Kristiawan M, Sobolik V, Al-Haddad M, Allaf K (2008) Effect of pressure-drop rate on the isolation of cananga oil using instantaneous controlled pressure-drop process. Chem Eng Process Process Intensif 47:66–75

Kulisic T, Radonic A, Katalinic V, Milos M (2004) Use of different methods for testing antioxidative activity of oregano essential oil. Food Chem 85:633–640

Kusuma HS, Mahfud M (2016) Chemical composition of essential oil of Indonesia sandalwood extracted by microwave-assisted Hydrodistillation. AIP Conf Proc 2016-1755

Legault J, Dahl W, Debiton E, Pichette A, Madelmont JC (2003) Antitumor activity of balsam fir oil: production of reactive oxygen species induced by-humulene as possible mechanism of action. Planta Med 69:402–407

Liu ZL, Ho SH (1999) Bioactivity of the essential oil extracted from Evodia rutaecarpa hook f. et Thomas against the grain storage insects, Sitophilus zeamais Motsch. And Tribolium castaneum (Herbst). J Stored Prod Res 35:317–328

Lopez P, Sanchez C, Battle R, Nerin C (2005) J Agric Food Chem 53:6939–6946

Lucci P, Saurina J, Núñez O (2017) Trends in LC-MS and LC-HRMS analysis and characterization of polyphenols in food. Trends Analytical Chem 88:1–24

Masango P (2005) Cleaner production of essential oils by steam distillation. J Clean Prod 13:833–839

Masotti V, Juteau F, Bessière JM, Viano J (2003) Seasonal and phenological variations of the essential oil from the narrow endemic species Artemisia molinieri and its biological activities. J Agri Food Chem 51:7115–7121

Meyer-Warnod B (1984) Natural essential oils: extraction processes and application to some major oils. Perfumer Flavorist 9:93–104

Milner JA (2001) A historical perspective on garlic and cancer. J Nutr 131:1027S–1031S

Misra P, Kumar A, Khare P, Gupta S, Kumar N, Dube A (2009) Pro-apoptotic effect of the landrace Bangla Mahoba of Piper betle on Leishmania donovani may be due to the high content of eugenol. J Med Microbial 58:1058–1066

Moirangthem DS, Laishram S, Rana VS, Borah JC, Talukdar NC (2015) Essential oil of Cephalotaxus griffithii needle inhibits proliferation and migration of human cervical cancer cells: involvement of mitochondria-initiated and death receptor-mediated apoptosis pathways. Nat Prod Res 29:1161–1165

Monzote L, Nance MR, Garcia M, Scull R, Setzer WN (2011) Comparative chemical, cytotoxicity and antileishmanial properties of essential oils from Chenopodium ambrosioides. Nat Prod Commun 6:281–286

Monzote L, García M, Pastor J, Gil L, Scull R, Maes L, Cos P, Gille L (2014) Essential oil from Chenopodium ambrosioides and main components: activity against Leishmania, their mitochondria and other microorganisms. Exp Parasitol 136:20–26

Moon T, Wilkinson JM, Cavanagh HM (2006) Antiparasitic activity of two Lavandula essential oils against giardia duodenalis, trichomonas vaginalis and Hexamita inflata. Parasitol Res 99:722–728

Moss M, Cook J, Wesnes K, Duckett P (2003) Aromas of rosemary and lavender essential oils differentially affect cognition and mood in healthy adults. Int J Neurosci 113:15–38

Moyler DA (1993) Extraction of essential oils with carbon dioxide. Flav Frag J 8:235–247

Ngoh SP, Choo LEW, Pang FY, Huang Y, Kini MR, Ho SH (1998) Insecticidal and repellent properties of nine volatile constituents of essential oils against the American cockroach, Periplaneta americana (L.). Pestic Sci 54:261–268

Noudogbessi J, Yedomonhan H, Alitonou G, Chalard P, Figueredo G, Adjalian E, Avlessi F, Chalchat J, Sohounhloué D (2012) Physical characteristics and chemical compositions of the essential oils extracted from different parts of Siphonochilusaethiopicus (Schweinf.) BL Burtt (Zingiberaceae) harvested in Benin. J Chem Pharm Res 4:4845–4851

O'Bryan CA, Pendleton SJ, Crandall PG, Ricke SC (2015) Potential of plant essential oils and their components in animal agriculture-in vitro studies on antibacterial mode of action. Front Vet Sci 2:35

Olmo-García L, Kessler N, Neuweger H, Wendt K, Olmo-Peinado JM, Fernández-Gutiérrez A, Baessmann C, Carrasco-Pancorbo A (2018) Unravelling the distribution of secondary metabolites in Olea europaea L.: exhaustive characterization of eight olive-tree derived matrices by complementary platforms (LC-ESI/APCI-MS and GC-APCI-MS). Molecules 23:2419

Paduch R, Kandefer-Szerszeń M, Trytek M, Fiedurek J (2007) Terpenes: substances useful in human healthcare. Arch Immunol Ther Exp 55:315–327

Paik SY, Koh KH, Beak SM, Paek SH, Kim JA (2005) The essential oils from Zanthoxylum schinifolium pericarp induce apoptosis of HepG2 human hepatoma cells through increased production of reactive oxygen species. Biol Pharm Bull 28:802–807

Pavela R (2015) Essential oils for the development of eco-friendly mosquito larvicides: a review. Ind Crop Prod 76:174–187

Pereira CG, Meireles MAA (2007) Economic analysis of rosemary, fennel and anise essential oils obtained by supercritical fluid extraction. Flav Frag J 22:407–413

Pizzale L, Bortolomeazzi R, Vichi S, Überegger E, Conte LS (2002) Antioxidant activity of sage (*Salvia officinalis* and S fruticosa) and oregano (*Origanum onites* and *O indercedens*) extracts related to their phenolic compound content. J Sci Food Agri 82:1645–1651

Pourmortazavi SM, Hajimirsadeghi SS (2007) Supercritical fluid extraction in plant essential and volatile oil analysis. J Chromatogr A 1163:2–24

Prakash P, Gupta N (2005) Therapeutic uses of Ocimum sanctum Linn (Tulsi) with a note on eugenol and its pharmacological actions: a short review. Ind J Physio Pharmaco 49:125–131

Priestley CM, Burgess IF, Williamson EM (2006) Lethality of essential oil constituents towards the human louse, Pediculus humanus, and its eggs. Fitoterapia 77:303–309

Queiroz RM, Takiya CM, Guimarães LPTP, Rocha GDG, Alviano DS, Blank AF, Alviano CS, Gattass CR (2014) Apoptosis-inducing effects of Melissa officinalis L. essential oil in glioblastoma multiforme cells. Cancer Investig 32:226–235

Rai M and Carpinella MC (2006) Naturally occurring bioactive compounds. Advances in phytomedicine Vol.3. First edition

Rattan RS (2010) Mechanism of action of insecticidal secondary metabolites of plant origin. Crop Prot 29:913–920
Raut JS, Karuppayil SM (2014) A status review on the medicinal properties of essential oils. Ind Crop Prod 62:250–264
Reed JC (2003) Apoptosis-targeted therapies for cancer. Cancer Cell 3:17–22
Reverchon E, Senatore F (1992) Isolation of rosemary oil: comparison between hydrodistillation and supercritical CO2 extraction. Flav Frag J 7:227–230
Rivera Calo J, Crandall PG, O'Bryan CA, Ricke S (2015) Essential oils as antimicrobials in food systems- a review. Food Control 54:111–119
Russo R, Corasaniti MT, Bagetta G, Morrone LA (2015) Exploitation of cytotoxicity of some essential oils for translation in cancer therapy. Evid Based Complement Alternat Med 2015:397821
Sadgrove N, Jones GA (2015) Contemporary introduction to essential oils: chemistry, bioactivity and prospects for Australian agriculture. Agri 5:48–102
Sahraoui N, Vian MA, Bornard I, Boutekedjiret C, Chema F (2008) Improved microwave steam distillation apparatus for isolation of essential oils: comparison with conventional steam distillation. J Chromatogr A 1210:229–233
Saito K, Matsuda F (2010) Metabolomics for functional genomics, systems biology, and biotechnology. Annu Rev Plant Biol 61:463–489
Salgueiro L, Martins A, Correia H (2010) Raw materials: the importance of quality and safety. A review. Flavour Fragr J 25:253–271
Samber N, Khan A, Varma A, Manzoor N (2015) Synergistic anti-candidal activity and mode of action of Mentha piperita essential oil and its major components. Pharm Boil 53:1496–1504
Sanli A, Karadogan T (2017) Geographical impact on essential oil composition of endemic Kundmannia anatolica hub.-Mor. (Apiaceae). Afr J Tradit Complementary Altern Med 14:131–137
Schauer N, Semel Y, Roessner U, Gur A, Balbo I, Carrari F et al (2006) Comprehensive metabolic profiling and phenotyping of interspecific introgression lines for tomato improvement. Nat Biotechnol 24:447–454
Schnitzler P, Schön K, Reichling J (2001) Antiviral activity of Australian tea tree oil and eucalyptus oil against herpes simplex virus in cell culture. Pharmazie 56:343–347
Schuhmacher A, Reichling J, Schnitzler P (2003) Virucidal effect of peppermint oil on the enveloped viruses herpes simplex virus type 1 and type 2 in vitro. Phytomedicine 10:504–510
Sell CS (2006) The chemistry of fragrances: from perfumer to consumer. Royal Society of Chemistry, London, UK
Senorans FJ, Ibanez E, Cavero S, Tabera J, Reglero G (2000) Liquid chromatographic–mass spectrometric analysis of supercritical-fluid extracts of rosemary plants. J Chromatogr A 870:491–499
Sereshti H, Rohanifar A, Bakhtiari S, Samadi S (2012) Bifunctional ultrasound assisted extraction and determination of *Elettaria cardamomum* Maton essential oil. J Chromatogr A 1238:46–53
Serpa RTHFC, Sousa AKL, Coelho-de-Souza PJC, Leal AN, Cardosa JH (1997) Phytother Res 11:552
Sertel S, Eichhorn T, Plinkert PK, Efferth T (2011) Cytotoxicity of Thymus vulgaris essential oil towards human oral cavity squamous cell carcinoma. Anticancer Res 31:81–87
Shah AJ, Gilani AH, Abbas K (2011) Studies on the chemical composition and possible mechanisms underlying the antispasmodic and bronchodilatory activities of the essential oil of Artemisia maritima L. Arch Pharm Res 34:1227
Sharma N, Aranjo M, Wu M, Quqish J, Charles C (2010) Int Dent J 60:175–180
Silva SLD, Figueiredo PM, Yano T (2007) Cytotoxic evaluation of essential oil from *Zanthoxylum rhoifolium* lam. Leaves. Acta Amazon 37:281–286
Singh N (2006) Drug resistance mechanisms in clinical isolates of Leishmania donovani. Ind J Med Res 123:411
Soeur J, Marrot L, Perez P, Iraqui I, Kienda G, Dardalhon M, Meunier JR, Averbeck D, Huang ME (2011) Selective cytotoxicity of Aniba rosaeodora essential oil towards epidermoid cancer cells through induction of apoptosis. Mutat Res 718:24–32

Sokmen A, Gulluce M, Akpulat HA, Daferera D, Bektas Tepe B, Polissiou M, Sokmen M, Sahin F (2004) The in vitro antimicrobial and antioxidant activities of the essential oils and methanol extracts of endemic thymus spathulifolius. Food Control 15:627–634

Sonwa MM (2000) Isolation and structure elucidation of essential oil constituents. Comparative study of the oils of Cyperus alopecuroides, Cyperus papyrus, and Cyperus rotundus. Ph.D. degree,, vol 56. University of Hamburg, Germany, p 321

Starmans DA, Nijhui HH (1996) Extraction of secondary metabolites from plant material: a review. Trends Food Sci Tech 7:191–197

Stevensen C (1994) The psychophysiological effects of aromatherapy massage following cardiac surgery. Complement Ther Med 2:27–35

Storz P (2005) Reactive oxygen species in tumor progression. Front Biosci 10:1881–1896

Svoboda KP, Greenaway RI (2003) Investigation of volatile oil glands of *Satureja hortensis* L. (summer savory) and phytochemical comparison of different varieties. Internat J Aroma 13:196–202

Tian J, Huang B, Luo X, Zeng H, Ban X, He J, Wang Y (2012) The control of aspergillus flavus with Cinnamomum jensenianum hand.-Mazz essential oil and its potential use as a food preservative. Food Chem 130:520–527

Tolouee M, Alinezhad S, Saberi R, Eslamifar A, Zad SJ, Jaimand K, Taeb J, Rezaee MB, Kawachi M, Shams-Ghahfarokhi M (2010) Effect of Matricaria chamomilla L. flower essential oil on the growth and ultrastructure of aspergillus Niger van Tieghem. Int J Food Microbiol 139:127–133

Tomaino A, Cimino F, Zimbalatti V, Venuti V, Sulfaro V, Pasquale AD, Saija A (2005) Influence of heating on antioxidant activity and the chemical composition of some spice essential oils. Food Chem 89:549–554

Trombetta D, Castelli F, Sarpietro MG, Venuti V, Cristani M, Daniele C, Saija A, Mazzanti G, Bisignano G (2005) Mechanisms of antibacterial action of three monoterpenes. Antimicrob Agents Chemother 49:2474–2478

Turek C, Stintzing FC (2013) Stability of essential oils: a review. Compr Rev Food Sci Food Saf 12:40–53

Turina AV, Nolan MV, Zygadlo JA, Perillo MA (2006) Natural terpenes: self-assembly and membrane partitioning. Biophys Chem 122:101–113

Turner GW, Gershenzon J, Croteau RB (2000) Distribution of peltate glandular trichomes on developing leaves of peppermint. Plant Phy 124:655–664

Ultee A, Kets EP, Smid EJ (1999) Mechanisms of action of carvacrol on the food-borne pathogen Bacillus cereus. Appl Environ Microbiol 65:4606–4610

Valente J, Zuzarte M, Liberal J, Gonçalves MJ, Lopes MC, Cavaleiro C, Cruz MT, Salgueiro L (2013) Margotia gummifera essential oil as a source of anti-inflammatory drugs. Ind Crop Prod 47:86–91

Veras HNH, Rodrigues FFG, Colares AV, Menezes IRA, Coutinho HDM, Botelho MA (2012) Synergistic antibiotic activity of volatile compounds from the essential oil of Lippia sidoides and thymol. Fitoterapia 83:508–512

Vian MA, Fernandez X, Visinoni F, Chemat F (2008) Microwave hydrodiffusion and gravity, a new technique for extraction of essential oils. J Chromatogr A 1190:14–17

Vinatoru M (2001) An overview of the ultrasonically assisted extraction of bioactive principles from herbs. Ultrason Sonochem 8(3):303–313

Wang L, Weller CL (2006) Recent advances in extraction of nutraceuticals from plants. Trends Food Sci Tech 17:300–312

Woronuk G, Demissie Z, Rheault M, Mahmoud S (2011) Biosynthesis and therapeutic properties of Lavandula essential oil constituents. Planta Med 77:7–15

Yahyaraeyat R, Khosravi A, Shahbazzadeh D, Khalaj V (2013) The potential effects of Zataria multiflora Boiss essential oil on growth, aflatoxin production and transcription of aflatoxin biosynthesis pathway genes of toxigenic aspergillus parasiticus. Braz J Microbiol 44:649–655

Yonekura-Sakakibara K, Fukushima A, Saito K (2013) Transcriptome data modeling for targeted plant metabolic engineering. Curr Opin Biotechnol 24:285–290

Yonekura-Sakakibara K, Saito K (2009) Functional genomics for plant natural product biosynthesis. Nat Prod Rep 26:1466–1487

Zuzarte M, Vale-Silva L, Gonçalves M, Cavaleiro C, Vaz S, Canhoto J, Pinto E, Salgueiro L (2012) Antifungal activity of phenolic-rich Lavandula multifida L. essential oil. Eur J Clin Microbial Infect Dis 31:1359–1366

Antioxidant Activity of Essential Oils: A Mechanistic Approach

3

Bijendra Kumar Singh and Akash Maurya

Abstract

Essential oils (EOs) have long been extracted using a variety of techniques from different aromatic and therapeutic plants. Because of their antioxidant quality and attractive fragrances, they have found broad utilisation in various fields as antioxidant, antibacterial, and antifungal agent in pharmaceutical, cosmetic, and food sectors. Antioxidants are a group of compounds that can prevent oxidation by acting as reducing agents, pro-oxidant inactivators, free radical scavengers, and radical species quenchers. Reactive oxygen species (ROS) interact with any material that may be oxidised and modify the structural makeup of the material as a result, causing oxidative damage. There are plenty of approaches available for the estimation of antioxidant activities of Eos; ABTS (2,2′-azino-bis(3-ethylbenzothiazoline-6-sulfonic acid) and DPPH (1,1-diphenyl-2-picrylhydrazine) assays are frequently used methods. Since EOs contain a variety of active components, their antioxidant mode of action cannot be achieved by a single method of action. The EOs and its nanoencapsulated products are more effective as free radical scavengers in nature due to slow release, longer shelf life, and easily available in the medium without direct interference with food matrix. The application of EOs as antioxidants in real food systems have been associated with some hurdles, which could be overcome using the encapsulation technology and advanced food packaging systems. In this chapter, the antioxidant properties of EOs, estimation of antioxidant activities, mechanisms involved in antioxidant activity, encapsulation process with enhanced activity and applicability, its application in real food systems, and future possibilities have been discussed.

B. K. Singh (✉)
Department of Botany, Feroze Gandhi College, Raebareli, Uttar Pradesh, India

A. Maurya
Department of Botany, Shri Murli Manohar PG College, Ballia, Uttar Pradesh, India

B. Prakash et al. (eds.), *Plant Essential Oils*,
https://doi.org/10.1007/978-981-99-4370-8_3

Keywords

Reactive oxygen species · Essential oils · Antioxidants · Nanoencapsulation · Natural preservatives

3.1 Introduction

A class of substances known as antioxidants are capable of inhibiting the oxidation process by acting as free radical scavengers, reducing agents, inactivators of pro-oxidants, and quenchers of radical species. Antioxidants are substances, when present in foods or the body in very small amounts, which slow down, stop, or stop oxidative processes that result in food quality deterioration or the development and spread of degenerative illnesses in the organisms. Numerous organic antioxidants, such as polyphenols (anthocyanins, flavanols, and flavones), carotenoids, phenolic acids, tannins, lignin, and vitamins, are found in plants (Loi and Paciolla 2021; Arias et al. 2022). Due to its capacity to lower oxidative stress, several studies demonstrate that antioxidants are crucial for preserving human health as well as for preventing and treating illnesses. In order to ensure the quality of functional foods and, more critically, to determine how effective food antioxidants are in preventing and treating illnesses associated with oxidative stress, it is crucial to measure the antioxidant activity/capacity of foods and biological samples (Munteanu and Apetrei 2021; Siano et al. 2023).

There are numerous origins of ROS production. Different processes to control ROS production and availability, such as localised and compartmentalised creation as well as the activation of sinks (detoxifying factors) and redox relaying, are in place to keep their levels at physiological concentrations. The presence of free Fe^{2+} ions, which are engaged in the Fenton reaction, which produces the most reactive ROS form hydroxyl radicals with a single unpaired electron and the ability to react with almost all biomolecules, has a substantial impact on ROS reactivity. Due to the high reactivity of hydroxyl radicals, cellular problems such as altered protein structures, lipid peroxidation, and membrane disintegration occurs (Demidchik 2015; Acosta-Casique et al. 2023). One of the primary factors in food degradation is oxidation. For instance, the oxidation of oil results in both poisonous and low-molecular-weight molecules with little taste. It also has other negative consequences, such as vitamin deterioration and important fatty acid loss, which reduces nutritional values (Umaraw et al. 2020). The pursuit of efficient antioxidants from natural resources as alternative solutions to suppress food deterioration is now centred on edible plants because they have fewer side effects than the synthetic chemicals used in today's food products, which is associated with growing consumer interest in natural food additives (Valdivieso-Ugarte et al. 2019). EOs are liquid compositions of volatile chemicals that are frequently derived from aromatic plants. Natural antioxidants including polyphenols, carotenoids, and vitamins may come from plants and various plant components like flowers, stems, and roots. The biological benefits of EOs have been recognised, including their antiviral, antibacterial, anti-inflammatory, and antioxidant activities (Tit and Bungau 2023). The use of natural and safer preservatives is on the rise, giving food an impression of being natural origin. EOs have received a lot of interest for usage as food preservatives. The popularity of EOs as a natural preservative is increasing since last decade due to growing concern of health-related issues caused by synthetic preservatives (Burt 2004; Wu et al. 2019a, b; Falleh et al. 2020).

There are numerous techniques that may be used to assess the transfer of the hydrogen atom or the transfer of electrons from antioxidants to free radicals directly. The antioxidant activities described for this technique category are typically linked to their ability to scavenge certain radical species, some of which may be synthetic and insignificant to biology. Some of these approaches have the drawback of not accurately reflecting the circumstances in oxidised food or in vivo cases (Munteanu and Apetrei 2021). ROS interact with other components that are susceptible to oxidation and undergo structural changes to cause oxidative degradation. The category of ROS includes hydroxyl radicals (OH^-), superoxide radicals (O_2^{2-}), hydrogen peroxide (H_2O_2), and alkoxyl (RO^-). EOs slow down the beginning of antioxidant chains, free radical scavengers, hydrogen abstraction, and quenchers of singlet oxygen production by a variety of direct and indirect mechanisms (Rodriguez-Garcia et al. 2016).

For EOs to be used as food preservatives, precise knowledge of how food matrix components affect Eos' antibacterial activity is needed. Despite their remarkable preservation and antioxidant efficacy, EOs have limited employment in the food and agricultural industries because of their volatility, hydrophobicity, and negative impacts on food organoleptic properties (Prakash et al. 2015; Falleh et al. 2020; Singh and Dubey 2023). It has been consistently justified to use encapsulation as a technical substitute to mask the extremely intense aroma of some of the EOs and increase its shelf life. By guaranteeing their equal dispersion in the media, EO encapsulation provides an effective method to boost their bioactivity and bioavailability. The exterior protective layer utilised in the encapsulation functions as a hydrophilic barrier to EOs to get over their hydrophobic restriction (Nehme et al. 2021; López-Gómez et al. 2023). By preventing exposure and degradation of EOs and their bioactive ingredients by forming a physical barrier and by facilitating their regulated release, nanoencapsulation looks to be a promising method to eliminate these restrictions. This results in increased bioavailability and efficacy. More focus should be placed on these particular parameters in order to achieve better nanoencapsulation and ultimately to their application into the food systems because a large number of intrinsic factors can affect the successful nanoencapsulation of EOs and their bioactive constituents into polymeric systems (Chaudhari et al. 2021; Singh et al. 2022a, b; Chaudhari et al. 2023). A lot of research has been done regarding the encapsulation materials, factors influencing the process, and behaviour of nanoencapsulated EOs in the real food systems. But there is need of more research to be carried out regarding the hurdles involved in the application of natural antioxidants such as EOs in real food systems and their large-scale practical applications, which is the demand of present-day generations and surely going to increase several folds in future.

3.2 Reactive Oxygen Species (ROS) and Antioxidants

The term 'ROS' refers to a class of molecules generated via reduction-oxidation (redox) processes or electronic excitation and derived from molecular oxygen. Several kinds of 'reactive species' have piqued attention in biology and medicine (Hawkins and Davies 2019). Their names reflect the characteristics of the reactive atoms. There are several origins of ROS production. Different strategies to control ROS production and availability, such as localised and compartmentalised creation as well as the activation of sinks (detoxifying factors) and redox relays, are used to keep their levels at physiological concentrations (Sies and Jones 2020; Zeng et al. 2023). ROS are a normal by-product of cellular aerobic metabolism in a biological setting and are primarily produced by mitochondrial respiration (Mailloux 2020). Free radicals are made up of atoms or molecules that have an unpaired electron in their outer shell, which makes them unstable and extremely reactive, or, to put it another way, makes them prone to 'stealing' electrons from other molecules (Pham-Huy et al. 2008). In addition, environmental pollution, exposure to xenobiotic compounds, pesticides, and heavy metals, drug and alcohol usage, cigarette use, extended use of some medicines, and a number of other lifestyle choices adversely influence ROS production. Recently, free radicals have become more prevalent due to pollution, stress, and eating commercial food products (Al-Gubory 2014; Khalid et al. 2023).

There are plenty of research material available depicting ROS as a signal factor taking part in defence mechanism and signal transduction, and considered as core contributors in complex signalling cascade. In reaction to stressors or other environmental perturbations, physiological targets of oxidants operate as molecular redox switches in signal transduction at different levels of cell control (Mittler et al. 2011; Sies and Jones 2020). Generally, the functions of ROS in signalling are correlated with tightly controlled site- and time-specific modifications of ROS levels, whereas unregulated ROS accumulation as a result of either consisted or defective catabolism, or their combined effect, is correlated with harmful oxidative effects that can result in cellular damage or even cell death (Mittler 2017). Through oxidative processes, namely those involving protein thiol groups or iron-containing clusters, ROS can alter the structures and behaviours of proteins. However, the harmful effects of accumulating ROS are also used as chemical defence within immunological repertoires of various species against pathogens, such as the ROS burst that causes plant cell death during the plant hypersensitive response to biotrophic infections (Eckardt 2017; Janků et al. 2019).

Food rotting mostly results from chemical reactions that happen when food is exposed to air or when spoilage bacteria are present. ROS interact with any substance that can be oxidised and undergo structural changes as a result to produce oxidative damage (Gómez-Estaca et al. 2014). Additionally, ROS cause a number of unwanted and substantial changes in food, including a reduction in nutritional content and a loss of colour, taste, and odour. This process in packaged foods is typically thought to be started by a change in the oxidation state of the metals already present in the food and/or exposure to ultraviolet (UV) light, like sunlight. This causes subsequent hydrogen extrapolation from fats, which in turn triggers subsequent reactions in nearby fat molecules (Shahidi and Ambigaipalan 2015). Moisture can contribute ROS formed from the food product's water content, which is a determining factor in oxidation. Additionally, when oxidative reactions develop over time, a number of compounds (such as sugars and short-chain fatty acids) are made more accessible to spoilage microbes, making the food surface more inviting to them. Foods are also sensitive to oxidation, enzymatic activity, and the release of ROS (Carpena et al. 2021).

Antioxidants are substances that have beneficial antioxidant effects due to their ability to provide electrons to neutralise free radicals and stop oxidation processes. Through a variety of processes, antioxidants scavenge ROS and help to convert it to less reactive and more stable forms (Pateiro et al. 2018). In order to be close to the places where ROS are produced, enzymatic antioxidants are localised in a variety of subcellular locales. Non-enzymatic antioxidants can come from inside the body or be obtained from food (Shields et al. 2021). There are around more than 6000 naturally occurring chemicals that belong to the huge and varied family of secondary metabolites known as polyphenols, which have important biological functions linked to their antioxidant capabilities and free radical scavenging abilities (Buchanan et al. 2015; Tomás-Barberán et al. 2000). The highly reactive superoxide anion is changed into a less reactive hydrogen peroxide by the enzyme superoxide dismutase (SOD). Using transition metals in its active site, the enzyme catalyses the transfer of electrons through two redox processes. The enzyme known as superoxide dismutase (SOD) has a specific function in reducing the consequences of oxidative stress caused by the presence of free radicals. SOD is involved in catalysing the recombination reaction of the oxygen radicals (Getzoff et al. 1983; Munteanu and Apetrei 2021)).

In order to deal with hydrogen peroxide-producing processes, including those of the SODs, catalase (CAT) catalyses the conversion of hydrogen peroxide to water and molecular oxygen (Ighodaro and Akinloye 2018). The glutathione peroxidase (GPx) and glutathione reductase (Grx) systems must function at optimal glutathione (GSH) levels because GSH works to restore each enzyme to its active state. Additionally, GSH itself has the ability to function as an antioxidant, donating electrons via the sulfhydryl group to lower and detoxify ROS (Birben et al. 2012). In addition to being a precursor molecule for GSH, N-acetyl cysteine (NAC) also possesses its own thiol group, which enables it to take part in redox processes and give its electrons for the detoxification of ROS and the defence of sulfhydryl-containing proteins against oxidative damage.

The secondary metabolism in plants produces volatile aromatic molecules known as essential oils (EOs). Between a dozen and several hundred components make up each EO, the majority of which are terpenes and terpenoids, including oxygenated derivatives such as aldehydes, ketones, alcohols, ethers, esters, and epoxides (Bajpai and Baek 2016). Another biological characteristic of significant interest is the antioxidant activity of EOs, which may help protect food from the harmful effects of oxidants (Valdivieso-Ugarte et al. 2019). Antioxidants have the function of neutralising free radicals in biological cells, which have a detrimental effect on living things. Bioactive EOs have been included into active food packaging due to their distinctive scent, tastes, and natural antibacterial properties to increase the shelf life and safety of perishable food. They have been categorised as natural preservatives, culinary flavourings, and therapeutic treatments because of plants' antiviral, antimicrobial, and insecticidal properties (Bonda et al. 2020; Ni et al. 2021).

3.3 Antioxidant Activity of EOs

Recently, numerous studies have investigated the antioxidant activities of EOs due to their chemical compositions. Moreover, Phenols and other secondary metabolites are linked with double bonds that are responsible for substantial antioxidant activities (Prasad et al. 2022; Tiwari et al. 2022). Therefore, aromatic plant EOs and their major constituents are widely used as an alternative food additive to avoid the degradation of the food products instead of suspected synthetic antioxidants such as butylated hydroxyanisole (BHA), butylated hydroxytoluene (BHT), and propyl gallate (PG) that pose potential adverse effects on human health (Maurya et al. 2021a; Al-Maqtari et al. 2021). EOs also exhibit important role in preventing ageing and variety of lifestyle-related diseases such as cancer, dysfunction, immune system decline, and heart disease due to closely associated with active oxygen and lipid peroxidation (Bhavaniramya et al. 2019). Thus, the EOs were recommended as good substitute of synthetic antioxidants, because the majority of EOs are generally recognised as safe (GRAS) (Singh et al. 2022a, b). In addition, most of the plant EOs are rich in oxygenated monoterpenes such as aldehydes (*Galagania fragrantissima*), alcohols (*Achillea fillipendulina*), ketones (*Artemisia rutifolia*, *Anethum graveolens*, and *Mentha longifolia*), and esters (*Salvia sclarea*). However, phenolic terpenoids (*Origanum tyttanthum* and *Mentha longifolia*) such as thymol and carvacrol exhibited strongest antioxidant activities in EOs, due to their phenolic structure (Bhavaniramya et al. 2019). Phenolic component has redox potential, thus, exhibited significant role in free radicals' neutralisation and peroxide decomposition. Additionally, some other EO components such as alcohols, ethers, ketones, aldehydes, and monoterpenes like citronellal, isomenthone, 1,8-Cineole, geranial/neral, menthone, linalool also play important role in the antioxidant activities (Angane et al. 2022).

Maurya et al. (2022) investigated the dose-dependent free radicals scavenging activity of *Carum carvi* EO (CcEO) by using DPPH radicals, and they revealed that, IC_{50} value of CcEO was 10.564 μL/mL. The free radical scavenging activity presented by CcEO was found better than some of the formerly reported EOs such as *Gaultheria fragrantissima* (13.09 μL/mL) and *Zingiber officinale* (14.44 μL/mL) (Ramsdam et al. 2021). Similarly, El-Baroty et al. (2010) reported the antioxidant activity of 45 EOs of various plant species against the DPPH radical. In this work, they evaluated that EOs of clove bud and cinnamon exhibited the higher inhibition rates (approx. 96%), and these EOs have more than 80% of eugenol in their chemical composition. Moreover, Chaudhari et al. (2020) measured the radical scavenging activity of *Pimenta dioica* EO (PDEO) by DPPH and ABTS assay. They found that, PDEO showed dose-dependent scavenging of free radicals and its IC_{50} value was 0.82 and 0.31 μL/mL for DPPH and ABTS, respectively. Thus, the lower IC_{50} value of PDEO strengthens its strong antioxidant potential and plays as natural eco-friendly antioxidant for food preservation by reducing the oxidative burden of food products. In addition, Singh et al. (2021a, b) evaluated the highest DPPH and ABTS radical scavenging activity that was exhibited in the sequence *Salvia sclarea* EO (SSEO) and Linalyl acetate (LA) combination (0.0024, 0.00082 μL/mL) > LA (0.0047, 0.00093 μL/mL) > SSEO (0.0051, 0.00109 μL/mL), respectively. In this study, they found that combined mixture of SSEO and LA revealed significant IC_{50} value as compared to some other previously investigated EOs such as *Stachys inflate*, *Satureja montana*, and *Satureja subspicata* EOs. The strong antioxidant potential corresponds to phenolic group in LA promoting the hydrogen donation and stability of phenoxyl radicals. Semeniuc et al. (2018) examined various EOs belonging to family Lamiaceae and Apiaceae, and found that thyme EO showed greater antioxidant activity due to the presence of phenolic rich compounds. Thus, the other plant species' EOs and their variety of bioactive components with strong antioxidant capacities are relevant to use as a natural substitute of synthetic antioxidant to reduce the oxidative deterioration of food and food items.

3.4 Estimation of Antioxidant Activity

Although ROS are challenging to detect precisely, especially in vivo, due to their short lifespan and strong reactivity, several techniques have been developed to quantify ROS directly (Griendling et al. 2016). Because it is challenging to measure ROS levels directly, researchers frequently measure ROS indirectly by evaluating oxidative damage to cell components or oxidative stress-related survival rates (Katerji et al. 2019). This consequence might also be caused by lower quantities of antioxidant or repair enzymes. Larger levels of oxidative damage are typically taken to mean increased levels of ROS. Increased amounts of ROS may be indicated by decreased tolerance to oxidative stress, but excessive levels of ROS can also promote the overexpression of antioxidant enzymes (Anaissi-Afonso et al. 2018; Ahmad and Suzuki 2019).

There are various analytical methods like DPPH (1,1-diphenyl-2-picrylhydrazine), ABTS (2,2′-azino-bis(3-ethylbenzothiazoline-6-sulfonic acid), phosphomolybdenum, reducing power and some other tests are known in identifying the antioxidant activity of EOs. Despite of merits and demerits of all analytical methods, it has been found that the most common and reliable methods are DPPH, ABTS, and reducing power assay methods (Munir et al. 2018; Vorobyova et al. 2019).

3.4.1 Reducing Power Assay

According to Oyaizu's approach, the reducing power of the samples is assessed. A phosphate buffer (5 ml, 0.2 M, pH 6.6) and potassium ferricyanide (5 ml, 1%) are combined with the sample in 1 ml of methanol, and the combination is then incubated at 50 °C for 20 min. The reaction mixture is then mixed with 5 ml of 10% trichloroacetic acid, centrifuged for 10 min at 3000 rpm. The solution's top layer (5 ml) is combined with ferric chloride (1 ml, 1%) and distilled water (5 ml), and is then tested for absorbance at 700 nm. An enhanced reducing power is shown by a greater absorption. In the process of developing corrosion inhibitors based on plant material, using an indicator of antioxidant activity as an indirect assessment of the inhibitory effect would save a lot of time and, more importantly, allow a greater proportion of the various types of plant material to be explored systematically (Maheshwari et al. 2011; Vorobyova et al. 2019).

3.4.2 DPPH Assay

The stable free radical DPPH interacts with substances that may give off an atom of hydrogen. This technique works by scavenging DPPH by including an antioxidant or radical species that changes the colour of the DPPH solution. One of the prominent techniques for determining a compound's capacity to scavenge free radicals is DPPH reactivity, which has been widely used to the antioxidants in fruits and vegetables. For plant samples, it is one of the most often used antioxidant tests. For this, the subsequent drop in absorption at 517 nm is used to calculate the antioxidant activity. It involves preparing a 0.1 mM DPPH solution in methanol and adding 4 ml of this solution, at different concentrations, to 1 ml of the sample solution in methanol ((Munir et al. 2018).

3.4.3 ABTS Assay

One of the most often used antioxidant tests for plant materials is the ABTS radical scavenging technique. The oxidation of ABTS with potassium persulfate produces the ABTS radical cation, and its decrease in the presence of antioxidants that donate hydrogen is detected spectrophotometrically at 734 nm. The enhanced procedure uses ABTS and potassium persulfate to react, producing a blue/green $ABTS^{+}$ chromophore. Radical scavenging technique 2,2′-azinobis(3-ethylbenzthiazoline-6-sulphonic acid), also known as ABTS, was created by Rice-Evans and Miller and then improved by Re et al. (1999). Metmyoglobin is activated by hydrogen peroxide in the presence of $ABTS^{+}$ to create a radical cation, which is the basis for the modification.

3.4.4 Hydroxyl Radical Scavenging Assay

The Fenton reaction, which takes place in the presence of reduced transition metal ions (such as Fe^{2+}) and H_2O_2, can result in the formation of hydroxyl radicals. Since the hydroxyl radical is highly reactive with sugars, amino acids, lipids, and even nucleotides, the antioxidant action of removing this radical is crucial. In this experiment, ascorbate-hydrogen peroxide produces hydroxyl radicals. Typically, gallic acid is employed as a control solution. In an ultraviolet-visible (UV-Vis) spectrophotometer, the mixture is heated for one hour at 37 °C before the absorbance is recorded at 510 nm (Diniz do Nascimento et al. 2020).

3.4.5 Cellular Antioxidant Activity Assay

This technique assesses the antioxidant activity using cell lines including Caco-2, HepG2, IPEC-J2, and MCF-7 cells as well as a biosensor (Wu et al. 2019a, b; Diniz do Nascimento et al. 2020). Dihydrodichlorofluorescein (DCFH2, fluorescence probe dye) is used in certain research as a redox sensor. When peroxyl (ROO$^\bullet$), produced by the breakdown of 2,2-azobis(2-methylpropionamidine) hydrochloride, is present, it oxidises to fluorescent dichlorofluorescein (DCF). The technique assesses a substance's capacity to prevent the oxidation of intracellular DCFH2, which may be detected by fluorescence. The amount of quercetin equivalents per mole is used to measure antioxidant activity. For one hour, the emitting fluorescence is measured every 5 min using absorbances of 485 nm and 535 nm (Amorati and Valgimigli 2015; Nascimento da Silva et al. 2016).

3.5 Mechanism of Antioxidant Activity of EOs

Preservation of food and food products faced a challenge due to oxidative deterioration, directly bound to the chemical composition of the food and food products (Pateiro et al. 2018). Oxidative deterioration is performed by the interaction of reactive oxygen species (ROS) with other component vulnerable to be oxidised and suffer structural changes in the process. Hydroxyl radicals (OH^-), superoxide radicals (O_2^{2-}), hydrogen peroxide (H_2O_2), and alkoxyl (RO^-) are classified under the category of ROS (Carpena et al. 2021). These ROS exhibited various undesirable and significant changes in foods such as reducing nutritional component, lowering their shelf life, loss of flavour, colour, and odour (Borges et al. 2019).

In these references, EOs and their active components act as promising natural antioxidant due to presence of various secondary metabolites, have inherent ability to reduce or stop the oxidation of lipids, and enhance the shelf life of food products at relatively low concentration. Indeed, EOs have multiple mechanism of direct and indirect actions to slow down the antioxidant chain initiation, free radical scavengers, hydrogen abstraction, and quenchers of singlet oxygen formation (Rodriguez-Garcia et al. 2016). In addition, due to the presence of various active component in EOs, their antioxidant mode of action cannot be performed in only single mode of action (Maurya et al. 2021b). Antioxidant mode of action of EOs occurs in three stages such as initiation, propagation, and termination. Hydroxyl group of EOs are generally the key site of hydrogen donation, inactivating the lipid peroxyl and lipid alkoxyl radicals (free radicals) produced from unsaturated fatty acids' oxidation

(Fig. 3.1) (Pateiro et al. 2018). Moreover, their redox activities are due to their property of electron donation to the free radicals, and reducing the other components from oxidising. In this way, phenolic compounds such as eugenol, carvacrol, and thymol can be used for the retardation of lipid oxidation of food and food products due to their redox potential (Jayasena and Jo 2014: Maqsood et al. 2014).

3.6 Antioxidant Activity of Nanoencapsulated Eos, Edible Films, and Coatings

EOs have an excellent source of phenolic components that exhibit antioxidant activity; however, their direct (free EOs) utilisation in food products has some limitations such as high volatility, lower antioxidant capacity compared to synesthetic, and also changes the organoleptic properties (Jamróz et al. 2018). Therefore, nanoencapsulation of EOs and their bioactive constitutes has been proven to be an effective and promising technique to enhance the chemical stability of components by preserving them from volatilisation, organoleptic properties, and extending their effectiveness (Ozogul et al. 2022). Thus, the nanoemulsions applied in food packaging have range of advantages such as enhancing the antioxidant and antimicrobial activity, minimising the contact with food ingredients, decreasing the transfer of important elements, and uniformity of food packaging (Al-Tayyar et al. 2020; Kocira et al. 2021). Moreover, antioxidant capacity of EOs can be demonstrated by their ability to act as oxygen scavengers and permit the diffusion of bioactive components into coated food items.

In this context, recently, Zheng et al. (2019) investigated that incorporation of eugenol and acorn starch in edible chitosan-based film significantly enhanced the antioxidant activity (approx. 86.77%). Su et al. (2020) reported that cinnamon EO (0.1–0.38%) is encapsulated into chitosan nanoparticles via oil-in-water emulsification and ionic gelation technique to develop biodegradable films. In this work they found that, percentage of radical scavenging activities of DPPH and hydroxyl radicals was in the ranges of 20.6 to 56.9% and 21.0 to 53.4%, respectively. However, DPPH and hydroxyl radicals' scavenging activities of free cinnamon EO were only 19.5 and 22.6%, respectively. In a similar way, Wu et al. (2019a, b) demonstrated the antioxidant activities of chitosan-based coating with liposomes that consist

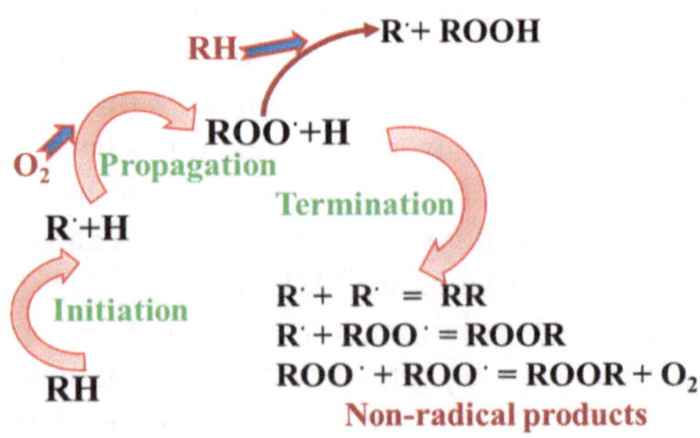

Fig. 3.1 Schematic diagram showing antioxidant mechanism of action

laurel EO (lip-LEO) and silver nanoparticle (AgNPs). They found that Lip-LEO-AgNPs (32.95 μg FeSO4/0.1 mL) had excellent antioxidant activity, which showed no important difference in redox potential in comparison with Lip-LEO (31.31 μg FeSO4/0.1 mL). In comparison to blank liposome, Lip-LEO-AgNPs and Lip-LEO had good antioxidant activity due to chemical composition of laurel EO such as eugenol, cajeputol, terpinene, and geraniol, which have significant ability to capture free radicals. In addition, Doost et al. (2019) examined the free radical scavenging ability of thymol by incorporation into quillaja saponin (QS) stabilised nanoemulsions that was ameliorated for the purpose of depressing chicken breast meat oxidation. Thus, the thymol molecules are more stable and soluble in delivery systems due to improved radical scavenging activity. Singh et al. (2022a, b) reported both the free and chitosan encapsulated nanoparticle antioxidant activity of *Aniba rosaeodora* EO (AREO) against DPPḤand ABTS$^{\cdot+}$ radicals. They found that IC50 value of free AREO showed 4.468 μL/mL against DPPḤ and 2.370 μL/mL against ABTS· + assay, while nanoencapsulated AREO (AREO-CsNe) exhibited enhanced free radical activity with IC50 3.792 μL/mL and 1.706 μL/mL against DPPH and ABTS^{+} respectively. In this way nanoencapsulated EOs improved the stability of number of oxidative-sensitive substances and reduce the volatilisation of their bioactive ingredients. Thus, the antioxidant activity of active components could be significantly protected by encapsulation technique (Zhaveh et al. 2015).

3.7 Antioxidant Activity of Essential Oils in Real Food Systems

One of the most significant factors in food deterioration is oxidation. It is a multilateral reaction that results in unfavourable modifications to the nutritional value, organoleptic standards, food value, and the production of potentially hazardous chemicals. Food that has experienced significant oxidation has serious flaws and cannot be consumed. While oxidation may be detected by colour changes and the emergence of bad odours in food during processing and/or storage, the modification of the food's primary constituents, such as lipids, is not always noticeable (Prakash et al. 2015; Falleh et al. 2020). At relatively low concentrations, antioxidant chemicals can stop, modify, or even completely stop oxidative processes. Accordingly, EOs and their components are crucial in promoting antioxidant activity.

The majority of the legal food additives are used for their preservation abilities, which are related to their acknowledged bioactivities. Acetic, malic, lactic, sorbic, and propionic acids, potassium and calcium acetates, carbon dioxide, benzoates, sorbates, propionates, nitrites, nitrates, and parabens are a few examples of additives with antimicrobial properties that can control food spoilage and/or prevent contamination by foodborne pathogens. Sulphites are the most common additives used to stop food from browning due to chemical or enzymatic processes (Wu et al. 2022). Despite the fact that all of these substances are legal to use in the food industry, the current fad is to swap out synthetic chemicals with natural ones. In this regard, using EOs is viewed as an alternative to using artificial additives (Carocho et al. 2014; Ribeiro-Santos et al. 2017).

Consumers' growing health consciousness has generated unfavourable attitudes of artificial food additives. As a result, the food business began using a large number of natural and secure EOs as part of an 'organic and green approach', as a new barrier to ensure the removal of pathogens from a particular food matrix (Falleh et al. 2019). EOs can function directly or indirectly in a variety of ways, such as by stopping the beginning of chains and scavenging free radicals (Rodriguez-Garcia et al. 2016). Both the European Commission (EC) and the Food and Drug Administration (FDA) in the United States authorised a number of EOs and EO components, classifying them as generally recognized as safe (GRAS) to be used as flavourings and/or preservatives in food items. Lavandin, menthol, rose, sage, oregano, cinnamon, basil, clove, coriander, nutmeg, ginger, and thyme EOs are among the list of crude EOs with the GRAS designation. The recognised EO components also contained substances that are all regarded as safe within the parameters of the recommended daily consumption, such as thymol, carvacrol, eugenol, linalool, carvone, vanillin, cinnamaldehyde, citral, and limonene (Hyldgaard et al. 2012; Falleh et al. 2020).

Free EOs have been blended with various biodegradable substances, including paper and edible films. Paper and cardboard have historically been utilised as packaging materials, but novel packaging solutions are constantly being developed, such as active cardboard trays covered with emulsions that include encapsulated EOs. These are frequently used as food and vegetable packaging materials (with waterproof layers) and are viewed as a replacement for the packaging industry's heavy reliance on plastics (Buendía et al. 2019; Derbassi et al. 2022). To create an aqueous dispersion for edible films, EOs are included into a hydrophilic matrix. As a result, the hydrophilic components of the film undergo structural rearrangement as the drying process causes the lipid droplets to get lodged in the matrix (Sánchez-González et al. 2011). Adding particular EOs (such as oregano, coriander, thyme, and clove) to meat as preservatives helps to eradicate numerous viruses and native flora responsible for deterioration. After ingestion, EO components, especially thymol and carnosic acid, may build up to sufficient amounts in the animal's muscle to operate as an efficient antibacterial agent (Burt 2004).

It is important to note that the source and makeup of EOs affect their impact on food preservation. Precise information about the impact of food matrix components on Eos' antibacterial activities is required for the application of EOs as food preservatives. The volatility, hydrophobicity, and detrimental effects on food organoleptic qualities drive EOs' limited employment in the food and agriculture industries despite their outstanding preservation effectiveness (Weisany et al. 2022; Polat and Aygun 2023). For a slow release of EO aroma suitable with food-based applications, particular delivery mechanisms are thus needed. Common EOs delivery methods include edible coating, active packaging, and nanoencapsulation. These methods facilitate EO dispersion while maintaining constant antibacterial activity and extending food shelf life. Additionally, regulated release behaviour exhibits improved diffusion kinetics and lessens the influence of EO on the organoleptic qualities of food systems (Singh et al. 2021a, b; Maurya et al. 2021a, b; Lim et al. 2023).

3.8 Conclusion and Future Perspectives

Growing health consciousness among consumers has led to negative views against artificial food additives. It has been widely researched and found to be promising that EOs might be used in active packaging technology. Because of their accessibility and vast variety of biological activity, EOs and their constituents are significant. Another benefit is that when employed in the precise quantities, they do not alter the flavour and intrinsic properties of stored food products, prolonging its shelf life. Food storage stability is determined by the amount of antioxidant present, which may be assessed using both direct and indirect approaches. To measure the antioxidant activity of EOs, a variety of techniques have been developed, and each technique employs a different strategy, indication, and analytical target. Using the most appropriate analytical techniques for each food product or antioxidant would be crucial for providing more precise information about the oxidation status and antioxidants' capacity. EOs embedded in nanoparticles are a relatively recent use in food packaging. With the regulated release of EOs, nanoencapsulation offers a longer and more efficient usage of the antibacterial and antioxidant impact while also limiting their strong aroma.

A fresh viewpoint on the organoleptic qualities of food may be offered by the use of EO blends that are created in line with the features of the stored products. Novel methods may be useful to safeguard and enhance EOs' features and biological activities due to their flimsiness under environmental challenges like temperature and light. Further research should concentrate on the mechanisms of action and synergistic effects between various EOs and their components. The EOs and their nanoencapsulated outcomes might be very effective as natural antioxidants in the present scenario of changing choices of the people and their heavy interest towards the natural products and preservatives. Thus, the antioxidant properties of EOs could be further enhanced by more suitable active packaging, synergism between involved components, and their more appropriate formulations using advanced techniques.

Acknowledgements Bijendra Kumar Singh is grateful to Head, Department of Botany, and Principal of Feroze Gandhi College, Raebareli, Uttar Pradesh, India, for the kind support and also thankful to the Council of Scientific and Industrial Research (CSIR) [grant no.: 09/013(0920)/2019-EMR-I], New Delhi, India, for providing research fellowship.

References

Acosta-Casique A, Montes-Alvarado JB, Barragán M, Larrauri-Rodríguez KA, Perez-Gonzalez A, Delgado-Magallón A, Maycotte P et al (2023) ERK activation modulates invasiveness and reactive oxygen species (ROS) production in triple negative breast cancer cell lines. Cell Signal 101:110487

Ahmad T, Suzuki YJ (2019) Juglone in oxidative stress and cell signaling. Antioxidants 8(4):91

Al-Gubory KH (2014) Environmental pollutants and lifestyle factors induce oxidative stress and poor prenatal development. Reprod Biomed Online 29(1):17–31

Al-Maqtari QA, Rehman A, Mahdi AA, Al-Ansi W, Wei M, Yanyu Z, Yao W et al (2021) Application of essential oils as preservatives in food systems: challenges and future prospectives–a review. Phytochem Rev:1–38

Al-Tayyar NA, Youssef AM, Al-Hindi RR (2020) Edible coatings and antimicrobial nanoemulsions for enhancing shelf life and reducing foodborne pathogens of fruits and vegetables: a review. Sustain Mater Technol 26:e00215

Amorati R, Valgimigli L (2015) Advantages and limitations of common testing methods for antioxidants. Free Radic Res 49(5):633–649

Anaissi-Afonso L, Oramas-Royo S, Ayra-Plasencia J, Martín-Rodríguez P, García-Luis J, Lorenzo-Castrillejo I, Machin F et al (2018) Lawsone, juglone, and β-lapachone derivatives with enhanced mitochondrial-based toxicity. ACS Chem Biol 13(8):1950–1957

Angane M, Swift S, Huang K, Butts CA, Quek SY (2022) Essential oils and their major components: an updated review on antimicrobial activities, mechanism of action and their potential application in the food industry. Foods 11(3):464

Arias A, Feijoo G, Moreira MT (2022) Exploring the potential of antioxidants from fruits and vegetables and strategies for their recovery. Innovative Food Sci Emerg Technol 102974:102974

Bajpai VK, Baek KH (2016) Biological efficacy and application of essential oils in foods-a review. J Essent Oil Bear Plants 19(1):1–19

Bhavaniramya S, Vishnupriya S, Al-Aboody MS, Vijayakumar R, Baskaran D (2019) Role of essential oils in food safety: antimicrobial and antioxidant applications. Grain Oil Sci Technol 2(2):49–55

Birben E, Sahiner UM, Sackesen C, Erzurum S, Kalayci O (2012) Oxidative stress and antioxidant defense. World Allergy Org J 5:9–19

Bonda AF, Regis L, Giovannelli L, Segale L (2020) Alginate/maltodextrin and alginate/shellac gum core-shell capsules for the encapsulation of peppermint essential oil. Int J Biol Macromol 162:1293–1302

Borges RS, Ortiz BLS, Pereira ACM, Keita H, Carvalho JCT (2019) Rosmarinus officinalis essential oil: a review of its phytochemistry, anti-inflammatory activity, and mechanisms of action involved. J Ethnopharmacol 229:29–45

Buchanan BB, Gruissem W, Jones RL (eds) (2015) Biochemistry and molecular biology of plants. John Wiley & Sons

Buendía L, Soto S, Ros M, Antolinos V, Navarro L, Sánchez MJ et al (2019) Innovative cardboard active packaging with a coating including encapsulated essential oils to extend cherry tomato shelf life. LWT 116:108584

Burt S (2004) Essential oils: their antibacterial properties and potential applications in foods—a review. Int J Food Microbiol 94(3):223–253

Carocho M, Barreiro MF, Morales P, Ferreira IC (2014) Adding molecules to food, pros and cons: a review on synthetic and natural food additives. Compr Rev Food Sci Food Saf 13(4):377–399

Carpena M, Nuñez-Estevez B, Soria-Lopez A, Garcia-Oliveira P, Prieto MA (2021) Essential oils and their application on active packaging systems: a review. Resources 10(1):7

Chaudhari AK, Das S, Singh BK, Dubey NK (2023) Green facile synthesis of cajuput (Melaleuca cajuputi Powell.) essential oil loaded chitosan film and evaluation of its effectiveness on shelf-life extension of white button mushroom. Food Chem 401:134114

Chaudhari AK, Singh VK, Das S, Dubey NK (2021) Nanoencapsulation of essential oils and their bioactive constituents: a novel strategy to control mycotoxin contamination in food system. Food Chem Toxicol 149:112019

Chaudhari AK, Singh VK, Dwivedy AK, Das S, Upadhyay N, Singh A, Dubey NK et al (2020) Chemically characterised Pimenta dioica (L.) Merr. Essential oil as a novel plant based antimicrobial against fungal and aflatoxin B1 contamination of stored maize and its possible mode of action. Nat Prod Res 34(5):745–749

Demidchik V (2015) Mechanisms of oxidative stress in plants: from classical chemistry to cell biology. Environ Exp Bot 109:212–228

Derbassi NB, Pedrosa MC, Heleno S, Carocho M, Ferreira IC, Barros L (2022) Plant volatiles: using scented molecules as food additives, vol 122. Trends in Food Science & Technology, p 97

Diniz do Nascimento L, Barbosa de Moraes AA, Santana da Costa K, Pereira Galúcio JM, Taube PS, Leal Costa CM, Guerreiro de Faria LJ et al (2020) Bioactive natural compounds and antioxidant activity of essential oils from spice plants: new findings and potential applications. Biomol Ther 10(7):988

Doost AS, Van Camp J, Dewettinck K, Van der Meeren P (2019) Production of thymol nanoemulsions stabilized using Quillaja Saponin as a biosurfactant: antioxidant activity enhancement. Food Chem 293:134–143

Eckardt NA (2017) The plant cell reviews plant immunity: receptor-like kinases, ROS-RLK crosstalk, quantitative resistance, and the growth/defense trade-off, vol 29, p 601

El-Baroty GS, Abd El-Baky HH, Farag RS, Saleh MA (2010) Characterization of antioxidant and antimicrobial compounds of cinnamon and ginger essential oils. Afr J Biochem Res 4(6):167–174

Falleh H, Ben Jemaa M, Djebali K, Abid S, Saada M, Ksouri R (2019) Application of the mixture design for optimum antimicrobial activity: combined treatment of Syzygium aromaticum, Cinnamomum zeylanicum, Myrtus communis, and Lavandula stoechas essential oils against Escherichia coli. J Food Proc Preserv 43(12):e14257

Falleh H, Jemaa MB, Saada M, Ksouri R (2020) Essential oils: a promising eco-friendly food preservative. Food Chem 330:127268

Getzoff ED, Tainer JA, Weiner PK, Kollman PA, Richardson JS, Richardson DC (1983) Electrostatic recognition between superoxide and copper, zinc superoxide dismutase. Nature 306:287–290

Gómez-Estaca J, López-de-Dicastillo C, Hernández-Muñoz P, Catalá R, Gavara R (2014) Advances in antioxidant active food packaging. Trends Food Sci Technol 35(1):42–51

Griendling KK, Touyz RM, Zweier JL, Dikalov S, Chilian W, Chen YR, Bhatnagar A et al (2016) Measurement of reactive oxygen species, reactive nitrogen species, and redox-dependent signaling in the cardiovascular system: a scientific statement from the American Heart Association. Circ Res 119(5):e39–e75

Hawkins CL, Davies MJ (2019) Detection, identification, and quantification of oxidative protein modifications. J Biol Chem 294(51):19683–19708

Hyldgaard M, Mygind T, Meyer RL (2012) Essential oils in food preservation: mode of action, synergies, and interactions with food matrix components. Front Microbiol 3:12

Ighodaro OM, Akinloye OA (2018) First line defence antioxidants-superoxide dismutase (SOD), catalase (CAT) and glutathione peroxidase (GPX): their fundamental role in the entire antioxidant defence grid. Alexandria J Med 54(4):287–293

Jamróz E, Juszczak L, Kucharek M (2018) Investigation of the physical properties, antioxidant and antimicrobial activity of ternary potato starch-furcellaran-gelatin films incorporated with lavender essential oil. Int J Biol Macromol 114:1094–1101

Janků M, Luhová L, Petřivalský M (2019) On the origin and fate of reactive oxygen species in plant cell compartments. Antioxidants 8(4):105

Jayasena DD, Jo C (2014) Potential application of essential oils as natural antioxidants in meat and meat products: a review. Food Rev Intl 30(1):71–90

Katerji M, Filippova M, Duerksen-Hughes P (2019) Approaches and methods to measure oxidative stress in clinical samples: research applications in the cancer field. Oxidative Med Cell Longev 2019:1

Khalid AD, Ur-Rehman N, Tariq GH, Ullah S, Buzdar SA, Iqbal SS, Sher F et al (2023) Functional bioinspired nanocomposites for anticancer activity with generation of reactive oxygen species. Chemosphere 310:136885

Kocira A, Kozłowicz K, Panasiewicz K, Staniak M, Szpunar-Krok E, Hortyńska P (2021) Polysaccharides as edible films and coatings: characteristics and influence on fruit and vegetable quality—A review. Agronomy 11(5):813

Lim DY, Lee JS, Lee HG (2023) Nano-encapsulation of a combination of clove oil and thymol and their application in fresh-cut apples and raw minced beef. Food Control 109683

Loi M, Paciolla C (2021) Plant antioxidants for food safety and quality: exploring new trends of research. Antioxidants 10(6):972

López-Gómez A, Navarro-Martínez A, Martínez-Hernández GB (2023) Effects of essential oils released from active packaging on the antioxidant system and quality of lemons during cold storage and commercialization. Sci Hortic 312:111855

Maheshwari DT, Kumar MY, Verma SK, Singh VK, Singh SN (2011) Antioxidant and hepatoprotective activities of phenolic rich fraction of Seabuckthorn (Hippophae rhamnoides L.) leaves. Food Chem Toxicol 49(9):2422–2428

Mailloux RJ (2020) An update on mitochondrial reactive oxygen species production. Antioxidants 9(6):472

Maqsood S, Benjakul S, Abushelaibi A, Alam A (2014) Phenolic compounds and plant phenolic extracts as natural antioxidants in prevention of lipid oxidation in seafood: a detailed review. Compr Rev Food Sci Food Saf 13(6):1125–1140

Maurya A, Kumar S, Singh BK, Chaudhari AK, Dwivedy AK, Prakash B, Dubey NK (2022) Mechanistic investigations on antifungal and antiaflatoxigenic activities of chemically characterised Carum carvi L. essential oil against fungal infestation and aflatoxin contamination of herbal raw materials. Nat Prod Res 36(17):4563–4568

Maurya A, Prasad J, Das S, Dwivedy AK (2021a) Essential oils and their application in food safety. Front Sustain Food Syst 5:653420

Maurya A, Singh VK, Das S, Prasad J, Kedia A, Upadhyay N, Dwivedy AK et al (2021b) Essential oil nanoemulsion as eco-friendly and safe preservative: bioefficacy against microbial food deterioration and toxin secretion, mode of action, and future opportunities. Front Microbiol 3326

Mittler R (2017) ROS are good. Trends Plant Sci 22(1):11–19

Mittler R, Vanderauwera S, Suzuki N, Miller GAD, Tognetti VB, Vandepoele K, Van Breusegem F et al (2011) ROS signaling: the new wave? Trends Plant Sci 16(6):300–309

Munir A, Sultana B, Bashir A, Ghaffar A, Munir B, Shar GA, Iqbal M et al (2018) Evaluation of antioxidant potential of vegetables waste. Pol J Environ Stud 27(2):947

Munteanu IG, Apetrei C (2021) Analytical methods used in determining antioxidant activity: a review. Int J Mol Sci 22(7):3380

Nascimento da Silva LC, Bezerra Filho CM, Paula RAD, Silva e Silva CS, Oliveira de Souza LI, Silva MVD, Figueiredo RCBQD et al (2016) In vitro cell-based assays for evaluation of antioxidant potential of plant-derived products. Free Radic Res 50(8):801–812

Nehme R, Andrés S, Pereira RB, Ben Jemaa M, Bouhallab S, Ceciliani F, Abdennebi-Najar L et al (2021) Essential oils in livestock: from health to food quality. Antioxidants 10(2):330

Ni ZJ, Wang X, Shen Y, Thakur K, Han J, Zhang JG, Wei ZJ et al (2021) Recent updates on the chemistry, bioactivities, mode of action, and industrial applications of plant essential oils. Trends Food Sci Technol 110:78–89

Ozogul Y, Karsli GT, Durmuş M, Yazgan H, Oztop HM, McClements DJ, Ozogul F (2022) Recent developments in industrial applications of nanoemulsions. Adv Colloid Interf Sci 102685:102685

Pateiro M, Barba FJ, Domínguez R, Sant'Ana AS, Khaneghah AM, Gavahian M, Lorenzo JM et al (2018) Essential oils as natural additives to prevent oxidation reactions in meat and meat products: a review. Food Res Int 113:156–166

Pham-Huy LA, He H, Pham-Huy C (2008) Free radicals, antioxidants in disease and health. Int J Biomed Sci: IJBS 4(2):89

Polat S, Aygun A (2023) Nanoencapsulation of essential oil-based packaging for shelf-life extension of foods. In: Nanotechnology interventions in food packaging and shelf life. CRC Press, pp 217–244

Prakash B, Kedia A, Mishra PK, Dubey NK (2015) Plant essential oils as food preservatives to control moulds, mycotoxin contamination and oxidative deterioration of Agri-food commodities–potentials and challenges. Food Control 47:381–391

Prasad J, Das S, Maurya A, Jain SK, Dwivedy AK (2022) Synthesis, characterization and in situ bioefficacy evaluation of Cymbopogon nardus essential oil impregnated chitosan nanoemulsion against fungal infestation and aflatoxin B1 contamination in food system. Int J Biol Macromol 205:240–252

Ramsdam MG, Chaudhari AK, Singh VK, Dkhar MS, Kayang H, Dubey NK, Prakash B (2021) Antifungal and antioxidant activity of plant based essential oils on aspergillus flavus link (Trichocomaceae) isolated from stored maize grains of Meghalaya. Arch Phytopathol Plant Protect 54(17–18):1405–1420

Re R, Pellegrini N, Proteggente A, Pannala A, Yang M, Rice-Evans C (1999) Antioxidant activity applying an improved ABTS radical cation decolorization assay. Free Radic Biol Med 26(9–10):1231–1237

Ribeiro-Santos R, Andrade M, de Melo NR, Sanches-Silva A (2017) Use of essential oils in active food packaging: recent advances and future trends. Trends Food Sci Technol 61:132–140

Rodriguez-Garcia I, Silva-Espinoza BA, Ortega-Ramirez LA, Leyva JM, Siddiqui MW, Cruz-Valenzuela MR, Ayala-Zavala JF et al (2016) Oregano essential oil as an antimicrobial and antioxidant additive in food products. Crit Rev Food Sci Nutr 56(10):1717–1727

Sánchez-González L, Chiralt A, González-Martínez C, Cháfer M (2011) Effect of essential oils on properties of film forming emulsions and films based on hydroxypropylmethylcellulose and chitosan. J Food Eng 105(2):246–253

Semeniuc CA, Socaciu MI, Socaci SA, Mureşan V, Fogarasi M, Rotar AM (2018) Chemometric comparison and classification of some essential oils extracted from plants belonging to Apiaceae and Lamiaceae families based on their chemical composition and biological activities. Molecules 23(9):2261

Shahidi F, Ambigaipalan P (2015) Phenolics and polyphenolics in foods, beverages and spices: antioxidant activity and health effects–A review. J Funct Foods 18:820–897

Shields HJ, Traa A, Van Raamsdonk JM (2021) Beneficial and detrimental effects of reactive oxygen species on lifespan: a comprehensive review of comparative and experimental studies. Front Cell Develop Biol 9:181

Siano F, Picariello G, Castaldo D, Cautela D, Caruso T, Vasca E (2023) Monitoring antioxidants by coulometry: quantitative assessment of the strikingly high antioxidant capacity of bergamot (Citrus bergamia R.) by-products. Talanta 251:123765

Sies H, Jones DP (2020) Reactive oxygen species (ROS) as pleiotropic physiological signalling agents. Nat Rev Mol Cell Biol 21(7):363–383

Singh BK, Dubey NK (2023) Bioactive components and biological properties of Zingiber officinale roscoe essential oil. In: Bioactives and pharmacology of medicinal plants. Apple Academic Press, pp 391–401

Singh BK, Tiwari S, Dubey NK (2021b) Essential oils and their nanoformulations as green preservatives to boost food safety against mycotoxin contamination of food commodities: a review. J Sci Food Agric 101(12):4879–4890

Singh BK, Tiwari S, Chaudhari AK, Maurya A, Das S, Singh VK, Dubey NK (2022a) Chitosan encompassed Aniba rosaeodora essential oil as innovative green candidate for antifungal and antiaflatoxigenic activity in millets with emphasis on cellular and its mode of action. Front Microbiol 2979

Singh BK, Tiwari S, Maurya A, Kumar S, Dubey NK (2022b) Fungal and mycotoxin contamination of herbal raw materials and their protection by nanoencapsulated essential oils: an overview. Biocatal Agric Biotechnol 39:102257

Singh VK, Das S, Kumar Dwivedy A, Kumar Chaudhari A, Upadhyay N, Dubey NK (2021a) Assessment of chemically characterized Salvia sclarea L. essential oil and its combination with linalyl acetate as novel plant-based antifungal, antiaflatoxigenic and antioxidant agent against herbal drugs contamination and probable mode of action. Nat Prod Res 35(5):782–787

Su H, Huang C, Liu Y, Kong S, Wang J, Huang H, Zhang B (2020) Preparation and characterization of cinnamomum essential oil–chitosan nanocomposites: physical, structural, and antioxidant activities. PRO 8(7):834

Tit DM, Bungau SG (2023) Antioxidant activity of essential oils. Antioxidants 12(2):383

Tiwari S, Singh BK, Dubey NK (2022) Aflatoxins in food systems: recent advances in toxicology, biosynthesis, regulation and mitigation through green nanoformulations. J Sci Food Agric 103:1621

Tomás-Barberán FA, Ferreres F, Gil MI (2000) Antioxidant phenolic metabolites from fruit and vegetables and changes during postharvest storage and processing. Stud Nat Prod Chem 23:739–795

Umaraw P, Munekata PE, Verma AK, Barba FJ, Singh VP, Kumar P, Lorenzo JM (2020) Edible films/coating with tailored properties for active packaging of meat, fish and derived products. Trends Food Sci Technol 98:10–24

Valdivieso-Ugarte M, Gomez-Llorente C, Plaza-Díaz J, Gil Á (2019) Antimicrobial, antioxidant, and immunomodulatory properties of essential oils: a systematic review. Nutrients 11(11):2786

Vorobyova VI, Skiba MI, Shakun AS, Nahirniak SV (2019) Relationship between the inhibition and antioxidant properties of the plant and biomass wastes extracts-A review. Int J Corros Scale Inhib 8(2):150–178

Weisany W, Yousefi S, Tahir NAR, Zadeh NG, McClements DJ, Adhikari B, Ghasemlou M (2022) Targeted delivery and controlled released of essential oils using nanoencapsulation: a review. Adv Colloid Interf Sci 102655:102655

Wu L, Zhang C, Long Y, Chen Q, Zhang W, Liu G (2022) Food additives: from functions to analytical methods. Crit Rev Food Sci Nutr 62(30):8497–8517

Wu Z, Tan B, Liu Y, Dunn J, Martorell Guerola P, Tortajada M, Ji P et al (2019a) Chemical composition and antioxidant properties of essential oils from peppermint, native spearmint and scotch spearmint. Molecules 24(15):2825

Wu Z, Zhou W, Pang C, Deng W, Xu C, Wang X (2019b) Multifunctional chitosan-based coating with liposomes containing laurel essential oils and nanosilver for pork preservation. Food Chem 295:16–25

Zeng FS, Yao YF, Wang LF, Li WJ (2023) Polysaccharides as antioxidants and prooxidants in managing the double-edged sword of reactive oxygen species. Biomed Pharmacother 159:114221

Zhaveh S, Mohsenifar A, Beiki M, Khalili ST, Abdollahi A, Rahmani-Cherati T, Tabatabaei M (2015) Encapsulation of Cuminum cyminum essential oils in chitosan-caffeic acid nanogel with enhanced antimicrobial activity against aspergillus flavus. Ind Crop Prod 69:251–256

Zheng K, Xiao S, Li W, Wang W, Chen H, Yang F, Qin C (2019) Chitosan-acorn starch-eugenol edible film: physico-chemical, barrier, antimicrobial, antioxidant and structural properties. Int J Biol Macromol 135:344–352

Mechanistic Investigation on Antibacterial Activity of Essential Oils against Resistant Bacteria Species

4

Bárbara Morandi Lepaus, Sara Jarske Geringer, Manueli Monciozo Domingos, Bárbara Santos Valiati, Daniel Sgrancio Uliana, Rhaiza Marcia Lopes Leal, Alessandra Peres Guimarães, and Jackline Freitas Brilhante de São José

Abstract

Consumption of contaminated food affects millions and may even lead to death. For this reason, foodborne outbreaks are considered a public health problem. Commonly, eating food contaminated with pathogenic bacteria involves treatment with antimicrobials, like antibiotics. However, there is the aggravating circumstance of food contamination by resistant bacteria species (RBS). Microbial resistance to antimicrobials is an ecological problem and appears when microorganisms adapt to continuous exposure to an antimicrobial agent. Thus, there is an interest in studying potential safe alternatives to conventional antimicrobials. Natural antimicrobials have been increasingly studied because they are compounds naturally present in plants. Essential oils (EOs) are considered a complex mixture of aromatic and volatile compounds and have the potential to be applied as natural additives due to their natural biological activities, such as antibacterial, antifungal, and antiparasitic. Different EOs have exhibited antimicrobial properties against Gram-positive and Gram-negative bacteria species. This chapter

B. M. Lepaus · M. M. Domingos · B. S. Valiati · A. P. Guimarães
Postgraduate Program in Nutrition and Health, Federal University of Espírito Santo, Espírito Santo, Brazil

S. J. Geringer · D. S. Uliana · R. M. L. Leal
Department of Integrated Education in Health, Federal University of Espírito Santo, Espírito Santo, Brazil

J. F. B. de São José (✉)
Professor in Postgraduate Program in Nutrition and Health, Department of Integrated Education in Health, Federal University of Espírito Santo, Espírito Santo, Brazil
e-mail: jackline.jose@ufes.br

B. Prakash et al. (eds.), *Plant Essential Oils*,
https://doi.org/10.1007/978-981-99-4370-8_4

describes some investigations about the antimicrobial effect of different EOs against RBS.

Keywords

Bacterial resistance · Antibacterial activity · Essential oil · Foodborne pathogens.

4.1 Introduction

Consumption of contaminated food affects millions and may even lead to death. For this reason, foodborne outbreaks are considered a public health problem. Among the microorganisms that contaminate food and can cause food outbreaks are species of *Bacillus*, *Campylobacter*, *Clostridium*, *Escherichia coli*, *Listeria monocytogenes*, *Salmonella*, and *Staphylococcus* (WHO 2022a, b).

Usually, ingesting food contaminated by pathogenic bacteria requires treatment with antimicrobials, like antibiotics. However, there is the aggravating factor of food contamination by resistant bacteria species (RBS). Microbial resistance to antimicrobials is an ecological issue and occurs when microorganisms adapt to continuous exposure to antimicrobial agents (Samtiya et al. 2022).

For this reason, there is a worldwide effort among researchers regarding the study of possible safe alternatives to conventional antimicrobials. On the other hand, natural antimicrobials have been increasingly studied because they are compounds naturally present in plants and can be extracted from different parts, like herbs, bark, buds, leaves, wood, roots, twigs, seeds, flowers, and fruits. Additionally, consumers are increasingly concerned and demanding the possibility of using safe and effective compounds.

Among these alternatives are essential oils (EOs). EOs are considered a complex mixture of aromatic and volatile compounds. They can have 20 to over 80 different compounds (Freitas and Cattelan 2018; Mani-López et al. 2018; Bouymajane et al. 2022; Jaouadi et al. 2023).

Due to their high ability to eliminate free radicals, EOs have several natural biological activities, such as antibacterial, antifungal, and antiparasitic (He et al. 2020). For this reason, new sources and applications of EOs are being increasingly studied. For example, different EOs have demonstrated high antimicrobial properties against Gram-positive and Gram-negative bacteria species (Albaqami et al. 2022; Khedhri et al. 2022; Almeida et al. 2023).

This chapter describes some investigations about the antimicrobial effect of different EOs against RBS. First, the concepts of microbial contamination are presented, followed by a topic about using natural compounds, specifically the EOs, as green alternatives to conventional antimicrobials.

4.2 Microbial Contamination and the Presence of Resistant Bacteria Species in Food

Food contamination has been recognized as a global challenge by the World Health Organization (WHO 2022a, b). Food can be contaminated with biological, chemical, or physical agents (Kamala and Kumar 2018). The biological contaminants are present in several sources throughout the supply chain, from farm to fork (São José et al. 2022). According to the Centers for Disease Control and Prevention of the United States (CDC), 250 food outbreaks were recorded between 2017 and 2020, resulting in 14,312 illnesses, 3517 hospitalizations, and 65 deaths. However, the actual number of cases may be underestimated since many people do not seek medical care, making it difficult to know which bacteria could be causing the disease (CDC 2022).

The occurrence of pathogens in food is dangerous because, usually, their presence in food does not guarantee changes in food's physicochemical and sensory aspects. *Bacillus cereus*, *Campylobacter jejuni*, *Clostridium botulinum*, *C. perfrigens*, *Escherichia coli*, *Listeria monocytogenes*, *Salmonella* spp., *Shigella* spp., *Staphylococcus aureus*, *Vibrio cholera*, and *Yersinia enterocolitica* are some of many examples of common pathogenic bacterial species that can contaminate food and cause disease outbreaks. Although many of these disease outbreaks are self-limiting diarrheal syndromes in healthy humans, in some cases, ingesting contaminated food can cause life-threatening illnesses that require immediate and specific antimicrobial treatment to treat the infections in humans (Villa et al. 2016; Samtiya et al. 2022). Antimicrobial drugs, like antibiotics, are also used globally to treat animal infections and prophylactically in production agriculture (Samtiya et al. 2022).

However, in recent decades, cases of food contamination by RBS have increased (Rahman et al. 2022), more frequently in low- and middle-income countries than in developed countries, but it is considered a serious issue worldwide. RBS, also called antimicrobial resistance species or resistance genes, emerge when microorganisms alter their physiology or genetic makeup after regular contact with antimicrobial agents (i.e., antibiotics) (Samtiya et al. 2022). Additionally, when they acquire resistance to antimicrobials, microorganisms can transmit this defense to other organisms by the horizontal transfer of resistance genes (Villa et al. 2016).

One of the mechanisms described for the antimicrobial resistance of bacteria is the efflux mechanism. These mechanisms transport proteins to eliminate toxic substances from the bacterial cell. The constant elimination of poisonous substances, like antimicrobials, makes them never remain at the proper concentration inside the cell to exert their antimicrobial action (Villa et al. 2016).

Nevertheless, the massive and indiscriminate use of existing antibiotics also contributes to the emergence of RBS. So, the increase in RBS and the scarcity of new antimicrobials require research on alternative strategies to deal with RBS (Chouhan et al. 2017).

Like other microorganisms, RBS can contaminate food at any step, from the field to retail. Therefore, it is very crucial to ensure food safety and not expose the consumer to hazards. For this, some points need attention. Firstly, it is necessary to identify the source of contamination, review the procedures adopted in production and handling, and apply strategies capable of reducing contamination. However, there is an increased demand for the study of new antimicrobial agents, so it is essential to understand and enable safe approaches to eliminate or reduce food contamination, especially by RBS.

4.3 Use of Natural Compounds as Green Alternatives to Conventional Antimicrobials

The use of natural compounds obtained naturally from plants has gained popularity because people are increasingly concerned about the possibility of using safe and effective compounds in many areas.

So, the biological properties of plants have been widely investigated since they often present low toxicity, besides a lot of potential pharmacological activities (e.g., antioxidant, antiseptic, antibacterial, antifungal, antiparasitic, and antiviral) (Seshadri et al. 2020). Among the new approaches of efficient natural antimicrobials against pathogens, including RBS, is the use of EOs. EOs, or aromatic or volatile oils, are a complex mixture of aromatic and volatile compounds in various plants (Table 4.1).

The EOs may contain between 20 and more than 80 compounds in different concentrations (Freitas and Cattelan 2018; Mani-López et al. 2018; Bouymajane et al. 2022; Jaouadi et al. 2023).

EOs are also known as bioactive secondary metabolites of plants, have low molecular weight, and have diverse chemical structures (Fig. 4.1). In addition, due to their hydrophobic nature, the compounds are soluble in organic solvents and have a lower density than water. For this reason, they are extracted by hydrodistillation, steam distillation, or solvent extraction methods and are commercialized like oily liquids (Chouhan et al. 2017; Freitas and Cattelan 2018).

These secondary metabolites can be extracted from vegetative parts of plants, like bark, leaves, roots, herbs, and wood, or reproductive parts as flowers and fruits (Ali et al. 2022).

Secondary metabolites are natural compounds synthesized by organisms or plants and are not essential to support growth and life (Pyne et al. 2019). They differ, for example, from primary metabolites like amino acids, nucleic acids, lipids, and carbohydrates. However, even though they are not essential for cell growth, these metabolites have biological activities, protect plants against many adverse environmental conditions, and have a survival function (Bhattacharya 2019; Mosunova et al. 2021). For this reason, EOs have an essential role in protecting plants against microorganisms, herbivores, and undesirable insects (Freitas and Cattelan 2018) because these organisms have a reduced appetite for plants that have these compounds (Chouhan et al. 2017).

Moreover, when exposed to higher temperature conditions, for example, plants accumulate more reactive oxygen species (ROS), which are very harmful to cell membranes and DNA. Therefore, in this high-temperature stress exposition, plant protection depends on increasing its capacity to scavenge the ROS and maintain the cell membrane stability. Thus, different metabolites, such as antioxidants (e.g., EOs), osmoprotectants, and heat shock proteins, are synthesized and accumulated by plants (Bhattacharya 2019).

Because of their high biological activity and radical-scavenging properties, EOs have been explored as a natural antimicrobial and antioxidant component in many industries, like the pharmacological, cosmetic, and food industries. In addition, EOs demonstrate broad antimicrobial spectra, acting against either Gram-positive or Gram-negative bacteria besides yeasts and molds (He et al. 2020; Albaqami et al. 2022; Khedhri et al. 2022; Almeida et al. 2023).

EOs are chemically derived from terpenes and their oxygenated derivatives, like terpenoids, which are aromatic and aliphatic acid esters and phenolic compounds (Chouhan et al. 2017). The biological properties of EOs are determined by their chemical composition, but one to three significant compounds are usually responsible for their antimicrobial activity (Freitas and Cattelan 2018; Mani-López et al. 2018).

Furthermore, the chemical composition, and consequently the biological properties of EOs, can be affected by environmental factors such as plant variety/cultivar, soil, climate (temperature and humidity), geographic location, maturity degree of the plant, and harvest season, as well as the method of obtaining and extracting.

For example, Kasrati et al. (2014) determined the volatile oil compounds of EOs from *Thymus saturejoides* collected from three places in Morocco. The sample collected from a more arid local presented carvacrol (45.3%), p-cymene (8.9%), linalool (8.4%), and borneol (7.5%) as the main constituents. On the other hand, EO from a medium arid site was mainly composed of carvacrol (26.5%), borneol (20.1%), camphene (8.0%), and terpinene (5.6%). At least, carvacrol (25.3%), borneol (19.7%), camphene (7.6%), and p-cymene (6.6%) were the main compounds identified in the less arid site sample. Moreover, the EO from the more arid place showed higher antioxidant capacity and antimicrobial activity.

In another study, Walasek-Janusz et al. (2022) investigated the chemical composition and variation in the antimicrobial activity of EOs from cultivars of *Lavandula*. Linalyl acetate and linalool were the main components of the EO in *L. angustifolia* and *L. intermedia*. Still, in the second variety, a high content of terpinen-4-ol (18.08%) was recorded, while in the other species, the content was less than 10%. However, the relationship analysis between the main components and the antimicrobial activity of EOs, demonstrated that linalool and terpinen-4-ol were the compounds potentially responsible for this activity.

In another research, three different species of *Curcuma* EO were evaluated (Albaqami et al. 2022). In the chemical composition analysis, *Curcuma longa*, *C. aromatica*, and *C. anguistifolia* showed different compound profiles (Table 4.1). Besides, *C. longa* EO demonstrated higher antioxidant and antibacterial activities.

In recent research, Jaouadi et al. (2023) demonstrated that the occurrence of carvacrol in 14 Tunisian *Thymus algeriensis* EO samples was positively correlated to geographic and climate conditions, being the evapotranspiration, summer drought

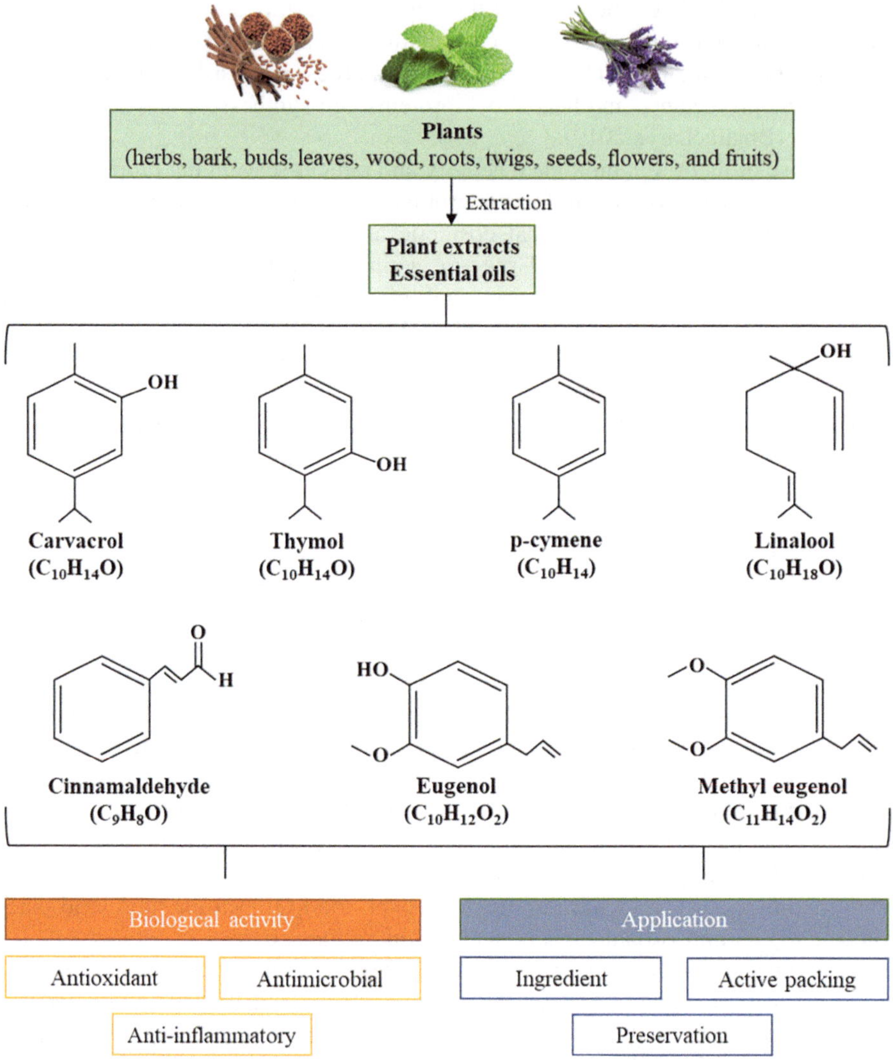

Fig. 4.1 Examples of some chemical structures of bioactive compounds present in essential oils and their applications

stress, and the temperature of the warmest months of the year. Furthermore, the antioxidant and anti-acetylcholinesterase capacity levels were linked to the chemical composition of EOs. More satisfactory results were described for EOs collected from the more arid zones, which had higher levels of carvacrol.

On the other hand, Marangoni et al. (2023) extracted the *Schinus terebinthifolia* EO from leaves and stems collected from six states of Brazil. α-pinene and limonene were predominant constituents of EO in five states, whereas α-phellandrene and limonene were the dominant components in the sixth place. Although the results

Table 4.1 Main components of various essential oils

The scientific name of the plant	Main components of EO[a]	Reference
Artemisia dracunculus L.	Methyl eugenol β-Phellandrene	(Sobieszczańska et al. 2020)
Atractylodes lancea (Thunb.) DC	β-Eudesmol Hinesol	(He et al. 2020)
Curcuma aromatic	Camphor 2-bornanone Curdione	(Albaqami et al. 2022)
Curcuma longa	α-Phellandrene 2-carene Eucalyptol	(Albaqami et al. 2022)
Curcuma anguistifolia	Eucalyptol Curzerenoneα-lemenone Longiverbenoneα-curcumene	(Albaqami et al. 2022)
Cinnamomum loureirii	Cinnamaldehyde Linalool	(Seshadri et al. 2020)
Eucalyptus sp.	Eucalyptol	(Khedhri et al. 2022)
Evolvulus alsinoides	Cis-α-Necrodol	(Seshadri et al. 2020)
Laurus nobilis	1,8-cineole α-Terpineol acetate Methyl eugenol	(Nabila et al. 2022)
Lavandula × *intermedia*	Linalool Linalyl acetate 1,8-cineol	(Ramić et al. 2022)
Lavandula sp.	Linalyl acetate Linalool	(Walasek-Janusz et al. 2022)
Ocimum basilicum L.	Eugenol Caryophyllene	(Cai et al. 2022)
Thymus vulgaris	Thymol p-cymene Linalool	(Benameur et al. 2019)
Thymus zygis subsp. *gracilis*	p-cymene Thymol	(Bouymajane et al. 2022)

[a]Main components refer to the compounds in higher amounts in the EO, according to the authors. EO: essential oil

have demonstrated that the EOs samples were different in their chemical composition, no difference in their anti-inflammatory and antihyperalgesic effects was observed.

So, the antimicrobial activity of EOs depends on the interaction between the compounds and not only on the chemical variation that each one presents.

4.4 Antimicrobial Activity of Essential Oils against Resistant Bacteria Species

Bacterial resistance is related to antimicrobial products' effectiveness and microorganism-resistant mechanisms. The resistance could be intrinsic and is related to bacteria's characteristics and tools, such as the generic efflux pump to remove antimicrobials from the cell, enzymatic inactivation, and permeability reduction. Genetic resistance can be obtained by horizontal gene transfer through bacteriophage, naked DNA, plasmids, integrons, or transposons. Furthermore, bacteria could use enzymes to modify drugs and plasmid-encoded efflux and unconventional metabolic pathways to modify the drug's target to decrease the effect of the antibiotic (El-Tarabily et al. 2021).

Considering the occurrence of microorganism resistance against conventional antimicrobials, there is a proposal to use EO. However, there is no consensus on the mechanism of microbial inactivation by which EOs inhibit cells (Clemente et al. 2016). In general, the lipophilic characteristic of EOs allows them to interact with the lipids in the cell membranes of bacteria. This interaction, consequently, affects cellular permeability, disturbing the structures and cellular functioning and causing cell death (Chouhan et al. 2017; He et al. 2020). However, the specific mechanism by which EOs exert antimicrobial activity depends on the oil and its chemical composition, the interaction between them, the concentration, besides the type of microorganism exposed.

For example, Gram-positive bacteria are more susceptible to EOs than Gram-negative bacteria. It occurs because Gram-negative species have a rigid and more complex outer membrane due to lipopolysaccharide (LPS), which interferes with the diffusion of EOs through the cell wall. In contrast, Gram-positive bacterial species are more sensitive to antimicrobial molecules because of lipoteichoic acid in the cell membranes (Yammine et al. 2022). However, according to Yammine et al. (2022), some studies indicated that Gram-negative bacteria are more sensitive to EOs, and different strains of the same bacterium may exert highly variable responses to EOs. Therefore, differences in the mechanism of action and antimicrobial capacity can be observed for the same microorganism subjected to different oils. Furthermore, the same oil can present different results against species of the same genus or microbial group (Clemente et al. 2016). Thus, it is essential to evaluate the antimicrobial activity of different EOs against different bacteria and strains of the same bacterium (Yammine et al. 2022).

Disc or well diffusion is one example of in vitro bioassays mostly applied to assess the antimicrobial capacity of EOs (Ali et al. 2022; Li et al. 2022). The disk-diffusion method, or plate diffusion, tests the efficiency of different antimicrobials in creating a halo of inhibition. First, the microorganism is inoculated onto the plate. Next, wells are made, and antimicrobials are deposited in them. Finally, the plates are incubated, and after this period, the halos, or zones of inhibition (ZI), are measured. The results are expressed as millimeters (mm) of inhibition, representing the part with no microbial growth (Fig. 4.2).

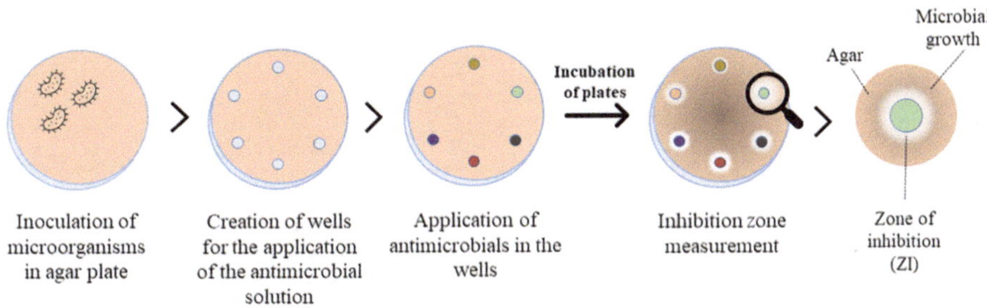

Fig. 4.2 Antimicrobial efficiency evaluation scheme by the disc or well diffusion method in plates

Besides, the minimum inhibitory concentration (MIC) is often evaluated. The MIC is the lowest concentration of some antimicrobial agents that completely inhibits the organism's growth in the bioassay. Time-kill assays also can be applied to comprehend the effect of EOs in microbial cells. Based on the period to produce an effective antimicrobial action, the EOs can be divided into compounds with slow activity and compounds that act faster (Chouhan et al. 2017).

EOs and their component parts are hydrophobicity, and they align between the fatty acid chains of cell membranes, destabilizing fats in the cellular membranes, and favoring the disruption of the cell structures. When membrane stability is lost, consequently, this fact tends to the destruction of microbial species owing to the leakage of essential elements from the microbial cell. Numerous substances regulate drug resistance by exploiting efflux pathways of various species of Gram-negative microbes (Hou et al. 2022; Yammine et al. 2022).

According to Hou et al. (2022), EOs disrupted the cell structure and occasioned the breakage of the cell membrane surface and increased permeation that disturbs the energy process, membrane transport, growth regulation, and other processes of microbial metabolism. Yammine et al. (2022) mentioned that another mechanism of action had been suggested for phenolic compounds in EOs, and this indicated that they interfere with specific proteins, inhibit flagella and toxins synthesis, suppress the expression of bacterial genes of biofilm formation, interfere in quorum sensing signaling, and stop cell division, thus leading to cell death.

Figure 4.3 illustrates some of the cellular changes in bacterial cells that can occur after exposure to EOs. These changes occurred after a cascade of reactions involving the entire bacterial cell. These modifications include the difference in the permeability of the membranes, loss of essential intracellular compounds (i.e., proteins, adenosine triphosphate [ATP], and DNA), inhibition of the energy generation mechanism, the release of enzymes and electrolytes, collapse of the cell, and finally, microbial death.

The improper application of antimicrobials conducts to drug resistance which affects animal and human health (Yasir et al. 2024). Next, some in vitro studies that evaluated the application of EO on different bacterial species that commonly contaminate food will be described.

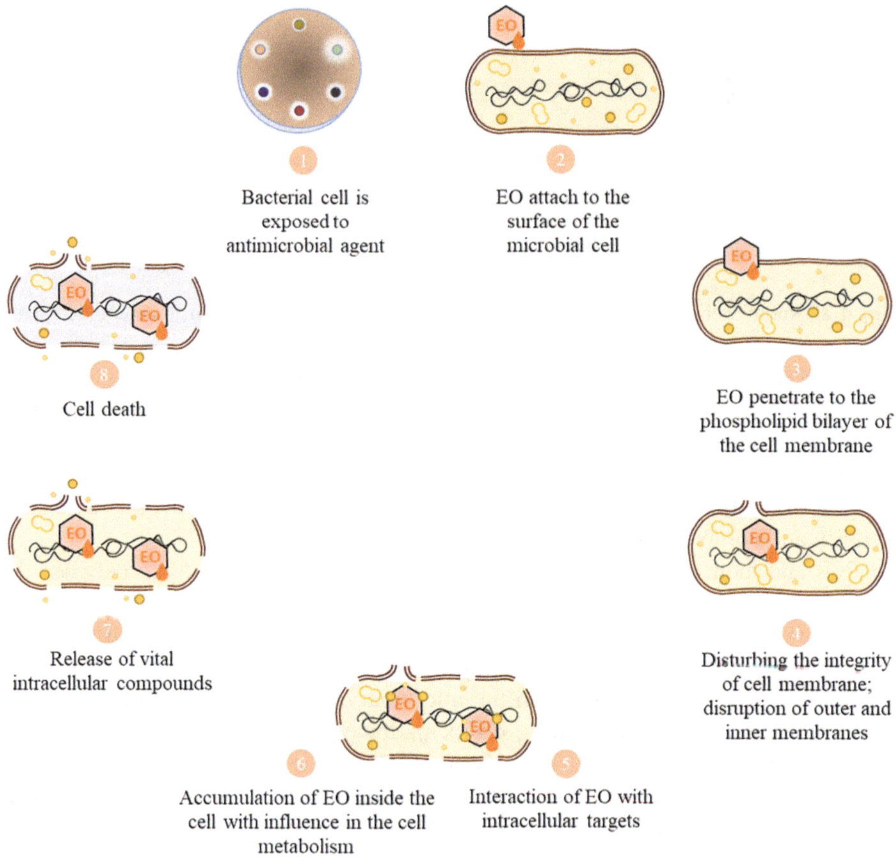

Fig. 4.3 Examples of mechanisms by which essential oils exert antimicrobial activity on bacterial cells

4.4.1 *Escherichia coli*

Escherichia coli is a resident bacterium of the gastrointestinal tract, but some strains, such as *Escherichia coli* (EPEC), enteroinvasive *Escherichia coli*, enterotoxigenic *Escherichia coli* (ETEC), enterohemorrhagic *Escherichia coli* (EHEC), enteroaggregative *Escherichia coli*, adherent *Escherichia coli*, and adherent-invasive *Escherichia coli*, could cause foodborne illnesses related to cause gastrointestinal and extra-intestinal disorders (Dávila-Aviña et al. 2020). Since it is a highly pathogenic bacterium and frequently related to foodborne outbreaks, several studies have tried to study the effects of different EOs on *E. coli* cells.

As mentioned in the section above, phenolic compounds presented in EOs have antibacterial activity. Thus, in this context, Dávila-Aviña et al. (2020) evaluated the effects of the phenolic compounds (tannic acid, gallic acid, methyl gallate, and epigallocatechin gallate) on the growth, swarming motility, biofilm formation, and expression of selected virulence genes of three *E. coli* pathotypes (EPEC, EHEC, and ETEC). All phenolic compounds were bactericidal with minimal bactericidal concentrations (0.07 to 2.1 mg/mL), and tannic acid was the most effective compound for inhibiting swarming motility (99.5–100%). For gallic acid, these authors observed reduced biofilm formation in all the tested strains.

Yasir et al. (2024) evaluated the antibacterial activity of EOs (thyme, oregano, lemongrass, mint, and rosemary) against multidrug-resistant foodborne pathogens isolated from raw milk and observed that EOs showed a broad range of antimicrobial activity against all bacteria except *Pseudomonas aeruginosa*. Furthermore, thyme was more effective against most of the multidrug resistant bacterial strains and formed the largest zone of inhibition (26 mm) against *Escherichia coli*.

In another research, a ZI of 30.1 mm was recorded after the exposure of *E. coli* to *Atractylodes lancea* EO. Moreover, the MIC and MBC (minimum bactericide concentration) values registered were 32.0 and 64.0 μg/mL (He et al. 2020).

Zhou et al. (2021) evaluated the EO of rhizomes of *Alpinia galanga* (AGREO) against the Enterohemorrhagic *E. coli* O157:H7 (EHEC O157) bacteria. The MIC and minimum bactericide concentration (MBC) values of AGREO against EHEC O157 were 2.5 mg/mL and 5.0 mg/mL, respectively. The reduction of EHEC O157 counts in the 1/2 MIC group reached 1.23–3.25 log CFU/mL, and the bactericidal effect was more marked when treated with AGREO at MIC.

Another study, performed by Benameur et al. (2019), investigated the susceptibility of *Enterobacteriaceae* producing blaESBL to Slovakian *Thymus vulgaris* EO (TVEO). The composition of TVEO was determined using a gas chromatograph-mass spectrometer. TVEO demonstrated excellent activity against *E. coli* S22/12, including blaESBL generating isolates, with ZI and MIC values ranging from 24 to 40 mm/10 μL.

The in vitro antibacterial activity of cinnamon, thyme, and clove EO against *E. coli* (ATCC 25922) was also reported, with MIC values ranging between 2.5 and 5 μL/mL (Oulkheir et al. 2017). Clove EO was used to determine antibacterial activity by the agar well diffusion method. It was found that clove oil successfully inhibited *E. coli* (MIC 2.5 mg/mL) (Nassan et al. 2015). Using disc diffusion and MIC assays, the cinnamon EO exhibited antibacterial activity against *E. coli* with an MIC value of 4.88 μg/mL (Atki et al. 2019).

The bacterial inhibitory activity of EO Linalool-loaded biocapsules against *E. coli* has been evaluated by the agar well diffusion method (Parasuraman et al. 2022). The pathogen exhibited maximum ZI of 17.23 and 13.42 mm for Linalool and linalool biocapsules at a concentration of (100 μg/mL), respectively. In addition, the MIC and MBC values were found to be 32 μg/mL and 64 μg/mL for linalool EO, respectively. However, the MIC and MBC values reported for linalool-loaded biocapsule were 128 μg/mL and 256 μg/mL, respectively.

The addition of thyme EO at supplementation levels of 0.6 and 0.9% in Tryptic Soy Broth broth showed a higher inhibitory effect against *E. coli* O157:H7 as compared to the addition at 0.3%. Both 0.6 and 0.9% levels showed no growth of viable cells of the pathogen on plates from 4 h and thereafter. The combined addition of EO at levels of 0.6% or 0.9% and nisin at 500 or 1000 IU/mL resulted in a robust inhibitory effect against *E. coli* O157:H7. Regarding antimicrobial activity in minced beef meat, the samples treated with EO at 0.6% plus nisin at 500 or 1000 IU/g presented populations of the pathogen significantly lower than the non-treated samples and the samples treated with either EO or nisin. In these samples, populations of *E. coli* O157:H7 were decreased to 3.1 log CFU/g on day 2 and kept almost unchanged up to the end of storage at 4 °C (Solomakos et al. 2008).

Huang et al. (2017) extensively investigated the antimicrobial effect of different EOs against *E. coli*. The disk diffusion assay showed ZI values (including 6 mm of the paper disk diameter) of 12, 35, 15, 10, 37, 10, and 78 mm for *L. cubeba*, clove, humifuse euphorbia herb, nutmeg, laurel leaf, thymifolious euphorbia, and *T. copticum*, respectively. The *T. copticum* EO showed the strongest and significantly higher antimicrobial activity than other EOs (ZI = 78 mm). Clove and laurel leaf EOs also inhibited *E. coli* cells (ZI >30 mm). The nutmeg EO showed the weakest antimicrobial activity (ZI <10 mm). Based on the results of the agar disk diffusion assay, clove, laurel leaf, and *T. copticum* EOs were selected for further analysis of their antimicrobial activity. The MIC values of these three EOs against *E. coli* were 0.5, 0.5, and 0.25 μL/mL, respectively. The MBC values of the above three EOs against *E. coli* were 0.5, 0.5, and 0.4 μL/mL, respectively. Overall, the *T. copticum* EO had the best antimicrobial activity against the bacterial strains tested.

When evaluating the antimicrobial susceptibility of *E. coli* ATCC 25922 and *E. coli* OI57:H7 CECT, 5947 to cinnamon and mustard EO, Clemente et al. (2016) reported MIC values of 25 and 200 (μg/mL) and 100 and 200 for the species, respectively. In the time-kill assay, cinnamon EO had the fastest killing effect on *E. coli* growth. The bactericidal effect was observed after 5 h of incubation, while for a mustard EO, the lethal effect was observed only after 23 h of incubation.

4.4.2 *Staphylococcus aureus*

Staphylococcus aureus is a Gram-positive bacteria that can survive under aerobic to facultative anaerobic conditions and is usually found in the human microbiota. *S. aureus* is able to cause infectious diseases, which comprise endocarditis, pneumonia, bacteremia, meningitis, and others (Kumar and Kiran Tudu 2023). Multidrug resistance by *S. aureus* signifies a rapidly growing public health concern because this bacteria is recognized as an opportunistic pathogen responsible for infections and high mortality and morbidity rates (Merghni et al. 2018). According to He et al. (2020), *S. aureus* is considered a model bacterial species to evaluate antibacterial mechanisms. Several studies have assessed the effect of essential oils on *S. aureus*.

According to Merghni et al. (2018), bacteria could develop various mechanisms to avoid the effects of antibiotics, and they comprise chemical changes and enzymatic inactivation, alteration of the target site, and decrease of intracellular antibiotics concentration by changes in membrane permeability or efflux pumps. Natural compounds, such as flavonoids, could affect microbial virulence components, reduce host ligand adhesion, and defuse bacterial toxins.

Clove essential oil was an option of natural antimicrobial, and eugenol is the main compound (Bai et al. 2023). According to Bai et al. (2023), this essential oil showed antibacterial activities against *S. aureus*, and the antibacterial mechanism was possibly associated with the destruction of the cell walls and membrane, the inhibition of biofilm formation, the oxidative stress, and the disruption of DNA synthesis.

The antibacterial activity and mechanism of cinnamon essential oil against *S. aureus* were evaluated (Zhang et al. 2016). These authors observed larger populations of sunken and malformed cells and swelling cell membrane-enclosed vacuoles after the application of cinnamon essential oil. Additionally, disruption of cell permeability was observed and caused leakage of electrolytes.

Nabila et al. (2022) used the disc diffusion technique to investigate the antibacterial activity of *Laurus nobilis* leaves EO against different species of *Staphylococcus*. *S. faecalis* and *S. aureus* demonstrated high antimicrobial resistance for the EOs tested, demonstrating a ZI of 13.6 and 11.2 mm, respectively, and MIC values of 0.25 mg/mL.

Al Zuhairi et al. (2020) demonstrated the antibacterial activity of *Rosmarinus officinalis* L. EO with ZI and MIC ranging from 7.00 ± 0.00 to 9.6 ± 0.32 mm and 0.06 ± 0.00 to 0.16 ± 0.07 mg/mL, respectively. The MIC and MBC values were identical in each test. The EOs MIC against *S. aureus* was 0.06 mg/mL, while the EOs maximum ZI against *S. aureus* was 9.5 mm.

Citrus medica L. var. *sarcodactylis* EO exhibited antibacterial activity against many pathogens, but the highest ZI was for *S. aureus* (19.2 mm) (Li et al. 2019). Ginger EO had a 17.1 mm ZI against *S. aureus*, with an MIC value of 1.0 mg/mL and MBC of 2.0 mg/mL (Wang et al. 2020b).

Oh et al. (2022) found that citronella, lemongrass, and oregano EO vapors had the lowest MIC (0.0781 μL/mL), and caraway seed, carrot seed, and kanuka EO vapors had the highest MIC (0.3125 μL/mL) against *S. aureus*. Citronella, lemongrass, oregano, may change, and thyme thymol EO vapors had the lowest minimal lethal concentration (MLC) (0.1563 μL/mL), and caraway seed and kanuka EO vapors had the highest MLC. Citronella vapor, in combination with lemongrass vapor, had the lowest fractional inhibitory concentration index (FICI) value (0.6250) when 1/8 MIC (0.0098 μL/mL) of citronella EO vapor was combined with 1/2 MIC (0.0391 μL/mL) of lemongrass EO vapor. A combination of citronella and oregano EO vapors showed partial synergism (FICI = 0.7500).

Still, in the study conducted by Oh et al. (2022), peppercorns were treated with EO vapors (0.0781, 0.1563, and 0.3125 μL/mL; 37 °C; 24 h). The initial population measured immediately after *S. aureus* inoculation onto the peppercorn's surface was 5.2 log CFU/g. After 24 h, the population decreased to 4.2 log CFU/g. Singular treatments with citronella, lemongrass, or oregano EO vapors at a concentration of 0.0781 μL/mL reduced the *S. aureus* populations to 3.1, 3.6, and 2.3 log CFU/g, respectively, showing significant differences compared to the control (4.2 log CFU/g). The population decreased significantly as the concentration of EO vapor increased. At the highest concentration of single EO vapor treatments (0.3125 μL/mL), the population of *S. aureus* decreased to <1.1–1.3 log CFU/g.

When citronella and lemongrass EO vapors were combined at a 1:4 vol. ratio and treated at a concentration of 0.0781 μL/mL, *S. aureus* populations were further reduced by 1.3 and 1.8 log CFU/g, respectively, compared to single treatment with either of the two EO vapors at 0.0781 μL/mL. Similar results were observed when citronella and lemongrass EO vapors were combined at a 1:2 vol. ratio. When citronella and oregano EO vapors were combined at a concentration of 0.0781 μL/mL, inhibition was greater than that resulting from treatment with citronella EO vapor alone ($P \leq 0.05$), but not oregano EO vapor alone (Oh et al. 2022).

A study investigated the antimicrobial efficacy of various EOs against antibiotic-resistant *S. aureus*. The isolates of *S. aureus* exhibited resistance to methicillin, cefoxitin, ampicillin, amoxicillin, penicillin, amoxicillin-clavulanate, ciprofloxacin, erythromycin, and gentamicin. The highest ZI was described for *Cinnamomum verum* (22.67 mm), followed by *Eucalyptus globulus* (18.67 mm) and *Syzygium aromaticum* (12.67 mm). Among the EO studied by the authors, *E. globulus* presented the lowest mean MIC value (0.33 mg/mL) (Ali et al. 2022).

Atractylodes lancea (Thunb.) DC, an important traditional herb used in eastern Asia, was studied as a source of EO for microbial inactivation of *S. aureus*. The ZI of the EO was 28.2 mm, whereas, for penicillin and streptomycin, sulfate was 28.6 and 11.5 mm, respectively. Besides, the MIC value for the EO was 64.0 μg/mL (He et al. 2020). Moreover, S. *aureus* was destroyed by different concentrations of A. lancea essential oil, and the authors suggested that the cell membrane might be the antibacterial target of *A. lancea* essential oil.

Eucalyptus globulus essential oil and 1,8-cineole were applied to control the antibiofilm and antiquorum sensing activities of methicillin-resistant *S. aureus* strains (Merghni et al. 2018). *Eucalyptus globulus* essential oil showed effective of antiquorum sensing action, even at a low concentration against *S. aureus*.

4.4.3 *Campylobacter* spp.

Campylobacter spp. is a Gram-negative, microaerophilic bacteria that present a spiral shape (Balta et al. 2021). *Campylobacter* spp. are present as normal enteric microbiota in animals (chicken, pigs, and cattle) but could affect humans through consumption of contaminated foods. *Campylobacter jejuni* and *C. coli* cause acute bacterial gastroenteritis and are considered emergent pathogens (Aslim and Yucel 2008; Ramatla et al. 2022).

Like other pathogenic bacteria, *Campylobacter* spp. Could acquire antibiotic resistance, and in this context, natural compounds could be an alternative (Aslim and Yucel 2008). Natural antimicrobials avoid the growth and development of pathogenic bacteria, but not sure how it happens. For *Campylobacter* spp., there is information that these compounds may have an anti-virulent effect. It is important to highlight that the chemical composition of plants is associated with antimicrobial function, as hydroxyl (-OH) could connect to bacterial cell membranes and destroy these structures, causing cell membrane lysis and microbial killing (Balta et al. 2021).

Aslim and Yucel (2008) evaluated the antimicrobial activities of an essential oil of *Origanum minutiflorum* against ciprofloxacin-resistant *Campylobacter* spp. The essential oil presented carvacrol (73.9%) and *p*-cymene (7.20%) as the most important components of *O. minutiflorum*. MIC values for bacterial strains, which were susceptible to the essential oil, were in the range of 7.8–800 μg/mL. The authors concluded that this essential oil presented a powerful antimicrobial action against all the tested ciprofloxacin-resistance *Campylobacter* spp.

Aiming to evaluate the anti-Campylobacter activities of the lavender EOs (*Lavandula × intermedia*), their MIC against *C. jejuni* NCTC 11,168 and *C. jejuni* 11168ΔluxS were determined. The most favorable effect was shown by samples from the group of EOs: EO 'Bila,' EO 'Budrovka' SN, e EO 'Budrovka'; the MICs for all EOs were 0.25 mg/mL. Lavandin EOs proved most effective against *C. jejuni* biofilm formation, whereas lavandin EO 'Budrovka' SN and 'Budrovka' reduced the biofilm formation by approximately 85% (Ramić et al. 2022).

In the study of Duarte et al. (2016), the antibacterial activity of coriander EO and its major compound, linalool, against *C. jejuni* and *C. coli* was assessed by the disc diffusion and vapor-phase methods. Both compounds, when tested at the pure state, inhibited the microbial growth of the strains all over the plate (ZI >85 mm) by diffusion in the agar medium. The antibacterial activity was also assessed by the broth microdilution method. The MIC values obtained for the linalool were equal to the ones obtained for the coriander oil, ranging between 0.5 and 1 μL/mL. All tested strains presented an MIC value of 0.5 μL/mL for both coriander oil and linalool, except for *C. jejuni 224,421*, which was resistant with an MIC of 1 μL/mL.

Thanissery et al. (2014) studied different oils against *C. jejuni* and *C. coli*. Thyme and clove EO showed the strongest antibacterial activity, resulting in no visible growth on the plates. Rosemary and orange EO obtained a ZI of 13.5 and 17.5 mm, respectively. The orange oil in the Thyme-orange combination (TOC) was associated with a decrease in the activity of thyme. However, TOC produced a ZI of 22 mm for *C. jejuni* and 21 mm for *C. coli*.

Using the disc diffusion assay, Kurekci et al. (2013) demonstrated that Leptospermum EO and Lemon Myrtle EO (LM) showed the highest activity toward *C. jejuni* and *C. coli*. Tea tree EO (TT) showed antimicrobial activity against both bacteria tested, with the ZI varying between 12.7 and 30 mm. It was also highly antimicrobial toward *C. coli* (30 mm) and *C. jejuni* ACM 3393 and C338 (29.3 and 26.7 mm, respectively). All formulations of neem EO formulated with other terpenes, as determined by the disc diffusion method, were antimicrobial toward *Campylobacter* spp. The broth dilution assay was used to assess the MIC of selected EOs and compounds. The MICs observed ranged from 0.001 to 1%. TT was the most effective with an MIC of 0.001%, followed by Leptospermum oil, LM (0.01%), terpinen-4-ol, and α-tops (0.06%) toward *C. jejuni*.

The antibacterial activity of *O. compactum* EO (OC), *R. officinalis* (RO), *M. pulegium* EO (MP), and *L. stoechas* EO (LS) against *Campylobacter* strains isolated from food was assessed using agar well diffusion assay for. The OC has very potent activity against all tested *C. jejuni*, and *C. coli* isolates, in which the measured ZI ranged from 15 mm to more than 80 mm and from 24 mm to 60 mm, respectively.

LS has shown a high anti-campylobacter effect which registered the highest ZI for *C. jejuni* (ranging from 48 to more than 80 mm), followed by *C. coli* (ranging from 24 to more than 80 mm), except *C. jejuni* isolated from raw milk (clinical mastitis case). The quantitative antibacterial effect of OC, RO, MP, and LS was evaluated in a liquid medium at final concentrations ranging from 0.063% (v/v) to >2% (v/v) using the microdilution technique. Overall, the MIC amounts were equal to MBC values indicating a bactericidal effect of three EOs, in particular in LS against *C. coli* and *C. jejuni*. RO had a bacteriostatic impact against the tested bacterial strains (el Baaboua et al. 2022).

4.4.4 *Salmonella* spp.

Salmonella is a Gram-negative bacteria associated with foodborne pathogens and causes more than 93.8 million cases of gastroenteritis every year. Salmonellosis is frequently associated with symptoms of diarrhea, fever, nausea, and abdominal discomfort (Tsai et al. 2017; Liu et al. 2022; Roasto et al. 2023).

Strains resistant to numerous antimicrobials have been described in human and poultry samples (Barbosa et al. 2020). Essential oils have exhibited bactericidal action against foodborne pathogens, such as *Salmonella* (Arellano et al. 2021). According to Barbosa et al. (2020), the association of antibiotics with essential oils could be an efficient strategy for the control of multiresistance and dose reduction. Next, some research results will be presented with the application of essential oils in the control of *Salmonella*.

Barbosa et al. (2020) evaluated the effect of the *Origanum vulgare* EO, thymol, and carvacrol against *Salmonella* Enteritidis ATCC 13076 to understand its antibacterial action mechanisms. According to the authors, protein expression indicated a stress response with variance expressions of chaperones and cellular protein synthesis. Thus, the effects of EO were especially related to thymol, and there was a disruption in protein regulation and DNA synthesis.

Thanissery et al. (2014) investigated the antimicrobial effect of rosemary, clove, and a mix of thyme-orange EO against different *Salmonella* species. Thyme oil had the highest mean of ZI (18.5 mm), followed by clove (13.8 mm) and rosemary (7.8 mm). An interesting observation was made after thyme-orange EO application, in which the combination resulted in a ZI of 20.5 mm, a higher value than the effects of individual EO, indicating a synergistic effect. The average MIC values for the oils of thyme, clove, and rosemary were 0.06, 0.11, and 0.88%, respectively. A higher concentration of orange oil (MBC > 1%) was required to inhibit *Salmonella*. The average concentration of the thyme-orange combination required for a similar bactericidal effect was 0.14%.

The antibacterial efficacy of three rosemary EOs (REOs) against *S. Typhimurium* varied, according to Hao et al. (2022), with halos ranging from 18.33 to 23.67 mm. The Dutch Mill Rosemary EO exhibited the greatest inhibitory impact on *S. Typhimurium* (ZI = 23.67 mm). Furthermore, the antibacterial properties of three REOs were compared using MIC. Algarve Rosemary EO and Dutch Mill Rosemary

EO MIC values were 5 mg/mL, whereas, for Majorca Pink Rosemary EO, the MIC value was 10 mg/mL. REOs completely inhibited *S. Typhimurium* growth at the MIC level in LB broth within 24 h. REO dosages of 1/16 MIC or 1/8 MIC demonstrated a slight inhibitory effect on the *pathogen* growth. In addition, the bacterial growth changed by REO at 1/2 MIC and 1/4 MIC compared to the control group.

The MIC of EO against *Salmonella* cells ranged from 0.62 μL/mL to values higher than 40 μL/mL; hence several EOs exhibited significant antibacterial activity, with strain-dependent sensitivity. In general, the most sensitive isolate was S. Derby, particularly susceptible to savory and thyme EOs (0.62 μL/mL). *Caryophillus aromatics*, *Cinnamomum zeylanicum*, and *Origanum vulgare* EOs showed similar efficacy against *S. Veneziana*, *S. Thompson*, and *S. Napoli*, with MIC values of 2.50, 1.25, and 5.00 μL/mL, respectively. Although both *Satureja montana* and *Thymus vulgaris* showed the lowest MIC value (0.62 μL/mL) against *S. Derby*, *Cinnamomum zeylanicum* showed the strongest broad-spectrum activity instead (MIC between 1.25 and 1.87 μL/mL) (Rossi et al. 2019).

The MIC and MBC values of thyme EO against *Salmonella Typhimurium* were described as 0.25 mg/mL and 0.5 mg/mL, respectively, by Lin et al. (2019), indicating an effective antibacterial agent against the bacteria. The bactericidal activity of thyme EO on *S. Typhimurium* growth was observed by a time-kill curve. After 0.5 h and 1 h of treatment, the total population of *S. Typhimurium* in the MIC group was reduced by about 99.23 and 99.89%, respectively, while the total population of *S. Typhimurium* was significantly reduced by about 99.99 and 99.999% in the MBC group. These results further revealed that thyme EO was an effective antibacterial agent against *S. Typhimurium*, showing a short period of inactivation.

Curcuma longa, *Curcuma aromatica*, and *Curcuma angustifolia* EOs have been applied against *Salmonella enterica* growth. The ZI reported for the three EO was 17.6, 16.1, and 15.5 mm, respectively. The MIC concentration (mg/mL) for *C. longa* was 0.625, whereas the other ones obtained a value of 1.250 (Albaqami et al. 2022).

Almeida et al. (2023) tested the antimicrobial activity of oregano, thyme, clove, cinnamon, and black pepper EOs, free or in combination, in *Salmonella* sp. Among the free application of EOs, cinnamon EO showed better inhibition efficiency, obtaining lower MIC and MBC values. Regarding the combination of the EOs, oregano and thyme presented a synergistic effect.

An example of a practical application is the use of essential oil microemulsions (suspensions of oregano oil at 0.1 and 0.3%, cinnamon oil at 0.3 and 0.5%, and lemongrass oil at 0.3 and 0.5%) on iceberg lettuce during 28-day storage at 4 °C. The authors verified that these compounds inactivate antibiotic-resistant *Salmonella* Newport with a reduction of 2.3–4.37 log CFU/g (Arellano et al. 2021).

4.4.5 *Pseudomonas* spp.

Pseudomonas is a Gram-negative bacterium, and an important species of this genus is the opportunistic pathogen *Pseudomonas aeruginosa. P. aeruginosa* demonstrates resistance to many antibiotics. Furthermore, *P. aeruginosa* is resistant to some EOs that are active against Enterobacterales and Gram-positive species (Pang et al. 2019; Tetard et al. 2022).

In their research, Kaskatepe et al. (2016) determined the antibacterial action of cinnamon oil against carbapenem-resistant nosocomial isolates of *Pseudomonas aeruginosa*. The antimicrobial activity of cinnamon oil was evaluated by the disk diffusion method. The zone diameter was determined as 35 mm for *P. aeruginosa*, and only one *P. aeruginosa* isolate was resistant to the antibacterial action of cinnamon brand 3. MIC was determined to be 0.0019 mL/mL. The authors concluded that cinnamon oil has the potential for use as an antimicrobial product in the pharmaceutical and food industry.

Yuan et al. (2023) evaluated rose essential oil on *Pseudomonas putida*, a common spoilage microorganism related to food contamination and foodborne illnesses. The MIC of rose essential oil against *P. putida* was 10 mg/mL, and the MBC was 10 mg/mL. The authors concluded that the treatment with this essential oil caused disruption of the integrity of bacterial cell walls and membranes, leakage of cell contents, and destroying cell morphological structure.

Clemente et al. (2016) tested the antimicrobial properties of mustard and cinnamon EO against different species of *Pseudomonas*. In general, *P. aeruginosa*, *P. fluorescens*, *P. putida*, and *P. carotovorum* were more sensitive to mustard than cinnamon EO. The MIC values reported after the antimicrobial activity of mustard EO were 12.5 μg/mL–25 μg/mL and 400 μg/mL–200 μg/mL, respectively. Regarding the killing kinetics of oils in P. putida cells, only 4 h was necessary to achieve a lethal effect of bacteria by mustard EO, whereas the killing time for cinnamon EO was 8 h. P. putida can degrade aromatic compounds, explaining why this pathogen was more resistant to cinnamon EO, that contains an aromatic compound (i.e., cinnamaldehyde) as the major component. In another way, the main component of mustard EO is allyl isothiocyanate, an aliphatic compound, producing a faster bactericidal activity against this bacterium.

P. vulgaris and *P. aeruginosa* were tested for antimicrobial resistance to *Atractylodes lancea* EO. Both species showed a similar ZI, reaching a value of 26.2 and 25.6 mm, respectively (He et al. 2020).

Eucalyptus EOs from seven different species were tested against *P. aeruginosa* growth. The MIC values ranged from 24 to 30 μL/mL, and all were higher than the MIC value obtained by tetracycline. The results also demonstrated that the EOs could inhibit the bacterial adhesion process, preventing the formation of biofilm (Khedhri et al. 2022). Similarly, Polito et al. (2022) investigated the phytotoxic and antimicrobial activity of the different varieties of *Eucalyptus* EO against *P. aeruginosa*. The MIC values varied from 28 to 30 μL/mL. In the antibiofilm formation assay, two doses were tested (10 and 20 μL/mL), and except for the *E. bicostata*, *E. obliqua*, and *E. pauciflora* at the lower concentration tested, all EOs proved capable of inhibiting biofilm formation.

Sobieszczańska et al. (2020) determined the antimicrobial effect of tarragon EO and isolated compounds (i.e., methyl eugenol and β-phellandrene) against *P. psychrophila* KM02, *P. orientalis* KM249, and *P. fluorescens* KM148. Concentrations of the sub-MIC of EO were 75.0 μL/mL for *P. psychrophila* KM02 and *P.*

fluorescens KM148; and 70.0 μL/mL for P. *orientalis* KM249. The sub-MIC for methyl eugenol was 10.0 μL/mL for *P. psychrophila* KM02, 12.0 μL/mL for *P. orientalis* KM249 and *P. fluorescens* KM148, and the sub-MIC for β-phellandrene was 8.0 μL/mL for *P. psychrophila* KM02 and *P. fluorescens* KM148; and 10 μL/mL for *P. orientalis* KM249.

To evaluate the antimicrobial effects of green pepper EO and β-caryophyllene against *P. aeruginosa* KM01, Myszka et al. (2019) analyzed microbial growth by measuring the optical density. The values of MIC for green pepper EO and β-caryophyllene were 140 and 30 nl ml^{-1}, respectively. Concentrations below the MIC of green pepper EO and β-caryophyllene (120 and 25 nl ml^{-1}, respectively) were shown to partly inhibit *P. aeruginosa* KM01 growth. To extend the results obtained in the *in vitro* experiments, they evaluated the true effects of green pepper EO and β-caryophyllene on the growth and virulence activity of exogenously provided *P. aeruginosa KM01* on model salmon-based products during storage. The results showed an inhibitory effect of examined substances on *P. aeruginosa* KM01 growth. This inhibition was independent of the storage period and equaled 95% for green pepper EO and 90% for β-caryophyllene. The green pepper EO added to the marinade also attenuated the virulence of *P. aeruginosa* KM01 on salmon meat, as evidenced by the decreases in pyocyanin production (80%) and protease (90%) and elastase activities (70%). A similar antivirulence activity was exhibited by β-caryophyllene added to the salmon marinade. In this work, pyocyanin production and protease and elastase activities were decreased by 88, 40, and 30%, respectively.

4.4.6 *Bacillus* spp.

Bacillus strains are Gram-positive spore-forming bacteria and can be found in different environments (Faille et al. 2014). *Bacillus* species related to foods are abundant, and their endospores can survive severe conditions and food processing. *Bacillus cereus* is a bacterium that causes foodborne diseases (Ayari et al. 2020). Generally, the symptoms associated with an illness caused by this bacteria are related to two types of toxins, causing vomiting and diarrhea, with a dose between 10^5 and 10^8 CFU/g (Jin et al. 2020).

The occurrence of multidrug resistance in pathogenic bacteria restricts the use of available antibiotics (Mujaddidi et al. 2021). Then, it is important to evaluate possible substitutes, such as essential oils.

Mustard and cinnamon EOs were applied to investigate the antibacterial effect against *Bacillus cereus*. Cinnamon proved to be more effective against the pathogen than mustard EO, demonstrating an MIC value of 100 μg/mL as opposed to 200 μg/mL. Furthermore, there were high differences in the MBC values between both EOs. Mustard EO showed a lower MBC value (800 μg/mL) than cinnamon EO, which presented twice as much (>1600 μg/mL). No MBC value was determined by the authors since all the concentrations tested exhibited no bactericidal effect (Clemente et al. 2016).

The antibacterial activity of *Atractylodes lancea* EO against *B. subtilis* and *B. cereus* was reported by (He et al. 2020). For both species, the ZI was equal to or higher than those observed for penicillin and streptomycin sulfate. The MIC value for *B. subtilis* was 32.0 μg/mL and for *B. cereus* 64.0 μg/mL.

The effects of Diallyl Disulfide (DADS) from garlic EO on the growth of *B. cereus* were determined and evaluated using the Gompertz model. In the control group, the growth curve included a lag time of 1.32 h, an exponential phase of 10 h, and a stationary phase with a *B. cereus* count of 7.86 log CFU/mL. When the concentration was ≥18 μg/mL, the lag time required for *B.cereus* growth increased with the rising concentration of DADS (0 μg/mL: 1.32 h; 18 μg/mL: 1.5 h; 30 μg/mL: 6.33 h; and 60 μg/mL: 7.34 h), and the maximum loads of bacteria in the stationary phase decreased (0 μg/mL:7.86 log CFU/mL; 18 μg/mL: 7.54 log CFU/mL; 30 μg/mL: 7.37 log CFU/mL; and 60 μg/mL: 7.20 log CFU/mL). After treatment with DADS, the production time for these two enterotoxins increased with the rising concentration of DADS (Nhe: 6–7 h; Hbl: 6.5–8 h). Meanwhile, the total counts of *B. cereus* required for toxin production also rose (for production of Nhe, the required count rose from 6.00 to 6.47 log CFU/g, and for the production of Hbl, the required count rose from 6.38 to 7.15 log CFU/g). The results indicated the effectiveness of DADS in inhibiting the enterotoxigenic nature of the bacteria, even at low concentrations (Jin et al. 2021).

Misaghi and Basti (2007) used the concept log P% in a factorial design study to quantify the effects of EO, Nisin (N), Temperature (T), and Day (D) on the growth response of *B. cereus* in BHI broth. A higher inhibitory effect of both *Z. multiflora* Boiss. EO (containing 71.12% carvacrol) and nisin on *B. cereus* were observed when used together than applied alone, even at sub-inhibitory concentrations. The growth of the test organism at combinations of EO = 0.005% and N = 1.5 μg ml^{-1}, and EO = 0.015% and N = 0.5 μg ml^{-1}, even at the optimum pH level (7.4) and storage temperature (30 °C) was strongly diminished (log P% = −3.85 and CN = 7.02×10^5 and log P% = −2.5 and CN = 3.5×10^5, respectively). The growth was completely inhibited at the combinations of EO ≥ 0.015% and N ≥ 1.5 μg ml^{-1}at the same pH (7.4) and T (30 °C) during the study (log P% = −4.54 and CN = 3.5×10^6) and results of plate count of the first three tubes of these combinations which inoculated with the highest level of inoculant (i.e., 10^5 CFU ml^{-1}) were < 1×10^1 CFU ml^{-1} (minimum limit of detection) after 43 days of storage in this study.

A study by Jin et al. (2020) aimed to apply garlic EO to control the *Bacillus cereus* content in white sufu, a traditional Chinese fermented soybean curd. After treatment with 32.5 μg/mL (0.5 MIC) of EO, the *B. cereus* content reduced at all processing stages, indicating the effectiveness of EO in inhibiting *B. cereus*. EO treatment reduced the count of *B. cereus* to a level of 4.58 log CFU/g, which should lower the risk of foodborne illness. In addition, the final content of *B. cereus* in the EO group also meets the relevant regulations of the Hong Kong Food Microbiological Guidelines on microbes in ready-to-eat foods (<5 log CFU/g). The total bacteria counts were similar in the EO and control groups at the tofu, pehtze, salted pehtze, and D15 sampling points, but after 15 days of ripening, the total bacterial counts in the EO group declined faster (from 6.27 to 4.77 log CFU/g) than in the control group (from 6.29 to 5.26 log CFU/g).

The contamination of food causes food spoilage and disease. So, contamination control is essential to eliminate bacteria. Usually, synthetic preservatives are applied, but there is an interest in new potential natural antimicrobials (Ayari et al. 2020). Ayari et al. (2020) evaluated the potential interactions between eight essential oils (Basil, Cinnamon, Eucalyptus, Mandarin, Oregano, Peppermint, Tea tree, and Thyme) against *Bacillus cereus* LSPQ 2872. Cinnamon/Thyme and Oregano/Thyme EO exhibited a synergistic effect against *B. cereus*.

4.4.7 *Listeria monocytogenes*

Bacteria genus *Listeria* includes 21 species, and one of them is *Listeria monocytogenes*. These bacteria are Gram-positive, short rods, facultative anaerobic and present a ubiquitous nature and survive severe conditions. Listeriosis causes a high death rate (Kayode and Okoh 2022).

According to Bermúdez-Capdevila et al. (2022), antimicrobial resistance in *L. monocytogenes* biofilms is a risk, and that is important to develop strategies for the control, mainly substances that do not cause the generation of resistance to the biocide itself or cross-resistance to antibiotics. So, in this context, essential oil could be an option. Below are some results of works with the application of essential oils against different strains of *L. monocytogenes*.

The antibacterial activity of *Thymus zygis* subsp. *gracilis* against different strains of *L. monocytogenes* was performed by disc diffusion method and broth micro-dilution assay. Results showed that the ZI ranged between 13.4 and 41.4 mm. The broth micro-dilution assay showed that the EO had the same MIC and MBC value (0.02%; v/v), demonstrating a bactericidal effect (MBC/MIC $\leq$4). The inhibition and eradication effects of the EO against *L. monocytogenes* biofilm were evaluated by crystal violet staining assay. The EO exhibited remarkable inhibitive ability against *L. monocytogenes* biofilm, with a MIC value of 0.02% after 24 h of treatment. There was also an eradication effect on preformed *L. monocytogenes* biofilm after 72 h of treatment. The microbiological analysis of smoked fish contaminated with *L. monocytogenes* and treated with the EO (0, 0.005, and 0.01%) during 30 days of storage at 8 °C shows the load of *L. monocytogenes* in the samples treated with 0.005 and 0.01% of TZG-EO reduced by approximately 2.77 and 2.00 4 $\log^{10}$ CFU/g respectively, after 30 days of storage (Bouymajane et al. 2022).

Another study also evaluated the antimicrobial effect of *T. zygis* EO against *L. monocytogenes* strains. The EO exhibited high antibacterial activity, with a ZI between 41.55 and 55.04 mm and an MIC of 0.05% (v/v) for three *L. monocytogenes* strains studied. Besides, the volatile compounds of the *T. zygis* EO also demonstrated high antimicrobial activity with ZI from 36.13 to 50.43 mm. The antibacterial activity was observed by the time-kill curves, showing that *T. zygis* EO has a bactericidal effect at 1× and 2 × MIC for all strains of *L. monocytogenes*. Further, after 8 h of incubation, significant reductions in the logarithmic bacterial counts were observed for *L. monocytogenes* LMG 16779 and LMG 16780 with 0.5 × MIC, 0.25 × MIC, and 0.125 × MIC of *T. zygis* EO, while only the sub-inhibitory concentration of 0.5 × MIC had a significant reduction in *L. monocytogenes* LMG 13305. The effect of *T. zygis* EO in the formation of biofilms by *L. monocytogenes* strains was also evaluated, and the biofilm formation was inhibited even at subinhibitory concentrations. The *T. zygis* EO showed biofilm inhibition values between 35 and 62.02% with the concentration of 0.5 × MIC and between 16.85 and 40.27% for the concentration of 0.25 × MIC, showing the potential of biofilm inhibition by the EO even at sub-inhibitory concentrations (Coimbra et al. 2022).

Carvalho et al. (2023) demonstrated that *Melissa officinalis* EO produced ZI of 38.4, 54.6, and 48.6 mm for *L. monocytogenes* LMG 13305, 16,779, and 16,780, respectively, while the values for its volatile compounds were 35.5, 51.1, and 46.1 mm. The MIC of the EO was 0.5 μL/mL for the three strains studied. Further, the anti-listerial activity was evaluated by time-kill curves, showing that the EO effectively controlled the growth of the three *L. monocytogenes* strains with a bactericidal effect of the EO at 4, 2, and 1 × MIC.

In solid medium and micro-atmosphere, the chitosan film added with Kernel EO almost didn't show any antimicrobial effect against *L. monocytogenes*. Increasing the EO concentration increased the effectiveness of the chitosan film, and the concentration of 0.5 and 1.0% obtained ZI values of 67.3 and 80.3 mm^2 in a solid medium. Besides, the growth reduction of 0.5 and 1.0% were 70.3 and 96.5% in the micro-atmosphere, respectively. In liquid media samples, the counts for *L. monocytogenes* were also gradually increased in the whole incubation time, and the counts for *L. monocytogenes* in liquid media samples treated by CS films were lower than that of the control sample. When applicating the chitosan film added with Kernel EO in sliced beef samples, the authors reported that the counts for *L. monocytogenes* in sliced beef samples treated by the films were lower than in the control sample (Wang et al. 2020a).

Carvalho et al. (2023) evaluated *Melissa officinalis* EO's antioxidant potential and antimicrobial activity against *L. monocytogenes*. The authors verified that pre-contact of the bacterium to subinhibitory concentrations of EO (0.125 μL/mL) did not cause high tolerance to stress conditions or cross-resistance with antibiotics. They also evaluated the use of EO in food models (lettuce, chicken, and milk) and watermelon juice, and the results indicated an antimicrobial activity in a lettuce leaf, further diminishing *L. monocytogenes* contamination.

Gottardo et al. (2022) evaluated the anti-*Listeria* potential of isolated and combined microencapsulated oregano and cinnamon essential oils in Italian salami. The 2, 5, and 10% of combined EO produced inhibition halos between 20 and 38 mm. According to the authors, the application of 2% microcapsule had a significant reduction of *L. monocytogenes*.

In a study, Bermúdez-Capdevila et al. (2022) proposed the application of new disinfectants with essential oils to biofilm control. The authors determined the MIC of four essential oils (cassia, clove, oregano, and cinnamon) in four *L. monocytogenes* strains. The authors observed that there were no differences in the biofilms before and after contact with cassia, except for the CECT 935 strain. Furthermore, they determined that cinnamon oil obtained a significantly higher MIC value (4562.5 ppm) than cassia oil (412.5 ppm).

4.5 Final Considerations

The bacterial resistance to different antibiotics boosted the need to find alternatives to these compounds. Essential oils have the potential to be applied in the control of resistant pathogenic bacteria with several mechanisms of action. These compounds could act in membrane integrity and permeability, disrupt cell membrane and respiratory processes, prevent biofilm development, and dysregulate quorum sensing interaction. However, there are variations in relation to the effects provided by these compounds. The different studies presented throughout this chapter indicated that there is variation in the diameter of inhibition zones formed by EOs. Different results could be associated with many factors, such as the composition of EOs, concentrations applied, and the target bacteria. The EOs can have the potential to be applied as an alternative antimicrobial option in cosmetics, food, and environmental areas.

References

Albaqami JJ et al (2022) Chemical composition and biological activities of the leaf essential oils of Curcuma longa, Curcuma aromatica and Curcuma angustifolia. Antibiotics 11(11):1–14. https://doi.org/10.3390/antibiotics11111547

Ali T et al (2022) Antibacterial activity of plant essential oils against indigenously characterized methicillin-resistant Staphylococcus aureus (MRSA). Tropical Biomed 39(1). https://doi.org/10.47665/tb.39.1.005

Almeida MJ et al (2023) Antimicrobial action of oregano, thyme, clove, cinnamon and black pepper essential oils free and encapsulated against foodborne pathogens. Food Control 144(109356):1–9. https://doi.org/10.1016/J.FOODCONT.2022.109356

Atki Y et al (2019) Antibacterial activity of cinnamon essential oils and their synergistic potential with antibiotics. J Adv Pharm Technol Res 10(2):63. https://doi.org/10.4103/japtr.JAPTR_366_18

Arellano S, Law B, Friedman M, Ravishankar S (2021) Essential oil microemulsions inactivate antibiotic-resistant salmonella Newport and spoilage bacterium lactobacillus casei on iceberg lettuce during 28-day storage at 4°C. Food Control 130:108209. https://doi.org/10.1016/j.foodcont.2021.108209

Aslim B, Yucel N (2008) In vitro antimicrobial activity of essential oil from endemic Origanum minutiflorum on ciprofloxacin-resistant campylobacter spp. Food Chem 107(2):602–606. https://doi.org/10.1016/j.foodchem.2007.08.048

Ayari S, Shankar S, Follett P, Hossain F, Lacroix M (2020) Potential synergistic antimicrobial efficiency of binary combinations of essential oils against Bacillus cereus and Paenibacillus amylolyticus-part A. Microb Pathog 141:104008. https://doi.org/10.1016/j.micpath.2020.104008

Baaboua A et al (2022) Evaluation of the combined effect of antibiotics and essential oils against campylobacter multidrug resistant strains and their biofilm formation. S Afr J Bot 150:451–465. https://doi.org/10.1016/j.sajb.2022.08.027

Bai J, Li J, Chen Z, Bai X, Yang Z, Wang Z, Yang Y (2023) Antibacterial activity and mechanism of clove essential oil against foodborne pathogens. LWT 173:114249. https://doi.org/10.1016/j.lwt.2022.114249

Balta I, Linton M, Pinkerton L, Kelly C, Stef L, Pet I, Stef D, Criste A, Gundogdu O, Corcionivoschi N (2021) The effect of natural antimicrobials against campylobacter spp. and its similarities to salmonella spp, listeria spp., Escherichia coli, vibrio spp., clostridium spp. and staphylococcus spp. Food Control 121:107745. https://doi.org/10.1016/j.foodcont.2020.107745

Barbosa LN, Alves FCB, Andrade BFMT, Albano M, Rall VLM, Fernandes AAH, Buzalaf MAR, de Leite A, de Pontes LG, dos Santos LD, Fernandes Junior A (2020) Proteomic analysis and antibacterial resistance mechanisms of salmonella Enteritidis submitted to the inhibitory effect of Origanum vulgare essential oil, thymol and carvacrol. J Proteome 214:103625. https://doi.org/10.1016/j.jprot.2019.103625

Bermúdez-Capdevila M, Cervantes-Huamán BRH, Rodríguez-Jerez JJ, Ripolles-Avila C (2022) Repeated sub-inhibitory doses of cassia essential oil do not increase the tolerance pattern in listeria monocytogenes cells. LWT 165:113681. https://doi.org/10.1016/j.lwt.2022.113681

Benameur Q et al (2019) Antibacterial activity of Thymus vulgaris essential oil alone and in combination with cefotaxime against blaESBL producing multidrug resistant Enterobacteriaceae isolates. Nat Prod Res 33(18):2647–2654. https://doi.org/10.1080/14786419.2018.1466124

Bhattacharya A (2019) High-temperature stress and metabolism of secondary metabolites in plants. In: Effect of high temperature on crop productivity and metabolism of macro molecules. https://doi.org/10.1016/b978-0-12-817562-0.00005-7

Bouymajane A et al (2022) Chemical profile, antibacterial, antioxidant, and anisakicidal activities of thymus zygis subsp. gracilis essential oil and its effect against listeria monocytogenes. Int J Food Microbiol 383:109960. Available at:. https://doi.org/10.1016/J.IJFOODMICRO.2022.109960

Cai M et al (2022) Antibacterial and antibiofilm activities of chitosan nanoparticles loaded with Ocimum basilicum L. essential oil. Int J Biol Macromol 202. https://doi.org/10.1016/j.ijbiomac.2022.01.066

Carvalho F et al (2023) Melissa officinalis essential oil as an antimicrobial agent against listeria monocytogenes in watermelon juice. Food Microbiol 109:104105. Available at:. https://doi.org/10.1016/J.FM.2022.104105

CDC (2022) Summary of possible multistate enteric (Intestinal) disease outbreaks in 2017–2020. Available at: https://www.cdc.gov/foodsafety/outbreaks/multistate-outbreaks/annual-summaries/annual-summaries-2017-2020.html

Chouhan S, Sharma K, Guleria S (2017) Antimicrobial activity of some essential oils—present status and future perspectives, Medicines 4(3). https://doi.org/10.3390/medicines4030058

Clemente I et al (2016) Antimicrobial properties and mode of action of mustard and cinnamon essential oils and their combination against foodborne bacteria. Innov Food Sci Emerg Technol 36:26. https://doi.org/10.1016/j.ifset.2016.05.013

Coimbra A et al (2022) Antimicrobial activity of thymus zygis essential oil against listeria monocytogenes and its application as food preservative. Innov Food Sci Emerg Technol 80(January):103077. https://doi.org/10.1016/j.ifset.2022.103077

Dávila-Aviña J et al (2020) Phenolics with bactericidal activity Alter motility and biofilm formation in Enterotoxigenic, Enteropathogenic, and Enterohemorrhagic Escherichia coli. Foodborne Pathog Dis 17(9):568–575. https://doi.org/10.1089/fpd.2019.2766

Duarte A et al (2016) Antioxidant properties of coriander essential oil and linalool and their potential to control Campylobacter spp. Food Control 61. https://doi.org/10.1016/j.foodcont.2015.09.033

El-Tarabily KA et al (2021) Using essential oils to overcome bacterial biofilm formation and their antimicrobial resistance. Saudi J Biol Sci 28(9):5145–5156. https://doi.org/10.1016/j.sjbs.2021.05.033

Faille C et al (2014) Sporulation of bacillus spp. within biofilms: A potential source of contamination in food processing environments. Food Microbiol 40:64–74. Available at:. https://doi.org/10.1016/j.fm.2013.12.004

Freitas IR, Cattelan MG (2018) Antimicrobial and antioxidant properties of essential oils in food systems—an overview. In: Microbial contamination and food degradation. https://doi.org/10.1016/b978-0-12-811515-2.00015-9

Gottardo FM et al (2022) Microencapsulated oregano and cinnamon essential oils as a natural alternative to reduce listeria monocytogenes in Italian salami. Food Bioscience 50:102146. https://doi.org/10.1016/j.fbio.2022.102146

Hao Y et al (2022) Label-free quantitative proteomics reveals the antibacterial mechanism of rosemary essential oil against salmonella enterica serovar typhimurium. Ind Crop Prod 189:115757. https://doi.org/10.1016/J.INDCROP.2022.115757

He F et al (2020) Antioxidant and antibacterial activities of essential oil from Atractylodes lancea rhizomes. Ind Crops Prod 153:112552. https://doi.org/10.1016/j.indcrop.2020.112552

Hou T et al (2022) Essential oils and its antibacterial, antifungal and anti-oxidant activity applications: a review. Food Bioscience 47:101716. https://doi.org/10.1016/j.fbio.2022.101716

Huang W et al (2017) Chemical analysis and in vitro antimicrobial effects and mechanism of action of Trachyspermum copticum essential oil against Escherichia coli. Asian Pac J Trop Med 10(7):663. https://doi.org/10.1016/j.apjtm.2017.07.006

Jaouadi R, Boussaid M, Zaouali Y (2023) Variation in essential oil composition within and among Tunisian Thymus algeriensis Boiss et Reut. (Lamiaceae) populations: effect of ecological factors and incidence on antiacetylcholinesterase and antioxidant activities. Biochem Syst Ecol 106(August 2022):104543. https://doi.org/10.1016/j.bse.2022.104543

Jin Z et al (2020) Inhibition of Bacillus cereus by garlic (Allium sativum) essential oil during manufacture of white sufu, a traditional Chinese fermented soybean curd. LWT 130:109634. https://doi.org/10.1016/j.lwt.2020.109634

Jin Z et al (2021) Diallyl disulfide, the antibacterial component of garlic essential oil, inhibits the toxicity of Bacillus cereus ATCC 14579 at sub-inhibitory concentrations. Food Control 126:108090. https://doi.org/10.1016/j.foodcont.2021.108090

Kamala K, Kumar VP (2018) Food products and food contamination. In: Microbial contamination and food degradation. https://doi.org/10.1016/b978-0-12-811515-2.00001-9

Kasrati A et al (2014) Antioxidative activity and synergistic effect of thymus saturejoides Coss. Essential oils with cefixime against selected food-borne bacteria. Ind Crop Prod 61:338. https://doi.org/10.1016/j.indcrop.2014.07.024

Kaskatepe B et al (2016) Antibacterial effects of cinnamon oil against carbapenem resistant nosocomial Acinetobacter baumannii and Pseudomonas aeruginosa isolates. Ind Crop Prod 81:191–194. https://doi.org/10.1016/j.indcrop.2015.11.058

Kayode AJ, Okoh AI (2022) Incidence and genetic diversity of multi-drug resistant listeria monocytogenes isolates recovered from fruits and vegetables in the eastern Cape Province, South Africa. Int J Food Microbiol 363:109513. https://doi.org/10.1016/j.ijfoodmicro.2021.109513

Kumar G, Kiran Tudu A (2023) Tackling multidrug-resistant Staphylococcus aureus by natural products and their analogues acting as NorA efflux pump inhibitors. Bioorg Med Chem 80:117187. https://doi.org/10.1016/j.bmc.2023.117187

Liu Y et al (2022) Inhibitory effects of clove and oregano essential oils on biofilm formation of salmonella Derby isolated from beef processing plant. LWT 162:113486. https://doi.org/10.1016/j.lwt.2022.113486

Khedhri S et al. (2022) Chemical composition, phytotoxic and antibiofilm activity of seven eucalyptus species from Tunisia, 27

Kurekci C et al (2013) Antimicrobial activity of essential oils and five terpenoid compounds against campylobacter jejuni in pure and mixed culture experiments. Int J Food Microbiol 166:450. https://doi.org/10.1016/j.ijfoodmicro.2013.08.014

Li YX et al (2022) Antimicrobial mechanisms of spice essential oils and application in food industry. Food Chemistry 382:132312. https://doi.org/10.1016/j.foodchem.2022.132312

Li ZH et al (2019) Antibacterial activity and mechanisms of essential oil from Citrus medica L. Var. Sarcodactylis. Molecules 24(8). https://doi.org/10.3390/molecules24081577

Lin L, Liao X, Cui H (2019) Cold plasma treated thyme essential oil/silk fibroin nanofibers against salmonella typhimurium in poultry meat. Food Packag Shelf Life 21:100337. https://doi.org/10.1016/j.fpsl.2019.100337

Mani-López E, Palou E, López-Malo A (2018) Biopreservatives as agents to prevent food spoilage. In: Microbial contamination and food degradation. https://doi.org/10.1016/b978-0-12-811515-2.00008-1

Marangoni JA et al (2023) Geographical variation in the chemical composition, anti-inflammatory activity of the essential oil, micromorphology and histochemistry of Schinus terebinthifolia Raddi. J Ethnopharmacol 301(115786):1–11. https://doi.org/10.1016/J.JEP.2022.115786

Merghni A et al (2018) Assessment of the antibiofilm and antiquorum sensing activities of Eucalyptus globulus essential oil and its main component 1,8-cineole against methicillin-resistant Staphylococcus aureus strains. Microb Pathog 118:74–80. https://doi.org/10.1016/j.micpath.2018.03.006

Misaghi A, Basti AA (2007) Effects of Zataria multiflora Boiss. Essential oil and nisin on Bacillus cereus ATCC 11778. Food Control 18(9):1043. https://doi.org/10.1016/j.foodcont.2006.06.010

Mosunova O, Navarro-Muñoz JC, Collemare J (2021) The biosynthesis of fungal secondary metabolites: from fundamentals to biotechnological applications. In: Encyclopedia of mycology. https://doi.org/10.1016/B978-0-12-809633-8.21072-8

Mujaddidi N et al (2021) Pharmacological properties of biogenically synthesized silver nanoparticles using endophyte Bacillus cereus extract of Berberis lyceum against oxidative stress and pathogenic multidrug-resistant bacteria. Saudi J Biol Sci 28(11):6432–6440. https://doi.org/10.1016/j.sjbs.2021.07.009

Myszka K et al (2019) Green pepper essential oil as a biopreservative agent for fish-based products: antimicrobial and antivirulence activities against Pseudomonas aeruginosa KM01. LWT 108:6–13. https://doi.org/10.1016/J.LWT.2019.03.047

Nabila B et al (2022) Chemical composition and antibacterial activity of the essential oil of Laurus nobilis leaves. Nat Prod Res 36:989. https://doi.org/10.1080/14786419.2020.1839450

Nassan MA et al (2015) Effect of clove and cinnamon extracts on experimental model of acute hematogenous pyelonephritis in albino rats: Immunopathological and antimicrobial study. Int J Immunopathol Pharmacol 28:60. https://doi.org/10.1177/0394632015572075

Oh J et al (2022) Inhibition of Staphylococcus aureus on a laboratory medium and black peppercorns by individual and combinations of essential oil vapors. Food Control 132:108487. https://doi.org/10.1016/j.foodcont.2021.108487

Oulkheir S et al (2017) Antibacterial activity of essential oils extracts from cinnamon, thyme, clove and geranium against a gram negative and gram positive pathogenic bacteria. J Dis Med Plants 3(1). https://doi.org/10.11648/j.jdmp.s.2017030201.11

Pang Z et al (2019) Antibiotic resistance in Pseudomonas aeruginosa: mechanisms and alternative therapeutic strategies. Biotechnol Adv 37(1):177–192. Available at:. https://doi.org/10.1016/j.biotechadv.2018.11.013

Parasuraman V et al (2022) Fabrication and bacterial inhibitory activity of essential oil linalool loaded biocapsules against Escherichia coli. J Drug Deliv Sci Technol 74:103495. Available at:. https://doi.org/10.1016/J.JDDST.2022.103495

Polito F et al (2022) Chemical composition and phytotoxic and Antibiofilm activity of the essential oils of Eucalyptus bicostata, E. gigantea, E. intertexta, E. obliqua, E. pauciflora and E. tereticornis. Plants 11(22):3017. https://doi.org/10.3390/plants11223017

Pyne ME, Narcross L, Martin VJJ (2019) Engineering plant secondary metabolism in microbial systems. Plant Physiol 179:844. https://doi.org/10.1104/pp.18.01291

Rahman M et al (2022) Contamination of fresh produce with antibiotic-resistant bacteria and associated risks to human health: A scoping review. Int J Environ Res Public Health 19(1). https://doi.org/10.3390/ijerph19010360

Ramatla T et al (2022) Prevalence and antimicrobial resistance profiles of campylobacter species in South Africa: a "one health" approach using systematic review and meta-analysis. Int J Infect Dis 125:294–304. https://doi.org/10.1016/j.ijid.2022.10.042

Ramić D et al (2022) Campylobacter jejuni biofilm control with Lavandin essential oils and by-products. Antibiotics 11(7). https://doi.org/10.3390/antibiotics11070854

Roasto M et al (2023) Salmonella enterica prevalence, serotype diversity, antimicrobial resistance and control in the European pork production chain. Trends Food Sci Technol 131:210–219. https://doi.org/10.1016/j.tifs.2022.12.007

Rossi C et al (2019) Salmonella enterica adhesion: effect of Cinnamomum zeylanicum essential oil on lettuce. LWT 111:16. https://doi.org/10.1016/j.lwt.2019.05.026

Samtiya M et al. (2022) Antimicrobial resistance in the food chain: trends, mechanisms, pathways, and possible regulation strategies, pp. 1–20. https://doi.org/10.3390/foods11192966

São José JFB, Faria-Silva L, Lepaus BM (2022) Decontamination of vegetables. In: Shah MA, Mir SA (eds) Microbial decontamination of food. Springer Nature Singapore, Singapore, pp 71–92. https://doi.org/10.1007/978-981-19-5114-5_3

Seshadri VD et al (2020) Essential oils of Cinnamomum loureirii and Evolvulus alsinoides protect guava fruits from spoilage bacteria, fungi and insect (Pseudococcus longispinus). Ind Crops Prod 154:112629. https://doi.org/10.1016/j.indcrop.2020.112629

Sobieszczańska N et al (2020) Tarragon essential oil as a source of bioactive compounds with anti-quorum sensing and anti-proteolytic activity against pseudomonas spp. isolated from fish–in vitro, in silico and in situ approaches. Int J Food Microbiol 331:108732. https://doi.org/10.1016/j.ijfoodmicro.2020.108732

Solomakos N et al (2008) The antimicrobial effect of thyme essential oil, nisin and their combination against Escherichia coli O157:H7 in minced beef during refrigerated storage. Meat Sci 80:159. https://doi.org/10.1016/j.meatsci.2007.11.014

Tetard A et al (2022) Exposure of pseudomonas aeruginosa to cinnamaldehyde selects multidrug resistant mutants. Antibiotics 11(12):1790. https://doi.org/10.3390/antibiotics11121790

Thanissery R, Kathariou S, Smith DP (2014) Rosemary oil, clove oil, and a mix of thyme-orange essential oils inhibit salmonella and Campylobacterin vitro. J Appl Poult Res 23:221. https://doi.org/10.3382/japr.2013-00888

Tsai H-C, Sheng L, Zhu M-J (2017) Antimicrobial efficacy of cinnamon oil against salmonella in almond based matrices. Food Control 80:170–175. https://doi.org/10.1016/j.foodcont.2017.04.045

Villa TG et al (2016) Resistant and emergent pathogens in food products. In: Antimicrobial food packaging. https://doi.org/10.1016/B978-0-12-800723-5.00002-4

Walasek-Janusz M et al (2022) Variation in the antimicrobial activity of essential oils from cultivars of Lavandula angustifolia and L. InterMedia 12(2955):1–13. https://doi.org/10.3390/agronomy12122955

Wang D et al (2020a) Incorporation of apricot (Prunus armeniaca) kernel essential oil into chitosan films displaying antimicrobial effect against listeria monocytogenes and improving quality indices of spiced beef. Int J Biol Macromol 162:838. https://doi.org/10.1016/j.ijbiomac.2020.06.220

Wang X et al (2020b) Antibacterial activity and mechanism of ginger essential oil against Escherichia coli and Staphylococcus aureus. Molecules 25(17). https://doi.org/10.3390/molecules25173955

WHO (2022a) Food safety, Food safety. Available at: https://www.who.int/news-room/fact-sheets/detail/food-safety

WHO (2022b) Foodborne diseases, Foodborne diseases. Available at: https://www.who.int/health-topics/foodborne-diseases#tab=tab_1

Yammine J et al (2022) Essential oils and their active components applied as: free, encapsulated and in hurdle technology to fight microbial contaminations. A review. Heliyon 8(12):e12472. https://doi.org/10.1016/j.heliyon.2022.e12472

Yasir M et al (2024) Anti-bacterial activity of essential oils against multidrug-resistant foodborne pathogens isolated from raw milk. Braz J Biol 84

Yuan Y et al (2023) Antibacterial mechanism of rose essential oil against pseudomonas putida isolated from white Hypsizygus marmoreus at cellular and metabolic levels. Ind Crop Prod 196:116523. https://doi.org/10.1016/j.indcrop.2023.116523

Zhang Y et al (2016) Antibacterial activity and mechanism of cinnamon essential oil against Escherichia coli and Staphylococcus aureus. Food Control 59:282–289. https://doi.org/10.1016/j.foodcont.2015.05.032

Zhou C et al (2021) Chemical composition, antibacterial activity and study of the interaction mechanisms of the main compounds present in the Alpinia galanga rhizomes essential oil. Ind Crops Prod 165:113441. https://doi.org/10.1016/j.indcrop.2021.113441

Zuhairi JJMJ et al (2020) Antioxidant, cytotoxic and antibacterial activity of Rosmarinus officinalis L. essential oil against bacteria isolated from urinary tract infection. Eur J Integr Med 38:101192. https://doi.org/10.1016/j.eujim.2020.101192

Essential Oils against Fruit Spoilage Fungi

5

Manueli Monciozo Domingos, Bárbara Santos Valiati, Daniel Sgrancio Uliana, Rhaiza Marcia Lopes Leal, Alessandra Peres Guimarães, Bárbara Morandi Lepaus, Sara Jarske Geringer, and Jackline Freitas Brilhante de São José

Abstract

Essential oils (EOs) are natural, low molecular weight secondary metabolites from aromatic plants. They are fat-soluble and soluble liquids in organic solvents, usually with lower density than water. EOs are widely known for their taste, aroma, and antimicrobial activity. Its volatile nature, attributed to secondary metabolites, has antioxidant and antimicrobial potential, representing approximately 90% of the composition. Other non-volatile components (acids, sterols, waxes), present in smaller amounts, can also significantly influence its bioactivity. Recent studies have shown antimicrobial, insecticide, or repellent action on pests using EOs. The use of essential oils proves to be a promising alternative to traditional artificial antifungals since they have effectiveness in the control of fungi related to food deterioration, low toxicity, act as preservatives in food. In this chapter, we will address the main fungi and yeasts responsible for the deterioration of fruits, the main essential oils used in food, and the application of essential oils against fruit-deteriorating fungi.

M. M. Domingos · B. S. Valiati · A. P. Guimarães · B. M. Lepaus
Postgraduate Program in Nutrition and Health, Federal University of Espírito Santo, Espírito Santo, Brazil

D. S. Uliana · R. M. L. Leal · S. J. Geringer
Department of Integrated Education in Health, Federal University of Espírito Santo, Espírito Santo, Brazil

J. F. B. de São José (✉)
Postgraduate Program in Nutrition and Health, Department of Integrated Education in Health, Federal University of Espírito Santo, Espírito Santo, Brazil
e-mail: jackline.jose@ufes.br

B. Prakash et al. (eds.), *Plant Essential Oils*,
https://doi.org/10.1007/978-981-99-4370-8_5

Keywords

Natural plant products · Antimicrobial activity · Essential oil · Antifungal · Biological control.

5.1 Introduction

Essential oils (EOs) are natural, low-molecular-weight secondary metabolites obtained from aromatic plants. They are aromatic, fat-soluble, and soluble liquids in organic solvents, usually with lower density than water (da Silva et al. 2021; Speranza and Corbo 2010). These substances can be extracted from parts such as leaves, branches, stems, seeds, and leaves (Arasu et al. 2019; Maia et al. 2015; Seshadri et al. 2020a). The main characteristics of these compounds are volatility and low solubility in water (Maia et al. 2015). In addition, some factors, such as genetic factors, climatic conditions, plant maturation stage, processing, concentration, and storage, can affect the composition and, consequently, the final quality of the product obtained (Barbosa 2010; Maia et al. 2015).

With the advance of globalization, there has been a change in the eating habits of populations, while the use of food additives and pesticides in food has intensified. However, consumers have shown concern about the possible harmful effects of these substances, and the uncontrolled use of these can provide the fungal resistance (e.g., Camele et al. 2012; de Corato et al. 2010; Seshadri et al. 2020a; da Silva et al. 2021). Thus, the use of less harmful antimicrobials is effective for the control of fungi and the promotion of food safety (Camele et al. 2012; da Silva et al. 2021). Therefore, these compounds have been widely used to increase the shelf life of processed and fresh foods and maintain sensorial characteristics due to their antifungal activity (Seshadri et al. 2020a; Speranza and Corbo 2010).

Furthermore, another critical factor is the control of post-harvest fungal deterioration. Fruits are highly perishable, and their properties corroborate that there are more significant losses and waste, in addition to their structure providing greater susceptibility to deterioration by fungi (de Botelho et al. 2022). Moreover, fruits have high water and moisture activity and various nutrients, which favor increased deterioration reactions (de Corato et al. 2010). Thus, care must be taken during cultivation, harvesting, and post-harvest to ensure the quality of the final product.

The use of essential oils proves to be a promising alternative to traditional artificial antifungals since they have effectiveness in the control of fungi related to food deterioration, low toxicity, act as preservatives in food, and are generally recognized as safe (GRAS) (Arasu et al. 2019; da Silva et al. 2021; Simas et al. 2017). Furthermore, several studies show the application of essential oils due to antimicrobial activity. The essential oils include clove oil, oregano, basil, thyme, rosemary, ginger, and cinnamon (Cui et al. 2020; Reis et al. 2020; Xu et al. 2022).

In this chapter, we will address the main fungi and yeasts responsible for the deterioration of fruits, the main essential oils used in food, and the application of essential oils against fruit-deteriorating fungi.

5.2 Food Spoilage Fungi

According to de Botelho et al. (2022), fungi are eukaryotic organisms described as heterotrophic, unicellular, and multicellular organisms. The difference between unicellular and multicellular is hyphae formation, a filamentous structure that forms the mycelium. Fungi produce spores that are related to the propagation of the species. These groups of microorganisms could be damaging to human health and cause mycoses and infectious diseases (de Botelho et al. 2022). In addition, there are fungi that cause food spoilage. Food, after being contaminated, can favor the multiplication of undesirable spoilage microorganisms causing damage to the product. Consequently, it causes economic losses and risks to the health of the consumer when mycotoxin-producing fungi occur (Bernardi et al. 2019; Santana Oliveira et al. 2019).

5.2.1 Contamination by Molds and Yeasts that Leads to Spoilage

Fruits and vegetables are essential sources of nutrients for humans; they are good sources of vitamins, minerals, phytochemicals, dietary fiber, and antioxidants. However, about 45% of these foods are wasted annually due to spoilage caused by contaminated environments, inadequate harvesting conditions, unsafe handling and storage processes, and incorrect display methods (Alegbeleye et al. 2022; Saleh and Al-Thani 2019). Also, according to the Food and Agriculture Organization (FAO), about 15 to 50% of the total amount of fruits and vegetables worldwide were lost during the post-harvest process in the year 2011 (Gunny et al. 2021), and 1.3 billion tons of food are lost every year due to spoilage (Kaczmarek et al. 2019).

Food spoilage has numerous socioeconomic implications, as it is related to food scarcity, waste, and hunger in some parts of the world, water stress, unnecessary loss of biodiversity, and increased emissions of greenhouse gases such as carbon dioxide and methane produced by spoilage, which contributes to the climate crisis (Alegbeleye et al. 2022).

Although there are microorganisms naturally present on the surface of fruits and vegetables, which do not cause disease and yet can act as a natural biological barrier to contamination by pathogens or spoilage agents, molds and yeasts, in general, are the leading cause of spoilage of raw fruits (Kaczmarek et al. 2019; Saleh and Al-Thani 2019). Molds can benefit human activities, as they are producers of enzymes, organic acids, and antibiotics and can be used in food production, for example. However, they can spoil various foods, from raw materials to finished products, causing changes in taste, odor, and spoilage. These degradative changes in food products due to microorganism contamination are responsible for losing food quality (Pandey et al. 2021; Quéro et al. 2019). The fermented aroma, sour taste, slimy surface, moisture, soft rot, discoloration, and even visible microbial growth characterize fungal contamination in many types of fruits (Kaczmarek et al. 2019).

Also, they are a significant threat to human health and safety, resulting in foodborne diseases due to the production of mycotoxins that can cause food poisoning and serious health problems to consumers, just as severe economic losses

(Alegbeleye et al. 2022; Atif et al. 2020a; Cai et al. 2022; Pandey et al. 2022). Furthermore, contamination by microorganisms and oxidation triggered by reactive oxygen species (ROS) are generally responsible for the spoilage of fresh vegetable foods and fruits during processing, transportation, and storage (Pandey et al. 2021).

The richness of nutrients, sugar, carbohydrates, free amino acids, and the high water activity make them suitable for microbial growth and survival. However, bacterial spoilage in this food category is limited due to their low pH since bacteria typically grow at neutral pH. Thus, spoilage of these foods is usually associated with aciduric bacteria, molds, and yeasts (Alegbeleye et al. 2022; Roumani et al. 2022).

As the physiological ripening process of fruits and vegetables progresses toward senescence and possible spoilage, metabolic and developmental activity is maintained. During this process, various physiological and compositional changes occur that make fresh fruits and vegetables a suitable substrate for the proliferation of microorganisms. However, the chances of microbial spoilage becoming established depend on certain factors, such as surface morphology and topography, exudates from the plant surface, stage of development, and post-harvest handling. As soon as foods are removed from their natural supply of nutrients, their quality begins to decline. Therefore, the period of ripening and senescence is the most susceptible to spoilage (Alegbeleye et al. 2022).

Spoilage microorganisms can be introduced into fruits and vegetables under different circumstances, in the farm, post-harvest handling, and processing, all of which can predispose these foods to spoilage. In addition, plants are usually grown in unprotected natural environments, where soils are teeming with life, making contamination possible (Alegbeleye et al. 2022).

Many soil microorganisms can be transferred to fruit or vegetable surfaces during food production or processing. Soil particles, dust, air, and water used for irrigation are common vectors and can carry fungal mycelium, spores, and live bacterial cells. Microorganisms can grow on fresh produce and have developed sophisticated biochemical mechanisms to break down food components, which provide them with the energy sources needed for growth (Kaczmarek et al. 2019).

Microbial contamination of fruits and vegetables can come from various sources, including the farm environment, post-harvest handling, raw materials, and contact with processing equipment. In addition, food contamination can occur after harvest and processing. Several examples occur, such as significantly higher bacterial counts after processing, such as shredding, cutting, slicing, rinsing, and packaging, as debris and microorganisms can accumulate on the equipment that performs these operations. If effective hygiene practices are not followed during fresh produce processing, new sources of contamination can be introduced (Kaczmarek et al. 2019; Seshadri et al. 2020a).

Furthermore, it is worth noting that specific post-harvest practices can promote the establishment of existing microorganisms or introduce new contamination (Alegbeleye et al. 2022). Thus, misbehaving practices in fruit handling, such as packing in plastic cages, stacking cages that cause pressure on the fruit leading to mechanical damage, high temperature during marketing, and pecking or biting by insects and wild animals that damage the outer cover of the products. This can cause damage such as wounds, abrasion, hematomas, perforations, cracks, breaks, or compression of the cellular and tissue structure that accelerate the contamination process and this favors the growth of microorganisms, such as fungi, prematurely. In addition, when the fruit is affected by molds and the fungal hyphae spread in the intercellular spaces and inside the parenchyma cells, they secrete enzymes that cause tissue maceration, eventually leading to fruit compromise (Alegbeleye et al. 2022; Moustafa 2018).

Applying good agricultural practices could reduce food loss from spoilage at all stages of vegetable production. The Federal Drug Administration (FDA, 2008) has a guide to minimize microbial food safety hazards for fresh vegetables (Guide to Minimize Microbial Food Safety Hazards of Fresh-cut Fruits and Vegetables) that recommends good practices for food producers, packers, and transporters to reduce the common microbiological hazards often associated with their operations (Kaczmarek et al. 2019).

It is also worth noting that several sources of contamination precede the harvest, such as seeds, agricultural water, soil, manure, biological alterations, livestock/wildlife, and humans. Contamination at this stage is crucial because it can determine the establishment of pathogens and spoilage within plant product tissues, making them impossible to remove by sanitizers or other decontamination strategies (Alegbeleye et al. 2022).

5.2.2 Main Molds and Yeasts that Spoilage Fruits

Plant-derived foods are often subject to spoilage by molds and yeasts, and contamination can occur from various sources described above (Table 5.1).

Members of the phylum *Ascomycota*, *Basidiomycota*, and *Zygomycota* are some of the most prevalent fungi. In addition, species of *Penicillium*, *Alternaria*, *Botrytis*, *Fusarium*, *Cladosporium*, *Phoma*, *Trichoderma*, *Lasiodiplodia*, *Aspergillus*, *Alternaria*, *Rhizopus*, *Aureobasidium*, and *Colletotrichum* are often responsible for spoilage of fresh produce and production of mycotoxins that impact product quality, can cause health risks to consumers, and socioeconomic losses (Alegbeleye et al. 2022; Kaczmarek et al. 2019). So, *Alternaria*, *Aspergillus*, *Penicillium*, and *Fusarium* are the main genera of spoilage fungi present in the postharvest stage (Atif et al. 2020a; Sanzani et al. 2016; Seshadri et al. 2020a), in which they secrete

Table 5.1 Fruit spoilage fungi and their effects

Fungi/Yeasts	Characteristics	Contaminated fruit	Reference
Genus *Alternaria*	Tenuazonic acid	Wide range of fruits, tomatoes	Moss (2008)
Genus *Aspergillus*	Black mold	Grapes, apricots	Alegbeleye et al. (2022)
Genus *Fusarium*	Fruit rot and crown rot	Banana	Alghuthaymi et al. (2020)
Genus *Penicillium*	Produced toxic patulin, blue and green rot of citrus	Pears and apples, citrus fruit	Neri et al. (2010)

a broad spectrum of enzymes such as amylases, cellulases, and pectinases that cause softening of plant tissues (Alegbeleye et al. 2022).

5.2.2.1 *Alternaria*

Alternaria is a fungal genus that has saprobic, endophytic, and pathogenic species and includes over 250 species. This group is related with a large range of serious plant pathogens that cause losses in large agricultural crops (de Botelho et al. 2022; de Lima et al. 2016).

One of the important species in this genus is *Alternaria alternata* that is a prevalent phytopathogen, causing serious foodborne spoilage and producing mycotoxins (Wang et al. 2019). Black spot caused by *Alternaria alternata* is associated to significant losses during storage and transportation of fruits (Guo et al. 2022). *Alternaria alternata* species are primarily responsible for causing brown spots on citrus fruits, also occur in a saprotrophic form on harvested fruits and vegetables, and can produce mycotoxins that exert poisonous effects after human consumption (Sanzani et al. 2016). The most important and frequent *Alternaria* mycotoxins are alternariol (AOH), and alternariol monomethyl ether (AME). The mycotoxins occur in derived from cereals, fruits, and vegetables, contaminating food and feedstuffs and could cause genotoxicity, cytotoxicity, mutagenicity, and carcinogenic (Wang et al. 2019).

5.2.2.2 *Aspergillus*

According to de Botelho et al. (2022), *Aspergillus* is an anamorph genus belonging to the phylum Ascomycota. There are more than 250 species of saprophytic mold in this genus, and they present asexual spores called aspergillum. *Aspergillus* can occur in different types of environments as they adapt to various growing conditions such as temperature from 6 to 55 °C and relatively low humidity. In food, they are frequently found in spoiled products, such as grains, cereals, vegetables, fruits, juices, and cakes (de Botelho et al. 2022).

Some species of the genus *Aspergillus* can affect entire plant organs and seeds and produce mycotoxins due to their secondary metabolism. Infection by these fungi usually occurs in the field, where the mycelium is confined to specific tissues (seed or fruit) (Sanzani et al. 2016).

One of the main species of this genus is *Aspergillus flavus*. This species is a facultative parasite that naturally occurs as a saprophytic soil fungus and contaminates various food crops such as cottonseed, maize, peanuts, and tree nuts (de Botelho et al. 2022; Chiotta et al. 2020; Tian et al. 2022). However, it is important to mention that *A. flavus* produces aflatoxin, which is a carcinogenic and mutagenic secondary metabolite and affects animal and human (Tian et al. 2022). For control *A. flavus* and aflatoxin contamination is important to avoid the growth of spores and mycelia and the inactivation of aflatoxins. Usually, to achieve these results it is necessary to use substances such as synthetic fungicides, ozone fumigation, irradiation, cooking processes, and controlled environmental conditions. However, many of these strategies are difficult, expensive, and inefficient. Then, considering this and the consumer's interest in ingesting products as natural as possible, essential oils could be a good option (Tian et al. 2022; de Botelho et al. 2022).

5.2.2.3 *Penicillium*

Penicillium is one of the most common genus belonging to Trichocomaceae and comprises 483 species (Sanzani et al. 2016; de Botelho et al. 2022). *Penicillium* occurs in a wide range of habitats, such as soil, vegetation, air, indoor environments, and various foodstuffs. They are decomposers of organic matter and destructive rot-causing agents in the food industry, producing a wide range of secondary metabolites (Sanzani et al. 2016). Also, some species are important in food industry to produce cheeses and fermented sausages and in medicine due to production of penicillin that is a secondary metabolite with biological activity (de Botelho et al. 2022).

However, *Penicillium* spp. are the major cause of deterioration and decomposition of many plant products after harvest, mainly fruits (El-Samawaty et al. 2021). Additionally, some species could produce toxic secondary metabolite as ochratoxin A (OTA) that widely contaminates food and feed (Ferrara et al. 2022). OTA could be synthesized by *Aspergillus*, *Penicillium*, *A. ochraceus*, *P. nordicum e P. verrucosum* and are founded in cereals, herbs, oilseeds, figs, beef jerky, fruits, and wine. The main toxic effects and diseases caused by this mycotoxin is kidney and liver damage, loss of appetite, nausea and vomiting, immune system suppression, and carcinogenic (El-Sayed et al. 2022).

Another species that could affect food is *Penicillium expansum* that causes blue mold rot and this problem is an international concern and is worldwide distributed. Furthermore, this is one of the most important post-harvest diseases of some fruits, such as apples. Various strains of *P. expansum* are able to produce patulin, a mycotoxin (Bahri et al. 2019; Luciano-Rosario et al. 2020).

5.2.2.4 *Fusarium*

Fusarium is a genus of molds that could be founded in different places as in soil, water, plants, and in the air. *Fusarium* species can adapt to a varied range of habitats but are mainly distributed in the tropical and subtropical regions (de Botelho et al. 2022; Perincherry et al. 2019). Some species of these genus are adaptable and resistant to antifungal compounds and many species are causing diseases in plants, such as wilt and canker (Perincherry et al. 2019). However, one of the main concerns regarding this group of fungi is the production of mycotoxins that could contaminate food products. Some known species that produce mycotoxins are *F. graminearum, F. oxysporum, F. sporotrichioides, F. verticillioides, and F. proliferatum* (El-Sayed et al. 2022; Perincherry et al. 2019).

The action of this fungus is characterized by the colonization of the host as biotrophic fungi, and then production of toxins and enzymes. *Fusarium* species produce mycotoxins and are most important in cereals (wheat, oats, barley, maize) and tropical fruit crops (banana and pineapple, lentils, tomato, and peas). Examples of species that produce mycotoxin are *Fusarium verticillioides* and *F. culmorum*. These species produce fumonisins that cause human encephalomalacia, pulmonary edema, carcinogenicity, neurotoxicity, liver damage, heart failure, and esophageal cancer (El-Sayed et al. 2022; Perincherry et al. 2019). *F. oxysporum* produces trichothecenes as T2 toxin and deoxynivalenol (DON) and these mycotoxins are found in cereals, feed, silage, legumes, fruits, and vegetables. These mycotoxins could cause immune suppression, cytotoxicity, skin necrosis, hemorrhage, anemia, granulocytopenia, oral epithelial lesions, hypotension, and coagulopathy (El-Sayed et al. 2022).

5.2.2.5 *Yeasts*

Yeasts are eukaryotic fungi, majority unicellular and reproduce by budding (Zhang et al. 2020). Among yeast species, the genera *Saccharomyces, Candida, Torulopsis*, and *Hansenula* have been the main ones associated with fruit and juice fermentation. Some other yeast species that can cause loss of product quality include *Rhodotorula mucilaginosa*, *Rhodotorula glutinis*, *Zygosaccharomyces bailii*, *Zygosaccharomyces bisporus*, and *Zygosaccharomyces rouxii* (Kaczmarek et al. 2019).

One of the major yeast genera is *Candida* that includes about 200 species. This genus could be found in distinct environments as in the human microbiota, in animals, in the skin, and mucous membranes. This group is considered opportunistic pathogenic fungi, and some species cause human infections (de Botelho et al. 2022).

As yeasts are ubiquitous, microorganisms could occur in raw materials that are utilized for food manufacture. Usually these microorganisms are associated to inappropriate practice in food production that facilitated the cross contamination. Spoilage caused by yeasts can generally be related to the action of enzymes (cellulases, lipases, and proteases), gas production, discoloration, and off-flavors in food. The most common species in spoilage fruit are *Hanseniaspora*, *Pichia*, and *Wickerhamomyces* (Osimani et al. 2022).

5.3 Plant Essential Oils

5.3.1 Definition

Essential oils (EOs) are natural products of several volatile molecules (El Asbahani et al. 2015; Huang et al. 2021). These compounds are widely known for their taste, aroma, and antimicrobial activity (Salvi et al. 2022). EOs are commonly distributed in Myrtaceae, Lamiaceae, Rutaceae, Umbelliferae, Lauraceae, Zingiberaceae, Asteraceae, and other plants (Ni et al. 2021). EOs are extracted from different plant parts, such as a leaf, flower, root, stems, and others non-woody organs (Arasu et al. 2019; Ni et al. 2021). Obtaining methods can be by distillation, mechanical pressure, or solvent extraction (Hou et al. 2022). The method can influence the yield, quality, and final product composition (Skendi et al. 2022).

5.3.2 Characteristics

Essential oils are formed by different substances, such as aldehydes, terpenes, phenolics, esters, alcohols, and ketones (Arasu et al. 2019). Its volatile nature, attributed to secondary metabolites, has antioxidant and antimicrobial potential (Salvi et al. 2022), representing approximately 90% of the composition. Other non-volatile components (acids, sterols, waxes), present in smaller amounts, can also significantly influence its bioactivity (Luque de Castro et al. 1999).

According to Ni et al. (2021), phenolic compounds are the most important for the antibacterial activity of EOs. Terpenes, phenols, esters, aromatic compounds, aldehydes, or terpenoids cause inhibition of microorganisms' growth in food and preventing spoilage of food.

Among the volatile substances found, monoterpenoids and sesquiterpenoids are the main constituents of EOs (Fotsing Yannick Stephane and Kezetas Jean Jules 2020). Terpenes and other compounds give EOs antimicrobial, antioxidant, and antifungal properties (Arasu et al. 2019). Terpenoids can affect the fatty acids of the cell membrane of microorganisms, thus altering their permeability and, consequently, causing leakage of intracellular substances (Ju et al. 2019). Phenolic compounds, conversely, have important antibacterial activities, such as disruption of the cytoplasmic membrane, changes in the flow of electrons, and coagulation of cell contents (Dhifi et al. 2016). It is important to highlight that the compound action in the essential oil will also depend on the type of target microorganisms, due to their composition and membrane thickness (Ju et al. 2019).

Additionally, other functional groups present in EOs have become important as hydrocarbons (α-pinene and sabinene), oxides (linalool oxide, and cineol), lactones (citroptene), esters (eugenol acetate), alcohols (linalool and geraniol), phenols (thymol and eugenol), ketones (camphor, and carvone), and aldehydes (citral, citronellal, and cinnamaldehyde) (Ni et al. 2021).

Recent studies have shown antimicrobial, insecticide, or repellent action on pests using EOs (Barros et al. 2022; Seshadri et al. 2020a; Wang et al. 2022; Xu et al. 2022). These positive results indicate that EOs can be used as natural preservative agents in the food industry. In addition, their use can attract consumers to be aware of food safety and reduce chemical agents in the food chain (Miao et al. 2020; Li et al. 2022).

5.3.3 Main Plants' Essential Oils Applied in Food Processing

There Are Many EOs that Are Tested in Food in Studies. Among them Are Lavender EO, Thyme EO, Peppermint EO, Cajuput EO, Cinnamon EO, Clove EO, Eucalyptus EO, and Sage EO (Wińska et al. 2019). Next, some Studies with the Application of these Different EOs Will Be Presented

Studies have shown that cinnamon essential oil is used alone, combined with other EOs, or with other preservation methods/agents for conservation and sanitization purposes. For example, Seshadri et al. (2020a) observed antimicrobial action against bacteria and fungi in vitro and in vivo analyses (in guavas) combining EOs extracted from the bark of *C. lourerii* and the plant *Evolvulus alsinoides*. In another study, using EO from *C. zeylanicum* and *Cymbopogon citratus* efficiently reduced fungal growth and total inhibition of the pathogens *C. musae*, *F. incarnatum, and F. verticillioides* (Kamsu et al. 2019). Xu et al. (2022) also observed significant antimicrobial effects when combining cinnamon EO with intermittent mild heating, resulting in a reduction of 4 log CFU/g in alfalfa seeds.

The use of EOs can also be used to delay post-harvest deterioration. For example, in in vivo tests, peppermint essential oil inhibited 100% of the development of *Colletotrichum gloeosporioides*, which is responsible for anthracnose in papayas (Ayón Reyna et al. 2022). Likewise, tomatoes use xanthan gum coating, whey protein, and clove EO that prolonged shelf life and maintenance of quality attributes, such as firmness, after 15 days of storage at 20 °C (Kumar and Saini 2021).

Research using clove, oregano, and thyme EO is also found in the literature. For example, Cui et al. (2020) observed a change in the morphology of biofilms when using clove EO, resulting in a 61.48% reduction in the metabolic activity of *E.coli*. This change may be associated with cell collapse and exudation of intracellular content caused by components present in clove EO. In another study, clove essential oil reduced 32.7% of *B. gladioli* cells, and this effect was potentiated by combining essential oil nanoemulsions in chitosan coating, achieving a 52% of reduction. The reduction was associated with the lipid structures of cell membranes disruption and mitochondrial membranes. Furthermore, this study observed better conservation of fresh Tremella fuciformis treated with clove oil and chitosan (Wang et al. 2022).

In vitro studies, the microencapsulated oregano essential oil could inhibit 100% of microorganisms on avocados. In vivo, a reduction in anthracnose lesions in the fruits was observed, and critical sensorial characteristics, such as color and firmness, were maintained. Another important finding in this research was the increase in the content of flavonoids and phenolic compounds in treated avocados (Colín-Chávez et al. 2022). Similar results in maintaining quality and delaying postharvest effects were also observed by Lee et al. (2022) when studying the EO effect of oregano on chitosan and cellulose coating in strawberries.

Sazvar et al. (2022) evaluated the effect of the essential oils (anise, chamomile, marjoram, black caraway, and thyme) on inhibiting the growth of *Alternaria*. These authors tested five concentrations (0, 200, 400, 600, and 800 μL/L) and observed that in vitro results revealed that by increasing the concentration of essential oils, their antifungal activity improved. However, the greatest inhibitory effect was associated to the use of thyme essential oil. When evaluating the application of the oil in vivo (fruit), the better result is related to lowest percentage of fruit weight loss, and this was observed with thyme oil at 600 μL/L and a 0.02% concentration.

Thyme essential oil can also be used as a repellent against pests that affect cereals during storage without changing sensory and quality parameters and can be used to replace chemical agents and reduce the environmental impact (Barros et al. 2022). Other essential oils used in food are presented in Table 5.2.

5.4 Application of Essential Oils against Fruit Spoilage Fungi

In the above section, we discussed about the application of different types of EOs. Essential oils have been used in the industry as a flavoring. Thus, in recent years, new research has shown that it can also be used for reducing microorganisms in food (Angane et al. 2022; Osimani et al. 2022).

Table 5.2 Examples of studies using essential oils in fruits

Matrix used	Type of essential oil	Application	Reference
Dragon fruit	Cinnamon oil and lemon grass oil and eucalyptus oil and clove oil, and rosemary oil	In vivo assay	Castro et al. (2017)
Tomato	*Adansonia digitata*	Sanitizer	Kayode et al. (2018)
Mangosteen	Peppermint oil and lime oil	Closed system	Owolabi et al. (2021)
Piper nigrum L. fruits	*Boswellia carterii*	Sanitizer	Prakash et al. (2014)
Avocado	*Lippia scaberrima*/spearmint oil	Commercial coatings	Regnier et al. (2010)
Papaya	Mint oil	Chitosan coating	Ayón Reyna et al. (2022)
Strawberry and cherry	*Zanthoxylum bungeanum*	Active package film	Zhang et al. (2022)

When it comes to vegetable food, there is a big concern about pathogen foodborne diseases since they are used to produce ready-to-eat food (Nikkhah et al. 2017; Nikkhah and Hashemi 2020; Shen et al. 2023). So, these oils can be applied to food in different ways; for example, sanitizers that can be sprayed on the food surface or essential oils can be used in edible coatings and films (Atif et al. 2020b; Shen et al. 2023). EO_S have already been used in different vegetable foods, such as raw vegetables and fruits or cereal-based foods and juices (Osimani et al. 2022).

According to Ni et al. (2021), these applications could be restricted due to the characteristics of EO such as high volatility, hydrophobicity, and ease of oxidation. However, there is a potential use when micro and nanotechnology were applied.

In this section, we will be present the application of EOs in food preservation and agricultural products.

5.4.1 Food Packaging: Edible Coatings and Films

Consumers demand healthy, fresh food that is as close to natural as possible, especially without the use of synthetic food additives. Food additives are purposely added to obtain technology and sensorial functions. Some of them could control microbial growth, chemical reactions, and promote an increase of shelf life of food products (Mesías et al. 2021). However, at same time, synthetic food additives could be a risk to human health and environment (Ni et al. 2021).

Currently, active food packaging could be a good option that unites an innovative method and possible ecological solution. This packaging is commonly applied with biodegradable materials and natural compounds. So, EOs could be used in films and coatings (Ni et al. 2021).

When it comes to edible coatings and films, there are differences between them, and it will influence industries' choices when applying them to food. Protein, starch, and polysaccharide are the most used ingredients in edible coatings. However, it is crucial that in the final product, the coat must have a particular texture to cover the fruits. In this case, the coats are thick liquids into which the fruit is dipped (Shen et al. 2023; Tripathi et al. 2021).

Edible films, on the other hand, are prepared as thin sheets. Therefore, a drawback is that it can be broken and let some portion of the food exposed, not conferring the protective goal (Pandey et al. 2022).

Edible coatings are often used in food. Usually, chitosan is one of the most popular ingredients in this material, mainly because it has antibacterial and antifungal properties. However, in recent studies, essential oils have been shown that it can improve the chitosan's effect on food (Anjum et al. 2020; Pandey et al. 2022; Shen et al. 2023).

Nanoemulsions are used for adding essential oils to edible coatings. The reason is that this process guarantees a stable emulsion that will not suffer during the storage period and break its connection (Anjum et al. 2020).

In a study with carboxymethyl cellulose and cardamom oil in a nanoemulsion edible coating in tomatoes, the edible coat was efficient to prolongate shelf life due to oxidative damage and bacterial growth (Das et al. 2022).

Long et al. (2022) analyzed chitosan/fennel seed essential oil/starch sodium octenyl succinate composite films in different formulations applied to apples. They found that the film in special one of the formulations had a powerful fungicidal effect against *B. cinerea*, *T. roseum*, and *P. expansum*. However, all six formulations used in the study showed antifungal activity in vitro.

Guo et al. (2022) produced a film material to preserve apricots by inhibiting the growth of *Alternaria alternata*. EO Oregano essential oil (OEO) (2%, 4%, 6% (w/v)) was added to chitosan-based films in the form of nanoemulsions. The 4% OEO nanoemulsion with 20 min of ultrasound treatment demonstrated the best antifungal action. The chitosan film combined with 1% (v/v) of 4% (w/v) OEO nanoemulsion significantly increased the antifungal properties. These authors observed also that CS-1%OEO films control the disease resistance of apricots from the black spot caused by *Alternaria alternata*.

According to Ni et al. (2021), EO affects the microstructure of the final packaging material and could contribute to increase the tensile, barrier, and optical properties. Furthermore, it is important to consider that composition of food could influence the migration rate of compounds.

5.4.2 Sanitizers

Sanitization is an important step for the microbiological control of fruits. The most common way to sanitize fruits and vegetables is by immersion in chlorinated compounds (de São José et al. 2018; de Moraes Motta Machado et al. 2022; Moncioso et al. 2021; Pelissari et al. 2021). However, since these chemicals are used mainly in food industries due to their efficiency and low costs, other additional treatments come to improve further functionality (Atif et al. 2020b; Kanetis et al. 2007). Furthermore, the chlorinated compounds are intended to be substituted due to potential harmful consequences on the environment and health (Moncioso et al. 2021; Pelissari et al. 2021).

Recently, many studies have shown, according to their methodology, possible use of essential oils as sanitizers, tested directly on the fruit or vegetable surface, or even as a possible compound to be used in a blend of other antimicrobial components (Antonioli et al. 2020; Atif et al. 2020b; Nikkhah et al. 2017; Nikkhah and Hashemi 2020; Seshadri et al. 2020b; Chen et al. 2022).

Studies proposing the use of essential oil to control fungi are scarce. However, we present this study to indicate that there is a trend toward the application of EOs as sanitizers. Soraggi Battagin et al. (2021) evaluated the clove essential oil as an alternative sanitizer for the disinfection of citrus fruit in packinghouses to control *Xanthomonas citri* subsp. *citri.* Clove EO was able to inhibit *X. citri* when used at 0.75% (v/v). In simulate packinghouse conditions, the sanitization process with 5% of the EO promoted an effect similar to sodium hypochlorite.

Atif et al. (2020a, b) have studied two essential oils from medicinal plants and protective properties in jack fruits. They sprayed the essential oils onto the jackfruit peel and the damage designed to the jackfruit peel. *Basilicum* oil has shown promising activity against *Rhizopus microsporus* MTCC383; mycelial growth was affected by *Basilicum* and V. zizanioides indicating that essential oils affected the developmental stages of fungi.

In a study with essential oils of *Cinnamomum loureirii* and *Evolvulus alsinoides*, researchers found that after applying essential oil emulsions to the guava peeled skin, after eight days, the oils were capable of inhibiting food spoilage fungus growth. (Seshadri et al. 2020b).

Moncioso et al. (2021) evaluated citric acid and clove essential oil as alternatives to chlorine compounds on sanitization of apples. The authors observed that the treatment with clove essential oil reduced the amount of the fungi group to 1.64 log CFU/g.

5.4.3 Agricultural Application

EOs have potential to control pests and diseases of crops as substitute of synthetic pesticides, and they could be safer to the humans and the environment (Torre et al. 2021).

Hashem et al. (2019) aimed to apply Thieves oil blend (six oils in one mixture) to protect peach fruit from spoilage during cold storage. The authors evaluated the effects on *Alternaria alternata, Fusarium oxysporum, Geotrichum candidum,* and *Monilinia laxa.* Thieves oil blend (2.0 mL/ L) inhibited fungal growth in vitro and reduced the disease rate to 12.0% and the disease severity index to 1.2 after 7 days at 27 °C *in peach fruit.*

Arasu et al. (2019) evaluated essential oil of four medicinal plants and protective properties in plum fruits against the spoilage bacteria and fungi. The minimum inhibitory concentration of the essential oils against fungi ranged from 1.1 ± 0.4 to 292 ± 3.2 μg/mL. *Allium sativum* presented the highest antifungal activity against

Penicillium notatum and inhibited the growth of *Aspergillus niger*, *Aspergillus flavus,* and *Rhizopus microsporus*. Authors concluded that *Allium sativum* EO could inhibit deterioration of plum fruit.

Torre et al. (2021) evaluated that basil aromatic oil was against three common storage fungi (*Fusarium oxysporum*, *Penicillium* spp., and *Colletotrichum gloeosporioides*), and also the effect of basil aromatic oil on the germination of commercial seeds (lettuce and tomatoes). The basil essential oil presented toxicity against the fungi *F. oxysporum*, *Penicillium* spp., and *C. gloeosporioides*. The authors observed also that tomato and lettuce seeds' germination were not significantly affected by the essential oils.

Xu et al. (2021) investigated the effectiveness of tea tree oil, thyme oil, rosemary oil, and lemon oil for controlling rot in post-harvest peaches (*Monilinia fructicola).* Tea tree oil had the antifungal action against in vitro and in inoculated peach fruit. According to these authors, tea tree oil impacts on the composition of cell membrane, causing alterations in mycelial morphology, membrane permeability, and concentrations of intracellular reactive oxygen species. Finally, the authors indicated this oil as a possible alternative for conventional fungicides applied to control rot in peach.

5.5 Limitations of the Use of EO

The structure and composition of a food can make it difficult and restrict the activity of essential oils since some foods require higher concentrations due to their matrix. In addition, many studies evaluate the use of essential oils in vitro. However, higher levels of essential oils are fundamental for the same effect to be performed on food (Speranza and Corbo 2010), which can have consequences for sensory aspects. Furthermore, the essential oil obtained depends on some aspects such as climate and maturation stage, which can interfere with the quality of the product to alter its constitution and, therefore, the activity of this compound.

5.6 Final Considerations and Future Perspectives

Essential oils presented potential in application against fruit spoilage fungi. The advances in studies in this area contribute to the issue of food safety from various aspects. The advantages include environmental and consumer health issues since they are compounds of natural origin. In addition, the antimicrobial potential of these oils contributes to reducing waste and fighting fruit spoilage, which has economic benefits. However, there are some challenges. For example, the use of essential oils on an industrial scale is hampered due to the low solubility and stability of these substances. More studies are required to unravel, elucidate, and improve the knowledge about essential oils and their mechanisms. Some applications, such as nanoemulsions, deserve more attention to expand their use in edible food coatings.

Furthermore, combining essential oils with other preservation techniques, for example, should be evaluated and used on an industrial scale.

References

Alegbeleye O, Odeyemi OA, Strateva M, Stratev D (2022) Microbial spoilage of vegetables, fruits and cereals. Appl Food Res 2(1):100122. https://doi.org/10.1016/j.afres.2022.100122

Alghuthaymi M, Alshehri WA, Al-Maary KS, Bahkali NA, AlKahtani MDF, Alarfaj AA, Alnadhari S et al (2020) Mycotoxigenicity of fusarium isolated from banana fruits: combining phytopathological assays with toxin concentrations. J King Saud Univ–Sci 32(2):1482–1485. https://doi.org/10.1016/j.jksus.2019.12.001

Angane M, Swift S, Huang K, Butts CA, Quek SY (2022) Essential oils and their major components: an updated review on antimicrobial activities, mechanism of action and their potential application in the food industry. Foods 11(3):464. https://doi.org/10.3390/foods11030464

Anjum MA, Akram H, Zaidi M, Ali S (2020) Effect of gum arabic and Aloe vera gel based edible coatings in combination with plant extracts on postharvest quality and storability of 'Gola' guava fruits. Sci Hortic 271:109506. https://doi.org/10.1016/j.scienta.2020.109506

Antonioli G, Fontanella G, Echeverrigaray S, Longaray Delamare AP, Fernandes Pauletti G, Barcellos T (2020) Poly(lactic acid) nanocapsules containing lemongrass essential oil for postharvest decay control: in vitro and in vivo evaluation against phytopathogenic fungi. Food Chem 326:126997. https://doi.org/10.1016/j.foodchem.2020.126997

Arasu MV, Vijayaraghavan P, Ilavenil S, Al-Dhabi NA, Choi KC (2019) Essential oil of four medicinal plants and protective properties in plum fruits against the spoilage bacteria and fungi. Ind Crop Prod 133:54–62. https://doi.org/10.1016/j.indcrop.2019.03.018

El Asbahani A, Miladi K, Badri W, Sala M, Addi EHA, Casabianca H, El Mousadik A et al (2015) Essential oils: from extraction to encapsulation. Int J Pharm 483(1–2):220–243. https://doi.org/10.1016/j.ijpharm.2014.12.069

Atif M, Ilavenil S, Devanesan S, AlSalhi MS, Choi KC, Vijayaraghavan P, Alfuraydi AA et al (2020a) Essential oils of two medicinal plants and protective properties of jack fruits against the spoilage bacteria and fungi. Ind Crop Prod 147:112239. https://doi.org/10.1016/j.indcrop.2020.112239

Atif M, Ilavenil S, Devanesan S, AlSalhi MS, Choi KC, Vijayaraghavan P, Alfuraydi AA et al (2020b) Essential oils of two medicinal plants and protective properties of jack fruits against the spoilage bacteria and fungi. Ind Crops Prod 147:112239. https://doi.org/10.1016/j.indcrop.2020.112239

Ayón Reyna LE, Uriarte Gastelum YG, Camacho Díaz BH, Tapia Maruri D, López López ME, López Velázquez JG, Vega García MO (2022) Antifungal activity of a chitosan and mint essential oil coating on the development of Colletotrichum Gloeosporioides in papaya using macroscopic and microscopic analysis. Food Bioprocess Technol 15(2):368–378. https://doi.org/10.1007/s11947-022-02764-w

Bahri BA, Mechichi G, Rouissi W, Ben Haj Jilani I, Ghrabi-Gammar Z (2019) Effects of cold-storage facility characteristics on the virulence and sporulation of Penicillium expansum and the efficacy of essential oils against blue mold rot of apples. Folia Horticulturae 31(2):301–317. https://doi.org/10.2478/fhort-2019-0024

Barbosa LN (2010) Propriedade antimicrobiana de óleos essenciais de plantas condimentares com potencial de uso como conservante em carne e hambúrguer bovino e testes de aceitação. p. 107

Barros FAP, Radünz M, Scariot MA, Camargo TM, Nunes CFP, de Souza RR, Gilson IK et al (2022) Efficacy of encapsulated and non-encapsulated thyme essential oil (Thymus vulgaris L.) in the control of Sitophilus zeamais and its effects on the quality of corn grains throughout storage. Crop Prot 153:105885. https://doi.org/10.1016/j.cropro.2021.105885

Bernardi AO, Garcia MV, Copetti MV (2019) Food industry spoilage fungi control through facility sanitization. Curr Opin Food Sci 29:28. https://doi.org/10.1016/j.cofs.2019.07.006

de Botelho AS, Ferreira OO, da Rocha Batista RJ, de Jesus Pereira Franco C (2022) Activity of essential oils against food spoilage fungi. In: Essential oils. Springer International Publishing, pp 43–70. https://doi.org/10.1007/978-3-030-99476-1_3

Cai M, Wang Y, Wang R, Li M, Zhang W, Yu J, Hua R (2022) Antibacterial and antibiofilm activities of chitosan nanoparticles loaded with Ocimum basilicum L. essential oil. Int J Biol Macromol 202:122–129. https://doi.org/10.1016/j.ijbiomac.2022.01.066

Camele I, Altieri L, de Martino L, de Feo V, Mancini E, Rana GL (2012) In vitro control of post-harvest fruit rot fungi by some plant essential oil components. Int J Mol Sci 13(2):2290–2300. https://doi.org/10.3390/ijms13022290

Castro JC, Endo EH, de Souza MR, Zanqueta EB, Polonio JC, Pamphile JA, Ueda-Nakamura T et al (2017) Bioactivity of essential oils in the control of Alternaria alternata in dragon fruit (Hylocereus undatus haw.). Ind Crop Prod 97:101–109. https://doi.org/10.1016/j.indcrop.2016.12.007

Chen H, Brashears MM, Zhong Q (2022) Sodium benzoate and sodium bisulfate as preservatives in apple juice and alternative sanitizers for washing cherry tomatoes. Int J Food Microbiol 372:109697. https://doi.org/10.1016/j.ijfoodmicro.2022.109697

Chiotta ML, Fumero MV, Cendoya E, Palazzini JM, Alaniz-Zanon MS, Ramirez ML, Chulze SN (2020) Toxigenic fungal species and natural occurrence of mycotoxins in crops harvested in Argentina. Rev Argent Microbiol 52(4):339–347. https://doi.org/10.1016/j.ram.2020.06.002

Colín-Chávez C, Virgen-Ortiz JJ, Miranda-Ackerman MA, Hernández-Cristóbal O, Martínez-Téllez MÁ, Esquivel-Chávez F, Gallegos-Santoyo NL (2022) Induction of defense mechanisms in avocado using Mexican oregano oil-based antifungal sachet. Future Foods 6:100171. https://doi.org/10.1016/j.fufo.2022.100171

de Corato U, Maccioni O, Trupo M, di Sanzo G (2010) Use of essential oil of Laurus nobilis obtained by means of a supercritical carbon dioxide technique against post harvest spoilage fungi. Crop Prot 29(2):142–147. https://doi.org/10.1016/j.cropro.2009.10.012

Cui H, Zhang C, Li C, Lin L (2020) Inhibition of Escherichia coli O157:H7 biofilm on vegetable surface by solid liposomes of clove oil. LWT 117:108656. https://doi.org/10.1016/j.lwt.2019.108656

Das SK, Vishakha K, Das S, Chakraborty D, Ganguli A (2022) Carboxymethyl cellulose and cardamom oil in a nanoemulsion edible coating inhibit the growth of foodborne pathogens and extend the shelf life of tomatoes. Biocatal Agric Biotechnol 42:102369. https://doi.org/10.1016/j.bcab.2022.102369

de Moraes Motta Machado MC, Lepaus BM, Bernardes PC, de São José JFB (2022) Ultrasound, acetic acid, and Peracetic acid as alternatives sanitizers to chlorine compounds for fresh-cut kale decontamination. Molecules 27(20):7019. https://doi.org/10.3390/molecules27207019

Dhifi W, Bellili S, Jazi S, Bahloul N, Mnif W (2016) Essential oils' chemical characterization and investigation of some biological activities: A critical review. Medicines 3(4):25. https://doi.org/10.3390/medicines3040025

El-Samawaty AE-RMA, El-Wakil DA, Alamery S, Mahmoud MMH (2021) Potency of plant extracts against Penicillium species isolated from different seeds and fruits in Saudi Arabia. Saudi J Biol Sci 28(6):3294–3302. https://doi.org/10.1016/j.sjbs.2021.02.074

El-Sayed RA, Jebur AB, Kang W, El-Demerdash FM (2022) An overview on the major mycotoxins in food products: characteristics, toxicity, and analysis. J Future Foods 2(2):91–102. https://doi.org/10.1016/j.jfutfo.2022.03.002

FDA (2008) The Food and Drug Administration, United States Department of Health and Human Services. Guide to Minimize Microbial Food Safety Hazards of Fresh-cut Fruits and Vegetables.

Ferrara M, Perrone G, Gallo A (2022) Recent advances in biosynthesis and regulatory mechanisms of principal mycotoxins. Curr Opin Food Sci 48:100923. https://doi.org/10.1016/j.cofs.2022.100923

Fotsing Yannick Stephane F, Kezetas Jean Jules B (2020) Terpenoids as important bioactive constituents of essential oils. In: Essential oils–bioactive compounds, new perspectives and applications. IntechOpen. https://doi.org/10.5772/intechopen.91426

Gunny AAN, Fang LP, Misnan NM, Gopinath SCB, Salleh NHM, Hashim RHR, Mat MHC (2021) Microwave-assisted solvent-free extraction of essential oil from Coleus aromaticus: anti-phytopathogenic potential for fruit post-harvesting. 3Biotech 11(4):166. https://doi.org/10.1007/s13205-021-02701-2

Guo S, Li T, Chen M, Wu C, Ge X, Fan G, Li X, Zhou D, Mi L, Zhao X, Yang T (2022) Sustainable and effective chitosan-based edible films incorporated with OEO nanoemulsion against apricots' black spot. Food Control 138:108965. https://doi.org/10.1016/j.foodcont.2022.108965

Hashem M, Alamri SAM, Alqahtani MSA, Alshehri SRZ (2019) A multiple volatile oil blend prolongs the shelf life of peach fruit and suppresses postharvest spoilage. Sci Hortic 251:48–58. https://doi.org/10.1016/j.scienta.2019.03.020

Hou T, Sana SS, Li H, Xing Y, Nanda A, Netala VR, Zhang Z (2022) Essential oils and its antibacterial, antifungal and anti-oxidant activity applications: A review. Food Biosci 47:101716. https://doi.org/10.1016/j.fbio.2022.101716

Huang X, Lao Y, Pan Y, Chen Y, Zhao H, Gong L, Xie N et al (2021) Synergistic antimicrobial effectiveness of plant essential oil and its application in seafood preservation: A review. Molecules 26(2):307. https://doi.org/10.3390/molecules26020307

Ju J, Chen X, Xie Y, Yu H, Guo Y, Cheng Y, Qian H et al (2019) Application of essential oil as a sustained release preparation in food packaging. Trends Food Sci Technol 92:22–32. https://doi.org/10.1016/j.tifs.2019.08.005

Kaczmarek M, Avery S, Singleton I (2019) Microbes associated with fresh produce: Sources, types and methods to reduce spoilage and contamination. Adv Appl Microbiol 107:29–82. https://doi.org/10.1016/bs.aambs.2019.02.001

Kamsu NP, Tchinda SE, Tchameni NS, Jazet DPM, Madjouko MA, Youassi Youassi O, Sameza ML et al (2019) Antifungal activities of essential oils of cinnamon (Cinnamomum zeylanicum) and lemongrass (Cymbopogon citratus) on crown rot pathogens of banana. Ind Phytopathol 72(1):131–137. https://doi.org/10.1007/s42360-018-0104-1

Kanetis L, Förster H, Adaskaveg JE (2007) Comparative efficacy of the new postharvest fungicides Azoxystrobin, Fludioxonil, and Pyrimethanil for managing citrus green Mold. Plant Dis 91(11):1502–1511. https://doi.org/10.1094/PDIS-91-11-1502

Kayode RM, Azubuike CU, Laba SA, Dauda AO, Balogun MA, Ajala SA (2018) Chemical composition and anti-microbial activities of the essential oil of Adansonia digitata stem-bark and leaf on post-harvest control of tomato spoilage. LWT 93:58–63. https://doi.org/10.1016/j.lwt.2018.03.014

Kumar A, Saini CS (2021) Edible composite bi-layer coating based on whey protein isolate, xanthan gum and clove oil for prolonging shelf life of tomatoes. Measurement: Food 2:100005. https://doi.org/10.1016/j.meafoo.2021.100005

Lee D, Shayan M, Gwon J, Picha DH, Wu Q (2022) Effectiveness of cellulose and chitosan nanomaterial coatings with essential oil on postharvest strawberry quality. Carbohydr Polym 298:120101. https://doi.org/10.1016/j.carbpol.2022.120101

Li Y, Erhunmwunsee F, Liu M, Yang K, Zheng W, Tian J (2022) Antimicrobial mechanisms of spice essential oils and application in food industry. Food Chem 382:132312. https://doi.org/10.1016/j.foodchem.2022.132312

de Lima CB, Rentschler LLA, Bueno JT, Boaventura AC (2016) Plant extracts and essential oils on the control of *Alternaria alternata*, Alternaria dauci and on the germination and emergence of carrot seeds (Daucus carota L.). Ciência Rural 46(5):764–770. https://doi.org/10.1590/0103-8478cr20141660

Luciano-Rosario D, Keller NP, Jurick WM (2020) *Penicillium expansum*: biology, omics, and management tools for a global postharvest pathogen causing blue mould of pome fruit. Mol Plant Pathol 21(11):1391–1404. https://doi.org/10.1111/mpp.12990

Long H, Bi Y, Pu L, Xu W, Xue H, Fu G, Prusky D (2022) Preparation of chitosan/fennel seed essential oil/starch sodium octenyl succinate composite films for apple fruit preservation. LWT 167:113826. https://doi.org/10.1016/j.lwt.2022.113826

Luque de Castro MD, Jiménez-Carmona MM, Fernández-Pérez V (1999) Towards more rational techniques for the isolation of valuable essential oils from plants. TrAC Trends Anal Chem 18(11):708–716. https://doi.org/10.1016/S0165-9936(99)00177-6

Maia TF et al (2015) Atividade antifúngica de óleos essenciais de plantas, Revista Brasileira de Produtos Agroindustriais. Revista Brasileira de Produtos Agroindustriais 17(1):105–116

Mesías FJ, Martín A, Hernández A (2021) Consumers' growing appetite for natural foods: perceptions towards the use of natural preservatives in fresh fruit. Food Res Int 150:110749. https://doi.org/10.1016/j.foodres.2021.110749

Miao P, Chen S, Li J, Xie X (2020) Decreasing consumers' risk perception of food additives by knowledge enhancement in China. Food Qual Prefer 79:103781. https://doi.org/10.1016/j.foodqual.2019.103781

Moncioso GL, Lepaus BM, Oliveira JSC, José JFBS (2021) Citric acid and clove essential oil as alternatives to chlorine compounds on sanitization of apples. Rev Bras Cienc Agrar 16(2):1–7. https://doi.org/10.5039/agraria.v16i2a9055

Moss MO (2008) Fungi, quality and safety issues in fresh fruits and vegetables. J Appl Microbiol 104(5):1239–1243. https://doi.org/10.1111/j.1365-2672.2007.03705.x

Moustafa SMN (2018) Role of hydrolytic enzymes of some fungi in rotting fruits and a possibility of retard spoilage. J Pure Appl Microbiol 12(3):1525–1535. https://doi.org/10.22207/JPAM.12.3.57

Neri F, Donati I, Veronesi F, Mazzoni D, Mari M (2010) Evaluation of Penicillium expansum isolates for aggressiveness, growth and patulin accumulation in usual and less common fruit hosts. Int J Food Microbiol 143(3):109–117. https://doi.org/10.1016/j.ijfoodmicro.2010.08.002

Ni Z-J, Wang X, Shen Y, Thakur K, Han J, Zhang J-G, Hu F, Wei Z-J (2021) Recent updates on the chemistry, bioactivities, mode of action, and industrial applications of plant essential oils. Trends Food Sci Technol 110:78–89. https://doi.org/10.1016/j.tifs.2021.01.070

Nikkhah M, Hashemi M (2020) Boosting antifungal effect of essential oils using combination approach as an efficient strategy to control postharvest spoilage and preserving the jujube fruit quality. Postharvest Biol Technol 164:111159. https://doi.org/10.1016/j.postharvbio.2020.111159

Nikkhah M, Hashemi M, Habibi Najafi MB, Farhoosh R (2017) Synergistic effects of some essential oils against fungal spoilage on pear fruit. Int J Food Microbiol 257:285–294. https://doi.org/10.1016/j.ijfoodmicro.2017.06.021

Osimani A, Garofalo C, Harasym J, Aquilanti L (2022) Use of essential oils against foodborne spoilage yeasts: advantages and drawbacks. Curr Opin Food Sci 45:100821. https://doi.org/10.1016/j.cofs.2022.100821

Owolabi IO, Songsamoe S, Matan N (2021) Combined impact of peppermint oil and lime oil on Mangosteen (Garcinia Mangostana) fruit ripening and mold growth using closed system. Postharvest Biol Technol 175:111488. https://doi.org/10.1016/j.postharvbio.2021.111488

Pandey AK, Chávez-González ML, Silva AS, Singh P (2021) Essential oils from the genus thymus as antimicrobial food preservatives: Progress in their use as nanoemulsions-a new paradigm. Trends Food Sci Technol 111:426–441. https://doi.org/10.1016/j.tifs.2021.02.076

Pandey VK, Islam RU, Shams R, Dar AH (2022) A comprehensive review on the application of essential oils as bioactive compounds in Nano-emulsion based edible coatings of fruits and vegetables. Appl Food Res 2(1):100042. https://doi.org/10.1016/j.afres.2022.100042

Pelissari EMR, Covre KV, do Rosario DKA, de São José JFB (2021) Application of chemometrics to assess the influence of ultrasound and chemical sanitizers on vegetables: impact on natural microbiota, salmonella Enteritidis and physicochemical nutritional quality. LWT 148:111711. https://doi.org/10.1016/j.lwt.2021.111711

Perincherry L, Lalak-Kańczugowska J, Stępień Ł (2019) Fusarium-produced mycotoxins in plant-pathogen interactions. Toxins 11(11):664. https://doi.org/10.3390/toxins11110664

Prakash B, Mishra PK, Kedia A, Dubey NK (2014) Antifungal, antiaflatoxin and antioxidant potential of chemically characterized Boswellia carterii Birdw essential oil and its in vivo practical applicability in preservation of Piper nigrum L. fruits. LWT Food Sci Technol 56(2):240–247. https://doi.org/10.1016/j.lwt.2013.12.023

Quéro L, Girard V, Pawtowski A, Tréguer S, Weill A, Arend S, Cellière B et al (2019) Development and application of MALDI-TOF MS for identification of food spoilage fungi. Food Microbiol 81:76–88. https://doi.org/10.1016/j.fm.2018.05.001

Regnier T, Combrinck S, du Plooy W, Botha B (2010) Evaluation of Lippia scaberrima essential oil and some pure terpenoid constituents as postharvest mycobiocides for avocado fruit. Postharvest Biol Technol 57(3):176–182. https://doi.org/10.1016/j.postharvbio.2010.03.010

Reis JB, Figueiredo LA, Castorani GM, Veiga SMOM (2020) Avaliação da atividade antimicrobiana dos óleos essenciais contra patógenos alimentares. Brazilian J Health Rev 3(1):342–363. https://doi.org/10.34119/bjhrv3n1-025

Roumani F, Gómez S, Rodrigues C, Barros-Velázquez J, Garrido-Maestu A, Prado M (2022) Development and evaluation of a real-time fluorescence, and naked-eye colorimetric, loop-mediated isothermal amplification-based method for the rapid detection of spoilage fungi in fruit preparations. Food Control 135:108784. https://doi.org/10.1016/j.foodcont.2021.108784

Saleh I, Al-Thani R (2019) Fungal food spoilage of supermarkets' displayed fruits. Veterinary World 12(11):1877–1883. https://doi.org/10.14202/vetworld.2019.1877-1883

Salvi P, Kumar G, Gandass N, Kajal, Verma A, Rajarammohan S, Rai N et al (2022) Antimicrobial potential of essential oils from aromatic plant Ocimum sp.; A comparative biochemical profiling and in-silico analysis. Agronomy 12(3):627. https://doi.org/10.3390/agronomy12030627

Santana Oliveira I, da Silva Junior AG, de Andrade CAS, Lima Oliveira MD (2019) Biosensors for early detection of fungi spoilage and toxigenic and mycotoxins in food. Curr Opin Food Sci 29:64–79. https://doi.org/10.1016/j.cofs.2019.08.004

Sanzani SM, Reverberi M, Geisen R (2016) Mycotoxins in harvested fruits and vegetables: insights in producing fungi, biological role, conducive conditions, and tools to manage postharvest contamination. Postharvest Biol Technol 122:95–105. https://doi.org/10.1016/j.postharvbio.2016.07.003

São José JFB, Medeiros HS, de Andrade NJ, Ramos AM, Vanetti MCD (2018) Effect of ultrasound and chemical compounds on microbial contamination, physicochemical parameters and bioactive compounds of cherry tomatoes. Ital J Food Sci 30

Sazvar E, Jahani M, Aminifard MH, Hosseini SA (2022) In vitro and in vivo control of *Alternaria alternata* in barberry (*Berberis vulgaris*) by some essential oils. Erwerbs-obstbau 64(3):413–423. https://doi.org/10.1007/s10341-022-00645-2

Soraggi Battagin T, Nicolas Caccalano M, Dilarri G, Felipe Cavicchia Zamuner C, Alleoni N, Leonardo Saldanha L, Bacci M, Ferreira H (2021) Syzygium aromaticum (clove) essential oil: an alternative for the sanitization of citrus fruit in packinghouses. J Food Proc Preserv 45(9). https://doi.org/10.1111/jfpp.15496

Seshadri VD, Balasubramanian B, Al-Dhabi NA, Esmail GA, Arasu MV (2020a) Essential oils of Cinnamomum loureirii and Evolvulus alsinoides protect guava fruits from spoilage bacteria, fungi and insect (Pseudococcus longispinus). Ind Crop Prod 154:112629. https://doi.org/10.1016/j.indcrop.2020.112629

Seshadri VD, Balasubramanian B, Al-Dhabi NA, Esmail GA, Arasu MV (2020b) Essential oils of Cinnamomum loureirii and Evolvulus alsinoides protect guava fruits from spoilage bacteria, fungi and insect (Pseudococcus longispinus). Indus Crops Prod 154:112629. https://doi.org/10.1016/j.indcrop.2020.112629

Shen C, Chen W, Li C, Aziz T, Cui H, Lin L (2023) Topical advances of edible coating based on the nanoemulsions encapsulated with plant essential oils for foodborne pathogen control. Food Control. https://doi.org/10.1016/j.foodcont.2022.109419

da Silva BD, Bernardes PC, Pinheiro PF, Fantuzzi E, Roberto CD (2021) Chemical composition, extraction sources and action mechanisms of essential oils: natural preservative and limitations of use in meat products. Meat Sci 176:108463. https://doi.org/10.1016/j.meatsci.2021.108463

Simas DLR, de Amorim SHBM, Goulart FRV, Alviano CS, Alviano DS, da Silva AJR (2017) Citrus species essential oils and their components can inhibit or stimulate fungal growth in fruit. Ind Crop Prod 98:108–115. https://doi.org/10.1016/j.indcrop.2017.01.026

Skendi A, Irakli M, Chatzopoulou P, Bouloumpasi E, Biliaderis CG (2022) Phenolic extracts from solid wastes of the aromatic plant essential oil industry: potential uses in food applications. Food Chem Adv 1:100065. https://doi.org/10.1016/j.focha.2022.100065

Speranza B, Corbo MR (2010) Essential oils for preserving perishable foods: possibilities and limitations. In: Application of alternative food-preservation technologies to enhance food safety and stability, pp 35–37

Tian F, Woo SY, Lee SY, Park SB, Zheng Y, Chun HS (2022) Antifungal activity of essential oil and plant-derived natural compounds against aspergillus flavus. Antibiotics 11(12):1727. https://doi.org/10.3390/antibiotics11121727

Torre R, Pereira EAD, Nascimento RV, Guedes TF, de Faria PRS, de Alves MS, de Souza MAA (2021) Agroecological approach to seed protection using basil essential oil. Ind Crop Prod 171:113932. https://doi.org/10.1016/j.indcrop.2021.113932

Tripathi AD, Sharma R, Agarwal A, Haleem DR (2021) Nanoemulsions based edible coatings with potential food applications. Int J Biobased Plastics 3(1):112–125. https://doi.org/10.1080/24759651.2021.1875615

Wang H, Ma Y, Liu L, Liu Y, Niu X (2022) Incorporation of clove essential oil nanoemulsion in chitosan coating to control Burkholderia gladioli and improve postharvest quality of fresh Tremella fuciformis. LWT 170:114059. https://doi.org/10.1016/j.lwt.2022.114059

Wang L, Jiang N, Wang D, Wang M (2019) Effects of essential oil Citral on the growth, mycotoxin biosynthesis and transcriptomic profile of Alternaria alternata. Toxins 11(10):553. https://doi.org/10.3390/toxins11100553

Wińska K, Mączka W, Łyczko J, Grabarczyk M, Czubaszek A, Szumny A (2019) Essential oils as antimicrobial agents—myth or real alternative? Molecules 24(11):2130. https://doi.org/10.3390/molecules24112130

Xu Y, Wei J, Wei Y, Han P, Dai K, Zou X, Jiang S, Xu F, Wang H, Sun J, Shao X (2021) Tea tree oil controls brown rot in peaches by damaging the cell membrane of Monilinia fructicola. Postharvest Biol Technol 175:111474. https://doi.org/10.1016/j.postharvbio.2021.111474

Xu Y, Xu J, Yang G, Guan X, Li R, Wang S (2022) Combined effects of intermittent radio frequency heating with cinnamon oil vapor on microbial control and quality changes of alfalfa seeds. Int J Food Microbiol 367:109586. https://doi.org/10.1016/j.ijfoodmicro.2022.109586

Zhang H, Zhang C, Wang X, Huang Y, Xiao M, Hu Y, Zhang J (2022) Antifungal electrospinning nanofiber film incorporated with Zanthoxylum bungeanum essential oil for strawberry and sweet cherry preservation. LWT 169:113992. https://doi.org/10.1016/j.lwt.2022.113992

Zhang X, Li B, Zhang Z, Chen Y, Tian S (2020) Antagonistic yeasts: A promising alternative to chemical fungicides for controlling postharvest decay of fruit. J Fungi 6(3):158. https://doi.org/10.3390/jof6030158

Essential Oils: A Natural Weapon against Mycotoxins in Food

6

Anjana Tiwari, Parshant, and Ravindra Shukla

Abstract

Essential oils are aromatic liquids that are isolated from plant parts and thus are extremely volatile and scented. These volatile oils are responsible for the different scents that plants emit. The major mycotoxin groups, including aflatoxin, fumonisin, ochratoxin (OT), zearalenone (ZEN), and deoxynivalenol (DON) can induce many harmful health effects, including allergies, cancer, and immunosuppression, owing to the intake of infected food. Currently, it is known that the majority of synthetic chemicals used as preservatives pose a danger to individuals and harm the environment. In this regard, utilizing diverse plant products, particularly essential oils (EOs) and their bioactive constituents, has been considered a green strategy and a safer alternative to synthetic chemicals due to their long-standing traditional use. Essential oils have several modes of action that prevent the growth of fungus and the production of mycotoxin, including changed fungal growth rates, disruption of cell permeability, disruption of the electron transport chain, alteration of gene expression patterns, and metabolic activities. This chapter aims to summarize the different recent studies on the effect of essential oils in inhibiting the growth of mycotoxigenic fungus, eliminating mycotoxins, and their mode of action.

Keywords

Essential oils · Aromatic compounds · Mycotoxins · Aflatoxin, Ochratoxin

---

A. Tiwari · Parshant · R. Shukla (✉)
Laboratory of Bio-resource Technology, Department of Botany, Indira Gandhi National Tribal University, Amarkantak, Madhya Pradesh, India
e-mail: ravindra.shukla@igntu.ac.in

B. Prakash et al. (eds.), *Plant Essential Oils*,
https://doi.org/10.1007/978-981-99-4370-8_6

6.1 Introduction

Foodborne diseases caused by consuming contaminated food are a major public health concern around the world. According to the World Health Organization (WHO), 31 foodborne hazards globally contributed to 600 million foodborne illnesses and 420,000 deaths (Havelaar et al. 2015; WHO 2022). Foodborne illnesses are caused by a variety of pathogens, including fungi, viruses, and bacteria. These pathogens can contaminate food at any point during its production, processing, or distribution (Hemalata and Virupakshaiah 2016). Toxins produced by certain fungi cause the most common fungal foodborne diseases. These toxins, known as mycotoxins, can induce diarrhea, abdominal pain, nausea, vomiting, and fever, among other symptoms. In severe cases, mycotoxin poisoning can lead to liver damage, kidney failure, and even death (Dhakal and Sbar 2021). More than 400 different types of mycotoxins have been identified, but the most common mycotoxins that cause foodborne illness are aflatoxins, ochratoxin-A, fumonisin, zearalenone, patulin, and trichothecenes (Ünüsan 2019). Consumption of mycotoxin-contaminated food or feed can cause acute (e.g., turkey X syndrome, human ergotism, stachybotryotoxicosis) or chronic (e.g., cancer induction, kidney toxicity, immune suppression) toxicity in humans and livestock due to their inherent carcinogenic, mutagenic, teratogenic, and immuno-suppressive characteristics (Bennett and Klich 2003; Chaudhari et al. 2021). Mycotoxins are low-molecular-weight (MW ~ 700 Da) secondary metabolites secreted by many filamentous fungi belonging to the genera *Aspergillus*, *Penicillium*, and *Fusarium* that are highly toxic to animals and humans (Liew and Mohd-Redzwan 2018; Alshannaq and Yu 2017). They are grown under favorable conditions (temperature, moisture, water activity, and relative humidity) between 10 to 40 °C and with a pH range of 4 to 8 (Bhat et al. 2010). Depending on the fungal species, mycotoxins can appear in both temperate and tropical climates. These mycotoxins are commonly found in dried fruits, nuts, coffee, oil seeds, cereals, cocoa, spices, beans, dried peas, and fruits, especially apples. Mycotoxins may also be identified in beer and wine due to the use of contaminated raw materials during their production, such as barley, cereals, grapes, etc. (Turner et al. 2009).

To ensure the consumption of toxin-free food with good nutritional values for human health, it is essential to maintain food quality. So, the best approach to maintain the quality of food and stop it from microbial deterioration is to use preservation techniques. Nowadays, there are many different types of preservation techniques available that can be used to preserve food commodities for a very long time, either by using traditional or modern preservation techniques (Sharif et al. 2017). Additional food preservatives, which can be divided into artificial (synthetic) and natural preservatives, are used in some of these preservation methods. Several synthetic preservatives such as sulfur dioxide, sulfites, sodium nitrite, sodium benzoate, benzoates, sorbates, formaldehyde, imidazoles, pyrrolidines, and thiocyanates, etc. have been employed to control microbial contamination of food items (Maurya et al. 2021a, b). A few of them are poisonous, and many others could have fatal adverse effects. Artificial preservatives can lead to major health problems like cancer, hyperactivity, hypersensitivity, allergy, asthma, and other respiratory and respiratory-related illnesses (Kumari et al. 2019). Customers are generally aware of the health risk posed by the use of artificial preservatives. As a result, the demand for natural food preservatives has significantly increased. In this regard, the use of various plant products, particularly essential oils (EOs) and their bioactive compounds, has been recognized as a green strategy and a safer alternative to synthetic antifungal and mycotoxin treatments (Shukla et al. 2012; Chaudhari et al. 2019).

The word "Quinta essentia" was first used to describe the active component of a drug by Paracelsus von Hohenheim in the sixteenth century. The word "essential" is derived from the Latin essentia (Guenther and Althausen 1948). Essential oils are secondary metabolites produced by many aromatic plant parts, including flowers, buds, leaves, stems, twigs, seeds, fruits, roots, bark, or wood. They are stored in secretory cells, cavities, canals, epidermal cells, or glandular trichomes. They are natural molecules with a complex mixture of chemical structures that are liquid, volatile, rarely colored, soluble in lipids and organic solvents, and have a density often less than that of water (Bakkali et al. 2008; Bouyahya et al. 2019). Essential oils are produced by two natural biochemical processes that involve several enzymatic reactions. The precursors of essential oil biosynthesis, isopentenyl diphosphate (IPP), and its isomer, dimethylallyl diphosphate (DMAPP), are produced in the cytoplasm and plastids via the mevalonic acid (MVA) and methyl-D-erythritol-4-phosphate (MEP) pathways (Rehman et al. 2016). Terpenes (pinene, myrcene, limonene, terpinene, and *p*-cymene), the prominent ingredient of essential oils are hydrocarbons made up of several isoprene (C5H8) units, while terpenoids (oxygen-containing hydrocarbons) are modifications of terpenes with different functional groups and moved or removed oxidized methyl groups at various positions (Masyita et al. 2022). In general, terpenoids possess greater antimicrobial activity than terpenes (Burt 2004). EOs have been widely used successfully in indigenous systems and are thought to have antimicrobial properties. Since the Middle Ages, essential oils have been used for a variety of purposes, including bactericidal, virucidal, fungicidal, antiparasitic, insecticidal, medicinal, and cosmetic ones (Prakash et al. 2015). Due to their natural origin, EOs and their components are regarded as user-friendly, friendly to the environment, and often exempt from the Environmental Protection Agency's standards for toxicity data (Prakash et al. 2010). Cinnamon, fennel, rosemary, oregano, thyme, palmarosa, clove, and eucalyptus have been revealed to be the most efficacious essential oils against mycotoxigenic fungi and their mycotoxins in research published over the past 10 years. Essential oils can prevent the growth of fungi and the production of mycotoxin through a variety of mechanisms, including altered fungal growth rate, extended lag phase, disruption of cell permeability, disruption of the electron transport chain, altering gene expression patterns, and metabolic processes (Mirza Alizadeh et al. 2022). A fungus's ability to produce toxins depends not only on its ability to grow but also on the fungistatic and fungicidal substances that may have an impact on its invasion and colonization. Extensive research publications, reviews, and reports have been written about the fungal species that contaminate food and feed; nevertheless, a large portion of the data is either limited to one type of mycotoxin or the data is fragmented. In light of this, we attempted to provide a thorough overview of essential oils against the mycotoxigenic fungus, the diversity of mycotoxins, associated health concerns for humans and livestock, etc. in the present book chapter.

6.2 Food Spoiling Mycotoxins and their Types

Mycotoxins are secondary, stable, and physiologically active metabolites that frequently contaminate agricultural goods. Along with significant financial losses, they also represent a long-term, hidden risk to the health of both humans and animals. Currently, more than 400 mycotoxins have been identified and described as toxic, non-volatile, and relatively low-molecular-weight secondary metabolites produced by certain filamentous fungi, such as species of *Aspergillus*, *Alternaria*, *Fusarium*, and *Penicillium* (Gurikar et al. 2022; Venkatesh et al. 2017). Many species of mycotoxins are commonly found in cereals, herbal materials, fruits, and spices (Chaudhari et al. 2021). According to Food and Agriculture Organization (FAO) statistics, approximately one-fourth (25%) of all agricultural products are harmed by mycotoxin-producing fungi worldwide. The most prevalent mycotoxin, aflatoxin, has significant toxicity and carcinogenic potential and is present in a wide range of foods and feeds. Besides aflatoxins, other major toxins are trichothecenes, fumonisins, ochratoxin, and zearalenone, which are present in grains used as food and other agricultural products. Mycotoxins found in contaminated food products and animal feeds can have many harmful effects on human and animal health, including cancer, immunosuppression, gastrointestinal, estrogenic, and kidney diseases (Tola and Kebede 2016; Neme and Mohammed 2017; Gurikar et al. 2022) (Table 6.1).

6.3 Aflatoxins

Aflatoxins are highly toxic secondary metabolites produced by *Aspergillus flavus*, *A. parasiticus*, and *A. nominus* (Rocha et al. 2023). These fungi grow on a wide variety of foods such as wheat, walnuts, and other dry fruits, legume seeds, corn, cotton, peanuts, and tree nuts that cause serious threats to human and animal health through various complications such as hepatotoxicity, teratogenicity, and immunotoxicity (Shukla et al. 2008; Kumar et al. 2017). The four major aflatoxins, B1, B2, G1, and G2, were isolated based on their fluorescence under UV light and identified by thin-layer chromatography (Baranyi 2013). AFTs-M1 and AFTs-M2 are produced by several metabolic processes from animals and animal products. The chemical structure of aflatoxins are shown in Fig. 6.1. The International Agency for Research on Cancer (IARC) classifies AFB1 as a Group 1 carcinogen, with high risks of hepatocellular carcinoma (HCC) in individuals exposed to aflatoxins, whereas AFM1 is classified as a "possible carcinogen" in Group 2B (Alshannaq and Yu 2017). The toxigenic capacities of pathogenic branches within an aflatoxigenic species differ considerably on a mycological and quantitative scale. Cytochrome P450 enzymes convert aflatoxin into reactive 8, 9-epoxide forms that can bind to DNA and proteins. It is generally understood that reactive aflatoxin epoxide

Table 6.1 Mycotoxin production in the food system and its adverse effects

Sr. no.	Mycotoxins	Common fungal sp.	Model foods (common name)	Health effects	Reference
1.	Aflatoxins: AFB1, AFB2, AFG1, AFG2	*Aspergillus flavus, A. parasiticus*, and *A. nomius*	Wheat, walnut, corn, cotton, peanuts, etc.	Teratogenicity, hepatotoxicity, immunotoxicity, etc.	Kumar et al. (2017)
2.	Fumonisins: FB1, FB2, FB3	*Fusarium moniliforme, F. proliferatum*, and *F. verticillioides*	Peanut, maize, grape, wheat, rye, barley, maize, oats, millet, etc.	Hepatocarci-noma, nephrotoxicity, adenocarcinoma, etc.	Kamle et al. (2019); Soriano et al. (2005)
3.	Ochratoxins (OTA, OTB, OTC)	*Aspergillus* sp., *Penicillium* sp.	Dry beans, maize, wheat, barley, oats, green coffee beans, etc.	Embryotoxic, teratogenic, immunotoxic, and nephrotoxicity	Reddy and Bhoola (2010)
4.	Zearalenone (ZEA)	*Fusarium graminearum, F. equiseti, F. cerealis, F. culmorum*	Rice, wheat, soybeans, corn, barley, millet, etc.	Genotoxic, teratogenic, immunotoxic, carcinogenic, hemotoxic, and hepatotoxic properties	Rogowska et al. (2019)
5.	Patulin	*Penicillium*, sp., *aspergillus* sp., and *Byssochlamys* sp.	Apple, apricot, black mulberry, corn, figs, pear, peach, pineapple, etc.	Immunotoxicity, gastrointestinal, hepatotoxicity, and neurological problems	Mahato et al. (2021)
6.	Ergot alkaloids: Ergometrine, ergocornine, ergosine, ergotamine, ergocryptine, ergocristine	*Claviceps purpurea, Balansia* sp., *Epichloe* sp., *Periglandula* sp., and *aspergillus fumigates*	Cereals like rye, wheat, millet triticale, oats, barley, etc.	Neurotoxicity, hallucinations, abdominal pain, insomnia, burning sensations of the skin, endocrine disruption in animals, convulsions, abortion, suppression of lactation, etc.	Arcella et al. (2017): Mulac and Humpf (2011)

(continued)

Table 6.1 (continued)

Sr. no.	Mycotoxins	Common fungal sp.	Model foods (common name)	Health effects	Reference
7.	Citrinin	*Aspergillus carneus A. terreus*, *A. niveus*, *Penicillium citrinum*, *P. verrucosum*, *P. expansum*, and *Monascus ruber*	Cereals (maize, rye, rice, barley, wheat, corn, oats), oilseeds (e.g., sunflower), and spices (e.g., turmeric, black pepper, cumin, cardamom, fennel, and coriander)	Nephrotoxicity and genotoxicity, teratogenic and embryotoxic effects	Kamle et al. (2022)
8.	Alternario, alternariol monomethyl ether (AME), and tentoxin (TEN)	*Alternaria alternata*, *A. brassicae*, *A. arborescens, A. radicina*, *A. infectoria*, *A. tenuissima*, and *A. brassicicola*	Cereals, cauliflowers, tomatoes, apples, oilseeds, grapes, oil crops, lemons, sunflower seeds, oranges, cucumbers, melons, peppers, and tangerines	Carcinogenicity, mutagenicity, sphingolipid metabolism disruption, induction of DNA strand breaks, and photophosphorylation	Escrivá et al. (2017)
9.	Trichothecenes: (T-2 toxin, HT-2 toxin, Neosolaniol (NEO), Diacetoxyscirpenol (DAS), Monoacetoxy-scirpenol (MAS), Verrucarol (VER) and Deoxynivalenol (DON))	*Fusarium culmorum* and *F. graminearum*, *Stachybotrys atra*, *Myrothecium* sp. And *Trichothecium* sp.	Cereals (wheat, corn, oats, and barley) seeds of rye, safflower, and in feed mixtures.	Inhibit the activity of the peptidyl transferase enzyme, as well as the initiation, elongation, and termination of eukaryotic protein synthesis	Polak-Śliwińska and Paszczyk (2021); da Rocha et al. (2014)
10.	Nivalenol/ deoxynivalenol	*Fusarium Graminearum*, and *F. culmorum*	Grains in the field, including corn, wheat, oats, barley, rice, and others	Immunotoxicity, hemotoxicity, edema, diarrhea, headaches, and other symptoms.	Sobrova et al. (2010); Kumar et al. (2022)

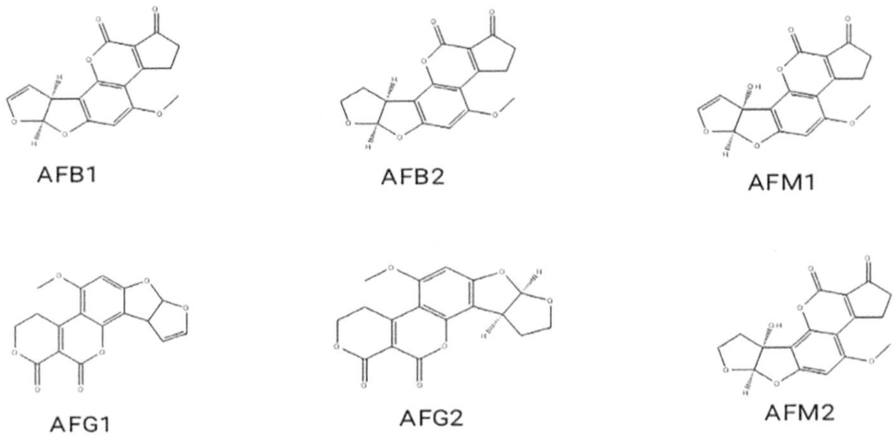

Fig. 6.1 Chemical structures of aflatoxins

interacts with the N7 site of guanines. The cytosol and microsomes contain a reactive glutathione S-transferase system that catalyzes the conjugation of activated aflatoxins with reduced glutathione, resulting in aflatoxin excretion (Bennett and Klich 2003).

6.4 Fumonisins

Fusarium mycotoxins are chemically and thermally stable secondary metabolites produced by *Fusarium verticillioides* (*Fusarium moniliforme*), *Fusarium proliferatum*, and other related species. *Fusarium* species produce mycotoxins such as deoxynivalenol, nivalenol, zearalenone, T-2 toxin, trichothecenes, and fumonisins (Ji et al. 2019; Rheeder et al. 2002). Fumonisins are found in corn, sorghum, millet, and other agricultural products. The Intergovernmental Agency for Research on Cancer (IARC) classified maize-derived toxins as category 2B (possibly carcinogenic to humans) in 1993 due to the typically higher presence of fumonisins in maize. Fumonisin also shows hepatotoxic, nephrotoxic, atherogenic, immunosuppressive, and embryotoxic effects other than its carcinogenic effect (Nair 1998). Since 1988, 28 fumonisin analogs have been identified, and they have been classified into four main groups (A, B, C, and P). The most abundant naturally occurring fumonisins are the FB analogs, which include the toxicologically significant FB1, FB2, and FB3. When cultured on corn, rice, or in a liquid medium, FB1 (Fig. 6.2) predominates and is usually found at higher levels (70 to 80%) than FB2 (15 to 25%) and FB3 (3 to 8%) (Rheeder et al. 2002). In 2014, fumonisin B1 (FB1), FB2, and FB3 were found in 98.1% of corn products collected in Shandong Province, China (Li et al. 2015).

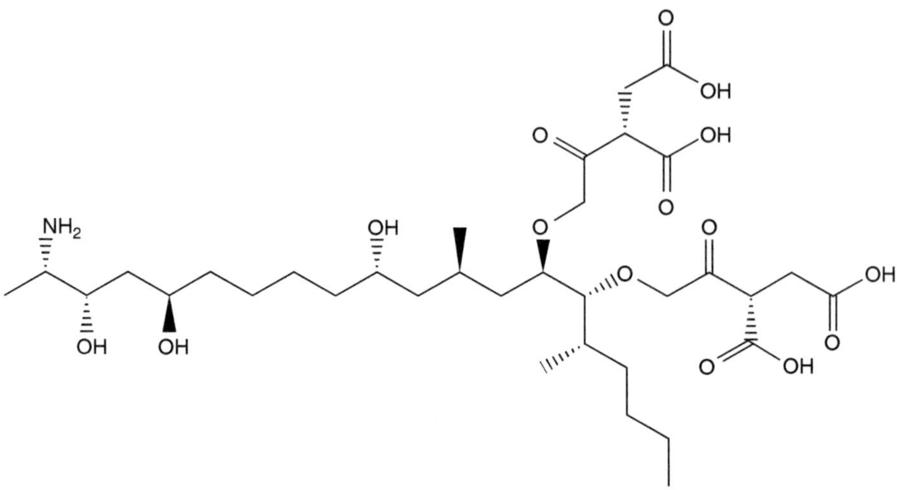

Fig. 6.2 Fumonisin B1

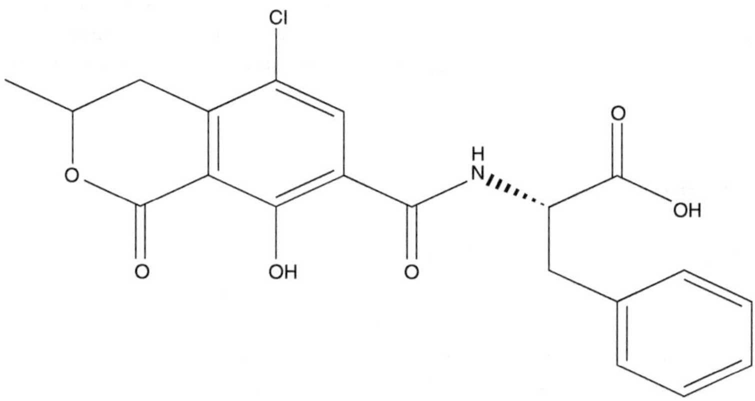

Fig. 6.3 Ochratoxin A

6.5 Ochratoxins

Aspergillus and *Penicillium* species produce ochratoxin as a secondary metabolite. The optimum temperature for the growth of *A. ochraceous* is 24 to 37 °C (pH is 3 to 10), and that of *P. verrucosum* is 20 °C (pH is 6.0 to 7.0) (Reddy and Bhoola 2010). These ochratoxins, ochratoxin A, ochratoxin B, and ochratoxin C, are all present in the environment. Ochratoxin A (Fig. 6.3) is a prominent toxin, and the International Agency for Research on Cancer categorized it as a category 2B likely human carcinogen in 1993 based on evidence that it could cause cancer in animals. Ochratoxin A is a naturally occurring mycotoxin produced by several fungi, including

Penicillium verrucosum, Aspergillus carbonarius, *A. ochraceus*, and *A. niger*. It can be found in a wide range of agricultural commodities worldwide, such as cereal grains, dried fruits, wine, and coffee (Reddy and Bhoola 2010; Bui-Klimke and Wu 2015; WHO,, and IARC 1993, and Kőszegi and Poór 2016). Ochretoxin A can cause a variety of health problems in animals and humans, including nephrotoxic, hepatotoxic, embryotoxic, teratogenic, neurotoxic, immunotoxic, genotoxic, and carcinogenic effects in many species (Malir et al. 2016). Spectroscopy techniques are widely used in the detection of toxic and harmful components in food products (Kumar et al. 2020).

6.6 Zearalenone

Zearalenone is a potent estrogenic mycotoxin produced by the *Fusarium* species, mainly *F. graminearum*, *F. culmorum*, *F. cerealis*, *F. equiseti*, *F. crookwellense*, *F. semitectum*, *F. verticillioides*, *F. sporotrichioides*, *F. oxysporum*, and *F. acuminatum* (Zhang et al. 2018). This mycotoxin is commonly found in cereals such as barley, maize, oats, sorghum, and wheat (Bhat et al. 2010). Zearalenone (Fig. 6.4) is a thermostable (160 °C) weakly polar compound with a molar mass of 318.364 g/mol that dissolves in various alkaline solutions such as benzene, acetonitrile, acetone, or alcohols (Ropejko and Twarużek 2021). McNutt et al. described a typical hyperestrogenism syndrome in pigs in 1928, due to the consumption of spoiled corn. Hyperestrogenism manifests as improper breastfeeding, aberrant udder or mammary gland growth, extended estrus, anestrus, changes in libido, infertility, and a higher frequency of pseudopregnancy. In humans, it can bind to alpha and beta estrogen receptors and disrupt the functioning of the endocrine system (Kuiper-Goodman et al. 1987). Additionally, zearalenone is genotoxic, immunotoxic, toxic to the reproductive system, and toxic to the development of the immune system. The interference with blood coagulation also causes hemato-toxic effects (Zinedine et al. 2017).

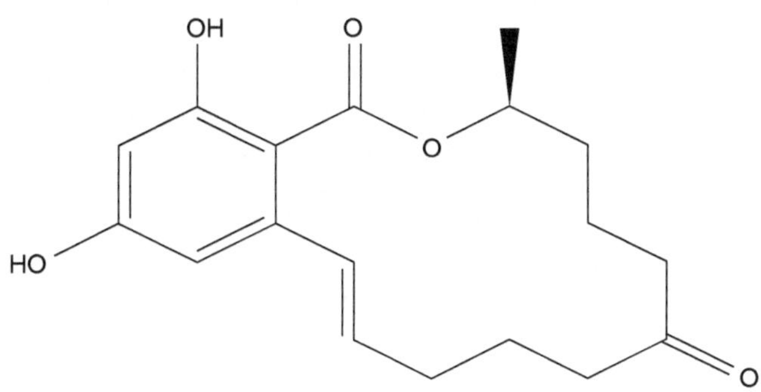

Fig. 6.4 Zearalenone

Fig. 6.5 Patulin

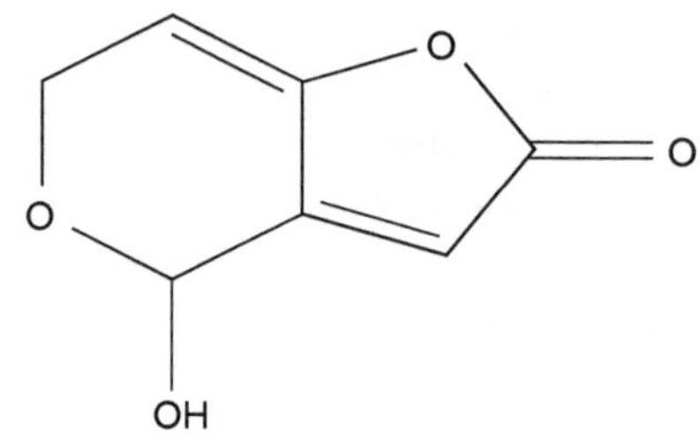

6.7 Patulin

Patulin (PAT) is a mycotoxin discovered in 1940 and produced as a secondary metabolite by numerous fungal species such as *Penicillium*, *Aspergillus*, and *Byssochlamys*. Chemically, PAT [4-Hydroxy-4H-furo(2,3-C)-pyron-2(6H)-1] is a polyketide lactone, which is a colorless, crystalline, water-soluble substance (Sajid et al. 2019; Pal et al. 2017). Patulin toxin was shown to be produced greater, respectively, at pH levels of 2.5 and 4, and storage temperatures of 20 to 25 °C. In most cases, it is associated with food products that have fungal infections. Patulin is generally found on fruits like apples, pears, peaches, and grapes but is mainly produced by the genus *Penicillium* with the species *Penicillium expansum*, which is capable of contaminating pome fruits, especially in apples and apple-based products (Baert et al. 2007). PAT (Fig. 6.5) was classified as non-carcinogenic in Group 3 by the International Agency for Cancer Research (IARC). Initially tested as an antibiotic, patulin was subsequently shown to be dangerous to humans and to produce symptoms including nausea, vomiting, ulceration, and hemorrhage. Glutathione is considered the scavenger of PAT-induced toxicity (Alshannaq and Yu 2017).

6.8 Ergot Alkaloid

The genus *Claviceps* produces the mycotoxins known as ergot alkaloids, which are predominantly produced by one type of fungus and infect a variety of cereals. Alkaloids found in ergot, including ergotamine (Fig. 6.6), ergocrystinine, ergometrimine, ergonovin, and ergocrystin, are toxic. One of the most significant species, *Claviceps purpurea*, mostly affects monocotyledonous plants. The crops infected by Ergot alkaloid (EAs) include rye, barley, wheat, millet, oats, and triticale, with rye having the highest rates of fungal infection (Krska et al. 2008; Agriopoulou 2021). Ergot alkaloids are consumed by humans through food. After consuming foods contaminated with EAs, it can cause ergotism, an illness causing strange hallucinations, convulsions, agalactia, burning sensations, vasoconstriction, and gangrenous loss of limbs, which are some of the symptoms in humans (Gurikar et al. 2022). Purified ergot alkaloids crystallized as translucent compounds that are soluble in organic solvents and buffers as well as inorganic solvents like acetonitrile and methanol. The two types of ergotism are gangrenous and convulsive, respectively.

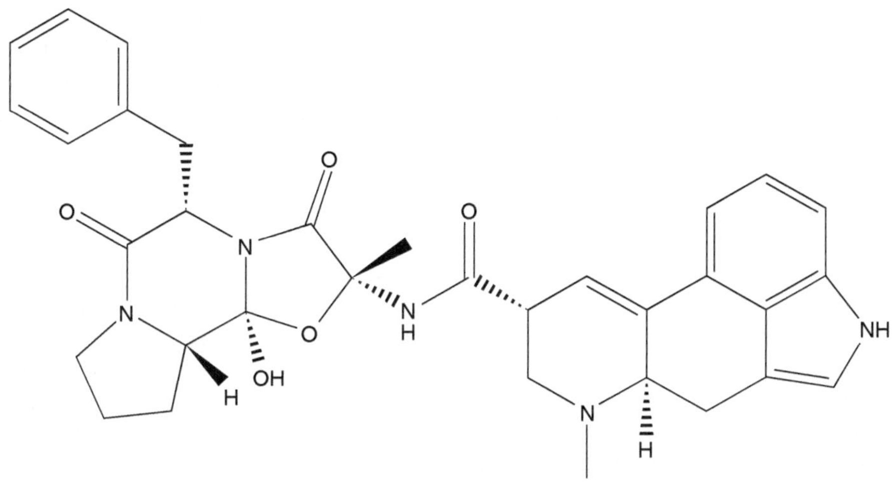

Fig. 6.6 Ergotamine

Convulsive ergotism affects the neurological system, whereas gangrenous ergotism affects the blood supply. Animals are not immune to ergotism, which is a serious threat to veterinarians and can affect a variety of species, including cattle, sheep, pigs, and poultry (Pleadin et al. 2019).

6.9 Effect of Aromatic Bioactive Compounds (Essential Oils) on Mycotoxins

Essential oils are naturally occurring volatile and aromatic compounds produced by aromatic herbs or shrubs for their requirements, such as defense or attracting pollinators, rather than nutrition. Aromatic plants, which are often found in temperate and tropical regions, generate essential oils, a mixture of different constituents. They typically possess a lower density compared to water and are liquid, transparent, sparingly colored, lipid-soluble, and soluble in organic solvents. Essential oils are synthesized by plant parts such as buds (*Syzygium aromaticum*), flowers (*Thymus vulgaris*), leaves (*Eucalyptus globulus, Callistemon lanceolatus*), stems (*Cordia trichotoma)*, seeds (*Pimpinella anisum)*, fruits (*Juniperus communis*), root or rhizome (*Acorus calamus, Zingiber officinale*), wood or bark (*Santalum austrocaledonicum*) and accumulate in cells, secretory cavities, or glandular hairs of plants (Bakkali et al. 2008; Roh et al. 2011; Shukla et al. 2011; Shukla et al. 2013; Bilia et al. 2014). Essential oils may be extracted using various methods. One of the most popular methods is steam- or hydrodistillation, which allows for the separation of mildly volatile, water-immiscible substances at low temperatures (Chamorro et al. 2012). Essential oils are complex natural mixtures that consist of about 20–60 components at quite different concentrations; two or three major components are present at high concentrations (20–70%) compared to other components that are present in trace amounts (Chouhan et al. 2017). The main components found in essential oils are typically responsible for their biological effects (Shukla et al. 2009). They have been extensively used as bactericidal, virucidal, fungicidal, antiparasitic, insecticidal, and therapeutic agents (Shaaban 2020). For example, 37 compounds are identified in the oregano essential oil (OEO), of which carvacrol (30.73%) and thymol (18.81%) are the two major constituents (Chen et al. 2022). In the past 10 years, numerous studies have been done on the antimycotic effects of essential oils. These studies have demonstrated the effect of EOs on the restriction or decrease of mycotoxin synthesis. Bocate et al. (2021) investigated the efficacy of garlic essential oil (GEO) to hinder the production of *Aspergillus parasiticus*, *Fusarium verticillioides*, and *Gibberella zeae*, which are responsible for producing the aflatoxins AFB1, ZEA, and, FB1 in stored corn kernels. According to the findings, GEO can prevent the growth of mycotoxigenic fungus. *A. parasiticus*, *G. zeae*, and *F. verticillioides* all exhibiting minimum inhibitory concentrations (MICs) of 0.0086, 0.069, and 0.0086 mg/mL, respectively. *Penicillium verrucosum* and *P. griseofulvum* were discovered in lentils and dry grapes (*Vitis vinifera* L.) in northern Morocco. Chidi et al. (2020) investigated the antifungal activities of *Melaleuca alternifolia* essential oil (EO) against both of these isolates. The results showed that adding *M. alternifolia* EO at various dosages completely decreased the amount of terrestric acid and ochratoxin-A produced by *P. griseofulvum* and *P. verrucosum*. In an in vitro study, Ferdes et al. (2017) investigated the effectiveness of five EOs (sage, rosemary, anise, quinoa, and savory) against *Aspergillus niger*, *Aspergillus oryza*, *Fusarium oxysporum*, and *Mucor pusillus*. Savoury oil treated with a concentration of 20 g/mL was efficient on *A. oryza*, *A. niger*, *F. oxysporum*, and *M. pusillus*. The essential oils of quinoa, sage, and rosemary have fungicidal properties against *A. oryza*, *A. niger*, and *F. oxysporum* at concentrations of 10 and 20 g/mL. Zhou et al. (2018) studied the effects of exogenous essential oil decanal on the growth and patulin production of *P. expansum*. According to the results, 0.12 g/L decanal was considered remarkably to restrict the growth of *P. expansum* in vitro, while 0.24 g/L decanal effectively prevented blue mold rot on apple and pear fruit. According to Ferreira et al. (2013), curcumin and the essential oil of *Curcuma longa* have antiaflatoxigenic properties at concentrations ranging from 0.01 to 5.0%. More than 96% of AFB1 and AFB2 production was suppressed by 0.5% of the essential oils of *C. longa* and curcumin (Table 6.2).

Tian et al. (2011) explored the dill essential oil derived from the seeds of *Anethum graveolens* to access its antifungal activity in vitro and in vivo against toxigenic strains of *Aspergillus flavus*, *A. niger*, *A. oryzae*, and *Alternaria alternata*. They observed that the mycelial growth of *A. flavus* (88.9%), *A. niger* (94.4%), *A. oryzae* (88.9%), and *A. alternata* (83.3%) was significantly reduced by the EO at a concentration of 120μL/mL. Kiran et al. (2016) explore the efficacy of *Cinnamomum zeylanicum* essential oil (CZEO) against *Aspergillus flavus* and aflatoxin B1 secretion, its functional properties, and mode of action. CZEO exhibited absolute fungitoxicity against *A. flavus*, and the MIC of the species fluctuated between 0.25 and 6.0μL/mL. Furthermore, CZEO drastically lowered the generation of aflatoxin B1 and inhibited it at 0.3μL/mL. Li et al. (2016) studied the inhibitory effects of *Litsea*

Table 6.2 Example of some essential oils (EOs) showing inhibitory effect against mycotoxigenic fungi

Source plants of EOs	Common name of the plants	Major constituents	Inhibited fungal species	Findings	References
Allium sativum	Garlic	Diallyl disulfide,2-vinyl-1,3-dithiane, diallyl trisulfide, Methyl allyl trisulfide	*Aspergillus parasiticus*, *fusarium verticillioide,and Gibberella zeae*	GEO reduces the growth of AFB1, FB1, or ZEA from stored kernels of corn	Bocate et al. (2021)
Cananga odorata	Ylang-Ylang	Linalool, germacrene-D, thymol, limonene, 1,8-cineole, α-phellandrene	*Fusarium graminearum*	2.5 mg/g of COEO limits the growth of *F. graminearum* in stored maize kernels	Kalagatur et al. (2018)
Melaleuca alternifolia	Tea tree	Terpinene-4-ol	*Penicillium griseofulvum*, *Penicillium verrucosum*	EO completely inhibits the growth of *P. verrucosum* and griseofulvin at 2.75%	Chidi et al. (2020)
Callistemon lanceolatus	Bottlebrush	1,8-cineole, γ-terpinene, α-pinene	A. *Absidia ramosa*, *aspergillus*, *fusarium*, *Alternaria*, *Penicillium*, *Dreschelera*, *Chetomium* spp.	CLEO and 1,8-cineole at 0.54 and 0.9 mg/mL, respectively, caused 100% inhibition of AFB1 production	Shukla et al. (2012)
Melaleuca rhaphiophylla	Swamp paperbark	Terpinolene, γ-terpinene, α-terpinene, 1,8-cineole, terpinen-4-ol	*A. flavus*, *A. niger*, *A. nomius*, *F. graminearum*	EO inhibits the 90% growth of selected fungal strains	Zimmermann et al. (2022)
Cymbopogon flexuosus	Lemongrass	E-citral, α-sinensal	*Aspergillus flavus*	CFEO, citral, geraniol, eugenol, α-pinene, and linlool have respective inhibition of AFB1 at 0.8, 0.4, 0.6, 0.2, and 0.6µL/mL	Kumar et al. (2009)

(continued)

Table 6.2 (continued)

Source plants of EOs	Common name of the plants	Major constituents	Inhibited fungal species	Findings	References
Xylopia aethiopica	Guinea pepper	α-Pinene, β-Pinene, β-Phellandrene, Z-γ-bisabolene	*Aspergillus Niger, fusarium oxysporium, A. flavus, A. fumigatus and A. versicolor*	Two fungal strains at 300 ppm conc. Of XAEO showed remarkable inhibition	Tegang et al. (2018)
Trachyspermumammi	Carom seeds	Thymol, P-cymene, γ-Terpinene, β-Pinene	*A. flavus, A. Niger, A. fumigatus, C. albicans, C. tropicalis, C. utilis*	TAEO (11.2 μg/mL) was significantly effective against *Candida* and *Aspergillus* spp.	Dutta et al. (2020)
Illicium verum	Star anise	Trans-anethole, estragole, D-limonene	*Aspergillus flavus*	The growth of *A. flavus* in lotus seeds is inhibited by IVEO at concentrations of 3.6 L/mL in vitro and 6.0 L/g in vivo	Li et al. (2020)
Citrus reticulata	Mandarin	Limonene	*Aspergillus Niger*	*A. niger*'s growth was prevented by the EO concentration of 30.72 mL/L	Abdel-Aziz et al. (2019)
Citrus sinensis	Sweet orange	Dl-limonene	*Aspergillus flavus*	CSEO and dl-limonene completely inhibited AFB1 production at 500 and 250 ppm, respectively	Singh et al. (2010)
Lippia alba	Bushy mat grass	Geranial, Neral	*Mycotoxigenic* fungi (*aspergillus, fusarium, Penicillium*) and *Rhizoctonia, Trichoderma* spp., etc.	LAEO, and their major components inhibited aflatoxin B1 production at 0.6 to 1.0μL/mL	Shukla et al. (2009)
Zingiber officinale	Ginger	α-Zingiberene, citral	*Fusarium verticillioides*	ZOEO exhibited the inhibitory effect at conc. Of 4000 and 2000μg/mL	Yamamoto-Ribeiro et al. (2013)

Source plants of EOs	Common name of the plants	Major constituents	Inhibited fungal species	Findings	References
Cymbopogon khasans	Jamrosa	Z-citral, linalyl acetate	*Aspergillus flavus*	CKEO suppressed AFB1 at 0.4μL/mL, whereas, linalyl acetate and Z-citral at 0.7–1.0μL/mL	Mishra et al. (2012a, b)
Pimenta dioica	Jamaica pepper	γ-Terpinene, α-terpineol, β-linalool	*Aspergillus flavus*	PDEO at varying conc. 2.5 μL/mL and 1.5 μL/mL completely retards the growth of *A. flavus*	Chaudhari et al. (2019)
Cinnamomum camphora, *Alpinia galanga*	Camphor and galangal, respectively	Fenchone, camphene, α-thujene, α-pinene	*Aspergillus flavus*	The combination of two EOs completely checked aflatoxin B1 synthesis at 250 ppm	Srivastava et al. (2008)
Mentha rotundifolia	Apple mint	Piperitenone oxide, caryophyllene oxide, cis-cinerolone	*Fusarium culmorum*	MREO effectively inhibits the growth of targeted fungi	Yakhlef et al. (2020)
Melissa officinalis	Lemon balm	P-mentha-1,2,3-triol, P-menth-3-en-8-ol, Pulegone	*Penicillium expansum*, *Botrytis cinerea, and Rhizopus stolonifer*	MOEO (10–160μL) obstructs the growth of selected fungal strains	El Ouadi et al. (2017)
Myristica fragrans	Nutmeg	Elemicin, myristicine, thujanol	*Aspergillus flavus*	MFEO observed at 2.75 mg/mL caused a 100% reduction of ergosterol content	Das et al. (2020)
Cinnamomum glaucescens	Sugandhkokila	2-Propenoic, acid1,8-cineole	*Aspergillus flavus*	CGEO inhibited the growth of fungi at 4.5 and 3.5μL/mL	Prakash et al. (2013)
Ocimum gratissimum L.	African basil	Methyl cinnamate, γ-terpinene	*Aspergillus flavus*	OGEO strongly inhibited aflatoxin production at 0.6μL/mL and 0.5μL/mL, respectively	Prakash et al. (2011)

(continued)

Table 6.2 (continued)

Source plants of EOs	Common name of the plants	Major constituents	Inhibited fungal species	Findings	References
Cinnamomum zeylanicum	Cinnamon	2-methoxy-3-(2-propenyl), caryophyllene	*Aspergillus flavus*	CZEO exhibited inhibition of all test molds at 0.25 to 0.6μL/mL and 0.3μL/mL	Kiran et al. (2016)
Anethum graveolens	Dill	Carvone, limonene, apiol	*Aspergillus flavus*, *A. niger*, *A. oryzae*, and *Alternaria alternata*	AGEO (2.0μL/mL) showed complete growth inhibition of all examined toxin-producing fungi	Tian et al. (2011)
Cinnamomum zeylanicum, *Curcuma longa*, *Zingiber officinale*, *Ocimum basilicum*, and *Cymbopogon martini*	Cinnamon, Turmeric, Ginger, Common basil, Palmarosa, respectively	Cinnamaldehyde, Ar-turmerone, Eugenol, Geranyl, proprionate, Geranyl acetate	*Aspergillus ochraceous*, *Penicillium verrucosum*	CZEO and CMEO inhibited the growth of OTA production at 1500 and 2500 mg/g, respectively	Kalagatur et al. (2020)
Rosmarinus officinalis	Rosemary	1.8 cineole, camphor, α-Pinene	*Fusarium verticillioides*	ROEO at 150μg/mL significantly reduced the growth of fumonisin	da Silva Bomfim et al. (2015)
Curcuma longa	Turmeric	Ar-turmerone	*Fusarium graminearum*	CLEO at 3500 and 3000μg/mL inhibited zearalenone	Kalagatur et al. (2018)
Caesulia axillaris	Pink node flower	Dl-limonene, euasarone	*Aspergillus flavus*	Aflatoxin B1 production was inhibited at 0.8μL/mL of EO	Mishra et al. (2012a, b)
Lantana indica	Shrub verbenas	α-Humulene, delta-3-carene, sabinene	*Aspergillus flavus* and 11 other spp.	LIEO at 0.75 mg/mL caused complete inhibition of AFB1	Kumar et al. (2010)
Carum carvi, *Coriandrum sativum*	Caraway and Coriander	Carvone, Linalool	*Aspergillus flavus*	EOs obstruct the growth of fungal biomass at 1000μg/mL	Lasram et al. (2019)

Source plants of EOs	Common name of the plants	Major constituents	Inhibited fungal species	Findings	References
Zanthoxylum alatum	Winged prickly ash	Linalool, methyl cinnamate	*Aspergillus flavus*	ZAEO at 1.25μL/mL and 2.5μL/mL protect 86.33% *Piper nigrum* L. fruits against fungi	Prakash et al. (2012)
Ageratum conyzoides	Chick weed	Precocene II, precocene, cumarine	*Aspergillus flavus*	Aflatoxin synthesis was inhibited by ACEO at 0.10 g/mL	Nogueira et al. (2010)
Boswellia carterii	Moxor	Ziranyl-acetate	*Aspergillus flavus*	BCEO protected *Piper nigrum* fruit storage against fungal damage by 65.38%	Prakash et al. (2014)
Matricaria chamomilla	Chamomile	α-Bisabolol, trans-farnesol, α-bisabolol oxide A	*Aspergillus Niger*	The maximal growth inhibition by MCEO was 92.50% at the highest conc. Of 1000 g/mL	Tolouee et al. (2010)
Thymus vulgaris, Satureja hortensis, Syzygium aromaticum	Thyme, summer savory, and clove, respectively	Thymol, carvacrol, eugenol, β-caryophyllene	*Aspergillus flavus*	EOs showed potential inhibition at 350–500 ppm in tomato paste	Omidbeygi et al. (2007)
Illicium verum	Star anise	Trans-anethole, caryophyllene, limonene	*Aspergillus flavus*, *A. parasiticus*, and *fusarium verticillioides*	At 200 ppm, IVEO inhibited all the fungi	Aly et al. (2016)
Rosmarinus officinalis, and Trachyspermumcopticum	Rosemary and carom seeds, respectively	Piperitone, α-pinene, thymol, P-cymene	*Aspergillus parasiticus*	TCEO showed more potency than ROEO against *A.parasiticus* and ROEO slightly inhibited the aflatoxin at 450 ppm	Rasooli et al. (2008)

(continued)

Table 6.2 (continued)

Source plants of EOs	Common name of the plants	Major constituents	Inhibited fungal species	Findings	References
Syzygium Aromaticum, *Rosmarinus officinalis*	Clove and rosemary, respectively	Eugenol, eucalyptol	*Aspergillus Luchuensis*	ROEO inhibited the growth of mycelia 36.6% and SAEO 100% inhibited the fungal mycelia growth at 200 μL/L	Laaziz et al. (2022)
Lippia origanoides	Brazilian lippia	Carvacrol, P-cymene, γ-terpinene	*Aspergillus flavus*, *aspergillus carbonarius*, *aspergillus ochraceus*	LOEO at different concentrations (0.24 L/mL, 0.98 L/mL, 0.98 L/mL) inhibited all fungi	Brandão et al. (2021)

cubeba essential oil (LC-EO) on *Aspergillus flavus* and aflatoxin B1 production in licorice. The minimum inhibitory concentration zone (MIC) and minimal fungicidal concentration (MFC) zone of LC-EO were 0.5 and 1.0μL/mL, respectively. The mycelia growth and aflatoxin B1 accumulation were completely inhibited when licorice was treated with 5.0μL/mL for 20 days. Gemeda et al. (2014) studied the effect of some EOs, viz., *Cymbopogon martinii, Foeniculum vulgare*, and *Trachyspermum ammi*, against toxigenic strains of *Aspergillus* species (*A. niger and A. flavus*). *T. ammi* oil showed the highest antifungal activity and could inhibit mycelial growth at 1 μL/mL. Ozcakmak et al. (2017) compared the effects of mint, garlic, sage, and wild oregano EOs on the growth of *Penicillium verrucosum* and its ochratoxin A production. Sage and mint essential oils reduced considerable toxin production, although wild oregano and garlic essential oils wholly prevented them, at quantities of 0.5 and 0.25%, accordingly. To control four corn fungi, *A. flavus* (CCC116–83 and BXC01), *P. oxalicum* (083296), and *P. minioluteum* (BXC03), which produce mycotoxin in maize, Camiletti et al. (2014) evaluated the antifungal activity of the essential oils of oregano varieties, mint varieties, and rosemary plants grown in Argentina. They noted that oregano EO was most effective against mycotoxigenic fungus, while others had neither antifungal nor toxin inhibitory activity. Pérez-Izquierdo et al. (2022) evaluated the essential oil from hydrodistillation of *Cistus ladanifer* and found a strong antifungal effect on all four species of phytopathogenic fungi (*Rhizoctonia solani, Fusarium oxysporum* sub sp., *radicis-lycopersici*, and *Cryphonectria parasitica*) and on oomycete (*Phytophthora cinnamomi*), when tested in vitro. All four species had their sporulation suppressed by EO. Marín et al. (2004) studied the effect of essential oils (cinnamon, clove, oregano, palmarosa, and lemongrass) on zearalenone and deoxynivalenol production by *Fusarium graminearum* in non-sterilized maize grain. The production of ZEA was significantly impacted by all parameters, including essential oils, temperature, a_w, treatment timing, and their interactions. In findings, they noticed that clove, lemongrass, and palmarosa oils reduced ZEA production when compared to the controls at 0.950 a_w/30 °C; however, cinnamon and oregano essential oils had no effect. Lee et al. (2008) explored the efficacy of 11 Myrtaceae essential oils *(Eucalyptus citriodora, E. smithii, E. globulus, E. radiata, E. dives, E. polybractea, Melaleuca dissitiflora, M. quinquenervia, M. uncinata, M. linariifolia, and Leptospermum petersonii)* and their components against three phytopathogenic fungi (*Phytophthora cactorum, Cryphonectria parasitica, and Fusarium circinatum*). As per the report, only three plant species have shown antifungal efficacy in in vitro fungicidal activity. Essential oils from *L. petersonii* had the strongest fungicidal effects (46.2%) in a test against *P. cactorum*, followed by essential oils from *M. quinquenervia* and *E. citriodora* (35.4 and 33.6%). Gakuubi et al. (2017) investigated the antifungal activity of *Eucalyptus camaldulensis* essential oil against maize infecting and fumonisins producing *Fusarium* spp., viz., *F. solani, F. verticillioides*, and *F. proliferatum*. After five days of incubation, the EO effectively inhibited fungal growth formation at a dosage of 7–8 L/mL in all of the tested pathogen species. Xing et al. (2014a, b) studied the degradation effect of essential oils on FB1. They found that eugenol oil, eucalyptus oil, cinnamon oil, anise oil, citral, and camphor oil were quite effective in reducing FB1. Furthermore, the optimal conditions for 94.06% FB1 reduction by EOs were 280 mg/mL on 120 h of incubation at 30 °C. The feasibility of using EOs and natural compounds used against human pathogenic fungal strains using conventional and non-traditional is highlighted by Abd Rashed et al. (2021) in a review. They explained that EOs with high monoterpene content have significant antifungal potential. Several species, including *Cymbopogon* sp., *Thymus* sp., *Lavandula* sp., and *Salvia* sp., were discovered to have excellent antifungal and toxin-inhibitory activity.

6.10 Mode of Action of Essential Oils against Mycotoxigenic Fungi and their Toxins

Aromatic oils extracted from various aromatic herbs or shrubs exhibit intense antimicrobial or antifungal properties. Essential oils are low-molecular-weight, lipophilic molecules that are liquid, lipid-soluble, and volatile (Bakkali et al. 2008; Basak and Guha 2018). Due to the lipophilic nature of essential oils, which enables them to cross the fungal cellular membrane, they may modify how permeable the membrane is to cations like H^+ and K^+. This influences the flow of protons, modifies cellular pH, and has an impact on the chemical components of cells as well as their activity. The interruption of membrane permeability leads to an imbalance in intracellular osmotic pressure, which degrades intracellular organelles including Ca^{2+} ion channels, proton pumps, and ATP (adenosine triphosphate) pools, destroying intracellular organelles, proton pumps, and diminishing membrane potential. The fluidity of plasma membranes could change, which can hamper cytochrome C pathways, affect protein metabolism, and depress calcium ion concentration, among other things. As a result, EOs may damage the protein, lipid, and nucleic acid composition of cells. Permeability of the inner and outer mitochondrial membranes may result in apoptosis or cell death. This leads to harmful morphological and ultrastructural changes that are irreversible i.e., spore germination inhibition, decreased mycelial growth, inhibition of energy release, and changes in gene expression patterns and metabolic processes (Basak and Guha 2018; Swamy et al. 2016; Bakkali et al. 2005; Mani-López et al. 2021; and Andrade-Ochoa et al. 2021). Das et al. studied the mode of action of Myristica fragrans-derived EOs against *A. flavus* and aflatoxin B1 (AFB1) contamination of stored scented rice (*Oryza sativa*) varieties in 2020. They demonstrate that MFEO reduced the ergosterol content of the fungal plasma membrane, increased cellular ion leakage, and decreased the amount of methylglyoxal, the substance that stimulates the production of aflatoxin. Similar studies by Maurya et al. (2021a, b) examine the antifungal mode of action of *Carum carvi* essential oil (CCEO) against the *Aspergillus flavus* (AF-LHP-WS-4) producing aflatoxin B1 strain. CCEO inhibited fungal growth by inducing the efflux of essential cellular ions (viz., K^+, Ca^{2+}, and Mg^{2+}) and inhibiting ergosterol and cellular methylglyoxal (MG) biosynthesis. Under light and scanning electron microscopy, Soylu et al. (2006) investigated the antifungal activities of essential oils extracted from the aerial parts of aromatic plants such as oregano

(*Origanum syriacum* var. bevanii), thyme (*Thymbra spicata* subsp. spicata), lavender (*Lavandula stoechas* subsp. stoechas), and rosemary (*Rosmarinus officinalis*). The images demonstrated significant morphological changes in the pathogen hyphae, including protoplast leakage, vacuolations, hyphal shrivelling, and cytoplasmic coagulation. These changes were found in pathogen hyphae that had been exposed to both the volatile and contact phases of oil. Ramsdam et al. (2021) investigated the ability of five medicinal aromatic plant essential oils, namely *Gaultheria fragrantissima*, *Curcuma longa*, *Zingiber officinale*, *Artemisia nilagirica*, and *Litsea cubeba*, to inhibit the growth of toxigenic *Aspergillus flavus* isolated from stored maize in Meghalaya. They evaluated that *Litsea cubeba* EO had the highest antifungal activity (0.8μL/mL) among the essential oils, and as shown by scanning electron microscopy images, *A. flavus* treated with the oils had damaged hyphae and conidiophores. Ahmad et al. (2013) assessed the in vitro synergistic antifungal activity of thymol and carvacrol with fluconazole against susceptible and resistant *Candida albicans* and also explained the synergistic antifungal effect with the azole antimycotic fluconazole by inhibiting the over-expression of efflux-pump genes CDR1 and MDR1. Both monoterpenes were highly effective at blocking drug efflux transporter pumps (70–90%) in *Candida* spp. Ahmad et al. (2011) elucidated that carvacrol and thymol have strong fungicidal effects against 11 fluconazole-sensitive and resistant *Candida* isolates. These natural isopropyl cresols appear to affect membrane integrity and impair ergosterol production as their primary mode of action (Table 6.3).

According to Rammanee and Hongpattarakere (2011), lime essential oil completely inhibited the growth and aflatoxin production of *A. flavus* at a dosage of 2.25 mg/mL, whereas kaffir and acid lime essential oils significantly reduced the aflatoxin production of *A. flavus* and *A. parasiticus*. The effects of acid lime essential oil on target cell damage were examined using transmission electron microscopy. Essential oil-treated cells had observable alterations to the plasma and nucleus membranes, including loss of cytoplasm, vacuole fusion, and separation of the fibrillar layer. Wang et al. (2018) reported that *A. ochraceus* exposed to 0.4 mmol/L of cinnamaldehyde showed harmful morphological and ultrastructural modifications that were irreversible, including folding of the cell, loss of cell wall integrity, disruption of the plasma membrane, destruction of the mitochondria, and the absence of intracellular organelles. The investigated regulatory and biosynthetic genes, such as pks, laeA, nrps, veA, and velB, were significantly downregulated in the presence of cinnamaldehyde. Singh et al. (2019) reported that chemically characterized nanoencapsulated *Ocimum sanctum* essential oil (OSEO) decreased the ergosterol content, increased the leakage of vital cellular ions, and also decreased the methylglyoxal content (an aflatoxin-inducing substrate). *Boswellia serrata* essential oil was tested in vitro and in a viable maize plant by Venkatesh et al. (2017) to determine its antifungal and anti mycotoxigenic properties. *B. serrata* essential oil completely inhibited the formation of aflatoxin B1 and fumonisin B1, and the ergosterol content fell dramatically as the concentration of essential oil increased. Bennis et al. (2004) exposed yeast cells to thymol and eugenol (phenolic major components of thyme and clove essential oils), which caused lysis as evidenced by

Table 6.3 Mode of action of essential oils against species of mycotoxigenic fungi and their toxins

Targeted fungi/toxins	Essential oils/ constituent	Mode of action/observation	Reference
Aspergillusflavus and AFB_1	*Melaleuca cajuputi*	Preventing the synthesis of AFB1 by inhibiting intracellular methylglyoxal, ergosterol synthesis, leakage of cellular components, and mitochondrial membrane damage potential	Chaudhari et al. (2022)
Aspergillus Niger	*Matricaria chamomilla*	The plasma membrane was damaged and separated from the cell wall, the cytoplasm was entirely depleted and the intracellular organelles were disorganized	Tolouee et al. (2010)
AspergillusNiger	*Thymus x-porlock, Thymus eriocalyx*	Irreversible disruption to the cellular organelles, cell membrane, and cell wall	Rasooli et al. (2006)
Fusarium solani	**Thymol and salicylic acid**	The structure and function of mitochondria were adversely affected, and the integrity and permeability of the cell membrane were degraded	Kong et al. (2021)
Monilinia fructicola	Tea tree oil, lemon oil, rosemary oil, and thyme oil	TTO altered mycelial morphology, intracellular reactive oxygen species (ROS) levels, and membrane permeability in post-harvest peaches	Xu et al. (2021)
Penicillium italicum	**Citral**	Disruption of integrity and permeability of cell membrane, alteration of extracellular pH, the release of cell constituent, and potassium ion leakage, as well as a decrease in ergosterol contents and total lipid	Tao et al. (2014a)
Penicillium italicum, Penicillium digitatum	*Citrus reticulata*	Changes in extracellular conductivity and total lipid content, the release of cell components, and cytotoxicity were caused by disruptions in cell membrane integrity and cell component leakage	Tao et al. (2014b)

Table 6.3 (continued)

Targeted fungi/toxins	Essential oils/ constituent	Mode of action/observation	Reference
Fusarium graminearum	*Humulus lupulus* (hop)	Changes in the chitin and total lipid composition of the outer cell membrane, disruption of the cytoplasmic membrane, and prevention of mycelial growth and spore germination	Jiang et al. (2023)
Aspergillus Niger	*Melaleuca alternifolia*	Down-regulation of important genes that are expressed concerning metabolic pathways like glycolysis, pentose phosphate, and phenylpropanoid was significantly induced	Kong et al. (2020)
Candida species	*Ocimum sanctum*	Disruption of membrane integrity and reduction in **ergosterol biosynthesis**	Khan et al. (2010)
Raffaeleaquercus-mongolicae, and *Rhizoctonia solani*	*Cinnamomum verum* **and** *Cymbopogon citratus*	Disruption of cell membranes and the generation of reactive oxygen species (ROS)	Lee et al. (2020)
Aspergillus flavus and AFB1	*Caesulia axillaris*	Increased the shelf life of herbal raw materials by reducing lipid peroxidation and biodeterioration	Mishra et al. (2012a, b)
Alternaria alternata	**Cinnamaldehyde**	Loss of cell membrane integrity, reduction in total lipid content, the release of intracellular components, leakage of electrolytes, and reduction in total lipid content	Xu et al. (2018)
Fusarium verticillioides	**Cinnamon oil and cinnamaldehyde**	Irreversible harmful morphological and ultrastructural changes, such as loss of integrity of the cell wall, lack of cytoplasmic contents, destroyed mitochondria, ruptured plasma membranes, and cell folding	Xing et al. (2014a, b)

(continued)

Table 6.3 (continued)

Targeted fungi/toxins	Essential oils/ constituent	Mode of action/observation	Reference
Aspergillus flavus	*Ageratum conyzoides*	Degradation in the surrounding fibrils as well as in the ultrastructure, which was more visible in the endomembrane system, such as in mitochondria	Nogueira et al. (2010)
Candida albicans, C. tropicalis, C. glabrata, Saccharomyces cerevisiae, Zygosaccharomyces parabailii, Debarymyces hansenii	*Chrysanthemum morifolium*	Cytoplasmic membrane permeability was disrupted, and mitochondrial membrane potential and DNA binding were also affected	Zhan et al. (2021)
Fusarium verticillioides	*Rosmarinus officinalis*	This causes the cell wall to be damaged and the loss of essential cellular elements, which in turn prevents the synthesis of ergosterol and fumonisins	da Silva Bomfim et al. (2015)
Candida spp.	*Laurus nobilis*	Negatively affects *C. albicans* biofilm adhesion and formation by interfering with cell wall production and membrane ionic permeability	Peixoto et al. (2017)
Aspergillus flavus	*Curcuma longa*	Aflatoxin production is suppressed by mycotoxin gene silencing, which disrupts plasma membrane integrity, and mitochondrial dysfunction	Hu et al. (2017)
Geotrichum citri-aurantii	Cinnamaldehyde	Interfering the build of the cell wall and therefore may lead to the damage of cell wall permeability and integrity	OuYang et al. (2019)
Trichophyton. Rubrum, T. tonsuran, Microsporumgypseum T. Mentagrophytes	*Foeniculum vulgare*	Damages the plasma membrane, and intracellular organelles and also inhibit the activities of a mitochondrial enzyme, such as malate dehydrogenase, succinate dehydrogenase, and ATPase	Zeng et al. (2015)

Table 6.3 (continued)

Targeted fungi/toxins	Essential oils/ constituent	Mode of action/observation	Reference
Aspergillus flavus, *aspergillus ochraceus*	Geraniol and citral	Citral showed antimycotic action mainly by inhibiting the genes involved in sporulation and proliferation. Geraniol inhibited *A. flavus* by increasing intracellular ROS generation, while it exhibited toxicity toward *A. ochraceus* via altering cell membrane permeability	Tang et al. (2018)
Aspergillus flavus and AFB_1	Cinnamaldehyde, citral, and eugenol	Gene transcription levels that are involved in the production of aflatoxin were decreased	Liang et al. (2015)
Candida albicans	Tea tree, thyme, peppermint, and clove oils	Damage to the mitochondria has lethal effects on *Candida albicans*, and also has cytotoxic and genotoxic effects	Rajkowska et al. (2016)

the release of substances that absorb at 260 nm. A scanning electron microscope analysis of the findings revealed significant surface damage in the treated cells. The findings revealed that the antifungal efficiency of both of these components involves modification of the membrane and cell wall of the yeast (Fig. 6.7).

6.11 Conclusion

The essential oils and their bioactive components have significant antifungal and anti-mycotoxigenic properties at the cellular and molecular levels against the major group of fungi responsible for food deterioration. The most commonly used essential oils have an effect on *Aspergillus* spp., *Penicillium* spp., and *Fusarium* spp., which were the most employed genera and species in the last decade for mycotoxigenic fungi, mycotoxins, and their modes of action. In terms of mode of action, various mechanisms involve extending the lagged phase, modifying the fungal growth rate, inhibiting the cell membrane and cell permeability, and disrupting the enzymatic cell system. The current reports suggest that developing aflatoxin-resistant varieties using a green transgenic strategy would be facilitated by their mode of action in AF inhibition via methylglyoxal. Despite the several significant characteristics of essential oils, it should be emphasized that their organoleptic effects can be problematic. In this sense, nanoencapsulation can be a possible approach to dealing with these issues because the process can limit their loss as well as enhance the stability of the antifungal and anti-mycotoxigenic potential of EOs and their bioactive components and may reduce their interaction with food while retaining their original organoleptic properties.

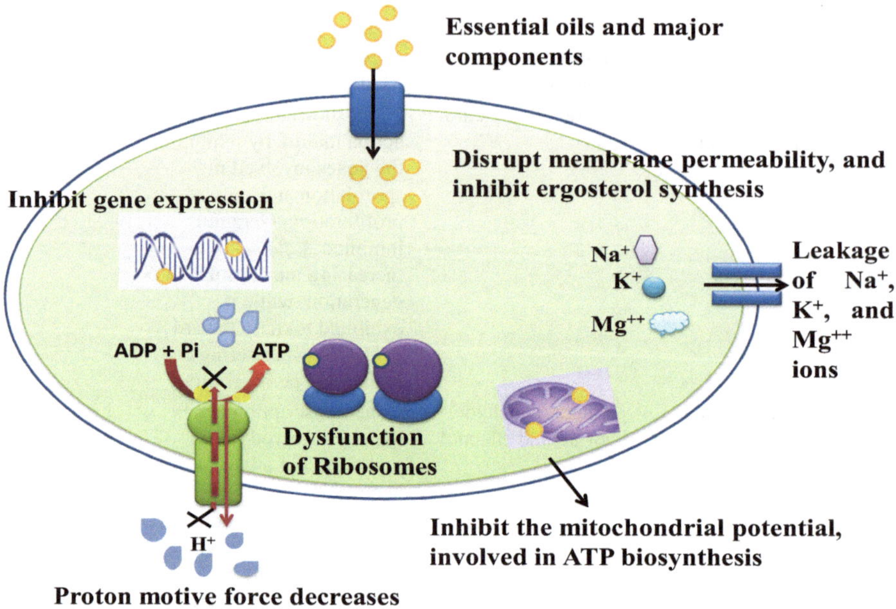

Fig. 6.7 Schematic of essential oil's principle of action against mycotoxigenic fungi

References

Alshannaq A, Yu JH (2017) Occurrence, toxicity, and analysis of major mycotoxins in food. Int J Environ Res Public Health 14(6):632

Abd Rashed A, Rathi DNG, Ahmad Nasir NAH, Abd Rahman AZ (2021) Antifungal properties of essential oils and their compounds for application in skin fungal infections: conventional and nonconventional approaches. Molecules 26(4):1093

Aly SE, Sabry BA, Shaheen MS, Hathout AS (2016) Assessment of antimycotoxigenic and antioxidant activity of star anise (Illicium verum) in vitro. J Saudi Soc Agric Sci 15(1):20–27

Ahmad A, Khan A, Manzoor N (2013) Reversal of efflux mediated antifungal resistance underlies synergistic activity of two monoterpenes with fluconazole. Eur J Pharm Sci 48(1–2):80–86

Abdel-Aziz MM, Emam TM, Elsherbiny EA (2019) Effects of mandarin (*Citrus reticulata*) peel essential oil as a natural antibiofilm agent against *aspergillus Niger* in onion bulbs. Postharvest Biol Technol 156:110959

Ahmad A, Khan A, Akhtar F, Yousuf S, Xess I, Khan LA, Manzoor N (2011) Fungicidal activity of thymol and carvacrol by disrupting ergosterol biosynthesis and membrane integrity against *Candida*. Eur J Clin Microbiol Infect Dis 30(1):41–50

Agriopoulou S (2021) Ergot alkaloids mycotoxins in cereals and cereal-derived food products: characteristics, toxicity, prevalence, and control strategies. Agronomy 11(5):931

Andrade-Ochoa S, Chacón-Vargas KF, Sánchez-Torres LE, Rivera-Chavira BE, Nogueda-Torres B, Nevárez-Moorillón GV (2021) Differential antimicrobial effect of essential oils and their main components: insights based on the cell membrane and external structure. Membranes 11(6):405

Bakkali F, Averbeck S, Averbeck D, Zhiri A, Idaomar M (2005) Cytotoxicity and gene induction by some essential oils in the yeast *Saccharomyces cerevisiae*. Mutat Res 585(1–2):1–13

Bennett JW, Klich M (2003) Mycotoxins. Clin Microbiol Rev 16(3):497–516

Burt S (2004) Essential oils: their antibacterial properties and potential applications in foods—a review. Int J Food Microbiol 94(3):223–253

Bui-Klimke TR, Wu F (2015) Ochratoxin A and human health risk: a review of the evidence. Crit Rev Food Sci Nutr 55(13):1860–1869

Bhat R, Rai RV, Karim AA (2010) Mycotoxins in food and feed: present status and future concerns. Compr Rev Food Sci Food Saf 9(1):57–81

Bakkali F, Averbeck S, Averbeck D, Idaomar M (2008) Biological effects of essential oils–a review. Food Chem Toxicol 46(2):446–475

Baert K, Devlieghere F, Flyps H, Oosterlinck M, Ahmed MM, Rajković A, De Meulenaer B et al (2007) Influence of storage conditions of apples on growth and patulin production by *Penicillium expansum*. Int J Food Microbiol 119(3):170–181

Bennis S, Chami F, Chami N, Bouchikhi T, Remmal A (2004) Surface alteration of *Saccharomyces cerevisiae* induced by thymol and eugenol. Lett Appl Microbiol 38(6):454–458

Bouyahya A, Abrini J, Dakka N, Bakri Y (2019) Essential oils of Origanum compactum increase membrane permeability, disturb cell membrane integrity, and suppress quorum-sensing phenotype in bacteria. J Pharm Anal 9(5):301–311

Bilia AR, Guccione C, Isacchi B, Righeschi C, Firenzuoli F, Bergonzi MC (2014) Essential oils loaded in nanosystems: a developing strategy for a successful therapeutic approach. Evid Based Complement Alternat Med 2014:2014

Baranyi N (2013) Current trends in aflatoxin research. Acta Biologica Szegediensis 57(2):95–107

Basak S, Guha P (2018) A review on antifungal activity and mode of action of essential oils and their delivery as nano-sized oil droplets in food system. J Food Sci Technol 55(12):4701–4710

Bocate KP, Evangelista AG, Luciano FB (2021) Garlic essential oil as an antifungal and anti-mycotoxin agent in stored corn. LWT 147:111600

Brandão RM, Ferreira VRF, Batista LR, Alves E, Santiago WD, Barbosa RB, Cardoso MDG et al (2021) Antifungal activity and the effect of the essential oil of *Lippia origanoides* Kunth on *aspergillus* mycotoxins production. Aust J Crop Sci 15:1005–1012

Chidi F, Bouhoudan A, Khaddor M (2020) Antifungal effect of the tea tree essential oil (*Melaleuca alternifolia*) against *Penicillium griseofulvum* and *Penicillium verrucosum*. J King Saud Univ-Sci 32(3):2041–2045

Chamorro ER, Zamb6n SN, Morales WG, Sequeira AF, Velasco GA (2012) Study of the chemical composition of essential oils by gas chromatography. In: Gas chromatography in plant science, wine technology, toxicology and some specific applications, vol 1. IntechOpen, pp 307–324

Chaudhari AK, Singh VK, Das S, Kujur A, Dubey NK (2022) Unveiling the cellular and molecular mode of action of *Melaleuca cajuputi* Powell. Essential oil against aflatoxigenic strains of *aspergillus flavus* isolated from stored maize samples. Food Control 138:109000

Chouhan S, Sharma K, Guleria S (2017) Antimicrobial activity of some essential oils—present status and future perspectives. Medicines 4(3):58

Chen J, Liu Y, Ma G, Yang F, Zhan Z, Guan L, Jin L et al (2022) Piperonyl butoxide synergizes the larvicidal activity of *Origanum vulgare* essential oil and its major constituents against the larvae of *Aedes albopictus* and *Culex pipiens quinquefasciatus*. J Asia Pac Entomol 102025

Chaudhari AK, Dwivedy AK, Singh VK, Das S, Singh A, Dubey NK (2019) Essential oils and their bioactive compounds as green preservatives against fungal and mycotoxin contamination of food commodities with special reference to their nanoencapsulation. Environ Sci Pollut Res 26(25):25414–25431

Chaudhari AK, Singh VK, Das S, Dubey NK (2021) Nanoencapsulation of essential oils and their bioactive constituents: A novel strategy to control mycotoxin contamination in food system. Food Chem Toxicol 149:112019

Camiletti BX, Asensio CM, Pecci MDLPG, Lucini EI (2014) Natural control of corn postharvest fungi *aspergillus flavus* and *Penicillium* sp. using essential oils from plants grown in Argentina. J Food Sci 79(12):M2499–M2506

Dhakal A, Sbar E (2021) Aflatoxin toxicity. In: StatPearls [Internet]. Stat Pearls Publishing

Das S, Kumar Singh V, Kumar Dwivedy A, Kumar Chaudhari A, Upadhyay N, Singh A, Dubey NK et al (2020) Assessment of chemically characterised *Myristica fragrans* essential oil against fungi contaminating stored scented rice and its mode of action as novel aflatoxin inhibitor. Nat Prod Res 34(11):1611–1615

da Silva Bomfim N, Nakassugi LP, Oliveira JFP, Kohiyama CY, Mossini SAG, Grespan R, Machinski M Jr et al (2015) Antifungal activity and inhibition of fumonisin production by *Rosmarinus officinalis* L. essential oil in *fusarium verticillioides* (Sacc.) Nirenberg. Food Chem 166:330–336

Dutta S, Kundu A, Saha S, Prabhakaran P, Mandal A (2020) Characterization, antifungal properties and in silico modelling perspectives of *Trachyspermum ammi* essential oil. Lwt 131:109786

da Rocha MEB, Freire FDCO, Maia FEF, Guedes MIF, Rondina D (2014) Mycotoxins and their effects on human and animal health. Food Control 36(1):159–165

European Food Safety Authority, Arcella D, Gómez Ruiz JÁ, Innocenti ML, Roldán R (2017) Human and animal dietary exposure to ergot alkaloids. EFSA J 15(7):e04902

El Ouadi Y, Manssouri M, Bouyanzer A, Majidi L, Bendaif H, Elmsellem H, Hammouti B et al (2017) Essential oil composition and antifungal activity of *Melissa officinalis* originating from North-Est Morocco, against postharvest phytopathogenic fungi in apples. Microb Pathog 107:321–326

Escrivá L, Oueslati S, Font G, Manyes L (2017) Alternaria mycotoxins in food and feed: an overview. J Food Qual 2017:1

Ferdes M, Al Juhaimi F, Özcan MM, Ghafoor K (2017) Inhibitory effect of some plant essential oils on growth of *aspergillus Niger, aspergillus oryzae, Mucor pusillus* and *fusarium oxysporum*. S Afr J Bot 113:457–460

Ferreira FD, Kemmelmeier C, Arrotéia CC, da Costa CL, Mallmann CA, Janeiro V, Machinski M Jr et al (2013) Inhibitory effect of the essential oil of *Curcuma longa* L. and curcumin on aflatoxin production by *aspergillus flavus* link. Food Chem 136(2):789–793

Guenther E, Althausen D (1948) The essential oils, vol 1. Van Nostrand, New York, p 81

Gurikar C, Shivaprasad DP, Sabillón L, Gowda NN, Siliveru K (2022) Impact of mycotoxins and their metabolites associated with food grains, vol 6. Grain & Oil Science and Technology, p 1

Gakuubi MM, Maina AW, Wagacha JM (2017) Antifungal activity of essential oil of *Eucalyptus camaldulensis* dehnh. Against selected *fusarium* spp. Int J Microbiol 2017

Gemeda N, Woldeamanuel Y, Asrat D, Debella A (2014) Effect of *Cymbopogon martinii, Foeniculum vulgare*, and *Trachyspermum ammi* essential oils on the growth and mycotoxins production by *aspergillus* species. Int J Food Sci 2014

Hemalata VB, Virupakshaiah DBM (2016) Isolation and identification of food borne pathogens from spoiled food samples. Int J Curr Microbiol App Sci 5(6):1017–1025

Havelaar AH, Kirk MD, Torgerson PR, Gibb HJ, Hald T, Lake RJ, World Health Organization Foodborne Disease Burden Epidemiology Reference Group (2015) World health organization global estimates and regional comparisons of the burden of foodborne disease in 2010. PLoS Med 12(12):e1001923

Hu Y, Zhang J, Kong W, Zhao G, Yang M (2017) Mechanisms of antifungal and anti-aflatoxigenic properties of essential oil derived from turmeric (*Curcuma longa* L.) on *aspergillus flavus*. Food Chem 220:1–8

Ji F, He D, Olaniran AO, Mokoena MP, Xu J, Shi J (2019) Occurrence, toxicity, production and detection of *fusarium* mycotoxin: a review. Food Production, Processing and Nutrition 1(1):1–14

Jiang H, Zhong S, Schwarz P, Chen B, Rao J (2023) Antifungal activity, mycotoxin inhibitory efficacy, and mode of action of hop essential oil nanoemulsion against *fusarium graminearum*. Food Chem 400:134016

Kőszegi T, Poór M (2016) Ochratoxin A: molecular interactions, mechanisms of toxicity and prevention at the molecular level. Toxins 8(4):111

Kumar P, Mahato DK, Sharma B, Borah R, Haque S, Mahmud MC et al (2020) Ochratoxins in food and feed: occurrence and its impact on human health and management strategies. Toxicon 187:151–162

Kong Q, An P, Xu Z, Zhang R, Qi J, Ren X (2020) New insights into the alleviating role of *Melaleuca alternifolia* oil on metabolites pathway disorder of grapes caused by *aspergillus Niger*, verified by corresponding key genes expression. Food Chem 327:127083

Khan A, Ahmad A, Akhtar F, Yousuf S, Xess I, Khan LA, Manzoor N (2010) Ocimum sanctum essential oil and its active principles exert their antifungal activity by disrupting ergosterol biosynthesis and membrane integrity. Res Microbiol 161(10):816–823

Kamle M, Mahato DK, Gupta A, Pandhi S, Sharma N, Sharma B et al (2022) Citrinin mycotoxin contamination in food and feed: impact on agriculture, human health, and detection and management strategies. Toxins 14(2):85

Kumar P, Mahato DK, Kamle M, Mohanta TK, Kang SG (2017) Aflatoxins: A global concern for food safety, human health and their management. Front Microbiol 7:2170

Kamle M, Mahato DK, Devi S, Lee KE, Kang SG, Kumar P (2019) Fumonisins: impact on agriculture, food, and human health and their management strategies. Toxins 11(6):328

Kumari PK, Akhila S, Rao YS, Devi BR (2019) Alternative to artificial preservatives. Syst Rev Pharm 10:99–102

Krska R, Stubbings G, Macarthur R, Crews C (2008) Simultaneous determination of six major ergot alkaloids and their epimers in cereals and foodstuffs by LC–MS–MS. Anal Bioanal Chem 391(2):563–576

Kong J, Xie Y, Yu H, Guo Y, Cheng Y, Qian H, Yao W (2021) Synergistic antifungal mechanism of thymol and salicylic acid on *fusarium solani*. LWT 140:110787

Kalagatur NK, Mudili V, Kamasani JR, Siddaiah C (2018) Discrete and combined effects of Ylang-Ylang (*Cananga odorata*) essential oil and gamma irradiation on growth and mycotoxins production by *fusarium graminearum* in maize. Food Control 94:276–283

Kiran S, Kujur A, Prakash B (2016) Assessment of preservative potential of *Cinnamomum zeylanicum* Blume essential oil against food borne molds, aflatoxin B1 synthesis, its functional properties and mode of action. Innovative Food Sci Emerg Technol 37:184–191

Kalagatur NK, Gurunathan S, Kamasani JR, Gunti L, Kadirvelu K, Mohan CD, Siddaiah C et al (2020) Inhibitory effect of *C. zeylanicum*, *C. longa*, *O. basilicum*, *Z. officinale*, and *C. martini* essential oils on growth and ochratoxin A content of *A. ochraceous* and *P. verrucosum* in maize grains. Biotechnology Reports 27:e00490

Kuiper-Goodman T, Scott P, Watanabe H (1987) Risk assessment of the mycotoxin zearalenone. Regul Toxicol Pharmacol 7(3):253–306

Kumar A, Shukla R, Singh P, Dubey NK (2009) Biodeterioration of some herbal raw materials by storage fungi and aflatoxin and assessment of *Cymbopogon flexuosus* essential oil and its components as antifungal. Int Biodeterior Biodegradation 63(6):712–716

Kumar A, Shukla R, Singh P, Dubey NK (2010) Efficacy of extract and essential oil of *Lantana indica* Roxb. Against food contaminating moulds and aflatoxin B1 production. Int J Food Sci Technol 45(1):179–185

Kumar P, Mahato DK, Gupta A, Pandey S, Paul V, Saurabh V, Agriopoulou S et al (2022) Nivalenol mycotoxin concerns in foods: An overview on occurrence, impact on human and animal health and its detection and management strategies. Toxins 14(8):527

Lasram S, Zemni H, Hamdi Z, Chenenaoui S, Houissa H, Tounsi MS, Ghorbel A (2019) Antifungal and antiaflatoxinogenic activities of *Carum carvi* L., *Coriandrum sativum* L. seed essential oils and their major terpene component against *aspergillus flavus*. Ind Crop Prod 134:11–18

Liew WPP, Mohd-Redzwan S (2018) Mycotoxin: its impact on gut health and microbiota. Front Cell Infect Microbiol 60

Li F, Jiang D, Zheng F, Chen J, Li W (2015) Fumonisins B1, B2 and B3 in corn products, wheat flour and corn oil marketed in Shandong province of China. Food Additives & Contaminants: Part B 8(3):169–174

Laaziz A, El Hammoudi Y, Qjidaa S, Hajjaj H, Haesaert G, Bouseta A (2022) Activity of essential oils from *Syzygium aromaticum* and *Rosmarinus officinalis* against growth and ochratoxin A production by aspergillus tubingensis and *aspergillus luchuensis* from Moroccan grapes. Phytopathol Mediterr 16:299–310

Li Y, Wang Y, Kong W, Yang S, Luo J, Yang M (2020) *Illicium verum* essential oil, a potential natural fumigant in preservation of lotus seeds from fungal contamination. Food Chem Toxicol 141:111347

Liang D, Xing F, Selvaraj JN, Liu X, Wang L, Hua H et al (2015) Inhibitory effect of cinnamaldehyde, citral, and eugenol on aflatoxin biosynthetic gene expression and aflatoxin B1 biosynthesis in *aspergillus flavus*. J Food Sci 80(12):M2917–M2924

Lee YS, Kim J, Shin SC, Lee SG, Park IK (2008) Antifungal activity of Myrtaceae essential oils and their components against three phytopathogenic fungi. Flavour Fragr J 23(1):23–28

Li Y, Kong W, Li M, Liu H, Zhao X, Yang S, Yang M (2016) *Litsea cubeba* essential oil as the potential natural fumigant: inhibition of *aspergillus flavus* and AFB1 production in licorice. Ind Crop Prod 80:186–193

Lee JE, Seo SM, Huh MJ, Lee SC, Park IK (2020) Reactive oxygen species mediated-antifungal activity of cinnamon bark (*Cinnamomum verum*) and lemongrass (*Cymbopogon citratus*) essential oils and their constituents against two phytopathogenic fungi. Pestic Biochem Physiol 168:104644

Mishra PK, Shukla R, Singh P, Prakash B, Dubey NK (2012a) Antifungal and antiaflatoxigenic efficacy of *Caesulia axillaris* Roxb. Essential oil against fungi deteriorating some herbal raw materials, and its antioxidant activity. Ind Crop Prod 36(1):74–80

Mishra PK, Shukla R, Singh P, Prakash B, Kedia A, Dubey NK (2012b) Antifungal, anti-aflatoxigenic, and antioxidant efficacy of Jamrosa essential oil for preservation of herbal raw materials. Int Biodeterior Biodegradation 74:11–16

Masyita A, Sari RM, Astuti AD, Yasir B, Rumata NR, Emran TB, Simal-Gandara J et al (2022) Terpenes and terpenoids as main bioactive compounds of essential oils, their roles in human health and potential application as natural food preservatives. Food Chem X:100217

Maurya A, Kumar S, Singh BK, Chaudhari AK, Dwivedy AK, Prakash B, Dubey NK (2021a) Mechanistic investigations on antifungal and antiaflatoxigenic activities of chemically characterised *Carum carvi* L. essential oil against fungal infestation and aflatoxin contamination of herbal raw materials. Nat Prod Res 1-6

Marín S, Velluti A, Ramos AJ, Marín S, Sanchis V (2004) Effect of essential oils on zearalenone and deoxynivalenol production by *fusarium graminearum* in non-sterilized maize grain. Food Microbiol 21(3):313–318

Malir F, Ostry V, Pfohl-Leszkowicz A, Malir J, Toman J (2016) Ochratoxin A: 50 years of research. Toxins 8(7):191

Maurya A, Prasad J, Das S, Dwivedy AK (2021b) Essential oils and their application in food safety. Frontiers in Sustainable Food Systems 5:653420

Mirza Alizadeh A, Golzan SA, Mahdavi A, Dakhili S, Torki Z, Hosseini H (2022) Recent advances on the efficacy of essential oils on mycotoxin secretion and their mode of action. Crit Rev Food Sci Nutr 62(17):4726–4751

Mahato DK, Kamle M, Sharma B, Pandhi S, Devi S, Dhawan K, Kumar P et al (2021) Patulin in food: A mycotoxin concern for human health and its management strategies. Toxicon 198:12–23

Mulac D, Humpf HU (2011) Cytotoxicity and accumulation of ergot alkaloids in human primary cells. Toxicology 282(3):112–121

Mani-López E, Cortés-Zavaleta O, López-Malo A (2021) A review of the methods used to determine the target site or the mechanism of action of essential oils and their components against fungi. SN Applied Sciences 3(1):1–25

Nair MG (1998) Fumonisins and human health. Ann Trop Paediatr 18(sup1):S47–S52

Neme K, Mohammed A (2017) Mycotoxin occurrence in grains and the role of postharvest management as a mitigation strategy. A review. Food Control 78:412–425

Nogueira JH, Gonçalez E, Galleti SR, Facanali R, Marques MO, Felício JD (2010) *Ageratum conyzoides* essential oil as aflatoxin suppressor of aspergillus flavus. Int J Food Microbiol 137(1):55–60

Ozcakmak S, Gul O, Dervisoglu M, Yilmaz A, Sagdic O, Arici M (2017) Comparison of the effect of some essential oils on the growth of *Penicillium verrucosum* and its Ochratoxin production. J Food Process Preserv 41(1):e13006

OuYang Q, Duan X, Li L, Tao N (2019) Cinnamaldehyde exerts its antifungal activity by disrupting the cell wall integrity of *Geotrichumcitri-aurantii*. Front Microbiol 10:55

Omidbeygi M, Barzegar M, Hamidi Z, Naghdibadi H (2007) Antifungal activity of thyme, summer savory and clove essential oils against *aspergillus flavus* in liquid medium and tomato paste. Food Control 18(12):1518–1523

Peixoto LR, Rosalen PL, Ferreira GLS, Freires IA, de Carvalho FG, Castellano LR, de Castro RD (2017) Antifungal activity, mode of action and anti-biofilm effects of *Laurus nobilis* Linnaeus essential oil against Candida spp. Arch Oral Biol 73:179–185

Prakash B, Mishra PK, Kedia A, Dubey NK (2014) Antifungal, antiaflatoxin and antioxidant potential of chemically characterized *Boswellia carterii* Birdw essential oil and its in vivo practical applicability in preservation of *Piper nigrum* L. fruits. LWT-Food Sci Technol 56(2):240–247

Prakash B, Kedia A, Mishra PK, Dubey NK (2015) Plant essential oils as food preservatives to control moulds, mycotoxin contamination and oxidative deterioration of Agri-food commodities–potentials and challenges. Food Control 47:381–391

Prakash B, Singh P, Mishra PK, Dubey NK (2012) Safety assessment of *Zanthoxylum alatum* Roxb. Essential oil, its antifungal, antiaflatoxin, antioxidant activity and efficacy as antimicrobial in preservation of Piper nigrum L. fruits. Int J Food Microbiol 153(1–2):183–191

Pérez-Izquierdo C, Serrano-Pérez P, del Carmen Rodríguez-Molina M (2022) Chemical composition, antifungal and phytotoxic activities of *Cistus ladanifer* L. essential oil and hydrolate. *Biocatalysis and agricultural.* Biotechnology 45:102527

Prakash B, Shukla R, Singh P, Kumar A, Mishra PK, Dubey NK (2010) Efficacy of chemically characterized *Piper betle* L. essential oil against fungal and aflatoxin contamination of some edible commodities and its antioxidant activity. Int J Food Microbiol 142(1–2):114–119

Polak-Śliwińska M, Paszczyk B (2021) Trichothecenes in food and feed, relevance to human and animal health and methods of detection: A systematic review. Molecules 26(2):454

Pal S, Singh N, Ansari KM (2017) Toxicological effects of patulin mycotoxin on the mammalian system: an overview. Toxicology Res 6(6):764–771

Prakash B, Shukla R, Singh P, Mishra PK, Dubey NK, Kharwar RN (2011) Efficacy of chemically characterized Ocimum gratissimum L. essential oil as an antioxidant and a safe plant based antimicrobial against fungal and aflatoxin B1 contamination of spices. Food Res Int 44(1):385–390

Prakash B, Singh P, Yadav S, Singh SC, Dubey NK (2013) Safety profile assessment and efficacy of chemically characterized *Cinnamomum glaucescens* essential oil against storage fungi, insect, aflatoxin secretion and as antioxidant. Food Chem Toxicol 53:160–167

Pleadin J, Frece J, Markov K (2019) Mycotoxins in food and feed. Adv Food Nutr Res 89:297–345

Reddy L, Bhoola K (2010) Ochratoxins—food contaminants: impact on human health. Toxins 2(4):771–779

Rogowska A, Pomastowski P, Sagandykova G, Buszewski B (2019) Zearalenone and its metabolites: effect on human health, metabolism and neutralisation methods. Toxicon 162:46–56

Roh HS, Lim EG, Kim J, Park CG (2011) Acaricidal and oviposition deterring effects of santalol identified in sandalwood oil against two-spotted spider mite, *Tetranychus urticae* Koch (Acari: Tetranychidae). J Pest Sci 84(4):495–501

Ropejko K, Twarużek M (2021) Zearalenone and its metabolites—general overview, occurrence, and toxicity. Toxins 13(1):35

Rehman R, Hanif MA, Mushtaq Z, Al-Sadi AM (2016) Biosynthesis of essential oils in aromatic plants: A review. Food Rev Intl 32(2):117–160

Rasooli I, Fakoor MH, Yadegarinia D, Gachkar L, Allameh A, Rezaei MB (2008) Antimycotoxigenic characteristics of *Rosmarinus officinalis* and *Trachyspermum copticum* L. essential oils. Int J Food Microbiol 122(1–2):135–139

Rasooli I, Rezaei MB, Allameh A (2006) Growth inhibition and morphological alterations of aspergillus Niger by essential oils from *Thymus eriocalyx* and *thymus x-porlock*. Food Control 17(5):359–364

Ramsdam MG, Chaudhari AK, Singh VK, Dkhar MS, Kayang H, Dubey NK, Prakash B (2021) Antifungal and antioxidant activity of plant based essential oils on *aspergillus flavus* link (Trichocomaceae) isolated from stored maize grains of Meghalaya. Arch Phytopathol Plant Protect 54(17–18):1405–1420

Rajkowska K, Nowak A, Kunicka-Styczyńska A, Siadura A (2016) Biological effects of various chemically characterized essential oils: investigation of the mode of action against *Candida albicans* and HeLa cells. RSC Adv 6(99):97199–97207

Rammanee K, Hongpattarakere T (2011) Effects of tropical citrus essential oils on growth, aflatoxin production, and ultrastructure alterations of *aspergillus flavus* and *aspergillus parasiticus*. Food Bioprocess Technol 4(6):1050–1059

Rocha AR, Cardoso MS, Júnior JAS, Júnior EAG, Maciel LF, Menezes-Filho JA (2023) Occurrence of aflatoxins B1, B2, G1, and G2 in beers produced in Brazil and their carcinogenic risk evaluation. Food Control 145:109348

Rheeder JP, Marasas WF, Vismer HF (2002) Production of fumonisin analogs by fusarium species. Appl Environ Microbiol 68(5):2101–2105

Shukla R, Kumar A, Prasad CS, Srivastava B, Dubey NK (2008) Antimycotic and antiaflatoxigenic potency of *Adenocalymma alliaceum* Miers. On fungi causing biodeterioration of food commodities and raw herbal drugs. Int Biodeterior Biodegradation 62:348–351

Shukla R, Kumar A, Singh P, Dubey NK (2009) Efficacy of *Lippia alba* essential oil and its monoterpene aldehyde constituents against fungi isolated from some edible legume seeds and aflatoxin B1 production. Int J Food Microbiol 135:165–170

Shukla R, Singh P, Prakash B, Dubey NK (2013) Efficacy of *Acorus calamus* L. essential oil as a safe plant-based antioxidant, aflatoxin B 1 suppressor and broad spectrum antimicrobial against food-infesting fungi. Int J Food Sci Technol 48:128–135

Shukla R, Singh P, Prakash B, Kumar A, Dubey NK (2012) Antifungal, aflatoxin inhibition and antioxidant activity of *Callistemon lanceolatus* (Sm.) sweet essential oil and its major component 1,8-cineole against fungal isolates from chickpea seeds. Food Control 25:27–33

Shukla R, Singh P, Prakash B, Kumar A, Mishra PK, Dubey NK (2011) Efficacy of essential oils of *Lippia alba* (mill.) N. E. Brown and *Callistemon lanceolatus* (Sm.) sweet and their major constituents on mortality, oviposition and feeding behaviour of pulse beetle, *Callosobruchus chinensis* L. J Sci Food Agric 91:2277–2283

Sobrova P, Adam V, Vasatkova A, Beklova M, Zeman L, Kizek R (2010) Deoxynivalenol and its toxicity. Interdiscip Toxicol 3(3):94

Soriano JM, Gonzalez L, Catala AI (2005) Mechanism of action of sphingolipids and their metabolites in the toxicity of fumonisin B1. Prog Lipid Res 44(6):345–356

Swamy MK, Akhtar MS, Sinniah UR (2016) Antimicrobial properties of plant essential oils against human pathogens and their mode of action: an updated review. Evid Based Complement Alternat Med 2016:2016

Sharif ZIM, Mustapha FA, Jai J, Yusof NM, Zaki NAM (2017) Review on methods for preservation and natural preservatives for extending the food longevity. Chem Eng Res Bull 19:145–153

Sajid M, Mehmood S, Yuan Y, Yue T (2019) Mycotoxin patulin in food matrices: occurrence and its biological degradation strategies. Drug Metab Rev 51(1):105–120

Singh P, Shukla R, Kumar A, Prakash B, Dubey NK (2010) Chemical profile, antifungal, antiaflatoxigenic and antioxidant activity of *Citrus maxima* Burm. And *Citrus sinensis* (L.) Osbeck essential oils and their cyclic monoterpene, *DL*-limonene. Food Chem Toxicol 48:1734–1740

Singh VK, Das S, Dwivedy AK, Rathore R, Dubey NK (2019) Assessment of chemically characterized nanoencapuslated *Ocimum sanctum* essential oil against aflatoxigenic fungi contaminating herbal raw materials and its novel mode of action as methyglyoxal inhibitor. Postharvest Biol Technol 153:87–95

Shaaban HA (2020) Essential oil as antimicrobial agents: efficacy, stability, and safety issues for food application. In: Essential oils-bioactive compounds, new perspectives and applications, pp 1–33

Soylu E, Soylu S, Kurt S (2006) Antimicrobial activities of the essential oils of various plants against tomato late blight disease agent *Phytophthora infestans*. Mycopathologia 161(2):119–128

Srivastava B, Singh P, Shukla R, Dubey NK (2008) A novel combination of the essential oils of *Cinnamomum camphora* and *Alpinia galanga* in checking aflatoxin B 1 production by a toxigenic strain of aspergillus flavus. World J Microbiol Biotechnol 24:693–697

Tolouee M, Alinezhad S, Saberi R, Eslamifar A, Zad SJ, Jaimand K, Razzaghi-Abyaneh M et al (2010) Effect of *Matricaria chamomilla* L. flower essential oil on the growth and ultrastructure of *aspergillus Niger* van Tieghem. Int J Food Microbiol 139(3):127–133

Turner NW, Subrahmanyam S, Piletsky SA (2009) Analytical methods for determination of mycotoxins: a review. Anal Chim Acta 632(2):168–180

Tegang AS, Beumo TMN, Dongmo PMJ, Ngoune LT (2018) Essential oil of *Xylopia aethiopica* from Cameroon: chemical composition, antiradical and in vitro antifungal activity against some mycotoxigenic fungi. J King Saud Univ-Sci 30(4):466–471

Tian J, Ban X, Zeng H, Huang B, He J, Wang Y (2011) In vitro and in vivo activity of essential oil from dill (*Anethum graveolens* L.) against fungal spoilage of cherry tomatoes. Food Control 22(12):1992–1999

Tang X, Shao YL, Tang YJ, Zhou WW (2018) Antifungal activity of essential oil compounds (geraniol and citral) and inhibitory mechanisms on grain pathogens (*aspergillus flavus* and *aspergillus ochraceus*). Molecules 23(9):2108

Tao N, OuYang Q, Jia L (2014a) Citral inhibits mycelial growth of *Penicillium italicum* by a membrane damage mechanism. Food Control 41:116–121

Tao N, Jia L, Zhou H (2014b) Anti-fungal activity of *Citrus reticulata* Blanco essential oil against *Penicillium italicum* and *Penicillium digitatum*. Food Chem 153:265–271

Tola M, Kebede B (2016) Occurrence, importance and control of mycotoxins: a review. Cogent Food & Agriculture 2(1):1191103

Ünüsan N (2019) Systematic review of mycotoxins in food and feeds in Turkey. Food Control 97:1–14

Venkatesh HN, Sudharshana TN, Abhishek RU, Thippeswamy S, Manjunath K, Mohana DC (2017) Antifungal and antimycotoxigenic properties of chemically characterised essential oil of *Boswellia serrata* Roxb. Ex Colebr. Int J Food Prop 20(sup2):1856–1868

WHO (2022). https://www.who.int/news-room/fact-sheets/detail/food-safety, Accessed on 18 february, 2023

World Health Organization, International Agency for Research on Cancer (1993) Some naturally occurring substances: food items and constituents, heterocyclic aromatic amines and mycotoxins. IARC Monogr Eval Carcinog Risk Chem Hum 56

Wang L, Jin J, Liu X, Wang Y, Liu Y, Zhao Y, Xing F (2018) Effect of cinnamaldehyde on morphological alterations of *aspergillus ochraceus* and expression of key genes involved in ochratoxin A biosynthesis. Toxins 10(9):340

Xing F, Hua H, Selvaraj JN, Yuan Y, Zhao Y, Zhou L, Liu Y (2014a) Degradation of fumonisin B1 by cinnamon essential oil. Food Control 38:37–40

Xu L, Tao N, Yang W, Jing G (2018) Cinnamaldehyde damaged the cell membrane of *Alternaria alternata* and induced the degradation of mycotoxins in vivo. Ind Crop Prod 112:427–433

Xing F, Hua H, Selvaraj JN, Zhao Y, Zhou L, Liu X, Liu Y (2014b) Growth inhibition and morphological alterations of *fusarium verticillioides* by cinnamon oil and cinnamaldehyde. Food Control 46:343–350

Xu Y, Wei J, Wei Y, Han P, Dai K, Zou X, Shao X et al (2021) Tea tree oil controls brown rot in peaches by damaging the cell membrane of *Monilinia fructicola*. Postharvest Biol Technol 175:111474

Yamamoto-Ribeiro MMG, Grespan R, Kohiyama CY, Ferreira FD, Mossini SAG, Silva EL, Junior MM et al (2013) Effect of *Zingiber officinale* essential oil on *fusarium verticillioides* and fumonisin production. Food Chem 141(3):3147–3152

Yakhlef G, Hambaba L, Pinto DC, Silva AM (2020) Chemical composition and insecticidal, repellent and antifungal activities of essential oil of *Mentha rotundifolia* (L.) from Algeria. Ind Crop Prod 158:112988

Zhou T, Wang X, Ye B, Shi L, Bai X, Lai T (2018) Effects of essential oil decanal on growth and transcriptome of the postharvest fungal pathogen *Penicillium expansum*. Postharvest Biol Technol 145:203–212

Zimmermann RC, Poitevin CG, Bischoff AM, Beger M, da Luz TS, Mazarotto EJ, Zawadneak MA et al (2022) Insecticidal and antifungal activities of *melaleuca rhaphiophylla* essential oil against insects and seed-borne pathogens in stored products. Ind Crop Prod 182:114871

Zeng H, Chen X, Liang J (2015) In vitro antifungal activity and mechanism of essential oil from fennel (*Foeniculum vulgare* L.) on dermatophyte species. J Med Microbiol 64(1):93–103

Zhan J, He F, Cai H, Wu M, Xiao Y, Xiang F, Li S et al (2021) Composition and antifungal mechanism of essential oil from *Chrysanthemum morifolium* cv. Fubaiju. J Funct Foods 87:104746

Zhang GL, Feng YL, Song JL, Zhou XS (2018) Zearalenone: a mycotoxin with different toxic effect in domestic and laboratory animals' granulosa cells. Front Genet 9:667

Zinedine A, Fernández-Franzón M, Mañes J, Manyes L (2017) Multi-mycotoxin contamination of couscous semolina commercialized in Morocco. Food Chem 214:440–446

Essential Oils against the Bio-Deteriorating Insect Pests of Stored Food Commodities

7

Akash Kedia

Abstract

Plant-based essential oils (EOs) are emerging as promising alternatives to chemical insecticides worldwide against stored food insect pests. EOs are plant secondary metabolites containing a mixture of various aromatic and aliphatic compounds that play crucial role in plant defense and signaling processes. Coleoptera followed by Lepidoptera are the most destructing insect orders causing huge loss to stored food items worldwide. A plethora of research and review articles have been published reporting the efficacy of different EOs against them in various methods like contact toxicity, fumigation toxicity, repellent activity, oviposition deterrent activity, ovicidal activity, larvicidal activity, pupaecidal activity and antifeedant activity. However, most of the studies focused only on one or two insect pests or one or two methods. The present chapter aims to furnish short information on various important stored product insect pests as well as efficacy of different EOs against them in various ways as analyzed in different research studies. Further, future research studies concerning the use of EOs as insecticides should focus on detailed investigations in the real food system and their safety profile, an area that needs more research input.

Keywords

Essential oil · stored food · Coleoptera · Lepidoptera · Insecticide

A. Kedia (✉)
Department of Botany, Govt. General Degree College Mangalkote,
Purba Bardhaman, West Bengal, India

B. Prakash et al. (eds.), *Plant Essential Oils*,
https://doi.org/10.1007/978-981-99-4370-8_7

7.1 Introduction

Essential oils (EOs) are gaining increased interest to be developed as botanical insecticides for stored pest management in both developing and developed countries. The increased insect resistance, and eco-toxicological, environmental and social consequences of the commonly applied synthetic chemical insecticides in agriculture have led researchers to investigate EOs as viable eco-friendly alternatives of hazardous chemical insecticides. Further, the worldwide availability and relative cost-effectiveness of EOs make them a suitable alternative of synthetic insecticides.

EOs are plant secondary metabolites composed of a mixture of hundreds of aromatic and aliphatic compounds with the dominance of monoterpenes, sesquiterpenes and their oxygenated derivatives and they play very important roles in plant defense and signaling processes (Zuzarte and Salgueiro 2015). EOs are produced by different plant organs like leaves, flowers, buds, fruits, seeds, bark, wood, rhizomes and roots. Some angiospermic families such as Asteraceae, Cupressaceae, Lamiaceae, Lauraceae, Rutaceae, Myrtaceae, Piperaceae, Apiaceae and Poaceae show increased accumulation of EOs in their glandular trichomes, secretory cavities and resin ducts. The main components of EOs, monoterpenes and sesquiterpenes are synthesized via either the methylerythritol 4-phosphate (MEP) pathway or the mevalonate-dependent (MVA) pathway in the cytoplasm and plastid (Hüsnü and Buchbauer 2015). The most common methods for extraction of EOs from raw plant materials are hydro-distillation, steam distillation and mechanical processes as considered in the European Pharmacopoeia and the International Standard Organization on Essential oils (ISO 9235:2013) (Zuzarte and Salgueiro 2015). However, some other modern methods frequently used for extraction of EOs are solvent extraction, microwave-assisted extraction, ultrasonic extraction, Soxhlet extraction, subcritical or superheated water extraction (SCWE) and supercritical fluid extraction. The composition of EOs is strongly subjected to variations according to their geographic origin, physiological status (i.e., flowering, vegetative etc.) of the plants, the part of the plant from which EOs were extracted, method of extraction etc. that can significantly alter the quality and the quantity of EO components and thus the toxicity of the EOs (Campolo et al. 2018).

EOs have a huge potential to be developed as plant-based alternative insecticide against different stored product insects as supported by a plethora of literature, however, information of various storage insects as well as efficacy of EOs against them in a single article is limited. The aim of this chapter is to provide short information on various important stored product insect pests and to analyze research studies on the use of essential oils against them in various aspects.

Coleoptera (beetles) and Lepidoptera (moths and butterflies) are the two major groups of insects that are responsible for post-harvest deterioration of stored food commodities. They can infest the crops both in the field and in store. Crop damage by Coleoptera is done by both larvae and adults while Lepidoptera damage is done mainly by the larvae. These insect pests of storage food commodities not only cause serious damage to agricultural products worldwide during their storage but also provide suitable medium for other contaminants such as bacteria, fungi and mites.

The following is a short description of most common post-harvest storage insect pests causing significant damage to stored food commodities worldwide.

7.2 Coleoptera

This is the largest order of insects that contains the most common and important stored product insects. In adults, the front pair of wings is modified into hard elytra. They inhabit a wide variety of habitats and can be found almost everywhere. They can be primary, i.e., able to attack intact grains; while others are secondary pests, i.e., attack already damaged grains or grain products.

7.2.1 Curculionidae (Snout Beetles)

The adults of this family are characterized by the presence of elongated downward-curved snout (rostrum). This family contains world's most common and destructive stored grain pests comprising *Sitophilus oryzae* (L.) (rice weevil), *S. zeamais* Motsch. (maize weevil) and *S. granarius* (L.) (granary/wheat weevil). These are the major insect pests of stored wheat, rice, maize and barley; however, they are able to develop on all cereals, dried cassava and other processed food products. Adult females create a small hole in intact grain, lay eggs inside and seal the hole with a secretion. Pupation takes place inside the grain and adults chew their way out through the outer layer of the grain. An adult female can lay more than 500 eggs and live for 6 months or more (Mound 1989).

7.2.2 Bruchidae (Seed Beetles)

This family includes short, stout-bodied beetles with a short forewing not reaching the tip of the abdomen. Adults have relatively long antennae. Larvae feed inside stored dry grains, mainly legumes. This family contains several important field and stored crop pests. The most destructive insect pest from this family is *Callosobruchus* sp. Adults have a short life span of about 12 days and do not feed. They show the polyphenism, i.e., two life forms can be seen: the active (flying) form and the flightless form. The adult females lay about 100 eggs glued to the seed surface or pods. Larvae tunnel inside the seed where the entire development takes place. *C. maculatus* (Fabricius) (cowpea weevil or pulse beetle) and *C. chinensis* (Adzuki bean beetle) are the major field-to-storage insect pest of pulses with broad host range and may cause up to 100% destruction of seeds within 3–4 months (Kedia et al. 2015a, b). Other species such as *C. rhodesianus*, *C. subinnotatus*, *Acanthoscelides obtectus* Say (*Bruchus obtectus* Say) (American bean weevil), *Caryedon serratus* (Olivier) (groundnut borer) are also important pests of legume seeds.

7.2.3 Tenebrionidae (Darkling Beetles)

The adults of this family are black or dark brown and are characterized by the tarsi of the hind leg with only four segments. *Tribolium castaneum* Herbst. (red-rust flour beetle) and *T. confusum* J. du Val (confused flour beetle) from this family are probably the most serious secondary pests of all plant commodities in store (flour mills, grain bins, empty cargo containers, storage units and retail stores) throughout the world causing significant damage and weight loss of stored grains (Ismail 2018). Damage is done by both larvae and adults generally to broken grains rather than intact grains. The adult females lay small, cylindrical, white eggs scattered in the product. Larvae are yellowish with a pale brown head, and they live inside grains until pupation. Females can lay up to 1000 eggs and live for a year or more. Another important pest from this family is *Tenebrio molitor* (L.) (the yellow mealworm beetle). It has worldwide distribution and feeds on a wide range of cereal products, plant and animal materials.

7.2.4 Bostrichidae (Branch and Twig Borers)

Adults of this family are elongated with the head bent down ventrally to the thorax and contain rasp-like hooks on the pronotum. This family includes two serious stored grain pests: *Rhizopertha dominica* (Fabricius) (lesser grain borer) and *Prostephanus truncatus* (Horn) (larger grain borer). *R. dominica* attacks on a large number of cereal grains during storage including wheat, barley, rice, oats, cassava, flour and other cereal products (Filomeno et al. 2020). Females can lay up to 500 eggs on the grain surface. Larvae feed either externally or inside the grain and pupation takes place within the eaten grain. Both adults and larvae eat the endosperm resulting in powdered grains. *P. truncates* is a serious primary pest of maize and dried cassava.

7.2.5 Laemophloeidae (Lined Flat Bark Beetles)

This family contains one common pest of stored grains: *Cryptolestes ferrugineus* (Stephens) (rusty grain beetle). Adults are small with relatively big head and prothorax. This insect is a secondary pest of stored grains, usually attacks the germs of broken or cracked grains, thus reducing germination (Pantenius 1988). Other species such as *C. pusillus* (Schonherr) and *C. pusilloides* (Steel and Howe) can also cause significant damage particularly in humid areas of the tropics.

7.2.6 Silvanidae (Silvan Flat Bark Beetles)

This family includes two important species: *Oryzaephilus surinamensis* (L) (saw-toothed grain beetle) and *O. mercator* (Fauvel) (merchant grain beetle). They infest a wide variety of stored grains, processed foodstuff and almost every product of vegetable origin (Abd El-Salam et al. 2019). *O. surinamensis* prefers cereals, seeds and nuts while *O. mercator* is more frequent on oil-seed products all over the globe. They enter through damaged grains and feed on the germ.

7.2.7 Dermestidae (Skin Beetles)

One of the world's most serious destructive stored product pest of grain and seeds from this family is *Trogoderma granarium* Everts (khapra beetle). Adults are oval, red brown in color with dark thorax. Adult females may lay up to 120 eggs within the stored products. Larvae are very hairy and bore into undamaged stored seeds. The larvae have the ability to fall under facultative diapause for several years and are tolerant to insecticidal treatments (Kavallieratos et al. 2020).

7.2.8 Anobiidae (the Wood Borers)

This family includes two important widespread storage pests: *Lasioderma serricorne* (Fabricius) (cigarette beetle) and *Stegobium paniceum* (Linnaeus) (drugstore beetle). *L. serricorne* is a common pest of stored cereals, cocoa beans, tobacco, ground nut, peas, beans, flours and other food items. Adults create holes on grains to mate and lay eggs inside. Newly hatched larvae feed on them and are responsible for most of the damage (Zhou et al. 2018). *S. paniceum* is less common and larvae are active feeders of stored biscuits, macaroni, dry fruits and other products.

7.2.9 Trogossitidae (Bark Gnawing Beetles)

The common pest in storehouse and granary from this family is *Tenebroides mauritanicus* (L.) (the cadelle). It primarily attacks on cereals, oilseeds and their products. Both adults and larvae feed directly on stored food. The larvae may tunnel in wooden walls of the store to create a pupation chamber.

7.3 Lepidoptera

Lepidoptera is the second most important order of stored product insect pests after Coleoptera. Adults are active flyers with two pairs of scaly wings and larvae possess well-developed mandibles and pseudopods (false legs) on some of the abdominal segments. They can attack crops in both the field and store. Adults generally attack ripening crop while larvae can be found in recently harvested stored grains.

7.3.1 Pyralidae (Snout Moths)

This is a large family containing two important stored product insect pests: *Ephestia* sp. and *Plodia* sp. *E. cautella* (Walker) (tropical warehouse moth) is a serious cosmopolitan pest of stored maize, wheat, dried fruits, beans, nuts, banana, groundnut and other grains. Adult females create holes in bags and lay sticky eggs within the substrate. Larvae cause considerable damage from webbing in the grain and on the surface of bags forming large lumps. *E. elutella* (Hub.) (warehouse moth) can be found on stored cocoa beans, grains, pulses, nuts, tobacco, coconut and dried fruits. *E. kuehniella* (Zeller) (Mediterranean flour moth) is a major pest of flour mills, grout mills, corn milling plants, bakeries and flour products. *P. interpunctella* (Hübner) (Indian meal moth) prefers meals and flours but can be found on stored raisins, nuts, pulses and cereals. The larvae of these insects feed on the germinal part of the grains causing food contamination with dead bodies, frass, exuviae and silk webbing (Mound 1989).

7.3.2 Gelechiidae (Twirler Moths)

This family contains two serious post-harvest pests *Sitotroga cerealella* and *Phthorimaea operculella*. *S. cerealella* (Olivier) (Angoumois grain moth) is a serious primary pest of maize, wheat and sorghum, both in the field and in stores, particularly the warmer parts of the world. Larvae feed and spend their entire life inside one grain leaving a hole after emergence (Bushra and Aslam 2014). *P. operculella* (Zeller) (potato tuberworm) is a cosmopolitan pest of potatoes, tomatoes and eggplants, both in the field and stores.

7.3.3 Acaridae (the Mites)

This is the family of mites including *Acarus siro* L. (flour mite), the cosmopolitan mite in foodstuff. *A. siro* can be found in granaries, feed mixing plants, threshing floors, stacks of hay and straw, dead organic matter, soil or plant residues and almost all products of plant or animal origin. Adult females lay large clutches of sticky eggs. The grains lose nutrients and germination ability, give a musty smell and unpleasant taste after contamination.

7.4 Efficacy of Essential Oils against Stored Product Insect Pests

To control these losses, mainly synthetic pesticides (gray chemicals) have been widely used throughout the world. However, due to adverse effects on non-target organisms and environment, resistance development among pests and high cost of synthetic insecticides, natural plant-based insecticides are getting preferred and research is going on for their efficacy in large scale. Recently essential oil-based products are gaining momentum in view of their negligible persistence in the nature, multiple modes of action, low toxicity, large-scale availability, renewable source and less chances of resistance development in pests (Chaudhari et al. 2021). A plethora of research and review articles have been published reporting the efficacy of EOs against stored insect pests, however, most of them have focused only on one or two insect pests. The reports concerning bio-efficacy of various EOs against various stored product insect pests through various ways are compiled in Table 7.1. The following section deals with insecticidal activity of different EOs against various stored insect pests that have been reported time to time through various methods.

7.4.1 Contact Toxicity

Contact toxicity is a way to kill pests when they come in contact with a chemical. Different EOs have been analyzed by different workers to record their activity as contact toxicant against various stored product insect pests. The methods most commonly used were residual film assay, impregnated paper assay, dipping method, direct topical application method etc. Abdelgaleil et al. (2016) evaluated the contact toxicity of 20 plant EOs against *S. oryzae,* and found strong insecticidal contact activity for *Artemisia judaica* (Asteraceae), *Callistemon viminalis* (Myrtaceae) and *Origanum vulgare* (Lamiaceae) EOs with LD_{50} value of 0.08, 0.09 and 0.11 mg/cm^2, respectively. Similar result was also observed for *Syzygium aromaticum* (Myrtaceae) and *Lavandula officinalis* (Lamiaceae) EO (LD_{50} values 0.04 and 0.07 mg/cm^2, respectively) (El-Bakry et al. 2016) and for *Coriandrum sativum* (Apiaceae), *Eucalyptus obliqua* (Myrtaceae) and *Pinus longifolia* (Pinaceae) EOs (LD_{50} values 36.68, 52.77 and 77.30 μg/cm^2, respectively) against *S. oryzae* (Rani 2012). In another study, some EOs showed similar contact insecticidal activity against *S. zeamais* (LD_{50} values for *Aster ageratoides* (Asteraceae) and *Dracocephalum moldavica* (Lamiaceae) were 27.16 and 22.10 μg/cm^2, respectively) (Chu et al. 2011, 2013) suggesting the reliability of these EOs at very low dosages for curculionid insects. In a similar study, Upadhyay et al. (2019) investigated the contact toxicity of *Melissa officinalis* (Lamiaceae) EO against *T. castaneum* and observed strong toxicity after 48 h of exposure at 0.157 μl/cm^2 air concentration. In a laboratory bioassay, Emamjomeh et al. (2021) estimated the contact LC_{50} of *Eucalyptus globulus* (Myrtaceae) and *Zataria multiflora* (Lamiaceae) EOs as 13.07 and 1.47 μl/l, respectively, against *E. kuehniella*. Researches showed the toxicity difference of different EOs on different insects could be affected by the penetration ability of the EO components, ability of insect to metabolize these components, thickness and the composition of cuticle in insects.

Table 7.1 Bio-efficacy of various EOs against various stored product insect pests as analyzed through various methods

EO	Test insects	Results	Reference
Contact toxicity			
Cupressus lusitanica *Eucalyptus saligna*	*T. castaneum* *A. obtectus* *S. zeamais*	5.0–65.0% mortality at 0.20% v/w EOs and 120 days grain storage	Bett et al. (2017)
Liriope muscari	*T. castaneum* *L. serricorne*	LD_{50} values 13.36 and 11.28 μg/adult, respectively	Wu et al. (2015)
Aloysia citriodora	*R. Dominica*	LC_{50} value 26.6 mg/cm^2	Benzi et al. (2009)
Chenopodium ambrosioides	*C. chinensis* *C. maculatus* *A. obtectus* *S. granarius* *S. zeamais* *P. truncatus*	80–100% mortality at 0.2 μl/cm^2 dose within 24 h except *C. maculatus* and *S. zeamais* (20 and 5% mortality, respectively)	Tapondjou et al. (2002)
Cinnamomum *cassia* *Cocholeria aroracia* *Brassica juncea*	*L. serricorne*	Over 90% mortality at 3 days after treatment	Kim et al. (2003)
Hyptis suaveolens *Ocimum canum*	*T. mauritanicus*	100 and 20% mortality, respectively, at 0.5 μl EO/g of peanut after 24 h	Adjou et al. (2019)
Cymbopogon martinii	*P. interpunctella*	LD_{50} value 22.8 μg/cm ^{2}	Jesser et al. (2017)
Fumigation toxicity			
Eucalyptus globulus	*P. interpunctella*	KT_{50} value 8.34 min.	Jesser et al. (2017)
Mentha spicata	*L. serricorne* *S. paniceum*	97.63 and 97.76% mortalities, respectively, at 20% v/v concentration after 24 h	Karakoç et al. (2018)
Rosmarinus officinalis	*O. surinamensis*	100% mortality at 0.15 μl/ml EO	Kiran and Prakash (2015)
Lippia palmeri	*S. zeamais* *P. truncatus*	92–100% mortality at 72 h with 1000 μl/l EO concentration	Martínez-Evaristo et al. (2015)
Artemisia vulgaris	*T. castaneum* *C. maculatus* *R. Dominica*	LC_{50} in the range of 52.47 to 279.86 μl/l air after 24 h	Sharifian et al. (2013)
Cuminum cyminum	*C. chinensis* *S. oryzae*	LC_{50} value 3.52 and 104.07 μl/l, respectively	Kedia et al. (2015)
Menthalongifolia	*T. castaneum* *T. confusum* *T. molitor* *O. surinamensis* *A. siro*	84.4, 42.2, 100, 100 and 87.8 mortalities, respectively, at 1000 ppm of EO	Kavallieratos et al. (2022)

(continued)

Table 7.1 (continued)

EO	Test insects	Results	Reference
Repellent activity			
Cupressus lusitanica *Eucalyptus saligna*	*T. castaneum* *A. obtectus* *S. zeamais*	34–52.4% repellency at 0.20% v/w EOs after 120 days grain storage	Bett et al. (2017)
Cuminum cyminum	*C. chinensis* *S. oryzae*	100 and 53% repellency, respectively at 12.5 μl/l air EO concentration	Kedia et al. (2015)
Lippia palmeri	*S. zeamais* *P. truncatus*	Total repellency and repellency index 0.15 at 24 h with 20 μl/l EO	Martínez-Evaristo et al. (2015)
Premna quadrifolia *P. angolensis*	*S. Cerealella*	96.84 and 91.55% repellency, respectively at 1% EO concentration	Adjalian et al. (2015)
Oviposition deterrence and ovicidal activity			
Cuminum cyminum	*C. chinensis* *S. oryzae*	>95 and > 75% deterrence, respectively at 100 μl/l air EO concentration	Kedia et al. (2015)
Cinnamomum verum *Citrus hystrix* *Cymbopogon nardus* *Euodia suaveolens* *Syzygium aromaticum*	*C. ferrugineus*	LC_{50} values for eggs 17, 12, 11, 16 and 11 ppm, respectively	Ikawati et al. (2020)
Eucalyptus Rosemary Lemon grass	*E. Cautella*	60.2, 33.6 and 46.7% egg hatching inhibition, respectively, at 5–15% EOs concentration	Al-Taie and Sabr (2018)
Majorana hortensis	*P. Operculella*	100% eggs failed to hatch at 0.1 ml dose of EO through contact mode	Abd El-Aziz (2011)
Larvicidal and pupaecidal activity			
Cinnamomum verum *Citrus hystrix* *Cymbopogon nardus* *Euodia suaveolens* *Syzygium aromaticum*	*C. ferrugineus*	LC_{50} values for larvae 24, 17, 22, 22 and 24 ppm and for pupae 9, 8, 8, 10 and 7 ppm, respectively	Ikawati et al. (2020)
Laurus nobilis *Salvia officinalis*	*T. granarium*	LC_{50} values 37.9 and 50.7 μl/l, respectively, for larvae	Tayoub et al. (2012)
Zataria multiflora	*E. Kuehniella*	LC_{50} values 0.61 μl/cm^2 for larvae after 72 h	Emamjomeh et al. (2017)

(continued)

Table 7.1 (continued)

EO	Test insects	Results	Reference
Eucalyptus Rosemary Lemon grass	*E. Cautella*	29.33, 40 and 17.33% mortality of young larvae and 39.77, 21.67 and 32.20% mortality of late larvae, respectively, at 5–15% concentration of EOs	Al-Taie and Sabr (2018)
Majorana hortensis	*P. Operculella*	100, 100 and 11.3% mortality of larva, prepupa and pupal stages after 24 h at 0.2 ml/10 g dose	Abd El-Aziz (2011)
Menthalongifolia	*T. castaneum* *T. confusum* *T. molitor* *O. surinamensis* *A. siro*	100, 100, 34.4, 100 and 67.8% mortalities to larvae, respectively, at 1000 ppm EO	Kavallieratos et al. (2022)
Antifeedant activity			
Cuminum cyminum	*C. chinensis* *S. oryzae*	100 and 97% FDI, respectively, at 100 μl/l air EO concentration	Kedia et al. (2015a, b)
Acorus calamus	*P. truncatus*	Feeding lowered to 50% compared to the control at 0.01% oil concentration	Schmidt and Streloke (1994)
Rosmarinus officinalis	*O. surinamensis*	100% antifeedant activity at 0.15 μl/ml EO	Kiran and Prakash (2015)
Premna quadrifolia *P. angolensis*	*S. Cerealella*	0.07 and 0.15% grain damage, respectively, at 15 μl/ml dose of EO	Adjalian et al. (2015)

7.4.2 Fumigation Toxicity

Fumigants are insecticides that act in the vapor or gaseous phase on the target pests. Being volatile in nature, EOs are well suited to be developed as plant-based fumigants. Extensive work has been done to assess the fumigation toxicity of EOs against various storage insects. The most followed method for fumigation toxicity of EOs was impregnated paper assay by using filter papers inside closed containers. *Ocimum gratissimum* (Lamiaceae) EO showed prominent fumigation toxicity against *S. oryzae*, *C. chinensis*, *R. dominica* and *O. surinamensis* (LC_{50} values 0.50, 0.20, 0.20 and 0.19 μl/l, respectively) but less toxicity toward *T. castaneum* (LC_{50} 24.9 μl/l) (Ogendo et al. 2008). In another study, *Artemisia sieberi* (Asteraceae) EO showed pronounced fumigation toxicity against *C. maculatus*, *S. oryzae* and *T. castaneum* (LC_{50} values 1.45, 3.86 and 16.75 μl/l air, respectively, after 24 h) (Negahban et al. 2007). Similarly, Naseri et al. (2017) investigated the fumigant toxicity of *A. sieberi* and *A. khorassanica* EOs against adults of *S. cerealella* and recorded LC_{50} values as 9.26 and 7.38 μl/l air concentrations, respectively. Research showed pronounced fumigation toxicity of EOs against various insect pests favoring their application for managing insect pest population in closed spaces such as storage bins or buildings. Currently, through microencapsulation method, some EOs have been encapsulated and showed increased activity by improving their handling, stabilization and controlled delivery. The application of EOs in post-harvest protection of stored food commodities would be economical as very low dose of EO may uniformly fumigate the commodities kept in large containers.

7.4.3 Repellent Activity

Repellents are chemicals that act locally or at a distance by providing a vapor barrier, deterring an insect from coming into contact to or landing over a surface. Hundreds of plant EOs have been investigated as potential sources of insect repellents, however, the studies mainly focused on Dipteran insects; the insect pests of stored food commodities (Coleoptera and Lepidoptera) have been less researched. The most commonly used methods were filter paper method, choice bioassay and olfactometer assay. In a study, Caballero-Gallardo et al. (2011) tested the repellent activity of *Lippia alba*, *Rosmarinus officinalis*, *Lepechinia betonicifolia* (Lamiaceae), *Tagetes lucida* (Asteraceae) and *Cananga odorata* (Annonaceae) EOs against *T. castaneum* and observed 96 ± 2, 92 ± 4, 92 ± 4, 90 ± 3 and 98 ± 2% repellent activity, respectively, after 4 h of exposure at 0.2 μl/cm^2 concentration. Similarly, Fogang et al. (2012) observed the 100% repellent activity of *Zanthoxylum xanthoxyloides* (Rutaceae) EO against *A. obtectus* at 0.501 μl/cm^2 air concentration. Nattudurai et al. (2017) observed 85.24 and 75.24% repellency against *C. maculates* and *S. oryzae*, respectively, at 25 μl/l air concentration of *Atalantia monophylla* (Rutaceae) EO after 3 h of exposure using a Y-tube glass olfactometer. In an another study, Mahdi and Behnam (2018) observed 49.99 and 58.33% repellency of *Citrus sinensis* (Rutaceae) EO against *R. dominica* and *L. serricorne*, respectively, after 3 h of exposure. Similarly, Ogendo et al. (2008) observed 78–93% repellency of *Ocimum gratissimum* (Lamiaceae) EO at 0.05–0.2% v/w concentration after 24 h against *C. chinensis* through choice bioassay in Petri plates.

7.4.4 Oviposition Deterrent and Ovicidal Activity

Oviposition deterrent is the property by which a chemical does not allow the females to deposit eggs on a surface. Ovicidal activity is the property by which a chemical kills the eggs by disrupting embryonic development. Several EOs have been reported to have oviposition deterrent and ovicidal activities against various stored product insect pests, however, reports are more for the insects that lay eggs over the seed surface and are clearly visible. Papachristos and Stamopoulos (2004) investigated *Lavandula hybrida*, *Rosmarinus officinalis* (Lamiaceae) and *Eucalyptus globulus* (Myrtaceae) EOs against *A. obtectus* and observed 27.1–29.9% oviposition deterrent and ovicidal activity at 197.2 μl/l air concentration. The activity increased on increasing time period. Mondal and Khalequzzaman (2009) observed the ovicidal activity of five EOs on *T. castaneum* eggs and found the strongest effect for *Elettaria cardamomum* (Zingiberaceae) while the lowest impact for *Azadirachta indica* (Meliaceae). Kedia et al. (2014) observed 98% oviposition deterrence and 100%

ovicidal performance of *Mentha spicata* EO against *C. chinensis* at 0.1 and 0.0125 μl/ml concentrations, respectively. In a study, Ayvaz et al. (2009) tested the ovicidal activity of five EOs against *P. interpunctella* and *E. kuehniella*. The highest egg mortality was observed from *Satureja thymbra* (Lamiaceae) EO (100%) and the lowest from *Citrus limon* (Rutaceae) EO (25–30%) for both the insects. EOs exhibited oviposition inhibition either due to killing females before laying their eggs or due to the failure of live females to lay many eggs. Further, the EO components may enter into the eggs through chorion and suppress embryonic development exhibiting their ovicidal activity. These properties of checking the pest population at the beginning of their life cycle would be advantageous in view of development of resistance in pests and can be recommended in food safety programs.

7.4.5 Larvicidal and Pupaecidal Activity

Most of the EO toxicity studies refer to adults. Being internal feeders, the toxicity of EOs toward larvae and pupae has been less investigated and these stages seemed to be more resistant than the mature adult stage. Kedia et al. (2015a, b) tested *Cuminum cyminum* (Apiaceae) seed EO against *C. chinensis* and *S. oryzae* immature stages and observed the early embryonic stages (eggs and neonate larvae) as more susceptible than the older stages (mature larvae and pupae). The toxicity of the EO against *C. chinensis* was 92% for LI/LII larvae, 53% for LIII/LIV larvae and 41% for pupae at 50 μl/l air concentration. Similarly against *S. oryzae*, EO showed 59% toxicity to larvae and 44% mortality to pupae at the same concentration. In another study, Polatoğlu et al. (2016) tested *Crithimum maritimum* (Apiaceae) EO against the larva of *O. surinamensis*, *S. granarius* and *S. oryzae* in Petri plates and observed 100% mortality at 100 μl/ml dose. Papachristos and Stamopoulos (2009) observed the effects of *Lavandula hybrida*, *Rosmarinus officinalis* (Lamiaceae) and *Eucalyptus globules* (Myrtaceae) EOs on the development, longevity and fecundity of *A. obtectus*. All EOs caused increased larval and pupal developmental time and reduced longevity and fecundity of the newly emerged female adults. Studies showed that EOs can penetrate the chorion and/or vitelline membrane, facilitating their diffusion to affect vital physiological and biochemical processes of different developmental stages of insects. During storage conditions, all developmental stages are normally present at a single time and thus the products showing toxicity to immature stages as well has an additional merit to protect food commodities.

7.4.6 Antifeedant Activity

The chemicals which control insect feeding (mainly the active larval stage) and cause death by starvation are called feeding deterrents. Certain EOs have been reported to control grain damage by checking insect feeding in terms of feeding deterrence index (FDI), weight loss of treated seeds and total seed damage. Liu and Ho (1999) tested the antifeedant activity of *Evodia rutaecarpa* (Rutaceae) EO and observed strong antifeedant action against larvae than adults (at a concentration of 0.75 and 1.5 mg/disc for *T. castaneum* and 1.5 and 2.2 mg/disc for *S. zeamais*, respectively for growth and food consumption). Kiran and Prakash (2015) reported complete feeding deterrence of *Gaultheria procumbens* (Ericaceae) EO at 58.62 and 2.71 μl/l air concentration against *S. oryzae* and *R. dominica*, respectively. Similarly, Shukla et al. (2011) observed 100 and 96.82% FDI of *Lippia alba* (Lamiaceae) and *Callistemon lanceolatus* (Myrtaceae) Eos, respectively, even after 24 months of storage against *C. chinensis*. In a study, Satureja hortensis (Lamiaceae) oil significantly decreased the relative growth rate and relative consumption rate of *P. interpunctella* larvae. Further at 2 μl/disk concentration, efficiency of conversion of ingested food (9.843%) was significantly low (Shahab-Ghayoor and Saeidi 2015). Plant products having feeding deterrent activity in general show high adult mortality, less oviposition, increased larvae mortality and low adult emergence. These properties of EOs make them a suitable choice of alternative insecticide for stored food commodities.

7.5 Mode of Insecticidal Action

In most of the studies, the mode of insecticidal activity of EOs have been reported as neurotoxic by either inhibiting acetylcholine esterase (AChE) or by blocking γ-amino butyric acid (GABA) and octopamine receptors. Some other studies also report the EO toxicity by altering enzymatic and nonenzymatic antioxidant defense systems. AChE inhibition is one of the most researched mechanism as the insect AChE differs from the mammalian system by only a single residue and can be used as a selective marker. Various EO components bind the catalytic site of AChE, reduce its activity that lead to the accumulation of acetylcholine at neuromuscular junctions which again in turn induces neuronal excitation, hyperactivity, paralysis and finally death of the insects occur (Abdelgaleil et al. 2009; Kiran and Prakash 2015). Octopamine and GABA receptors are second important targets next to AChE for various EOs. EO components may also bind with octopamine receptors causing increased intracellular cAMP (cyclic adenosine monophosphate) concentration and subsequent death (Kostyukovsky et al. 2002). Some studies also suggested the blockage of GABA receptors as another targets of EOs mediated toxicity (Chaudhari et al. 2021). In a study, Kiran et al. (2017) reported that the toxicity of EO can be assigned to the increase in reactive oxygen species (ROS), superoxide dismutase (SOD) and catalase (CAT) and reduction in glutathione (GSH/GSSG) ratio upon treatment. The depletion in glutathione level can cause oxidative burden resulting into damage to nucleic acids and lipoproteins, and ultimately cell death. However, further research is needed to elucidate the exact mechanism of EOs against stored product insect pests.

7.6 Conclusion and Future Challenges

Due to multiple modes of action, eco-friendly nature, renewable source and favorable safety profile, EOs would be the better alternative to the hazardous chemicals fumigants. To assess their practical application and effective formulation, the large scale testing in storage is needed. Because of growing consumer awareness and negative concerns toward synthetic chemicals, the use of plant-based natural EOs is becoming more popular in food security. Further, these products must be standardized and registered before use to ensure product safety and efficacy. Some of the EO-based pesticides are already available in Western market; however, their use is limited due to higher volatility, low persistence and rapid oxidation. Recently, nanotechnology's booming research trends show that EOs are encapsulated into edible secondary wall materials such as chitosan, gelatin, alginate, carrageenan, cyclodextrins etc. using different nanoencapsulation techniques such as ionic-gelation, spray drying or chilling, coacervation, electrospinning, emulsification etc. This technique not only solved the low persistence of EO but also caused increased efficacy and controlled release of EO at low concentrations, making their application easy. Further, detailed investigations are required for the efficacy of these products in the real food system and their safety profile, an area that needs more research insight.

References

Abd El-Aziz MF (2011) Bioactivities and biochemical effects of marjoram essential oil used against potato tuber moth *Phthorimaea operculella* Zeller (Lepidoptera: Gelechiidae). Life Sci J 8(1):288–297

Abd El-Salam AME, Abd Elwahed Salem S, Abdel-Rahman RS (2019) Fumigant and toxic activity of some aromatic oils for protecting dry dates from *Oryzaephilus surinamensis* (L.) (Coleoptera: Silvanidae) in stores. Bull Nat Res Cent 43(1):1–6

Abdelgaleil SA, Mohamed MI, Badawy ME, El-arami SA (2009) Fumigant and contact toxicities of monoterpenes to *Sitophilus oryzae* (L.) and *Tribolium castaneum* (Herbst) and their inhibitory effects on acetylcholinesterase activity. J Chem Ecol 35:518–525

Abdelgaleil SAM, Mohamed MIE, Shawir MS, Abou-Taleb HK (2016) Chemical composition, insecticidal and biochemical effects of essential oils of different plant species from northern Egypt on the rice weevil, Sitophilus oryzae L. J Pest Sci 89(1):219–229

Adjalian E, Sessou P, Odjo T, Figueredo G, Kossou D, Avlessi F, Menut C, Sohounhloué D (2015) Chemical composition and insecticidal and repellent effect of essential oils of two *Premna* species against *Sitotroga cerealella*. J Insects 2015:1–6

Adjou ES, Chougourou D, Soumanou MM (2019) Insecticidal and repellent effects of essential oils from leaves of *Hyptis suaveolens* and *Ocimum canum* against *Tenebroides mauritanicus* (L.) isolated from peanut in post-harvest. J Consum Prot Food Saf 14(1):25–30

Al-Taie MM, Sabr SH (2018) Evaluation of the ovicidal and larvicidal activity of essential oils of three plant species on fig moth, *Ephestia cautella* (Walker) (Lepidoptera: Pyralidae). J Entomol Zool Stud 6(1):1338–1342

Ayvaz AB, Karaborklu SA, Sagdic O (2009) Fumigant toxicity of five essential oils against the eggs of *Ephestia kuehniella* Zeller and *Plodia interpunctella* (Hübner) (Lepidoptera: Pyralidae). Asian J Chem 21(1):596–604

Benzi VS, Murray AP, Ferrero AA (2009) Insecticidal and insect-repellent activities of essential oils from Verbenaceae and Anacardiaceae against *Rhizopertha Dominica*. Nat Prod Commun 4(9):1286–1290

Bett PK, Deng AL, Ogendo JO, Kariuki ST, Kamatenesi-Mugisha M, Mihale JM, Torto B (2017) Residual contact toxicity and repellence of *Cupressus lusitanica* miller and *Eucalyptus saligna* smith essential oils against major stored product insect pests. Ind Crop Prod 110:65–74

Bushra S, Aslam M (2014) Management of *Sitotroga cerealella* in stored cereal grains: a review. Arch Phytopathol Pflanzenschutz 47(19):2365–2376

Caballero-Gallardo K, Olivero-Verbel J, Stashenko EE (2011) Repellent activity of essential oils and some of their individual constituents against *Tribolium castaneum* Herbst. J Agric Food Chem 59:1690–1696

Campolo O, Giunti G, Russo A, Palmeri V, Zappalà L (2018) Essential oils in stored product insect pest control. J Food Qual 2018:6906105

Chaudhari AK, Singh VK, Kedia A, Das S, Dubey NK (2021) Essential oils and their bioactive compounds as eco-friendly novel green pesticides for management of storage insect pests: prospects and retrospects. Environ Sci Pollut Res 28(15):18918–18940

Chu SS, Liu SL, Liu QZ, Jiang GH, Liu ZL (2013) Chemical composition and insecticidal activities of the essential oil of the flowering aerial parts of *Aster ageratoides*. J Serbian Chem Soc 78(2):209–216

Chu SS, Liu SL, Liu QZ, Liu ZL, Du SS (2011) Composition and toxicity of Chinese *Dracocephalum moldavica* (Labiatae) essential oil against two grain storage insects. J Med Plant Res 5(18):4621–4626

El-Bakry AM, Abdel-Aziz NF, Sammour EA, Abdelgaleil SAM (2016) Insecticidal activity of natural plant essential oils against some stored product insects and their side effects on wheat seed germination. Egypt J Biol Pest Control 26(1):83–88

Emamjomeh L, Imani S, Talebi Jahromi K, Moharramipour S (2021) Nanoencapsulation enhances the contact toxicity of *Eucalyptus globulus* Labill and *Zataria multiflora* Boiss essential oils against the third instar larvae of *Ephestia kuehniella* (Lepidoptera: Pyralidae). Int J Pest Manag 2021:1–9

Emamjomeh L, Imani S, Talebi KH, Moharramipour S, Larijani K (2017) Preparation of nanoemulsion formulation of essential oil of *Zataria multiflora* and comparison of contact toxicity with pure essential oil on *Ephestia kuehniella*. Entomol Phytopathol 85:181–190

Filomeno CA, Barbosa LCA, Teixeira RR, Pinheiro AL, de Sá FE, Ferreira JS, Picanço MC (2020) Chemical diversity of essential oils of Myrtaceae species and their insecticidal activity against *Rhyzopertha Dominica*. Crop Prot 137:105309

Fogang HPD, Womeni HM, Piombo G, Barouh N, Tapondjou LA (2012) Bioefficacy of essential and vegetable oils of *Zanthoxylum xanthoxyloides* seeds against *Acanthoscelides obtectus* (say) (Coleoptera: Bruchidae). J Food Prot 75:547–555

Hüsnü CB, Buchbauer G (2015) Handbook of essential oils: science, technology, and applications. CRC Press, Boca Raton, FL, USA

Ikawati S, Himawan T, Abadi AL, Tarno H (2020) Fumigant and feeding deterrent activity of essential oils against *Cryptolestes ferrugineus* (Stephens) (Coleoptera: Laemophloeidae). Biodiversitas 21(9):4301–4308

Ismail EH (2018) Toxicity, repellency and latent effects of some medicinal oils against *Tribolium confusum* and *T. castaneum* (Coleoptera: Tenebrionidae). J Entomol Zool Stud 6:1337–1347

Jesser EN, Werdin-González JO, Murray AP, Ferrero AA (2017) Efficacy of essential oils to control the Indian meal moth, *Plodia interpunctella* (Hübner) (Lepidoptera: Pyralidae). J Asia Pac Entomol 20(4):1122–1129

Karakoç ÖC, Alkan M, Şimşek Ş, Gökçe A, Halit ÇAM (2018) Fumigant activity of some plant essential oils and their consustent against to *Stegobium paniceum* and *Lasioderma serricorne* (Coleoptera: Anobidae). Plant Prot Bull 58(3):163–169

Kavallieratos NG, Boukouvala MC, Ntalli N, Skourti A, Karagianni ES, Nika EP, Kontodimas DC, Cappellacci L, Petrelli R, Cianfaglione K, Morshedloo MR, Tapondjou LA, Rakotosaona R, Maggi F, Benelli G (2020) Effectiveness of eight essential oils against two key stored-product beetles, *Prostephanus truncatus* (horn) and *Trogoderma granarium* everts. Food Chem Toxicol 139:111255

Kavallieratos NG, Nika EP, Skourti A, Xefteri DN, Cianfaglione K, Perinelli DR, Spinozzi E, Bonacucina G, Canale A, Benelli G, Maggi F (2022) Piperitenone oxide-rich *Mentha longifolia* essential oil and its nanoemulsion to manage different developmental stages of insect and mite pests attacking stored wheat. Ind Crop Prod 178:114600

Kedia A, Prakash B, Mishra PK, Chanotiya CS, Dubey NK (2014) Antifungal, antiaflatoxigenic, and insecticidal efficacy of spearmint (*Mentha spicata* L.) essential oil. Int Biodeterior Biodegradation 89:29–36

Kedia A, Prakash B, Mishra PK, Dwivedy AK, Dubey NK (2015a) Biological activities of *Cuminum cyminum* seed oil and its major components against *Callosobruchus chinensis* and *Sitophilus oryzae*. J Asia Pac Entomol 18:383–388

Kedia A, Prakash B, Mishra PK, Singh P, Dubey NK (2015b) Botanicals as eco friendly biorational alternatives of synthetic pesticides against *Callosobruchus* spp.(Coleoptera: Bruchidae)-a review. J Food Sci Technol 52(3):1239–1257

Kim SI, Park C, Ohh MH, Cho HC, Ahn YJ (2003) Contact and fumigant activities of aromatic plant extracts and essential oils against *Lasioderma serricorne* (Coleoptera: Anobiidae). J Stored Prod Res 39(1):11–19

Kiran S, Kujur A, Patel L, Ramalakshmi K, Prakash B (2017) Assessment of toxicity and biochemical mechanisms underlying the insecticidal activity of chemically characterized *Boswellia carterii* essential oil against insect pest of legume seeds. Pestic Biochem Physiol 139:17–23

Kiran S, Prakash B (2015) Toxicity and biochemical efficacy of chemically characterized *Rosmarinus officinalis* essential oil against *Sitophilus oryzae* and *Oryzaephilus surinamensis*. Ind Crop Prod 74:817–823

Kostyukovsky M, Rafaeli A, Gileadi C, Demchenko N, Shaaya E (2002) Activation of octopaminergic receptors by essential oil constituents isolated from aromatic plants: possible mode of action against insect pests. Pest Manag Sci 58:1101–1106

Liu ZL, Ho SH (1999) Bioactivity of the essential oil extracted from *Evodia rutaecarpa* hook f. et Thomas against the grain storage insects, *Sitophilus zeamais* Motsch. And *Tribolium castaneum* (Herbst). J Stored Prod Res 35:317–328

Mahdi KR, Behnam AB (2018) Fumigant toxicity and repellency effect of orange leaves *Citrus sinensis* (L.) essential oil on *Rhyzopertha Dominica* and *Lasioderma serricorne*. J Essent Oil Bear Plan 21:577–582

Martínez-Evaristo XC, Borboa-Flores J, Wong-Corral FJ, Cinco-Moroyoqui FJ, Del-Toro-Sánchez CL, González-de León A, Ortega-Nieblas MM, Cornejo-Ramirez YI, Osuna-Amarillas PS, Cabral-Torres FA (2015) Insecticidal efficacy and repellency of oregano (*Lippia palmeri*) essential oil for control of *Sitophilus zeamais* and *Prostephanus truncatus* in stored maize. Southwestern Entomol 40(4):713–720

Mondal M, Khalequzzaman M (2009) Ovicidal activity of essential oils against red flour beetle, *Tribolium castaneum* (Coleoptera: Tenebrionidae). J Biosci 17:57–62

Mound L (1989) Common insect pests of stored food products. Economic series no. 15 (7th Edn). British Museum (Natural History), London, UK, p 68

Naseri B, Abedi Z, Abdolmaleki A, Jafary-Jahed M, Borzoui E, Mansouri SM (2017) Fumigant toxicity and sublethal effects of *Artemisia khorassanica* and *Artemisia sieberi* on *Sitotroga cerealella* (Lepidoptera: Gelechiidae). J Insect Sci 15(5):100

Nattudurai G, Baskar K, Paulraj MG, Islam VIH, Ignacimuthu S, Duraipandiyan V (2017) Toxic effect of *Atalantia monophylla* essential oil on *Callosobruchus maculatus* and *Sitophilus oryzae*. Environ Sci Pollut Res 24:1619–1629

Negahban M, Moharramipour S, Sefidkon F (2007) Fumigant toxicity of essential oil from *Artemisia sieberi* Besser against three stored product insects. J Stored Prod Res 43:123–128

Ogendo JO, Kostyukovsky M, Ravid U, Matasyoh JC, Deng AL, Omolo EO, Kariuki ST, Shaaya E (2008) Bioactivity of *Ocimum gratissimum* L. oil and two of its constituents against five insect pests attacking stored food products. J Stored Prod Res 44(4):328–334

Pantenius C (1988) Storage losses in traditional maize granaries in Togo. Insect Sci Appl 9(6):725–735

Papachristos DP, Stamopoulos DC (2004) Fumigant toxicity of three essential oils on the eggs of *Acanthoscelides obtectus* (say) (Coleoptera: Bruchidae). J Stored Prod Res 40:517–525

Papachristos DP, Stamopoulos DC (2009) Sublethal effects of three essential oils on the development, longevity and fecundity of *Acanthoscelides obtectus* (say) (Coleoptera: Bruchidae). Hell Plant Prot J 2:91–99

Polatoğlu K, Karakoç ÖC, Yücel YY, Gücel S, Demirci B, Başer KHC, Demirci F (2016) Insecticidal activity of edible *Crithmum maritimum* L. essential oil against coleopteran and lepidopteran insects. Ind Crop Prod 89:383–389

Rani PU (2012) Fumigant and contact toxic potential of essential oils from plant extracts against stored product pests. J Biopest 5(2):120–128

Schmidt GH, Streloke M (1994) Effect of *Acorus calamus* (L.) (Araceae) oil and its main compound β-asarone on *Prostephanus truncatus* (Horn) (Coleoptera: Bostrichidae). J Stored Prod Res 30(3):227–235

Shahab-Ghayoor H, Saeidi K (2015) Antifeedant activities of essential oils of *Satureja hortensis* and *Fumaria parviflora* against Indian meal moth *Plodia interpunctella* Hübner (Lepidoptera: Pyralidae). Entomol Ornithol Herpetol 4(3):1–4

Sharifian I, Hashemi SM, Darvishzadeh A (2013) Fumigant toxicity of essential oil of Mugwort (*Artemisia vulgaris* L.) against three major stored product beetles. Arch Phytopathol Pflanzenschutz 46(4):445–450

Shukla R, Singh P, Prakash B, Kumar A, Mishra PK, Dubey NK (2011) Efficacy of essential oils of *Lippia alba* (mill.) NE Brown and *Callistemon lanceolatus* (Sm.) sweet and their major constituents on mortality, oviposition and feeding behaviour of pulse beetle, *Callosobruchus chinensis* L. J Sci Food Agric 91:2277–2283

Tapondjou LA, Adler CLAC, Bouda H, Fontem DA (2002) Efficacy of powder and essential oil from *Chenopodium ambrosioides* leaves as post-harvest grain protectants against six-stored product beetles. J Stored Prod Res 38(4):395–402

Tayoub GH, Odeh AD, Ghanem IY (2012) Chemical composition and fumigation toxicity of *Laurus nobilis* L. and *Salvia officinalis* L. essential oils on larvae of khapra beetle (*Trogoderma granarium* everts). Herba Pol 58(2):26–37

Upadhyay N, Singh VK, Dwivedy AK, Das S, Chaudhari AK, Dubey NK (2019) Assessment of *Melissa officinalis* L. essential oil as an eco-friendly approach against biodeterioration of wheat flour caused by *Tribolium castaneum* Herbst. Environ Sci Pollut Res 26:14036–14049

Wu Y, Zhang WJ, Wang PJ, Yang K, Huang DY, Wei JY, Tian ZF, Bai JF, Du SS (2015) Contact toxicity and repellency of the essential oil of *Liriope muscari* (DECN.) bailey against three insect tobacco storage pests. Molecules 20(1):1676–1685

Zhou J, Zou K, Zhang W, Guo S, Liu H, Sun J, Borjigidai A (2018) Efficacy of compounds isolated from the essential oil of *Artemisia lavandulaefolia* in control of the cigarette beetle, *Lasioderma serricorne*. Molecules 23(2):1–11

Zuzarte M, Salgueiro L (2015) Essential oils chemistry. In: de Sousa DP (ed) Bioactive essential oils and cancer. Springer, Cham, Switzerland, pp 19–61

Application of Plant Essential Oils in Pharma and Aroma Industries

8

Aleksandra Bogdanovic and Dejan Skala

Abstract

The medicinal plants perform wide applications due to essential oils and biologically active constituents which are utilized in various industries. Revealed biological activity of constituents in essential oils, such as antibacterial, antioxidant, anti-inflammatory, antifungal, antitumor, antiviral, and insecticidal led to their application in the role of importance in pharmacy, cosmetic, food, and aroma industries. This review is focused on recently applied extraction techniques followed by optimization of the processes to provide the most effective potential of essential oils in the therapeutic and pharmaceutical industries. Bioactive compounds and their maintenance in the industry are gaining attention due to their features to overcome synthetic and unsafe medicaments on the global market. Thus, the widespread application of bioactive compounds of plant essential oils requires scientific investigations of their isolation, an optimization of their yields, a concise analysis of their properties, formulations, and stability. Significant research is performed on the production and characterization of essential oils with the main goal to apply them as pharmaceutical constituents, as flavors which might be used for the preparation of some food, cosmetic, and aroma.

Keywords

Essential oils · Bioactive compounds · Biological activity, Industrial application

A. Bogdanovic (✉) · D. Skala
Faculty of Technology and Metallurgy, University of Belgrade, Belgrade, Serbia
e-mail: abogdanovic@tmf.bg.ac.rs

B. Prakash et al. (eds.), *Plant Essential Oils*,
https://doi.org/10.1007/978-981-99-4370-8_8

8.1 Introduction

Essential oils manifest wide spectrum of the biological activity which evaluate their spread application in the industry according to their therapeutic, preservative, medicinal, and favoring properties due to their constituents. The compositions of essential oils and extracts are usually complex due to the presence of highly functionalized chemical substances of different chemical classes which include terpenes (monoterpenes and sesquiterpenes), phenylpropanoids, anthocyanins, alkaloids, sapogenins, and saponins.

Since ancient times, alternative medicine applied plants as main treatment for wide spectrum of diseases, but nowadays modern and official commercial medicaments which are approved by FDA are derived and also accepted for treating various diseases. There are already on the global market active compounds and drugs that have successfully been applied in the industry production and commercialized from plant material such as aspirin from Willow bark, paclitaxel, podophyllotoxin, vinblastin, and scopolamine, a chemotherapy medication derived from *Taxus* spp. Proved and established natural products for industrial application and commercialization on global level require several approaches based on their isolation, optimization for more convenient extraction techniques with goal to achieve the high yield of active compounds, extract composition characterization, proven activity and metabolism mechanism, and convenient pharmaceutical formulation of the final products. The most important secondary metabolites due to their bioactivity are classified as terpenoids, phenols, and alkaloids. Alkaloids are the most significant constituents of essential oils, due to remained medicinal activity, therefore known as the most important source of phytomedicinal active compounds and therefore targeted at most for industrial production. Among alkaloids, there are many known bioactive compounds used and accepted as commercial medications such as morphine, quinine, vincristine, vincamine, as well caffeine, cocaine, nicotine, and methamphetamines. Terpenoids reveal antioxidant activity of importance such as retinol and β-carotene, including essential oils as carriers of fragrances, although including drugs (cannabinoids). Antioxidant properties are referred to phenols as well, characterized by the aromatic rings with hydroxyl groups. Among phenols, polyphenols which contain polymerized phenolic groups are revealed to possess antioxidant properties. Phenols more frequently find application in supplements or as pigments in food industry, such as resveratrol and anthocyanidins.

Essential oils as natural organic compounds remain valuable sources of bioactive compounds, as secondary metabolites synthesized in plant cells including enzymatic reactions. Insights into biosynthesis of essential oils, as well in their more priceless constituents, can improve extraction process of active compounds isolation and their valuable application.

The extraction techniques for essential oils isolation from plants are based on classic and traditional usage from ancient time as steam distillation and hydrodistillation, although extraction with organic solvents is still in common use. Development of green technology in achieving more cleaner final product is maintaining by supercritical fluid extraction processes.

A wide variety of essential oils from plants due to their traditional usage and even modern application requests have been of interest in our research file involving lemon balm (*Melissa officinalis* L.), fenugreek seeds (*Trigonella foenum-graecum* L.), pomegranate seeds (*Punica granatum* L.), *Helichrysum immortelle*, *Cannabis sativa* L. (family Cannabinaceae), sage leaves (*Salvia officinalis*), and wild bilberry (*Vaccinium myrtillus* L., Ericaceae).

The biological activity of essential oil constituents recommends their application due to their medical and therapeutical value in pharmaceutical industry as medicaments in treating diseases, as supplements, or as a skin cream for skin disorders (Bogdanovic 2022, Dima and Dima 2015). Primarily their use was as fragrances in aromatherapy, while according to their revealed antioxidative, anti-inflammatory, cell regeneration properties, antibacterial, antifungal, antitumor, antiviral, and insecticidal activity recommended their application in phytomedicine. Essential oils as well remain positive influence on hypercholesterolemia, hyperglycemia, and hypertensia, which led to precision investigation and their implementation as phytomedicaments. The antioxidative properties set essential oil application as conserving and preservation in either food or cosmetic industry. Regarding food industry, essential oils have spread application, although maintaining preserving properties and extending food shelf life, their role in food is to obtain natural flavor as well.

This manuscript reveals main active compounds in abovementioned plants which proved valuable interest in several year research of our team working at Belgrade University in Laboratory for High Pressure Processes of Faculty of Technology and Metallurgy (FTM), developing analytical chromatographic methods for their confirmation and quantifications at Department of Organic Chemistry FTM. Furthermore, several extraction methods were applied and compared to improve the yield of valuable biological active compounds including hydrodistillation, extraction with organic solvents, and supercritical CO_2 extraction. The obtained essential oils have been quantified, identificated, and correlation within active compounds and revealed and proven activity have been estimated and observed. Moreover, achieved optimized yield of essential oils with bioactive constituents was further addressed for their potential industrial application.

8.2 Biological Active Constituents of Essential Oils

The various plants are used from ancient times, primarily as a food, than spices, and as medicaments in alternative medicine. Their biological properties and application are usually known but nowadays, the requests of medicinal devices and industrial application of plants and their isolates in commercial medicine, cosmetic, and food industry require the proven bioactivity of phytopreparations with their exact quantification of composition. Especially since synthetic medicaments revealed negative side effects, the main attention based on "green technology" was focused on the preparations of bioactive compounds from plants using modern techniques of isolation without the use of organic solvents for their isolation. Moreover, the topic of commercial and official application nowadays in industry is based on proved bioactivity toward exact constituents, their quantity, formulation, and pathways of activity which can provide benefits with their implementation for wide usage.

The chemical composition of essential oils in plants is different, consists of polar and nonpolar compounds, with some major compounds responsible for biological activity. The chemical constituents can be in different quantity in essential oils, although very often they can interact and create synergistic effect resulting in stronger bioactivity. Regarding complex structure of essential oils, manifested bioactivity is under consideration whether it is from major bioactive compound or from synergism activity of many bioactive compounds. Terpenes and aromatic compounds are usually low molecular weight constituents of essential oils, while monoterpenes represent the main content of essential oils.

The chemical composition of essential oils depends on applied extraction methods and performed specific conditions of extraction, and therefore, by choosing and optimizing them is related to their specific application with certain biological activities. The active compounds of attractive plants in our investigation were presented in the Table 8.1.

Every plant is known to possess a wide range of biological activities related to their constituents. Their isolates consist of many different compound classes, among which the most common are terpenes and phenolic compounds, their aldehydes, and esters. Traditional application of plants uses whole plant, while official application maintains achievement and knowledge of major bioactive compounds, which can be isolated in higher yield with purpose of their more effective application. Achieving extracts abundant in major bioactive compounds is the challenge of nowadays technology processes and their more effective application in the industry. Classifications of major bioactive compounds in some plants with exact mechanism of activity lead to optimization of extraction methods which can obtain their higher content with more effective application and activity. According to the purpose of application of essential oils, specific biological activity which should remain, extraction of individual group of compounds was estimated.

"Fenugreek seeds" (*Trigonella foenum-graecum* L.), is known as abundant source of *steroidal sapogenins* as the main precursor of steroidal hormones which are official in its use in the pharmaceutical industry. Besides steroidal sapogenins of highest importancy and official application value in industry, fenugreek seed remains natural abundant source of a high amount of *protein*, *phytosterols*, *phytoestrogens*, *vitamins E* and *vitamin D,* and minerals and essential amino acids. Obtaining extracts abundant in steroidal sapogenins is of commercial importance and economic valuable requests in pharmaceutical industry. Synthetic synthesized steroidal sapogenins are not compatible as phytosapogenins, which have advantage in their application. Among several plants with a higher percentage of steroidal sapogenins, besides fenugreek, there is well known Dioscorea genus. The amount of steroidal sapogenins is higher in Dioscorea (1–8%), in addition to fenugreek (1–2%), while competitiveness of fenugreek gets advantage in its easier and faster cultivation and rapid growth (3–4 months) (Bogdanovic 2016b). Although steroidal sapogenins gain the most significant application as the main precursor of steroidal hormones, their application in lowering cholesterol, glycemia, and high blood pressure is

Table 8.1 Composition of active compounds in plants

Plant	Active compounds	Literature
Lemon balm, *Melissa officinalis* L.	Phenolic compounds, flavonoids, phenolic acids, and tannins	Bogdanovic (2016a)
Fenugreek, *Trigonella foenum-graecum* L.	Steroidal sapogenins and saponins (diosgenin, sarsapogenin, protodioscin, and yamogenin), phytosterols, phytoestrogens, trigonelline, vitamin E, and vitamin D	Bogdanovic (2016b, 2016c, 2020)
Pomegranate, *Punica granatum* L.	Unsaturated fatty acids, oleic acid, punicic acid, linoleic acid, linolenic acid, and phytoestrogens	Đurđević (2017)
Immortelle, *Helichrysum immortelle*	Chlorogenic acid, phloroglucinol alpha-pyrone compound arzanol, sesquiterpenes (curcumene, selinene isomers, nerol derivatives, bisabolol isomers, etc.), flavonoid, coumarin, phenolic acid, acetophenone, and α-pyrone derivative	Maksimovic (2013), Maksimovic et al. (2017), Tadic (2021a)
Cannabis sativa	Cannabinoids, omega-6, omega-3, essential fatty acids, and vitamin E	Tadic (2021a)
Sage, *Salvia officinalis*	Monoterpenes (pinene, thujone isomers, 1,8-cineole, camphor, and borneol), oxygenated sesquiterpenes (such as caryophyllene oxide and viridiflorol), diterpenes (manool, ferruginol isomers, and carnosol derivatives)	Glisic (2010, 2011), Maksimovic (2013), Djarmati (1991), Menaker (2004), El A. Hayouni (2008)
Wild bilberry, *Vaccinium myrtillus* L., Ericaceae	Anthocyanin, phenolic acids (chlorogenic being the most abundant), flavonoids (isoquercetin in the highest amount), and resveratrol in leaves extract, while in seeds oil the essential ω-3 and ω-6 fatty acids were determined in favorable ratio, almost being 1, vitamin E	Tadic (2021b)
Oregano, *Origanum vulgare* L.	Eugenol, carvacrol and thymol, thymoquinone, eugenyl acetate, and trans-caryophyllene	Ivanovic (2011), Morshedlooa (2018)

getting more attention as well. Beneficial effects on regulation of glycemia, cardiovascular system, high blood pressure, and cholesterolemia lead to phytopreparations for its applications in health protection. Within steroidal sapogenins there are several significant, as diosgenin, protodioscin, sarsapogenin, yamogenin, tigogenin, etc. (Bogdanovic 2020).

Diosgenin represents the most known and significant within them, according to its best possibility as the main precursor of steroidal hormones. It is also the most significant precursor in synthesis of steroidal drugs such as estrogen, progesterone, testosterone, contraceptive agents, and corticosteroids with high value and place taking in pharmaceutical market and industry. The therapeutical effect of steroidal sapogenins is well known: they possess anticancer, hypocholesterolemic, hypoglycemic, hypotension, and antidepressant activities. Fenugreek seeds maintain their application as phytochemicals in hypolipidemic activity in decreasing the serum's total cholesterol, LDL, VLDL cholesterol, and triglyceride levels in plasma.

Isolation of steroidal sapogenins from fenugreek seeds in the form of saponin glycoside requires pretreatment of seeds which include the process of deffating and acid hydrolysis of seeds followed by its extraction (Bogdanovic 2016b).

Besides steroidal sapogenins, fenugreek seeds became valuable source in extraction of other biological active compounds, as *vitamin E* and *vitamin D*, followed by *phytoestrogens*, *phytosterols*, and *saponins* with known activity, *ursolic* and *oleanolic acid* saponins. Oleanolic and ursolic acids have bioactivity to reduce high blood pressure, hypocholesterolemic activity, and antitumor activity (Bogdanovic 2016c).

Phytosterols contained in fenugreek seeds have been of importance due to their effect to lower cholesterol levels, their anti-atherosclerotic activity, and antioxidative effect.

Vitamin E according to its efficient antioxidant properties contained in fenugreek seed in a higher amount also represents the valuable source. It is used for the treatment of osteoporosis and plays a protective role to multiple sclerosis, although, its application is well known for the treatment of skin disorders and known supplements for newborn and maternity.

Although fenugreek seed is also a valuable source of phytoestrogens and unsaturated fatty acids, it finds its usefulness in food industry as famous spice of curry powder. As rich source of protein and vitamins and minerals, have been used as nutraceutical and supplement in food especially in poor countries and skinny kids. The fenugreek seed is known to enhance the quantity of milk during maternity; therefore, its application as supplement for increasing lactation in breastfeeding mothers is always a growing interest and request in human population.

Another plant which is abundant in bioactive constituents for nervous disorders and psychological problems, exhibit relaxing in case of insomnia, anxiety, headache, and vomiting, is lemon balm (*Melissa officinalis* L.). The use of lemon balm for nervous disorders and gastrointestinal disorders is approved by some official commissions for medical health such as German commission. Lemon balm is abundant in *phenolic compounds*, *phenolic acid*, *tannins*, *flavonoids*, *terpenoids,* etc. (Bogdanovic 2016a). Some bioactive compounds of interest are contained in lemon balm such as, *rosmarinic acid*, *ferulic*, *hydroxycinnamic* and *caffeic acid* as phenolic acids, *eugenol* as phenolic compound; *geraniol, neral, citronellal, citral* and *citronellol, (E)-caryophyllene, cubebene, viridifloro*as terpenes, and flavonoid compounds as *luteolin-7-O-glucoside, isoquercitrin, apigenin-7-O-glucoside, rhamnocitrin* (Bogdanovic 2016a). The listed compounds have represented calmative, gastric, antioxidative, carminative, antiseptic, and antimicrobial activity. The antioxidative activity studied in hypolipidemic effect in lowering cholesterol have determined specific compound recognized as the carrier of hypolipidemic activity *2-(3,4 -dihydroxyphenyl)-1,3-benzodioxole-5-aldehyde*. Achieving extracts abundant in target compounds with specific bioactive properties gained their more effective application in industry.

Optimizing extraction processes and techniques obtained more effective application of lemon balm in benefit of health.

Anthocyanins are the main phenolic compound known to their beneficial properties in health effects, such as hypoglycemic and hypolipidemic activity, cardiovascular disease treatment, and anti-inflammatory effect. Anthocyanins as promoters of antioxidant activity and reduction of oxidative stress are priceless in their application in human health. *Wild bilberry (Vaccinium myrtillus L.)* as abundant source of anthocyanins takes attention for their extraction. Anthocyanins, besides other phenolic compounds, are the major constituent of the wild bilberry achieving their extraction in higher yields. Food supplements abundant in anthocyanins find their application in lowering cholesterol, lowering high blood pressure, lowering levels of glucose, and treating cardiovascular disease (Tadic 2021a). Besides anthocyanins, bilberry as a rich source of flavonoids, phenolic acids, tannins reveal antioxidant properties and hypoglycemic activity (Tadic 2021a). Vitamin E is a well-valuable constituent of bilberry antioxidative properties and therefore, bilberry found increasing application in cosmetic and food industry. Skin preparations based on phenolic compounds which possess antioxidant activity are of interest and requests in cosmetic and pharmaceutical industry.

Lowering cholesterol and high blood pressure is connected to the constituents from plant *Cannabis sativa L. (family Cannabinaceae)*. Cannabis has high level of proteins, carbohydrates, mine, fiber, tocopherols, carotenoids, and some phenolic compounds. The therapeutic potential of *cannabinoids* as anti-inflammatory and antibacterial activity, antiepileptic, anticonvulsive, antineurodegenerative, and analgesic bioactivity has been of significant value. Application of cannabis oil have been known for eczema, dermatitis, psoriasis, lichen, and rosacea as dry skin conditions diseases, which is mostly due to the balanced content of polyunsaturated fatty acids. The pharmacological activities of cannabis have been attributed to the presence of cannabinoids. The main cannabinoid constituent is Δ9-tetrahydrocannabinol (Δ9-THC), while other important cannabinoids include cannabidiolic acid (CBDA), cannabidiol (CBD), cannabigerol (CBG), cannabichromene (CBC), cannabinol (CBN), cannabicyclol (CBL), and cannabidiol (CBND) (Fathordoobady et al. 2019, Tadic 2021a). Due to bioactive compounds of high value with a wide range of health benefits for humans, gaining extracts abundant in cannabionid compounds have been of interest and challenge. Efficient extraction which can afford maximal yield of cannabinoid from Cannabis can provide more efficient application in the food, pharmaceutical, nutraceutical, and cosmetic industry.

The *Helichrysum italicum*, known as *immortelle* plant, found their application due to their antimicrobial, antioxidant, anti-inflammatory, and anti-allergic activity (Tadic 2021a). The bioactivity of immortelle is most found in application of skin disorders such as eczema, psoriasis, and inflammations of the skin. High content of phenolic acids and flavonoids found in immortelle are responsible for beneficial effect on skin disorders. Content of coumarin, chlorogenic acid, acetophenone, α-pyrone, and arzanol afforded most significant bioactivity of immortelle such as anti-inflammatory and antioxidative activity (Tadic 2021a). The activity of chlorogenic acid has been of valuable application due to significant antioxidative properties which are manifested in decreasing lipid peroxidation. The topical application of immortelle extracts abundant in these bioactive compounds exhibited successful wound healing due to antioxidant potential.

Sage, Salvia officinalis, has been noticed for a wide range of applications due to its significant antioxidant, antibacterial, hypoglycemic, antidiabetic, anti-inflammatory, fungistatic, virustatic, and antidiabetic properties. The essential oil yield varies from the 0.39 mass% to 2.7 mass%. The composition of essential oil is mostly contained of monoterpenes, sesquiterpenes, diterpenes, triterpenes, esters, and waxes. Essential oil is abundant mostly in oxygenated monoterpenes, while amount of diterpene compounds depends from origin of the plant. Due to the exhibited bioactivity, the diterpenes are most valuable compounds as bearers of examined antioxidative properties (Glisic 2010, 2011). The major compounds present in essential oil are *α -thujone, manool, carnosol derivatives, camphor, viridiflorol*, and *1,8-cineole*. Extraction by supercritical process obtained extracts abundant in most antioxidative active compounds, diterpenes, *carnosol derivatives* (Glisic 2010).

Punica granatum L., known as *pomegranate*, found their application in protection of cardiovascular diseases due to ability to reduce total and HDL cholesterol. Pomegranate found their roles in hormone therapy in menopause treatment, as content of *phytoestrogens* are abundant in plant. As well, anticancer properties of plant are proved due to regulation of the tumor suppressor gene. Pomegranate is applied in skin treatment, especially in mature and wrinkles skin due to their ability of revitalization. Pomegranate seeds are abundant in fatty acids 65–80% among which the most significant due to its bioactivity is punicic acid. Considering all the beneficial properties of the oil rich in fatty acids, this study focuses on its extraction from pomegranate seed. The microwave pretreatment of supercritical extraction of pomegranate seeds obtained extracts were abundant in fatty acid content which was of interest according to the potential application of pomegranate seed oil in the treatment of cardiovascular and skin diseases, type II diabetes mellitus, and cancer. Sixteen fatty acids were detected in the composition of extracts obtained from pomegranate seeds; polyunsaturated fatty acids were the principal fatty acid class in range of 84.3–86.1% of total fatty acids. The dominant fatty acid extracted from pomegranate seed was *punicic acid* (60.0%). The punicic acid as the principal in fatty acid content of pomegranate seeds has functional meaning in its extraction and application due to its biological properties and health beneficial effects. Besides punicic acid, obtained extracts contained high amounts of *linoleic, oleic, palmitic, and stearic acid.* Higher carbon content fatty acids including arachidic acid, behenic acid, and lignoceric acid were also present in pomegranate seeds oil samples but in low amount (Djurdjevic 2017).

Wild bilberry fruits contain 15 major anthocyanins, a large group of water-soluble flavonoids that give fruits, consisting of anthocyanidin aglycones, which are formed as 3-O-glycosides attached by galactose, glucose, and arabinose. Fruits contain fiber, vitamins, and minerals, and contains high levels of polyphenols, flavonoids, anthocyanins, and other components that exhibit significant bioactivities. Phenolic compounds contain molecules with at least one hydroxyl group attached to an aromatic ring with revealed bioactivity with recommendation for health improvement (Altiok 2022, Kerbstadt 2015, Yang 2011, Paes 2014, Tadic 2021b).

Oregano (Origanum vulgare L.) is one of the most representative plants due to its significant bioactive components including *phenolic glucosides*, *flavonoids*, *sterols*, *triterpenes*, and *tannins* (Morshedlooa 2018). Essential oil of oregano is in wide application in food supplementation according to the strong antioxidant, antimicrobial, and antifungal activity (Sarikurkcu 2015). Significant bioactivity of oregano is correlated to the major compounds of content such as *carvacrol*, *linalyl acetate*, *(Z)-α-bisabolene*, *(E)-βcaryophyllene*, *caryophyllene oxide*, *germacrene D*, and *elemol* (Gulluce 2012). Chemical profile of oregano is abundant in compounds of valuable bioactivity; the monoterpenoids were mainly consist of carvacrol, linalyl acetate, p-cymene, γ-terpinene, trans-sabinene hydrate, (Z)-β-ocimene, and sabinene, while among the sesquiterpenoids (Z)-α-bisabolene, (E)-β-caryophyllene, caryophyllene oxide, germacrene D, elemol, and β-bisabolene were the most abundant compounds. The bioactive components were found in high amount in essential oil of oregano inducing significant bioactivity of oil in its application. Some constituents were in dependence of species of oregano and harvesting year, such as methyl-eugenol, δ-3-carene, and p-cymene as well as oxygenated monoterpenes (Ivanovic 2011).

8.3 Biological Activity of Essential Oils

In regard to the proved bioactivity of essential oils and developed extraction methods, their implementation in industry and commercial markets has been more represented. Application of plants has been known from traditional use, although there was no exact mechanism of activity and functionality of extracts. Developed extraction processes and preparation of various formulations for wide commercial application and production on global level has been performed. Nowadays, in commercial market there are plenty of supplements based on plants, which might be used in food and medical purposes. There are many official registered medicaments realized in pharmaceutical industry such as Aspirin, Paclitaxel, and Vinblastine. Global market of essential oils is in constant increase representing the market value of US$700 million per year (Dima and Dima 2015). There are predictions that worldwide essential oil market will cross 13 billion in the next year to the latest report of the Global Market Insights. According to consistence, mechanism of activity, functionality, proved bioactivity and formulations, extracts found place in fragrances, cosmetics, food and supplement at global market as well as product of medical and pharmaceutical industry.

The interest for some plants has been focused toward their bioactive constituents and their pharmacological bioactivity: lemon balm (*Melissa officinalis*), fenugreek (*Trigonella foenum-graecum*), pomegranate (*Punica granatum*), immortelle (*Helichrysum immortelle*), cannabis (*Cannabis sativa*), salvia (*Salvia officinalis*), wild bilberry (*Vaccinium myrtillus*), and oregano (*Origanum vulgare*).

Lemon balm (*Melissa officinalis*) among many wide bioactive properties has found popularity as bioactive plant against psychological disorders according to the Tibetan medicine from ancient time. Therefore, application of lemon balm in medicinal use mostly represented in calmative effect for nervous disorders, in treatment of insomnia, anxiety, hyperactivity, headache, melancholia, hysteria, and high blood pressure. Gastrointestinal disorders as vomiting, colic, cramps, dysmenorrheal, and infections can be successfully treated with melissa. Hypolipidemic and hypotension effect of melissa have been recently in research and application due to its antioxidative effect in free radical reactions and lipid peroxidation. Obtaining essential oils and extracts with antioxidative effect and its application in lowering high cholesterol levels and high blood pressure was aim of further research (Bogdanovic 2016a, Shidvar 2019, Bolkent 2005).

Fenugreek (*Trigonella foenum-graecum*) is the plant mostly used and known in Arabian countries, China, and India in first form as ingredient of spice carry and further due to its benefits of immunity of human health, promoting breastfeeding, nutrition satisfied, and benefits for either man or woman in their deficiency of hormones. Nowadays, the advantage of fenugreek in human health has been investigated in treating high blood pressure, high cholesterol levels in blood, diabetes, and impotency. The main goal was focused to obtain extracts with hypotension effect, hypochloremic activity, and hypoglycemic potential. Positive effect of fenugreek seeds on lipid serum levels and blood glucose was investigated with aim to define effective formulations for preparation and its application (Sharma 2014, Saadh 2020, Bogdanovic 2016b, 2016c). The most attention is based on impotency promoting supplements based on bioactive compounds of steroidal sapogenins which are steroidal precursor (Bogdanovic 2016b, 2016c, 2020). Supplements for impotency based on fenugreek extracts found their place on commercial market worldwide as steroidal sapogenins, mainly diosgenin are steroidal precursor. Many supplements for breastfeeding products contain fenugreek seeds which promote milk production caused by high level of phytoestrogens. Breastfeeding supplements are in the form of oriblets with fenugreek extract or in the form of tea (smashed fenugreek seeds). Fenugreek seed is still mostly used in the form of spice carry.

Pomegranate seeds are globally known for its antioxidative properties. However, the pomegranate waste is in juice industry by-product and it is used as raw material containing desired unsaturated fatty acids and phenols (Basiri 2015, Djurdjevic 2017). Waste is abundant in fiber, containing a high quantity of unsaturated fatty acids, phenolic content, phytoestrogens, vitamin E, and magnesium. Nowadays, there are more requests in human health for antioxidant supplements from natural sources, therefore extracts of pomegranate seeds found implementation in medical and food applications and are preferred by consumers than synthesized antioxidants, such as BHA and BHT.

Immortelle has been mostly used in skin disorders, respiratory and digestive disorders, and fever (Acimovic 2022, Maksimovic 2013, 2023). Skin inflammations and wounds have been successfully healed by treatment of immortelle oil. Due to anti-inflammatory, antimicrobial, and antioxidant activity, this plant considered benefits in its use in cosmetic products which became most promising in commercial application (Joze 2022, Maksimovic 2023). Skin treatment with immortelle oil has been gained popularity in its implementation. Phenolic compounds and terpenes have been responsible for bioactivity of immortelle oil. Insecticidal, antiallergenic, and anti-inflammatory effect of immortelle oil contributed in the treatment of allergies, inflammation, and infections.

Salvia remained synergistic effect with immortelle increasing bioactivity which recommended their application in dermatology disorders and treatment in skin hydration and antiaging wrinkles reduction (Maksimovic 2013). Salvia as aromatic plan approved antioxidant, antimicrobial, anti-inflammatory, and anticancer properties which diverse bioactivity was prescripted to the total polyphenolic compounds (Sidiropoulou 2022). The volatiles compounds, monoterpenes and sesquiterpenes, presented abundantly in plant are remained as main bearers of antioxidant activity which isolation and extraction from plant obtained increased activity. Mainly borneol, camphor, and 1,8-cineole as constituents of salvia are mostly responsible for antiviral activity (Santoyo 2014). Fractions abundant in these groups of compounds had increased bioactivity promising efficiency in their implementation (Glisic 2010, 2011, Maksimovic 2013, Aleksovski 2007).

Cannabis is still not approved legally in most of countries for commercial use mainly due to its psychoactive properties. Meanwhile, as more countries have changed their laws for cannabis application in official and medical use, more and more investigations were performed regarding the extraction of active compounds, isolation, purification, and formulation in more efficient implementation. Terpenes and non-psychoactive cannabinoids are present in higher quantity, and due to their synergistic effects their implementation in cosmetic and skin products is recommended. Topical cannabis implementation is mostly approved for skin disorders and treatment, such as dermatitis, eczema, itchy skin, systemic sclerosis, alopecia, acne, and antiaging skin care against wrinkles and for non-dermatologic conditions such as joint stiffness. In medicine implementation of psychoactive cannabinoid dominantly tetrahydrocannabinol (THC) was of interest treating several neuro and non-neuro disorders such as Alzheimer's disease and Parkinson's disease, neurological disorders as depression, epilepsy, anxiety, insomnia, as well in treatment in tumor disease (Jokic 2022, Pattnaik 2022, Tadic 2021a). Among cannabinoids potential bioactivity such as remained antioxidant, anti-inflammatory, and analgesic effects, the potential of cannabionoids in medicinal application has been recognized in its potential. Future clinical trials will obtain researches in its efficacy and safety in use (Pattnaik 2022, Tadic 2021a).

Wild Bilberry with high levels of polyphenols, flavonoids, anthocyanins, and other components that reveal significant bioactivities found implementation in medicine and food industry (Tadic 2021b). Anthocyanins, flavonoids, and polyphenols with remained antioxidative properties proved benefits in human health and their application in medical and food industry (Altiok 2022, Paes 2014, Tadic 2021b, Gustinelli 2018). The anthocyanins reveal positive effect in the treatment of vision problems, Alzheimer's disease, cardiovascular disorders, urinary disorders, and cancer. In regard to antioxidant properties, the fiber and vitamins of wild bilberry extract improve the overall health and immune system and have positive effect in the treatment of cardiovascular diseases, neuro disorders, and tumor disease (Altiok 2022).

Oregano worldwide application in food and medical industry is mostly based on approved antiviral, anti-inflammatory, antimicrobial, and antioxidant activity (Ivanovic 2011, Busatta 2017, Santoyo 2014). Benefits of oregano on human health have been used and applied mostly in treating urinary infections and common cold. Thymol and carvacrol as major constituents of oregano extracts are mainly responsible for significant manifested bioactivity of oregano (Ocaña-Fuentes 2010). Oregano manifested increased antiviral activity as result of synergism effect with salvia. It is influence of borneol, camphor, and 1,8-cineole as constituents of salvia which are mostly responsible for antiviral activity (Santoyo 2014). Positive effect of oregano extracts in their antimicrobial activity revealed the inhibition of *Helicobacter pillori* growth. The antioxidant and antibacterial properties of oregano extracts revealed significant activity in treatment against persistent bacteria *E. coli*, *Salmonella,* and *Klebsiella pneumonia* (Stamenic 2014). Oregano remained antiproliferative activity in tumor cells of HeLa, positive treatment in the neurodegenerative disorders, and prevention and treatment of atherosclerosis (Ocaña-Fuentes 2010) (Table 8.2).

8.4 Extraction Methods of Plant Essential Oils

Great effort is being dedicated to the investigation for more convenient, cheaper sources of natural phytomedicaments, which can be alternative to synthetic drugs (with plenty of side effects), as well as to the development and optimization of most efficient and convenient extraction techniques. Extraction with traditional techniques and conventional solvents are characterized by low selectivity and high temperatures, which could result in degradation of the desired compounds. Obtained extracts usually have solvent residues, which can be toxic and require purification and extended process of extraction which are more expensive and again can have some residue in traces which is not convenient in medicinal use. Therefore, a need for development of advanced and improved extraction techniques has been posted which will afford a higher selectivity, without toxic residues in obtained extracts for shorter time of extraction along with more economical process of extraction. Supercritical extraction (SCE) represents that advanced technique with ecological satisfied requests for medicinal use of obtained extracts according to the achieved clean extracts without toxic residues and no need for further purification and extended processes. SCE exhibit higher selectivity compared to conventional techniques of extraction, achieving extraction of targeted bioactive group of compounds. Selectivity of SCE toward some specific bioactive groups of compounds can be modulated by changing extraction conditions (pressure and temperature). Optimization of SCE condition can be used for obtaining the "clean" extracts with specific biological active compounds for most convenient economical and time-satisfied process. According to easy and total separation of supercritical fluid from the extract achieved only by lowering pressure which release CO_2 from extract, SCE presents type of extraction process which gave extracts without residual solvents. Such product of extraction is more convenient and safe for application in pharmaceutical, food, and cosmetics industry.

The selection of extraction method influences quantity and quality of extracts which consequently correlate biological activities. The conditions of applied method

Table 8.2 Biological activity of selected plants

Plant	Bioactivity	Literature
Melissa officinalis L.	Antioxidative, lipid lowering activity, antidiabetic, antihypertensive, anti-inflammatory, calmative	Bogdanovic (2016a)
Fenugreek seeds, (Trigonella foenum-graecum L.)	Antioxidative, anti-inflammatory, lipid lowering effect, lowering blood glucose, antihypertensive, menopause treatment, impotency treatment-steroid precursor, antitumor	Bogdanovic (2016b, 2016c 2020)
Pomegranate (*Punica granatum* L.)	Cardiovascular and skin diseases, type II diabetes mellitus, and cancer, antioxidative, antitumor	Đurđević (2017)
Immortelle (Helichrysum immortelle)	Antioxidant, antiviral, antimicrobial and antifungal behavior, and antiallergen agents, for the treatment of respiratory diseases and different skin disorders such as eczema, psoriasis, and inflammatory conditions of the skin, as well as for protection against lipid peroxidation caused by oxidative stress	Maksimovic (2013), Tadic (2021a, b)
Cannabis sativa	Reduce cholesterol, high blood pressure, eczema, dermatitis, psoriasis, lichen planus, and acne rosacea, antifungal activity, antimicrobial activity	Tadic (2021a)
Salvia officinalis	Antioxidant, antimicrobial, and anti-inflammatory properties of sage's extract or sage essential oil	Glisovic (2010, 2011), Maksimovic (2013), Djarmati (1991), Menaker (2004), El A. Hayouni (2008).
Wild bilberry, (*Vacciniummyrtillus* L., Ericaceae)	Antioxidative, anti-inflammatory, lipid-lowering effect, lowering blood glucose, antitumor	Tadic (2021b)
Oregano (*Origanum vulgare* L.)	Antioxidant, cure convulsive coughs, digestive disorders, menstrual problems, bronchitis and asthma, and frequently as a carminative, diaphoretic, expectorant, stimulant, anti-inflammatory and antimicrobial agent among other medical applications	Ivanovic (2011), Morshedlooa (2018), Sarikurkcu (2015), Mozaffarian (2013)

of extraction could be directly correlated to the composition of bioactive compounds as, for example, content of terpenes and terpenoids, phenol-derived aromatic components, and aliphatic components. Optimization of extraction condition also contributes to more efficient and effective extraction process resulted in desired quality and higher quantity of essential oils.

The chemical composition of essential oil depends also on the applied extraction method. The appropriate selection of extraction method is related to the further application of essential oil and its purpose in accordance to the biological activity and application in food, fragrance, repellent, and cosmetics of pharmaceuticals. In food industry, more frequent isolation process of essential oil is given to the usual traditional techniques as hydrodistillation, steam distillation, or organic solvent extraction, while the supercritical extraction is favorable for application in pharmaceutical industry according to higher requests for quality of active compounds (Table 8.3).

Hydrodistillation is traditional and classical method from ancient times for extraction of essential oils from various plants. It is relatively simple extraction process based on the extraction of water-soluble compounds from plants. Regarding the boiling point of water during extraction process, which affords cycling of vapor and water during extraction of bioactive compounds, besides many extracted compounds some of them potentially bioactive, can degrade at that temperature. Disadvantage of this traditional method is usually low yield and more complex separation of essential oils from not miscible water followed by lose some quantities of oils. In advantage of cheaper and simple equipment and simple process of hydrodistillation, a longer time of extraction process followed with low yield set up the new advanced methods for essential oil isolation. An organic solvent extraction took part in more applied extraction process as they achieve higher extraction yields. However, toxic solvent residues and need for purification of the extracts contributed to the higher costs and longer processes of organic solvent extraction. In regard to the residue of, for example, toxic solvents used for extraction (e.g., chlorinated hydrocarbons, even in traces after severe purification) in obtained extracts, make inconvenient extracts for medical use. Using environmental convenient solvents with safe use in medicine induced new extraction requests. The microwave-assisted extraction and extraction using ultrasounds present advanced methods which can reduce the extraction time and increase the extraction yield, and therefore, these methods are applied and used as pretreatment in extraction process. The microwave and ultrasound pretreatment cause break of the cell structure of the vegetable material and easier and rapid release of different compounds during extraction process.

Supercritical extraction remains advanced process of extraction which satisfied environmental impact for safe use of extracts in human health. This advanced extraction process achieved higher yields, shorter time of extraction, and selective extraction toward specific group of bioactive compounds. In order to improve extraction process for safer implementation in medicine, the supercritical CO_2 extraction (SCE CO_2) uses CO_2 as convenient and safe fluid in supercritical state as extraction solvent (supercritical conditions are pressure and temperature above critical conditions of CO_2). Carbon dioxide at this state is supercritical fluid, where is no exact difference within the liquid and the gaseous state of CO_2, extragent-supercritical fluid possesses properties of both liquid (density) and gas state (viscosity and diffusivity). Regarding the advantages of supercritical CO_2 as chemically inert, nontoxic, low cost, its implementation in extraction processes of bioactive compounds from essential plants is safe, economical, and environmentally satisfied. Supercritical extraction affords clean extracts from plants without toxic resides which make them convenient in food and medical industry. There is no need for further purification inducing shorter time of extraction with lower costs and economical process.

Table 8.3 Extraction methods of essential oils of selected plants

Method	Description of the method	Advantages	Disadvantages	Literature
Hydrodistillation	Plants are soaked in the water and heated until boiling. The water and compounds from plant evaporates, pass through a condenser, and recovered in boiling flask again as a liquid. Cycling of the water in boiling and evaporation collects and extracts vapor compounds from plant.	Easy and cheap method, easily applied and common.	Overheating induces degradation of some thermolabile bioactive compounds time, low yield, low extraction time, loss of polar molecules due to the use of water and harder separation.	Bogdanovic (2016a), Stamenic (2014)
Organic solvent extraction	The plant material is mixed with an organic solvent to the boiling point for some hours, extracting compounds from plant material (ethanol, hexan, and chloroform). Then the organic solvent is evaporated under vacuum.	Higher yield of extracted compounds, especially nonpolar, adding water to ethanol can induce higher extraction rate of polar compounds either	Purification and quality of extraction due to residue toxic solvents, long time of extraction and purification of organic solvent, more expensive process due to organic solvents and purifications	Bogdanovic (2016b, 2016c)

(continued)

Table 8.3 (continued)

Method	Description of the method	Advantages	Disadvantages	Literature
Supercritical extraction	Plant material is placed in a reactor vessel and extracted under process condition of optimized pressure, temperature, and flow of supercritical CO_2. The bioactive compounds are extracted in supercritical CO_2 and collected from vessel in clean extract, while CO_2 is separated as a gas in edition of temperature and pressure	Fast process, nontoxic, green chemistry extraction process, obtaining clean and safe extracts for pharmaceutical and cosmetic requests, selectivity, efficient extraction of nonpolar compounds, possible optimization in obtaining higher yield of specific target bioactive compounds	Equipment investment, favorable nonpolar compounds extraction, more complex processing with equipment	Tadic (2021a, b), Djurdjevic (2017), Maksimovic (2013, 2023), Bogdanovic (2016a, 2016b), Ivanovic (2011), Skala et al. (2002)
Microwave-assisted extraction	The samples are mixed with water and heated in a microwave oven previous to the other method of extraction which can be hydrodistillation, organic solvent or supercritical, extraction	The plant material in this pretreatment is well mixed and stirred, allows better extraction of compounds from plant material	Increased nonselectivity extraction, extraction of unwanted compounds	Djurdjevic (2017), Glisic (2011)

Selection of extraction method should produce extracts with valuable source of bioactive compounds from plant material. Extraction and separation methods should consider experimental conditions, achieved yields, number of identified compounds, isolation of specific compound, or group of compounds with significant bioactivity, time of extraction, quality of extracts, and economical satisfaction. The biological activities of extracts are usually screened and distinct in regard to applied conditions of extraction methods. Relation of biological activity of extracts and applied extraction methods gave insights in choosing more convenient method for extraction. Choosing preferable extraction method is related to the quality and achieved yield of extracts.

Melissa officinalis Extraction of essential oils from lemon balm was mostly realized by classical methods of extractions (hydrodistillation), which is followed by low yield of essential oils. In addition to the classical methods, the conventional methods of extraction gained increased use such as supercritical carbon dioxide extraction, liquid CO_2 extraction, and microwave-assisted extraction. The advantage of supercritical extraction is mainly in its selectivity and ability for specific bioactive compounds isolation. The two-step CO_2 extraction was applied with the purpose to achieve extraction of essential oil in the first step of extraction at 10 MPa and 40 °C, and extracts in the second step of extraction at 30 MPa and temperature: 25 °C, 40 °C, and 100 °C. The achieved yield was higher in second step of extraction at higher temperature and pressure, while chemical profile of obtained extracts and extracted compounds was very different. In the first step of extraction, monoterpenes and sesquiterpenes were mostly separated, while in the second step of extraction isolated compounds with greater molecular mass such as heavy alcohols and waxes, and phenolic compounds as eugenol and thymol. According to DPPH antioxidant activity of obtained extracts, the extracts obtained in the supercritical extraction at 30 MPa/40 °C and 30 MPa/100 °C revealed the highest activity. A higher antioxidant activity is a result of larger amount of heavier compounds (diterpenes and heavy alcohols) or increased amounts of thymol and eugenol (Bogdanovic 2016a).

Fenugreek Supercritical extraction was applied to obtain a maximal yield of extracts with the highest content of steroidal sapogenins as steroidal precursor and important pharmaceutical bioactive compounds in supplements for impotency as well as for hypolipidemic activity. The definition of optimal conditions for maximal contents of diosgenin as main representative and other steroidal sapogenins as sarsasapogenin, protodioscin, as well as oleanolic and ursolic acid sapogenins in obtained extracts was valuable in fenugreek application and implementation in commercial market. Optimal process conditions determined for achieving maximal extract yield and the content of steroidal sapogenins were 24.7 MPa and 38.15 °C. Separation of steroidal sapogenins from fenugreek plant material requires (previously mentioned) pretreatment of seeds prior to applied extraction (defatting and subsequent hydrolysis process). The pretreatment to the supercritical extraction remains acid hydrolysis, which leads to release of steroidal sapogenins bound from sugar chains in plant, while the defatting step enables obtaining more pure and quality SC-CO_2 extract. The maximum yield of isolated diosgenin from plant as main steroidal precursor of commercial value was about 1.3% of total extract (Bogdanovic 2016b, 2020).

Obtaining extracts abundant in phytosterols, vitamin E, and vitamin D was of interest as well. The optimal supercritical extraction conditions for achieving the maximal yield of valuable active compound such as phytosterols (16.28%), vitamin E (1.44%), and vitamin D (0.71%) were at slightly lower pressure (22.8–24.8 MPa) and temperature (39.9–42.0 C) (Bogdanovic 2016c).

Pomegranate The microwave pretreatment of seeds followed by supercritical extraction estimated benefits in increased yield, although in positive influence on profile and quantity of fatty acid content. Microwave pretreatment was performed at conditions of 100, 250, and 600 W during 2 and 6 min prior to the supercritical extraction. Classical method of solvent extraction with *n*-hexane in Soxhlet apparatus was performed in addition to supercritical extraction. Maximal oil yield with previous microwave pretreatment in *n*-hexane extraction (36.3%) was obtained with microwave radiation of 600 W for 6 min, while for SC CO_2 extraction the maximal oil yield (27.2%) was with 250 W for 6 min. Punicic acid, representative among fatty acids due to revealed bioactivity, was the most abundant fatty acid in pomegranate seed oil (>60%). Microwave pretreatment with supercritical extraction achieved increased yield of extracts, followed by the valuable profile and amount of fatty acids which are recommended in potential application in the treatment of cardiovascular disease, atherosclerosis, hyperlipidemia, diabetes, and tumor (Djurdjevic 2017).

Immortelle It represents abundant source of bioactive compounds, such as terpenes and phenolic compounds. Investigation and defining the conditions for extraction and separation of bioactive compounds from immortelle, as well as for maximal yield and number of bioactive compounds were performed at different pressure (9–26 MPa) and temperature (40–60 °C) and compared to the yield and composition of extracts. Average selectivity was achieved toward monoterpenes, sesquiterpenes, and diterpenes presented in immortelle plant, under these conditions. Similar composition profile in extracts was obtained by hydrodistillation. Immortelle is primarly abundant in sesquiterpenes such as curcumene and selinene isomers, nerol derivatives, and bisabolol isomers (Maksimovic 2013, 2023). Performed supercritical extraction with simultaneous extraction of sage and immortelle proved increased extraction and separation in either yield or amount of bioactive compounds confirming and revealing synergism of these two plants (Maksimovic 2013).

Cannabis The supercritical extraction is increased request as “green” and selective extraction method for extraction of cannabinoids and terpenes from industrial cannabis. The optimized supercritical conditions were defined in two ranges of pressure and temperature, to perform maximal solubility of specific cannabinoids and terpenes and their separation in obtained extracts (Tadic 2021a). Extraction of non-psychoactive cannabinoids in CO_2 compared to psychoactive cannabinoids THC extraction was estimated. The lower temperatures were more efficient for good solubility of non-psychoactive cannabinoids, while higher temperature was more convenient for psychoactive cannabinoids (THC) extraction. The optimal condition for maximal yield of extract was determined at 40 °C and 30 MPa. The first fraction of supercritical isolate was collected at 10 MPa and 40 °C, which contains mainly biological active volatile compounds recommended for application in cosmetic and food industry.

Wild bilberry Supercritical extraction has been used to improve extraction of valuable bioactive compounds from wild bilberry (total phenolics, tannins, and flavonoids). The bilberry was pretreated with ethanol as cosolvent promoting supercritical extraction of polyphenolic compounds. The most efficient extraction condition was estimated at 30 MPa and 40 °C, gaining the highest yield and most effective extracts (Tadic 2021b).

Sage (*Salvia officinalis*) The supercritical extracts were collected during extraction of bioactive compounds from sage, mainly diterpenes and sesquiterpenes which are responsible for antioxidant activity. Fractional extraction of plant at 10 MPa obtained extraction of lower molecular compounds, monoterpenes and oxygenated monoterpenes, while extraction at 15 MPa and higher pressure achieve extracts with higher molecular compounds such as diterpenes, mostly responsible for bioactivity (Glisic 2010). The fractional supercritical extraction with prior ultrasound pretreatment (USP) to SCE was also performed with the aim to promote extraction of higher molecular compounds (Glisic 2011). The sugars were isolated during USP, while extracts obtained after SCE were abundant in diterpenes and sesquiterpenes. Extracts obtained by USP and SC were compared to the essential oil obtained by hydrodistillation, revealing the most strongest antioxidant activity of supercritical extract. The diterpenes and sesquiterpenes from supercritical extract revealed higher activity in addition to the monoterpenes and sesquiterpenes from essential oil by hydrodistillation (Maksimovic 2013).

Oregano Supercritical extraction is analyzed with a goal to obtain the maximal yield of bioactive compounds which are bearers of antioxidant, antibacterial, and antiviral activity (Ivanovic 2011, Stamenic 2014). The fractional two-step CO_2 extraction at supercritical conditions was exhibited, the same applied as in the case of extraction of melissa. The first fraction collected during extraction represented the essential oil of oregano, while the second fraction was collected at 30 MPa and 100 °C containing the most effective antioxidant activity. Essential oil and extract obtained by supercritical extraction were compared to the essential oil obtained by simple hydrodistillation. Extract obtained by supercritical extraction has a higher antioxidant activity than essential oil collected after hydrodistillation. Furthermore, the supercritical extract showed also strong antibacterial activity toward staphylococci, including MRSA strain.

8.5 Industrial Application of Essential Oil and Bioactive Compounds from Plants

Plants have been used from ancient times as spices, food, medicaments, fragrance, preservative, insecticide, and massage. In addition to the previous use, nowadays application requires approved activity, exact mechanism of activity, most efficient method for obtaining bioactive compounds, optimization of extraction method, design of economical processes, and ecological satisfaction with most required quality and quantity of obtained isolates (essential oils, extracts, etc.) (Bogdanovic 2022). There is an increasing request for implementation of isolate from plants in the food, pharmaceutical, and cosmetic industry. The increasing interest for implementation of plants is due to their less side effects which might possess synthetic products, and due to their less-harmful residue of toxic chemicals. Essential oil and extracts from plant gained popularity due to biological properties with great variety in their content which manifested potential benefit of activity. Benefit of natural products in addition to the synthetic products is in their biocompatibility and in the meantime higher manifested bioactivity. As there is increased demand for implementation of natural products from consumers, industry must fulfill the requests of market and to secure stable and economical production based on optimized and efficient processes of production. Furthermore, produced extracts must be insecure as it depends on cultivation, which can be slow growth, low yield, severe conditions for growth of some plants, difficult handling of isolation processes to obtain natural products followed by satisfied quality and quantity. Industrial implementation of investigated plants was performed and obtained in optimized procedures, with most convenient methods with defined formulations of natural products in most effective sight, which is represented in Table 8.4.

8.5.1 Application in Pharmaceutical and Medical Industry

The recent increase in the demands for phytochemicals alternative medicines, compared to conventional therapy of synthetic medicaments, has contributed more research and investigation and many officially registered phytopreparations for the treatment of various health issues and disorders. The different manifested activity of extracts from plants are in relation to the specific bioactivity of some group of compounds in their content, which can be influenced by method of extraction and applied process conditions. Bacteria resistance to antibiotics resulted in requests for alternative antibacterial products from plant, among which, the oregano, the fenugreek seed, the wild bilberry, the sage, and melissa exhibited significant activity. Antioxidant activity as exhibited in fenugreek, oregano, melissa, pomegranate, and cannabis is in increased demand for lowering high cholesterol levels and high blood pressure. Extracts of fenugreek, melissa, pomegranate, and wild bilberry found implementation at market as drugs or supplements due to the hypoglycemic effect on diabetic, lowering levels of glucose. Anti-inflammatory activity of cannabis and immortelle resulted in cream preparation for skin hydration, antiaging treatment, and skin disorders such as eczema, psoriasis, cream for lichen, cream for acne rosacea, and antifungal activity cream. Used directly or included in drug formulations, oregano extracts are required for their pharmacological effects on the urinary system benefits and urinary inflammations.

Table 8.4 Industrial application of plants

Plant					Literature
	Spices	Food industry	Cosmetic industry	Pharmaceutical industry	
Melissa officinalis L.	Food ingredient, tea	Calmative tea, menopause treatment	Essential oil, fragrance for perfume, repellent	Skin cream, supplements for reducing cholesterol, supplements for reducing high pressure, supplements for diabetes	Bogdanovic (2016a)
Trigonella foenum-graecum L.	Curry powder	Nutraceutical supplement, impotency supplements, promoting breast milk feeding supplements, treatment for menopausal disorders, antidiabetic, lowering cholesterol, lipids levels supplements	Phytoestrogen preparation, skin cream recover	Steroidal hormones, supplements for reducing cholesterol and high blood pressure, supplements for diabetes	Bogdanovic (2016b, 2016c, 2020, 2022)
Punica granatum L.		Antioxidative supplements	Skin cream, antiaging cream, acne reduce tonic, eczema cream	Supplements for lowering cholesterol	Đurđević (2017)
Helichrysum immortelle			Antiaging cream, essential oil in massage, fragrances for perfumes	Antiallergen agents, treatment for respiratory diseases, cream for skin disorders such as cream for eczema, psoriasis, and inflammatory conditions of the skin	Maksimovic (2013), Tadic (2021a, b)

(continued)

Table 8.4 (continued)

Plant					Literature
	Spices	Food industry	Cosmetic industry	Pharmaceutical industry	
Cannabis sativa				Supplements for reducing cholesterol and high blood pressure, eczema cream for skin, cream for dermatitis, cream for psoriasis, cream for lichen planus, and cream for acne rosacea, antifungal activity cream	Tadic (2021a)
Salvia officinalis	Tea		Antimicrobial liquid for teeth	Antimicrobial oriblets for the neck	Glisovic (2010, 2011), Maksimovic (2013), Djarmati (1991), Menaker (2004), Hayouni (2008)
Wild bilberry, *Vaccinium myrtillus* L., Ericaceae	Odor addition for food			Supplements for lipid lowering effect, supplements for lowering blood glucose	Tadic (2021b)

(continued)

Table 8.4 (continued)

Plant					Literature
	Spices	Food industry	Cosmetic industry	Pharmaceutical industry	
Origanum vulgare L.			Essential oil	Essential oil for convulsive coughs, digestive disorders, menstrual problems, bronchitis and asthma diaphoretic, expectorant, stimulant, anti-inflammatory and antimicrobial agent, oriblets for urinary infections (Palisept)	Ivanovic (2011), Morshedlooa (2018), Sarikurkcu (2015), Mozaffarian (2013)

8.5.2 Application in Cosmetic Industry

Extracts of melissa, wild bilberry, pomegranate and immortelle are used as fragrances and perfumes for various cosmetic products, soaps, shampoos, conditioner, perfumes, and antiperspirants. Skin cream for hydration and antiaging effect contains immortelle essential oil. Antiaging ability is related to the antioxidant activity which reduces free radicals in skin aging. Cannabis oil is more frequent ingredient of cosmetic products, for skin hydration as well as for pharmacological preparation for skin disorders as well as immortelle and melissa due to their antioxidant, antimicrobial, and anti-inflammatory properties. Melissa and immortelle are also ingredients of repellent products due to their insecticidal essences.

8.5.3 Application in Food Industry

Antioxidative and antimicrobial properties of extracts from plants found implementation in food as preservatives, extending the storage stability of the food and inhibiting microbial growth and protecting food from oxidation. The extracts added to the food contributed to its taste too.

8.5.4 Application as Flavors and Fragrances

The melissa essential oil has valuable market price and the highest production requests worldwide. Essential oil of melissa is in use of perfumes, cosmetic, massage oil, while in medical purpose it has mainly calmative activity. Fenugreek is primarily used as spice and main ingredient of well-known ingredient in carry, with prior heritage in Indian food. Milled fenugreek seeds are part of content in medical supplement for promoting breastfeeding at mothers. Extracts from fenugreek is used as supplements for decreased impotency in men. Wild bilberry extracts are used as flavors in beverages, candies, gelatins, and food products. Sage, melissa, and immortelle are used in oral care products, as part of chewing gum or cleaning and refreshing products. Immortelle and melissa are often represented in perfume industry due to their bioactive volatile compounds with fragrances.

8.6 Conclusion

Use of various plants for producing so-called natural products are significantly integrated in the industry and global market due to their properties related to the antioxidant, antibacterial, anti-inflammatory, and antiviral activity. In addition to the synthetic products, biocompatibility of phytopreparations due to their nature contributes to a higher efficiency of their activity and implementation. In regard to the unsecure controlling growth conditions of plant, seeds, rise, and germination, the exact procedure of production methods for obtaining desired quantities of plants are one of requests defined by market demands. Furthermore, chosen methods for extraction and their optimization to satisfy quality and quantity of extracts and validation and standardization methods for development of plants and extracts taking into account their content of bioactive compounds for industry production and implementation is in increasing demand. The global market is going to prepare, in the close future, the new requests for plant applications and industrial production and commercialization of various natural products from plant material.

References

Aleksovski SA, Sovová H (2007) Supercritical CO2 extraction of Salvia officinalis L. J Supercrit Fluids 40(2):239–245. https://doi.org/10.1016/j.supflu.2006.07.006

Altiok E, Kacmaz S, Altiok D (2022) Extraction of bioactive compounds in wild bilberry (Vaccinium Myrtillus L.) in the eastern Black Sea region with different techniques Tr. J Nature Sci 11(2):130–136

Aćimović M, Zeremski T, Šovljanski O, Lončar B, Pezo L, Zheljazkov VD, Pezo M, Šuput D, Kurunci Z (2022) Seasonal variations in essential oil composition of immortelle cultivated in Serbia. Horticulturae 8:1183. https://doi.org/10.3390/horticulturae8121183

Asiri S (2015) Evaluation of antioxidant and antiradical properties of pomegranate (*Punica granatum* L.) seed and defatted seed extracts. J Food Sci Technol 52(2):1117–1123. https://doi.org/10.1007/s13197-013-1102-z

Bogdanovic A, Tadic V, Arsic I, Milovanovic S, Petrovic S, Skala D (2016a) Supercritical and high pressure subcritical fluid extraction from lemon balm (Melissa officinalis L., Lamiaceae). J Supercrit Fluids 107:234–242

Bogdanovic A, Tadic V, Stamenic M, Petrovic S, Skala D (2016b) Supercritical carbon dioxide extraction of Trigonella foenum-graecum L. seeds: process optimization using response surface methodology. J Supercrit Fluids 107:44–50

Bogdanovic A, Tadic V, Ristic M, Petrovic S, Skala D (2016c) Optimization of supercritical CO2 extraction of fenugreek seed (Trigonella foenum-graecum L.) and calculating of extracts solubility. J Supercrit Fluids 117:297–307

Bogdanovic A, Tadic V, Petrovic S, Skala D (2020) Supercritical CO_2 extraction of steroidal sapogenins from fenugreek (Trigonella foenum-graecum L.) seed. Chem Ind Chem Eng Q 26(2):171–182

Bogdanovic A, Tadic V, Petrovic S, Skala D (2022) Optimization of the supercritical carbon dioxide extraction of phytochemicals from fenugreek seeds, in the book. In: Ghosh D, Thakurdesai P (eds) Fenugreek: traditional and modern medicinal uses. Taylor & Francis Group. https://doi.org/10.1201/9781003082767

Bolkent S, Yanardag R, Karabulut-Bulan O, Yesilyaprak B (2005) Protective role of *Melissa officinalis* L. extract on liver of hyperlipidemic rats: a morphological andbiochemical study. J Ethnopharmacol 99(3):391–398. https://doi.org/10.1016/j.jep.2005.02.038

Busatta C, Barbosa J, Cardoso RI, Paroul N, Rodrigues M, de Oliveira D, de Oliveira JV, Cansian RL (2017) Chemical profiles of essential oils of marjoram (*Origanum majorana*) and oregano (*Origanum vulgare*) obtained by hydrodistillation and supercritical CO_2. J Essent Oil Res 29(5):367–374. https://doi.org/10.1080/10412905.2017.1340197

Dima C, Dima S (2015) Essential oils in foods: extraction, stabilization, and toxicity, current opinion in food. Science 5:29–35. https://doi.org/10.1016/j.cofs.2015.07.003

Djarmati RM, Jankov E, Scwirtlic B, Djulinac A (1991) Djordjevic, high antioxidant activity of extracts obtained from sage by supercritical CO2 extraction. J Am Oil Chem Soc 68:731–734

Đurđević S, Milovanović S, Šavikin K, Ristić M, Menković N, Pljevljakušić D, Petrović S, Bogdanović A (2017) Improvement of supercritical CO_2 and n-hexane extraction of wild growing pomegranate seed oil by microwave pretreatment. Ind Crops Prod 104:21–27

Fathordoobady F, Singh A, Kitts DD, Pratap Singh A (2019) Hemp (Cannabis sativa L.) extract: Antimicrobial properties, methods of extraction, and potential oral delivery. Food Rev Intl 35(7):664–684

Glisic S, Ivanovic J, Ristic M, Skala D (2010) Extraction of sage (Salvia officinalis L.) by supercritical CO2: kinetic data, chemical composition and selectivity of diterpenes. J Supercrit Fluids 52(1):62–70. https://doi.org/10.1016/j.supflu.2009.11.009

Glisic SB, Ristic M, Skala DU (2011) The combined extraction of sage (Salvia officinalis L.): ultrasound followed by supercritical CO2 extraction. Ultrason Sonochem 18(1):318–326. https://doi.org/10.1016/j.ultsonch.2010.06.011

Gulluce M, Karadayi M, Guvenalp Z, Ozbek H, Arasoglu T, Baris O (2012) Isolation of some active compounds from Origanum vulgare L. ssp vulgare and determination of their genotoxic potentials. Food Chem. 130(2):248–253

Gustinelli G, Eliasson L, Svelander C, Alminger M, Ahrné L (2018) Supercritical CO2 extraction of bilberry (Vaccinium myrtillus L.) seed oil: fatty acid composition and antioxidant activity. J Supercrit Fluids 135:91–97

Hayouni El A, Chraief I, Aderrabba M, Bouix M, Leveau JY, Mohammed H, Hamdi M, Tunisian Salvia officinalis L. and Schinus molle L (2008) Essential oils: their chemical compositions and their preservative effects against salmonella inoculated in minced beef meat. Int J Food Microbiol 125:242–251

Ivanovic J, Zizovic I, Ristic M, Stamenic M, Skala D (2011) The analysis of simultaneous clove/oregano and clove/thyme supercritical extraction. J Supercrit Fluids 55(3):983–991

Jažo Z, Glumac M, Drventić I, Žilić L, Dujmović T, Bajić D, Vučemilo M, Ivić E, Bektić S, Anačkov GT, Radan M (2022) The essential oil composition of *Helichrysum italicum* (Roth)

G. Don: influence of steam, hydro and microwave-assisted distillation. Separations 9:280. https://doi.org/10.3390/separations9100280

Jokic S, Jerkovic I, Pavic V, Aladic K, Molnar M, Jakovljevic Kovac M, Vladimir-Kneževic S (2022) Terpenes and cannabinoids in supercritical CO2 extracts of industrial hemp inflorescences: optimization of extraction, antiradical and antibacterial activity. Pharmaceuticals 15:1117. https://doi.org/10.3390/ph15091117

Kerbstadt S, Eliasson L, Mustafa A, Ahrné L (2015) Effect of novel drying techniques on the extraction of anthocyanins from bilberry press cake using supercritical carbon dioxide. Innovative Food Sci Emerg Technol 29:209–214. https://doi.org/10.1016/j.ifset.2015.02.002

Maksimovic S, Stankovic M, Roganovic S, Nesic I, Zvezdanovic J, Tadic V, Zizovic I (2023) Towards a modern approach to traditional use of Helichrysumitalicum in dermatological conditions: in vivo testing supercritical extract on artificially irritated skin. J Ethnopharmacol 301:115779. https://doi.org/10.1016/j.jep.2022.115779

Maksimovic S, Kesic Z, Lukic I, Milovanovic S, Ristic M, Skala D (2013) Supercritical fluid extraction of curry flowers, sage leaves, and their mixture. J Supercrit Fluids 84:1–12

Maksimovic S, Tadic V, Skala D, Zizovic I (2017) Separation of phytochemicals from Helichrysum italicum: an analysis of different isolation techniques and biological activity of prepared extracts. Phytochemistry 138:9–28

Menaker A, Kravets M, Koel M, Orav A (2004) Identification and characterization of supercritical fluid extracts from herbs. C R Chim 7:629–633

Morshedlooa MR, Salamib SA, Nazerib V, Maggic F, Craker L (2018) Essential oil profile of oregano (Origanum vulgare L.) populations grown under similar soil and climate conditions. Ind Crops Prod 119:183–190

Ocaña-Fuentes A, Arranz-Gutiérrez E, Señorans FJ, Reglero G (2010) Supercritical fluid extraction of oregano (Origanumvulgare) essentials oils: anti-inflammatory properties based on cytokine response on THP-1 macrophages. Food Chem Toxicol 48(6):1568–1575. https://doi.org/10.1016/j.fct.2010.03.026

Paes J, Dotta R, Barbero GF, Martínez J (2014) Extraction of phenolic compounds and anthocyanins from blueberry (Vacciniummyrtillus L.) residues using supercritical CO2 and pressurized liquids. J Supercrit Fluids 95:8–16. https://doi.org/10.1016/j.supflu.2014.07.025

Pattnaik F, Nanda S, Mohanty S, Dalai AK, Kumar V, Ponnusamy SK, Naik S (2022) *Cannabis*: chemistry, extraction and therapeutic applications. Chemosphere 289:133012. https://doi.org/10.1016/j.chemosphere.2021.133012

Rovetto LJ, Aieta NV (2017) Supercritical carbon dioxide extraction of cannabinoids from Cannabis sativa L. J Supercrit Fluids 129:16–27. https://doi.org/10.1016/j.supflu.2017.03.014

Saadh MJ (2020) Hypoglycemic and hypolipidemic activity of combined milk thistle and fenugreek seeds in alloxan-induced diabetic albino rats. Vet World 13(8):1732–1736. https://doi.org/10.14202/vetworld.2020.1732-1736

Santoyo S, Jaime L, García-Risco MR, Ruiz-Rodríguez A, Reglero G (2014) Antiviral properties of supercritical CO_2 extracts from oregano and sage. Int J Food Prop 17(5):1150–1161

Sarikurkcu C, Zengin G, Oskay M, Uysal S, Ceylan R, Aktumsek A (2015) Composition antioxidant, antimicrobial and enzyme inhibition activities of two *Origanum vulgare* subspecies (subsp. *vulgare* and subsp. *hirtum*) essential oils. Ind Crop Prod 70:178–184

Sharma MS, Choudhary PR (2014 Aug) Hypolipidemic effect of fenugreek seeds and its comparison with atorvastatin on experimentally induced hyperlipidemia. J Coll Physicians Surg Pak 24(8):539–542

Sidiropoulou E, Marugán-Hernández V, Skoufos I, Giannenas I, Bonos E, Aguiar-Martins K, Lazari D, Papagrigoriou T, Fotou K, Grigoriadou K, Blake DP, Tzora A (2022) In vitro antioxidant, antimicrobial, Anticoccidial, and anti-inflammatory study of essential oils of oregano, thyme, and sage from Epirus, Greece. Life (Basel) 12(11):1783. https://doi.org/10.3390/life12111783

Shidfar F, Safari M, Hosseini AF, Huseini HF, Heidari I, Rajab A (2019) Efficacy of *Melissa officinalis* L. (lemon balm) extract on glycemic control and cardiovascular risk factors in individuals with type 2 diabetes: A randomized, double-blind, clinical trial. Phytother Res 33(3):651–659. https://doi.org/10.1002/ptr.6254

Skala DU, Žižović, IT, Gavrančić, SS (2002) Supercritical fluid extraction: application in the food industry. Hemijska industrija 56(5):179–190

Stamenic M, Vulic J, Djilas S, Misic D, Tadic V, Petrovic S, Zizovic I (2014) Free-radical scavenging activity and antibacterial impact of Greek oregano isolates obtained by SFE. Food Chem 165:307–315. https://doi.org/10.1016/j.foodchem.2014.05.091

Tadic V, Žugic A, Martinovic M, Stankovic M, Maksimovic S, Frank A, Nešic I (2021a) Enhanced skin performance of emulgel vs. cream as systems for topical delivery of herbal actives (Immortelle Extract and Hemp Oil). Pharmaceutics 13:1919. https://doi.org/10.3390/pharmaceutics13111919

Tadic VM, Nešic I, Martinovic M, Rój E, Brašanac-Vukanovic S, Maksimovic S, Žugic A (2021b) Old plant, new possibilities: wild bilberry (Vaccinium myrtillus L., Ericaceae) in topical skin preparation. Antioxidants 10(465). https://doi.org/10.3390/antiox10030465

Yang B, Ahotupa M, Määttä P, Kallio H (2011) Composition and antioxidative activities of supercritical CO2-extracted oils from seeds and soft parts of northern berries. Food Res Int 44(7):2009–2017. https://doi.org/10.1016/j.foodres.2011.02.025

9 Application of Microbial Consortia and Biofertilizer to Improve the Quality and Yield of Essential Oils in Aromatic Plants

Weria Weisany, Esmail Khosropour, and Ayda Alavian

Abstract

The use of microbial consortia and biofertilizers has emerged as a promising approach for enhancing the yield and quality of essential oils in aromatic plants. This strategy leverages the benefits of microorganisms such as arbuscular mycorrhizal fungi (AMF) and plant growth-promoting rhizobacteria (PGPRs) to improve plant growth and the production of secondary metabolites. However, achieving optimal results also requires consideration of environmental factors such as light, temperature, humidity, and soil fertility, as well as appropriate cultivation techniques. Innovative systems such as aeroponic and hydroponic systems, as well as micropropagation, can provide even more favorable growth conditions for medicinal plants, resulting in the production of high-quality bioactive substances. Implementation of this approach has significant implications for the pharmaceutical and herbal industries, as well as for overall health and well-being. By leveraging the benefits of microbial consortia and biofertilizers, we can enhance the sustainability and productivity of agricultural practices while also promoting a healthier and more natural approach to medicine.

Keywords

Biofertilizer · Bioactive compounds · Medicinal plant · Mycorrhiza · Plant growth-promoting rhizobacteria

W. Weisany (✉)
Department of Agriculture and Food Science, Science and Research Branch, Islamic Azad University, Tehran, Iran

E. Khosropour
Department of Forestry and Forest Economics, University of Tehran, Tehran, Iran

A. Alavian
Sari Agriculture Sciences and Natural Resources University, Mazandaran, Iran

B. Prakash et al. (eds.), *Plant Essential Oils*,
https://doi.org/10.1007/978-981-99-4370-8_9

9.1 Introduction

Throughout history, medicinal plants have been an important source of therapeutic agents for treating various ailments and diseases, including diarrhea, fever, colds, and malaria. This has been highlighted by several studies, such as Dambisya and Tindimwebwa (2003); Ghiaee et al. (2014), Mathens and Bellanger (2010), and Titanji et al. (2008). The increase in population and the consequent rise in human needs to meet their needs has caused humans to exploit resources and facilities during the last century with the help of new technologies and relying on low-cost natural resources. The world's growing population will lead to several advancements, resulting in increase in food production during recent decades. Meanwhile, plant nutrition plays a key role in significantly increasing food supply (Mukhopadhyay et al. 2021). Increasing plant production has become possible through the use of man-made commercial fertilizers. Sustainable agriculture is a system that, while enjoying economic dynamism, can improve the environment and optimal use of available resources, and also play a significant role in providing human food needs and improving the quality of life of human societies. In addition, sustainable agriculture by observing the principles of ecology can increase the efficiency of resource use while creating a balance in the environment and provide a basis for productivity for a longer period of time (Polukhin and Panarina 2022). In sustainable agriculture, in addition to emphasizing on reducing the consumption of synthetic chemicals and pesticides, crop rotation, organic fertilizers, plant residues and agricultural wastes are used as a substitute for some of the chemical fertilizers, and the use of fertilizers is also optimal according to the plant's needs and production potential (Mukhopadhyay et al. 2021). In this way, the amount of damage or negative effects for natural resources and the environment will be reduced to a minimum. Long-term studies show that excessive use of chemical fertilizers reduces the performance of agricultural and medicinal plants. This reduction is due to soil acidification, reduction of soil biological activities, loss of soil physical properties, and absence of micronutrients in K, P, N fertilizers. In many cases, the use of chemical fertilizers causes environmental pollution and ecological damage, which also increases the cost of production. In order to reduce these risks, resources and institutions should be used that, in addition to providing the current needs of the plant, also ensure the sustainability of agricultural systems in the long term.

By producing humus, organic fertilizers reduce the adverse effects of chemical fertilizers and increase the efficiency of fertilizer use (Khosropour et al. 2023). Sustainable agriculture in the form of combined use of chemical and organic fertilizers as a suitable solution for the development of sustainable agriculture during the transition period from conventional agriculture to sustainable agriculture is effective for producing agricultural products and maintaining plant performance at an acceptable level (Khosropour et al. 2023). Organic materials are one of the important elements of soil fertility due to their beneficial effects on the physical, chemical, biological, and fertility properties of the soil. Organic fertilizers increase the organic matter and pH of the soil, and due to the improvement of the chemical properties of the soil, such as cation exchange capacity and increasing the activity of microorganisms and the amount of access to nutrients, they increase the fertility of the soil (Mukhopadhyay et al. 2021).

Plants in nature are associated with a vast number of beneficial microorganisms such as endophytic or symbiotic bacteria and fungi. These microorganisms have a crucial role in maintaining the health, development, and productivity of plants, and also play a significant part in regulating the synthesis of various metabolites. The importance of this relationship has been emphasized in several studies, including Ezzati Lotfabadi et al. (2021), Panke-Buisse et al. (2015), Castrillo et al. (2017), de Vries et al. (2020), Brader et al. (2014), and Compant et al. (2021).

Bioactive compounds, including carotenoids, essential oils, antioxidants, and flavors, are often incorporated into food products to enhance their flavor, aroma, texture, and overall appeal while also improving their nutritional and health benefits. Recent research has suggested that the inoculation of arbuscular mycorrhizal fungi (AMF) can help optimize the production of these important biomolecules (Hazzoumi et al. 2015; Oliveira et al. 2015). The symbiotic relationship between plants and AMF has been shown to boost the production of secondary compounds, which can increase the value of the plant's phytomass and its potential medicinal properties (Oliveira et al. 2013; Zitter-Eglseer et al. 2015).

This chapter aims to provide an overview of the current knowledge on the inoculation of AMF and PGPRs, their mechanisms, and their effects on the production of essential oils and bioactive compounds in aromatic plants. Additionally, this chapter highlights the benefits of using microbial consortia and biofertilizer management for enhancing the yield and quality of essential oils. Finally, the chapter identifies areas where further research is needed to gain a better understanding of the microbial communities and their mechanisms.

9.2 Biofertilizers

Fertilizers are materials that are used to achieve maximum production per unit area. However, the use of fertilizers should be able to improve the quality of agricultural products in addition to increasing production, and while not polluting the environment, it should reduce the accumulation of nitrate pollutants in the consuming organs of crops to the minimum possible, so as to ensure the health of humans and animals (Kumar et al. 2022). In the last few decades, the use of chemical inputs in agricultural lands has caused many environmental problems, including the pollution of water sources, the decline in the quality of agricultural products and food, and the reduction of soil fertility. Based on this, paying special attention to agriculture from the point of view of sustainability in order to preserve the environment and meet the needs of society in terms of food is necessary and unavoidable, and paying attention to alternative systems has recently attracted a large number of studies (Fasusi et al. 2021; Kumar et al. 2022).

Today, with the increase in agricultural and horticultural production to meet the growing needs of the expanding population, concern has been raised about the future of providing food and green space for the people (Mahapatra et al. 2022).

Pollution of water, soil, air, and soil erosion, resistance of pests to poisons, and spread of chemical fertilizers caused us to go back to the past and industrial crops in order to preserve resources. So, to produce healthy and clean products and as a result, healthy and cheerful people, we have no other way than biological agriculture (Dehsheikh et al. 2020). Considering the increasing demand for the consumption of biological agricultural products, which is based on the proper management of the soil and the growth environment of plants and trees, it is done in such a way that the balance between the required elements in the soil is not disturbed in the nutrition of plants and trees. During growth, there is no need to use poisons and pesticides, and instead of using chemical fertilizers, natural fertilizers such as leaf soil, algae, and animal and biological fertilizers should be used to feed the agricultural soil (Fasusi et al. 2021; Gurikar et al. 2022). If there is a need to fight against pests, instead of using chemical pesticides and poisons, biological methods such as effective microorganisms, ladybugs, bees and bacteria, or pest-resistant cultivars are used in cultivation, and in this type of agriculture, modified seeds are used (Kumar et al. 2022). Genetically modified and exposed to radiation is not used. From this point of view, the final product that reaches the consumer will be free of toxic and chemical residues and preservatives. On the other hand, quality food products, which are the product of biofertilizers, not only satisfy consumers, but also provide and guarantee their physical health (Gurikar et al. 2022). Advantages of using biological fertilizer are as follows:

1. Increasing the amount of nutrients and soil organic matter,
2. Reducing the need to use chemical fertilizers,
3. Increasing the absorption power of food by plants,
4. Prevent soil erosion,
5. The sulfur in the fertilizer improves soil salinity (Chojnacka et al. 2020).

Classification of biofertilizers according to the type of microorganisms:

1. Bacterial biofertilizers (rhizobium-azotobacter-azospirlium),
2. Fungal biofertilizers (mycorrhiza),
3. Algal biofertilizers (blue-green algae and Azolla),
4. Actinomycetes biofertilizers (Frankia).

9.3 Arbuscular Mycorrhizal Fungi (AMF)

One of the most intricate ecosystems on earth is the soil. Fungus is an essential biotic element of the soil that also keeps the plant-soil healthy. Fungi and vascular plants may cooperate to improve soil quality, according to paleontological and evolutionary evidence. Paleobotanists have demonstrated that these symbiotic connections before the development of terrestrial plants. They evolved alongside arbuscules in the Devonian period (400 Mya), hyphae and spores in the Ordovician period (460

Mya), and glomeromycota in the Ediacaran period (600 Mya). Primary associations in the soil environment are vesicular-arbuscular mycorrhizae (VAM), ectomycorrhizae (ECM), orchid mycorrhizae, ericoid mycorrhizae, ectendomycorrhizae (arbutoid), centianoid mycorrhizae, and monotropoid mycorrhiza. The endomycorrhizal association, one of the seven major types of mycorrhizae, is the most widespread and may be found in all kinds of soil (Gujre et al. 2021). VAM are known as arbuscular mycorrhizal fungi, or AMF, because they produce arbuscules and vesicles. AMF assists in nutrient balancing and enhances the efficiency of the soil system. It reproduces by means of small, multinucleate, asexual spores, the quantity of which is greatly enhanced by the mycelium. With the aid of AMF diversity, biodiversity, ecosystem variability, and productivity can all be preserved. Over 85% of plant families worldwide, across all sorts of environments, are colonized by mycorrhiza. Although they are not limited to it and can extend to 70 cm or even more, AMF typically appear in the topsoil zone of 0–20 cm (Sangwan and Prasanna 2022).

The population of AMF is at its peak in the spring, and the soil has an average (Bhardwaj et al. 2023). The diverse agricultural systems and related practices control the population of AMF, which are detritivores closely linked to organic carbon. Nonetheless, there is a notable rise in the AMF population with crops like sorghum, chickpeas, and maize, but oats, barley, and wheat only develop slowly. By destroying the soil aggregates and upsetting the mycelia network, practices like tillage, crop rotation, residual crop burning, inadequate drainage, fungicide sprays, and waterlogging have a detrimental impact on AMF. AMF can trap a considerable quantity of carbon through its exudations and turnover rates. Around the roots, a dense network of fungi creates an advantageous soil structure with a binding effect. Abiotic stress like heavy metals and water stress can be reduced by the fungus network inside the soil. With the help of the viscous glycoprotein glomalin, AMF promotes soil aggregation in the pedosphere. Its color is reddish-brown, and it has a powerful reinforcing impact that improves the general soil quality. Moreover, the main element of AMF called glomalin interacts with ions in the rhizosphere to increase metal binding.

9.4 Essential Oils

Due to rising consumer demand and interest in these plants for therapeutic purposes, the appeal of aromatic and medicinal plants continues to expand applications in the culinary and other human-made fields. Consumers are becoming more aware of the possibilities and advantages of aromatic and medicinal plants and their metabolites as they become more informed about issues related to food, nutrition, and health (Hanif et al. 2019). These plants generate a variety of secondary metabolites, including essential oils (EOs). EOs have a highly complex chemical makeup. EO constituents individually have beneficial uses in a variety of industries, including agriculture, the environment, and human health. EOs are effective complements to synthetic compounds which are widely utilized in the chemical industry. The term "essential oil" dates back to the sixteenth century and derives from the drug

Quinta Essentia, named by Paracelsus von Hohenheim of Switzerland (Brenner 1993). EOs are the substances obtained from a vegetable raw material, either by steam distillation or by mechanical processes from the epicarp of citrus, or "dry"distillation (Hanif et al. 2019). Inorganic solvents like water do not dissolve EOs; however, organic solvents do (ether, alcohol, and fixed oils). With the exception of vetiver, sassafras, and cinnamon, they are volatile liquids with a distinct smell and a density below unity. They are widely utilized in the cosmetics, aromatherapy, and perfume industries (Falleh et al. 2020). Aromatherapy is a therapeutic treatment that uses essential oils in baths, massages, and inhalations (volatile oils). A plant can govern and regulate its surroundings (play an ecological function) by using EOs, which can operate as chemical messages to deter predators, draw pollinating insects, prevent seed germination, and communicate with other plants. Moreover, EOs have insecticidal, deterring, and antifungal properties (Hanif et al. 2019).

9.5 The Effects of AMF on Essential Oil

The positive role of AMF in improving EO content and EO profile of various plants species has been addressed. The AMF can modify the biosynthesis pathway of EO production in plants and lead to significant changes in yield and compounds. The improvement of EO content by AMF has been presented in Table 9.1. In this regard, the improvement of EO content has been reported on *Mentha piperita (*Zare et al. 2023), *Lavandula angustifolia* (Pirsarandib et al. 2022), and *Cymbopogon citratus* (de Souza et al. 2022; Eke et al. 2020). Khalediyan et al. (2021) has shown that MF and PGPR can increase linalol, methyl chavicol, trans-geraniol, camphor, and limonene concentrations in basil EO and carvacrol, thymol, *p*-cymene, α-terpinene, and γ-terpinene in satureja EO components.

Several medicinal plants, including *Chlorophytum borivilianum, Dioscorea* spp., *Gymnema sylvestre, Glycyrrhiza uralensis, Libidibia ferrea, Ocimum basilicum, Satureja macrostema*, and *Salvia miltiorrhiza*, have shown a positive correlation between the biomass of arbuscular mycorrhizal fungi (AMF)-colonized plants and the concentration of secondary metabolites. Studies by Dave et al. (2011), Lu et al. (2015), Zimare et al. (2013), Chen et al. (2017), Silvia et al. (2014), Zolfaghari et al. (2013), Carreón-Abud et al. (2015), and Yang et al. (2017) have highlighted this relationship. However, for *Cynara cardunculus* colonized by *R. intraradices* and *F. mosseae*, a significant increase in yield was observed, but the concentrations of phenolics decreased (Colonna et al. 2016).

AMF can be beneficial for plants in various ways, including improving their nutrient uptake and indirectly affecting the concentration of secondary metabolites. For example, when *Glycyrrhiza uralensis* was grown under nutrient-deficient conditions, *F. mosseae* helped to increase its shoot and root biomass, root system architecture, and flavonoid accumulation (Chen et al. 2017). Furthermore, AMF can help micropropagated medicinal plants to survive and grow better when transferred from in vitro to ex vivo conditions. Studies have shown that when *Spilanthes acmella* and

Table 9.1 The effects of AMF on essential oil of different plant species

Authors	Plant	Results
Weisany et al. (2015)	*Anethum graveolens* L.	The presence of AM colonization resulted in a notable rise in EO yield when compared to plants that were not inoculated. Moreover, alterations in EO composition were observed in dill plants that were intercropped and inoculated with AM.
Sadegh Sadat Darakeh et al. (2021)	*Nigella sativa* L.	The utilization of bioorganic fertilizers can be an effective method of enhancing soil fertility and decreasing reliance on chemical inputs, ultimately leading to greater yields of medicinal and aromatic plants.
Zare et al. (2023)	*Mentha piperita*	Co-application of *glomus intraradices* and titanium dioxide nanoparticles (150 mg L^{-1}) improved EO content and amount of menthol, menthyl acetate, and 1, 8 cineole.
Darakeh et al. (2022)	*Nigella sativa* L.	The utilization of bio and organic fertilizers had an effect on the chemical composition of black cumin essential oil. The application of nutrients such as vermicompost, PGPRs, and AMF to the dillapiole plant resulted in percentage increases in EO composition.
Pirsarandib et al. (2022)	*Lavandula angustifolia*	AMF (*Funneliformis mosseae*) inoculation ameliorated soil heavy metals via improving growth and EO content.
Yilmaz and Karik (2022)	*Ocimum basilicum*	The AMF improve EO yield. Eugenol and Ɣ-cadinene, which are two of the highest ratio components of EO composition, were higher in the AMF.
de Souza et al. (2022)	*Cymbopogon citratus*	AMF improved plant growth and development and modified the content and composition of EOs.
Khalediyan et al. (2021)	*Ocimum basilicum*, *Satureja hortensis*	AMF and PGPR increased linalol, methyl chavicol, trans-geraniol, camphor and limonene concentrations in basil EO and carvacrol, thymol, *p*-cymene, α-terpinene, and γ-terpinene in satureja EO components.
Eke et al. (2020)	*Cymbopogon citratus*	Arbuscular mycorrhizal fungi improved antifungal potential of EO against *fusarium solani,* causing root rot in common bean.
Weisany et al. (2021)	*Coriandrum sativum* L.	The contents of limonene, apiole, coriander ether, n-dihydrocarvone, carvone, myristicin, and coriander apiole were enhanced in *glomus intraradices* treated.
Golubkina et al. (2020)	*Artemisia* dracunculus, *Hyssopus officinalis*	AMF inoculation resulted in a significant enhancement of EO yield.
Azizi et al. (2021)	*Common Myrtle*	AMF (*Funneliformis mosseae*, *Rhizophagus irregularis*) and PGPR (*Pseudomonas fluorescens*, *P. putida*) inoculation improved growth, biochemical traits and EO content under drought stress.
Kordi et al. (2020)	*Ocimum basilicum* L.	According to the findings, the greatest essential oil yield of sweet basil was obtained during the second harvest of the second year by using N_2-fixing bacteria along with 50% nitrogen chemical fertilizer in sole cropping, resulting in a yield of 30.8 kg h^{-1}.

(continued)

Table 9.1 (continued)

Authors	Plant	Results
Weisany et al. (2016a)	*Anethum graveolens* L.	The content of -phellandrene, limonene, −phellandrene, dill-ether, and carvone were enhanced in seed EO obtained from AM-inoculated and intercropped dill plants.
Weisany et al. (2016b)	*Trigonella foenum-graecum* L., *Coriandrum sativum* L., *and Nigella sativa* L.	AM colonization of fenugreek, coriander, and nigella plants resulted in alterations in the composition of essential oils.
Arpanahi et al. (2020)	*Thymus vulgaris*	AMF inoculation modulated drought through increasing EO production.
Eshaghi Gorgi et al. (2022)	*Melissa officinalis*	Co-application of AMF and PGPRs improve growth and EO content.
Arpanahi and Feizian (2019)	*Thymus vulgaris*	AMF with mild-moderate water stress, there is an increase in EO content but at severe water-stress essential oil content decreased. Interaction of AMF and water stress had significant effects on most components of EOs.
Pankaj et al. (2019)	*Cymbopogon martinii*	AMF significantly improved EO yield and affected EO profile.

Glycyrrhiza glabra plantlets were inoculated with *F. mosseae*, their survival rate was 100% and they exhibited improved growth and development in greenhouse and glasshouse conditions, whereas without AMF, the survival rate was only 60–70% (Yadav et al. 2012, 2013). Similarly, *Scutelleria integrifolia* seedlings inoculated with *C. etunicatum* had significantly increased height and fresh weight of shoots, roots, and seeds following micropropagation (Joshee et al. 2007).

According to Zeng et al. (2013), it is commonly believed that the increased levels of various secondary metabolites found in AMF-colonized plants are a result of several defense response pathways that are triggered. These pathways include the carotenoid pathway, the phenylpropanoid pathway, and alkaloid synthesis, which are known to promote signaling, stress tolerance, nutrient uptake, and resistance against biotic and abiotic stresses (Kaur and Suseela 2020). Despite this, the exact mechanisms through which AMF induce changes in the concentration of phytochemicals in plant tissues are not yet fully understood (Toussaint et al. 2007).

The production of terpenoids, phenolic compounds, and alkaloids in plants and how they are influenced by AMF has been the subject of many studies. Terpenoids, which are synthesized from isoprene units, are produced through two distinct pathways: the mevalonic acid (MVA) pathway and the methyleritrophosphate (MEP) pathway (Zhi et al. 2007). Various nutritional and non-nutritional factors have been suggested to account for the higher production of secondary metabolites in plants colonized by AMF, as outlined in Fig. 9.1, in studies by Kapoor et al. (2017), Sharma et al. (2017), and Dos Santos et al. (2021).

AMF and PGPRs can impact the production of secondary metabolites in medicinal plants through both nutritional and non-nutritional means. Nutritional factors

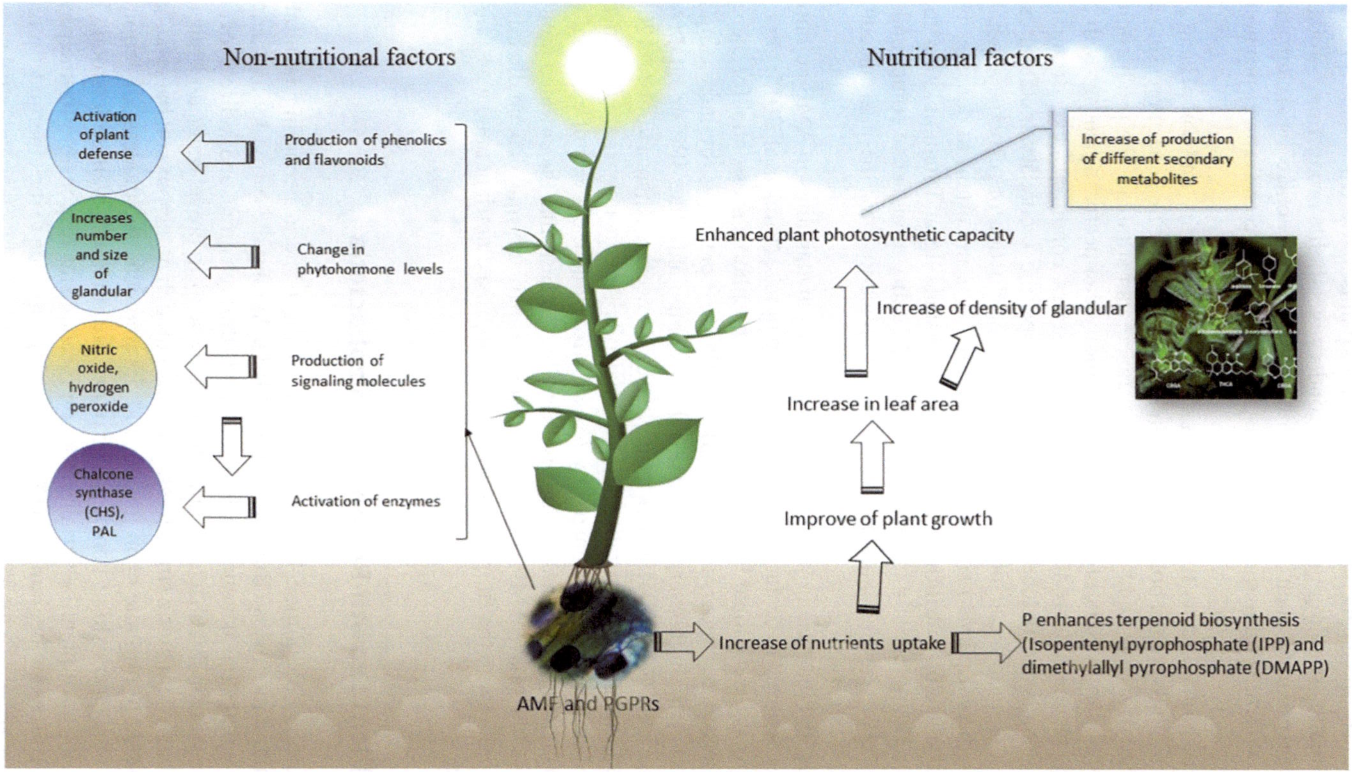

Fig. 9.1 The production of secondary metabolites, such as terpenoids, phenolics, and flavonoids, is affected by both nutritional and non-nutritional factors in plants that have been colonized by AMF and PGPRs

such as the availability of essential nutrients and micronutrients can influence the growth and secondary metabolite production in plants. AMF can enhance nutrient uptake in plants, resulting in increased secondary metabolite production. Similarly, PGPRs can promote plant growth and improve nutrient uptake by fixing atmospheric nitrogen, solubilizing phosphorus, and producing plant growth-promoting hormones.

Non-nutritional factors such as light, temperature, humidity, and soil pH can also influence secondary metabolite production in plants by altering gene expression and enzyme activity. High light intensity, for instance, can stimulate the production of certain secondary metabolites, while low humidity can decrease plant growth and secondary metabolite production. In addition, stress conditions like drought or pathogen attack can induce the production of specific secondary metabolites in plants as a defense mechanism. To optimize the production of secondary metabolites in medicinal plants using AMF and PGPRs, it is important to consider both nutritional and non-nutritional factors. By understanding and manipulating these factors, we can potentially increase the yield and quality of bioactive compounds in medicinal plants, which can have significant implications for the pharmaceutical and herbal industries.

Initially, the increase in production of secondary metabolites in AMF-colonized plants was believed to be due to improved nutrient uptake (Lima et al. 2015; Oliveira et al. 2015) as a nutritional factor. For example, phosphorus is essential for the synthesis of terpenoid precursors via the MVA and MEP pathways, by increasing the concentration of high-energy pyrophosphate compounds like IPP and DMAPP (Kapoor et al. 2002, 2004; Zubek et al. 2010). However, Khaosaad et al. (2006) discovered that the concentration of essential oils significantly increased in two *Origanum* sp. genotypes colonized by *F. mosseae*, while the levels of essential oils remained unchanged in plants treated with P. This suggests that the elevated production of essential oils in AMF-colonized *Origanum* sp. plants may depend directly on the association with the fungus rather than just improved nutrient uptake as a nutritional factor.

9.6 Plant Growth-Promoting Rhizobacteria

Plant-microbe coevolution has led to some of the bacteria becoming facultative intracellular endophytes (Bulgarelli et al. 2013). Among these free-living bacteria are PGPRs that exert beneficial effects on plants through direct and indirect mechanisms. Beneficial rhizobacteria have been utilized to improve water and nutrient uptake, abiotic and biotic stress tolerance. Even though numerous soil bacteria have been reported to promote plant growth and development, the mode(s) of action by which the bacteria exhibit beneficial activities are often not well understood. The molecular basis of plant-bacteria interaction mechanisms responsible for the physiological changes are beginning to be discerned, mainly due to the emerging "omics" approaches. PGPRs can affect the physiological and biochemical attributes in plants. The changes in secondary metabolite like essential oils is eminent in the interaction of plants and PGPRs. Table 9.1 shows the positive effect of PGPRs in improving essential oil quality and quantity.

9.6.1 Azotobacter

PGPR helps to replace chemical fertilizer for the sustainable agriculture production by fixing the atmospheric nitrogen and producing growth-promoting substances. Among the PGPR group, *Azotobacter* are ubiquitous, aerobic, free-living, and N_2-fixing bacteria commonly living in soil, water, and sediments. Being the major group of soilborne bacteria, *Azotobacter* plays different beneficial roles and is known to produce varieties of vitamins, amino acids, plant growth hormones, antifungal substances, hydrogen cyanide, and siderophores. The growth-promoting substances, such as indoleacetic acid, gibberellic acid, and arginine, produced by *Azotobacter* have direct influence on shoot and root length as well as seed germination of several agricultural crops. *Azotobacter* species are efficient in fixation of highest amount of nitrogen, production of indoleacetic acid and gibberellic acid, and formation of larger phosphate-solubilizing zone. Many species of *Pseudomonas*, *Bacillus*, and *Azotobacter* can grow and survive at extreme environmental conditions, namely, tolerant to higher salt concentration, pH values, and even at dry soils with maximum temperature. Different factors affect *Azotobacter* population in soil such as pH, phosphorus content, soil aeration, and moisture contents. *A. chroococcum* found tolerant to a maximum NaCl concentration of 6% with a temperature of 45 °C and also up to pH of 8. *Azotobacter* species such as *A. vinelandii*, *A. chroococcum*, *A. salinestris*, *A. tropicalis*, and *A. nigricans* are able to produce antimicrobial compounds which inhibit the growth of common plant pathogens, viz., *Fusarium*, *Aspergillus*, *Alternaria*, *Curvularia*, and *Rhizoctonia* species. Pesticides used to control pests, insects, and phytopathogens are known to cause direct effect on soil microbiological aspects, environmental pollution, and health hazards in all living beings of the soil ecosystem. The species of *Azotobacter* are known to tolerate up to 5% pesticide concentration and also to degrade heavy metals and pesticides. *A. chroococcum* and *A. vinelandii* proved their biodegradation efficiency of many commonly used pesticides, viz., endosulfan, chlorpyrifos, pendimethalin, phorate, glyphosate, and carbendazim.

9.6.2 *Pseudomonas* Sp.

Pseudomonas sp. is an aerobic, gram-negative, ubiquitous organism present in agricultural soils and well adapted to grow in the rhizosphere. This rhizobacterium possesses many traits to act as a biocontrol agent and to promote the plant growth ability. It grows rapidly in vitro and can be mass-produced. It rapidly utilizes seed and root exudates and colonizes and multiplies in the rhizosphere and spermosphere environments. In the plant rhizosphere, it produces a wide spectrum of bioactive metabolites, that is, antibiotics, siderophores, volatiles, and growth-promoting substances; competes aggressively with other microorganisms; and adapts to environmental stresses. In addition, pseudomonads are responsible for the *natural suppressiveness* of some soilborne pathogens. It suppresses the growth of pathogenic microorganisms by various mechanisms, namely, production of antibiotics, bacteriocins, siderophores, hydrolytic enzymes such as β-1,3-glucanase and chitinases, and other metabolites such as phytoalexins and induction of systemic resistance. In this chapter, the characteristics of *Pseudomonas* sp., plant growth-promoting properties, mechanisms of plant growth promotion, and induction of systemic resistance by plant growth-promoting rhizobacterium (PGPR) against diseases and insect and nematode pests have been reviewed. PGPR strains initiating induced systemic resistance against a wide array of plant pathogens causing fungal, bacterial, and viral diseases and insect and nematode pests are discussed. Synergistic effects of PGPR strain mixtures and PGPRs as endophytes are brought out. Modes of action of *Pseudomonas* against fungal pathogens have been explained. Plant-disease controls by *P. fluorescens* have been elaborated. Interaction of *P. fluorescens* with pesticides has been indicated. Formulation characteristics, its approved uses in India, methods of application, and data requirements for registration particularly in India are discussed. Factors affecting growth of biopesticides and future issues and research needs in biopesticides are discussed.

PGPRs can exert both direct and indirect effects on plants, as shown in Fig. 9.2. Direct effects involve the PGPRs' direct interactions with the plants, such as the production of plant growth hormones like auxins, cytokinins, and gibberellins, which can enhance plant growth and development. Additionally, PGPRs can secrete siderophores, iron-chelating compounds that facilitate iron uptake by plants, especially in iron-limited soils.

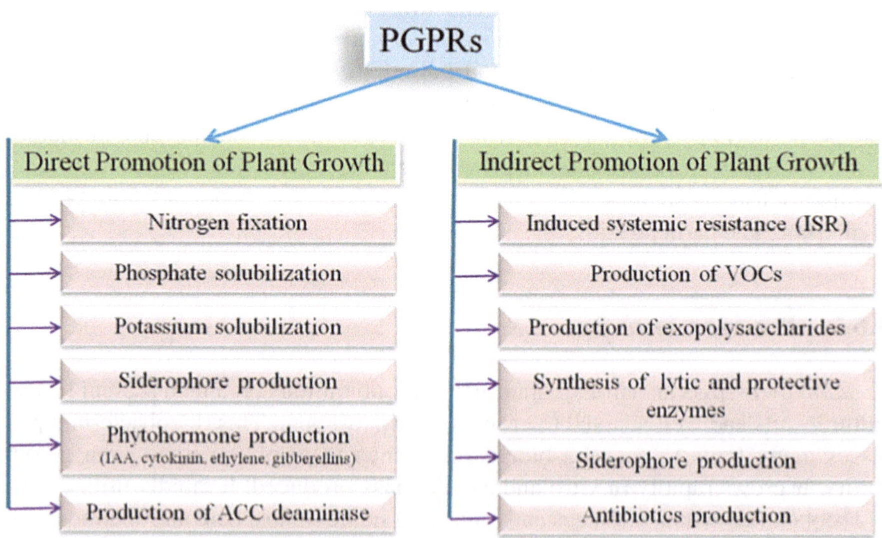

Fig. 9.2 Direct and indirect effects of PGPRs on plants

Indirect effects refer to the influence of PGPRs on other soil organisms, which subsequently affect plant growth and development. For example, PGPRs can stimulate the activity of other soil microorganisms, including mycorrhizal fungi, which can form mutually beneficial associations with plants to enhance nutrient uptake. PGPRs can also trigger systemic resistance in plants, making them more resistant to pathogens and pests. The direct and indirect effects of PGPRs on plants can significantly enhance plant growth, nutrient uptake, and overall health, which can have vital implications for agriculture and sustainable crop production.

9.7 The Effects of PGPRs on Essential Oil

The positive role of PGPRs in improving EO content and EO profile of some medicinal plants has been reported. The PGPRs can modify the biosynthesis pathway of EO production in plants and lead to significant changes in yield and compounds. The improvement of EO content by PGPRs has been presented in Table 9.2. Accordingly, Yilmaz & Karik reported that PGPRs inoculation under low drought stress-enhanced *Trachyspermum ammi* seeds' EO bioactivity. Sammak showed that the synergistic effect of AMF and *Pseudomonas fluorescens* improved growth and yield of *Thymus kotschyanus* EOs. Amini et al. (2020) noted that PGPRs improved EO yield and composition of *Dracocephalum moldavica* as affected by inoculation treatments under drought stress condition.

PGPRs have been found to positively affect the essential oil (EO) production of medicinal plants through various mechanisms.

Table 9.2 The effects of PGPRs on essential oil of different plant species from 2020 to 2023

Author/s	PGPRs	Results
Yilmaz and Karik (2022)	*Azotobacter chroococcum*	PGPRs inoculation under low drought stress-enhanced *Trachyspermum ammi* seeds' EO bioactivity.
Hatami et al. (2021)	*Pseudomonasfluorescens* and *P. putida*	Silicon nanoparticle-mediated seed priming and pseudomonas spp. inoculation-enhanced growth, antioxidant capacity, and EO yield in *Melissa officinalis* L.
Sammak et al. (2020)	*Pseudomonas fluorescens*	The synergistic effect of AMF and *Pseudomonas fluorescens* improved growth and yield of *Thymus kotschyanus* EOs.
Amini et al. (2020)	*Piriformospora indica*, *Pseudomonas fluorescens*	PGPRs improved EO yield and composition of *Dracocephalum moldavica* as affected by inoculation treatments under drought stress condition.
Dehsheikh et al. (2020)	N-fixing bacteria, P-solubilizing bacteria	Combined PGPRs improved soil microbial activity, essential oil quantity, and quality of Thai basil.
Mirzaei et al. (2020)	*Pseudomonas* sp., *Azotobacter* sp.	PGPR improved the plant growth and EO properties by increasing antioxidant capacity of lemongrass.

One of the main mechanisms is the ability of PGPRs to improve nutrient acquisition in plants, particularly the uptake of nitrogen, phosphorus, and other essential minerals. PGPRs can achieve this by solubilizing insoluble minerals, chelating minerals, and producing enzymes that facilitate nutrient uptake. By improving nutrient availability, PGPRs can support the growth and development of medicinal plants, leading to an increase in EO production.

PGPRs can also modulate plant hormone levels, including the production of phytohormones such as auxins, cytokinins, and gibberellins. These hormones are crucial in plant growth and development, and can influence the production of secondary metabolites such as EOs. For example, auxins have been shown to promote the biosynthesis of several EO components, while cytokinins can enhance the synthesis of certain aromatic compounds.

Furthermore, PGPRs can help to mitigate abiotic and biotic stress in medicinal plants. Stresses such as drought, salinity, and pathogen attack can negatively impact plant growth and development, as well as EO production. PGPRs can alleviate these stresses by inducing the expression of stress-responsive genes and producing compounds such as osmolytes, antioxidants, and siderophores that protect the plant from damage. In addition, PGPRs can interact with other microorganisms in the rhizosphere, including mycorrhizal fungi and other beneficial bacteria. These interactions can promote synergistic effects on plant growth and EO production, by improving nutrient acquisition, hormone levels, and stress tolerance.

Overall, the effects of PGPRs on EO production in medicinal plants are complex and multifaceted. The ability of PGPRs to improve nutrient uptake, induce hormone production, and modulate stress responses can enhance the growth and development of medicinal plants, leading to an increase in EO production and quality. Further research is needed to fully understand the mechanisms behind the effects of PGPRs on EO production, as well as to identify the optimal PGPR strains and application methods for different medicinal plant species.

9.8 Conclusion

In conclusion, the use of biofertilizers and microbial consortia is a promising and sustainable approach to enhance the growth and secondary metabolite production of medicinal plants. By harnessing the beneficial effects of microorganisms like AMF and PGPRs, we can improve plant resilience and increase the accumulation of active compounds. However, to achieve optimal results, it is crucial to also consider environmental factors such as light, temperature, humidity, and soil fertility, as well as cultivation techniques. By using innovative substrate-based and substrate-free systems like aeroponic and hydroponic systems, and micropropagation, we can create ideal growth conditions for medicinal plants and ensure the production of high-quality bioactive substances. This can have significant implications for the pharmaceutical and herbal industries, as well as for the overall health and well-being of society.

References

Amini R, Zafarani-Moattar P, Shakiba MR, Sarikhani MR (2020) Essential oil yield and composition of Moldavian balm (*Dracocephalum moldavica* L.) as affected by inoculation treatments under drought stress condition. J Essent Oil Bear Plants 23(4):728–742. https://doi.org/10.1080/0972060X.2020.1765851

Arpanahi AA, Feizian M (2019) Arbuscular mycorrhizae alleviate mild to moderate water stress and improve essential oil yield in thyme. Rhizosphere 9:93–96. https://doi.org/10.1016/j.rhisph.2019.05.007

Arpanahi AA, Feizian M, Mehdipourian G, Khojasteh DN (2020) Arbuscular mycorrhizal fungi inoculation improve essential oil and physiological parameters and nutritional values of *thymus daenensis* Celak and *Thymus vulgaris* L. under normal and drought stress conditions. Eur J Soil Biol 100:103217. https://doi.org/10.1016/j.ejsobi.2020.103217

Azizi S, Kouchaksaraei MT, Hadian J, Abad ARFN, Sanavi SAMM, Ammer C, Bader MKF (2021) Dual inoculations of arbuscular mycorrhizal fungi and plant growth-promoting rhizobacteria boost drought resistance and essential oil yield of common myrtle. For Ecol Manag 497:119478. https://doi.org/10.1016/j.foreco.2021.119478

Bhardwaj AK, Chandra KK, Kumar R (2023) Water stress changes on AMF colonization, stomatal conductance and photosynthesis of *Dalbergia sissoo* seedlings grown in entisol soil under nursery condition. For Sci Technol 19:1–13

Brader G, Compant S, Mitter B, Trognitz F (2014) Metabolic potential of endophytic bacteria. Curr Opin Biotechnol 27:30–37

Brenner DM (1993) Perilla: botany, uses and genetic resources. In: Janick J, Simon JE (eds) New crops. Wiley, New York, pp 322–328

Bulgarelli D, Schlaeppi K, Spaepen S, van Themaat EVL, Schulze-Lefert P (2013) Structure and functions of the bacterial microbiota of plants. Annu Rev Plant Biol 64:807–838

Carreón-Abud Y, Torres-Martínez R, Farfán-Soto B, Hernández-García A, Ríos-Chávez P, Bello-González MA, Martínez-Trujillo M, Salgado-Garciglia R (2015) Arbuscular mycorrhizal symbiosis increases the content of volatile terpenes and plant performance in *Satureja macrostema* (Benth.) Briq. Boletín Latinoamericano y del Caribe de Plantas Medicinales y Aromáticas 14(4):273–279. https://doi.org/10.13140/RG.2.1.3233.9449

Castrillo G, Teixeira PJ, Paredes SH, Law TF, de Lorenzo L, Feltcher ME, Finkel OM, Breakfield NW, Mieczkowski P, Jones CD, Paz-Ares J, Dangl JL (2017) Root microbiota drive direct integration of phosphate stress and immunity. Nature 543:513–518

Chen ML, Yang G, Sheng Y, Li PY, Qiu HY, Zhou XT, Huang LQ, Chao Z (2017) Glomus mosseae inoculation improves the root system architecture, photosynthetic efficiency and flavonoids accumulation of liquorice under nutrient stress. Front Plant Sci 8:931. https://doi.org/10.3389/fpls.2017.00931

Chojnacka K, Moustakas K, Witek-Krowiak A (2020) Bio-based fertilizers: a practical approach towards circular economy. Bioresour Technol 295:122223

Colonna E, Rouphael Y, De Pascale S, Barbieri G (2016) Effects of mycorrhiza and plant growth promoting rhizobacteria on yield and quality of artichoke. Acta Hortic 1147:43–50. https://doi.org/10.17660/ActaHortic.2016.1147.7

Compant S, Cambon MC, Vacher C, Mitter B, Samad A, Sessitsch A (2021) The plant endosphere world–bacterial life within plants. Environ Microbiol 23(4):1812–1829

Dambisya YM, Tindimwebwa G (2003) Traditional remedies in children around Eastern Cape. South Africa East Afr Med J 80(8):402–405

Darakeh SASS, Weisany W, Abdul-Razzak Tahir N, Schenk PM (2022) Physiological and biochemical responses of black cumin to vermicompost and plant biostimulants: arbuscular mycorrhizal and plant growth-promoting rhizobacteria. Ind Crop Prod 188:115557

Dave S, Das J, Tarafdar JC (2011) Effect of vesicular arbuscular mycorrhizae on growth and saponin accumulation in *Chlorophytum borivilianum*. Science Asia 37(2):165–169. https://doi.org/10.2306/scienceasia1513-1874.2011.37.165

de Souza BC, da Cruz RMS, Lourenço ELB, Pinc MM, Dalmagro M, da Silva C, Alberton O et al (2022) Inoculation of lemongrass with arbuscular mycorrhizal fungi and rhizobacteria alters plant growth and essential oil production. Rhizosphere 22:100514. https://doi.org/10.1016/j.rhisph.2021.100514

de Vries FT, Griffiths RI, Knight CG, Nicolitch O, Williams A (2020) Harnessing rhizosphere microbiomes for drought-resilient crop production. Science 368:270–274

Dehsheikh AB, Sourestani MM, Zolfaghari M, Enayatizamir N (2020) Changes in soil microbial activity, essential oil quantity, and quality of *Thai basil* as response to biofertilizers and humic acid. J Clean Prod 256:120439. https://doi.org/10.1016/j.jclepro.2020.120439

Dos Santos EL, Falcão EL, Barbosa da Silva FS (2021) Mycorrhizal technology as a bioinsumption to produce phenolic compounds of importance to the herbal medicine industry. Res Soc Dev 10(2). https://doi.org/10.33448/rsd-v10i2.15891

Eke P, Adamou S, Fokom R, Nya VD, Fokou PVT, Wakam LN et al (2020) Arbuscular mycorrhizal fungi alter antifungal potential of lemongrass essential oil against *fusarium solani*, causing root rot in common bean (*Phaseolus vulgaris* L.). Heliyon 6(12):e05737. https://doi.org/10.1016/j.heliyon.2020.e05737

Eshaghi Gorgi O, Fallah H, Niknejad Y, Barari Tari D (2022) Effect of plant growth promoting rhizobacteria (PGPR) and mycorrhizal fungi inoculations on essential oil in Melissa officinalis L. under drought stress. Biologia 77:11–20. https://doi.org/10.1007/s11756-021-00761-x

Ezzati Z, Weisany W, Abdul-razzak Tahir N (2021) Arbuscular mycorrhizal fungi species improve the fatty acids profile and nutrients status of soybean cultivars grown under drought stress. J Appl Microbiol 132:2177–2188

Falleh H, Jemaa MB, Saada M, Ksouri R (2020) Essential oils: A promising eco-friendly food preservative. Food Chem 330:127268. https://doi.org/10.1016/j.foodchem.2020.127268

Fasusi OA, Cruz C, Babalola OO (2021) Agricultural sustainability: microbial biofertilizers in rhizosphere management. Agriculture 11(2):163. https://doi.org/10.3390/agriculture11020163

Ghiaee A, Naghibi F, Esmaeili S, Mosaddegh M (2014) Herbal remedies connected to malaria like fever in Iranian ancient medicinal books. Iran J Parasitol 9(4):553–559

Golubkina N, Logvinenko L, Novitsky M, Zamana S, Sokolov S, Molchanova A, Caruso G et al (2020) Yield, essential oil and quality performances of Artemisia dracunculus, Hyssopus officinalis and Lavandula angustifolia as affected by arbuscular mycorrhizal fungi under organic management. Plan Theory 9(3):375. https://doi.org/10.3390/plants9030375

Gujre N, Soni A, Rangan L, Tsang DC, Mitra S (2021) Sustainable improvement of soil health utilizing biochar and arbuscular mycorrhizal fungi: A review. Environ Pollut 268:115549. https://doi.org/10.1016/j.envpol.2020.115549

Gurikar C, Sreenivasa MY, Gowda NN, Lokesh AC (2022) Azotobacter—A potential symbiotic rhizosphere engineer. In: Rhizosphere engineering. Academic Press, pp 97–112. https://doi.org/10.1016/B978-0-12-824508-6.00006-4

Hanif MA, Nisar S, Khan GS, Mushtaq Z, Zubair M (2019) Essential oils. In: Essential oil research: trends in biosynthesis, analytics, industrial applications and biotechnological production, vol 3-17. Springer, Cham. https://doi.org/10.1016/B978-0-12-818765-3.00001-1

Hatami M, Khanizadeh P, Bovand F, Aghaee A (2021) Silicon nanoparticle-mediated seed priming and pseudomonas spp. inoculation augment growth, physiology and antioxidant metabolic status in Melissa officinalis L. plants. Ind Crop Prod 162:113238. https://doi.org/10.1016/j.indcrop.2021.113238

Hazzoumi Z, Moustakime Y, Elharchli EH, Joutei KA (2015) Effect of arbuscular mycorrhizal fungi (AMF) and water stress on growth, phenolic compounds, glandular hairs, and yield of essential oil in basil (Ocimum gratissimum L.). Chem Biol Technol Agric 2:2–11. https://doi.org/10.1186/s40538-015-0033-6

Joshee N, Yadav AK, Mentreddy R (2007) Mycorrhizal fungi and growth and development of micropropagated Scutellaria integrifolia plants. Ind Crop Prod 25(2):169–177. https://doi.org/10.1016/j.indcrop.2006.09.005

Kapoor R, Anand G, Gupta P, Mandal S (2017) Insight into the mechanisms of enhanced production of valuable terpenoids by arbuscular mycorrhiza. Phytochem Rev 16:677–692. https://doi.org/10.1007/s11101-017-9508-5

Kapoor R, Giri B, Mukerji KG (2002) Glomus macrocarpum: a potential bioinoculant to improve essential oil quality and concentration in dill (Anethum graveolens L.) and Carum (Trachyspermum ammi (Linn.) Sprague). World J Microb Biot 18:459–463. https://doi.org/10.1023/A:1015262525430

Kapoor R, Giri B, Mukerji KG (2004) Improved growth and essential oil yield and quality in Foeniculum vulgare mill on mycorrhizal inoculation supplemented with P-fertilizer. Bioresour Technol 93(3):307–311. https://doi.org/10.1016/j.biortech.2003.10.016

Kaur S, Suseela V (2020) Unraveling arbuscular mycorrhiza-induced changes in plant primary and secondary metabolome. Meta 10(8):335. https://doi.org/10.3390/metabo10080335

Khalediyan N, Weisany W, Schenk PM (2021) Arbuscular mycorrhizae and rhizobacteria improve growth, nutritional status and essential oil production in Ocimum basilicum and Satureja hortensis. Ind Crop Prod 160:113163. https://doi.org/10.1016/j.indcrop.2020.113163

Khaosaad T, Vierheilig H, Nell M, Zitterl-Eglseer K, Novak J (2006) Arbuscular mycorrhiza alter the concentration of essential oils in oregano (Origanum sp., Lamiaceae). Mycorrhiza 16(6):443–446. https://doi.org/10.1007/s00572-006-0043-8

Khosropour E, Hakimi L, Weisany W (2023) Role of engineered nanomaterials in biotic stress management. Plant Biology, sustainability and climate change, Engineered Nanomaterials for Sustainable Agricultural Production, Soil Improvement and Stress Management. Elsevier, pp 257–272

Kordi S, Salmasi SZ, Kolvanagh JS, Weisany W, Shannon DA (2020) Intercropping system and N2 fixing bacteria can increase land use efficiency and improve the essential oil quantity and quality of sweet basil (Ocimum basilicum L.). Front Plant Sci 11:610026. https://doi.org/10.3389/fpls.2020.610026

Kumar S, Sindhu SS, Kumar R (2022) Biofertilizers: an ecofriendly technology for nutrient recycling and environmental sustainability. Curr Res Microb Sci 3:100094. https://doi.org/10.1016/j.crmicr.2022.100094

Lima KB, Riter Netto AF, Martins MA, Freitas MSM (2015) Crescimento, acúmulo de nutrientes e fenóis totais de mudas de cedro australiano (Toona ciliata) inoculadas com fungos micorrízicos. Ciência Florestal 25:853–862. https://doi.org/10.5902/1980509819663

Lu FC, Lee CY, Wang CL (2015) The influence of arbuscular mycorrhizal fungi inoculation on yam (Dioscorea spp.) tuber weights and secondary metabolite content. PeerJ 3:e1266. https://doi.org/10.7717/peerj.1266

Mahapatra DM, Satapathy KC, Panda B (2022) Biofertilizers and nanofertilizers for sustainable agriculture: Phycoprospects and challenges. Sci Total Environ 803:149990. https://doi.org/10.1016/j.scitotenv.2021.149990

Mathens A, Bellanger R (2010) Herbs and other dietary supplements: current regulations and recommendations for use to maintain health in the management of the common cold or other related infectious respiratory illnesses. J Pharm Pract 23:117–127

Mirzaei M, Ladan Moghadam A, Hakimi L, Danaee E (2020) Plant growth promoting rhizobacteria (PGPR) improve plant growth, antioxidant capacity, and essential oil properties of lemongrass (Cymbopogon citratus) under water stress. Iranian J Plant Physiol 10(2):3155–3166. https://doi.org/10.22034/ijpp.2020.1881571.1183

Mukhopadhyay R, Sarkar B, Jat HS, Sharma PC, Bolan NS (2021) Soil salinity under climate change: challenges for sustainable agriculture and food security. J Environ Manag 280:111736. https://doi.org/10.1016/j.jenvman.2020.111736

Oliveira MS, Campos MAS, Silva FSB (2015) Arbuscular mycorrhizal fungi and vermicompost to maximize the production of foliar biomolecules in Passiflora alata Curtis seedlings. J Sci Food Agric 95:522–528. https://doi.org/10.1002/jsfa.6777

Oliveira MS, Campos MA, Albuquerque UP, Silva FS (2013) Arbuscular mycorrhizal fungi (AMF) affect biomolecules content in Myracrodruon urundeuva seedlings. Ind Crop Prod 50:244–247. https://doi.org/10.1016/j.indcrop.2013.07.049

Pankaj U, Verma RS, Yadav A, Verma RK (2019) Effect of arbuscular mycorrhizae species on essential oil yield and chemical composition of palmarosa (Cymbopogon martinii) varieties grown under salinity stress soil. J Essent Oil Res 31(2):145–153. https://doi.org/10.1080/10412905.2018.1558801

Panke-Buisse K, Poole A, Goodrich J, Ley RE, Kao-Kniffin J (2015) Selection on soil microbiomes reveals reproducible impacts on plant function. ISME J 9:980–989

Pirsarandib Y, Hassanpouraghdam MB, Rasouli F, Aazami MA, Puglisi I, Baglieri A (2022) Phytoremediation of soil contaminated with heavy metals via arbuscular mycorrhiza (Funneliformis mosseae) inoculation ameliorates the growth responses and essential oil content in lavender (lavandula angustifolia L.). Agronomy 12(5):1221. https://doi.org/10.3390/agronomy12051221

Polukhin AA, Panarina VI (2022) Financial risk management for sustainable agricultural development based on corporate social responsibility in the interests of food security. Risks 10(1):17. https://doi.org/10.3390/risks10010017

Sadegh Sadat Darakeh SA, Weisany W, Diyanat M, Ebrahimi R (2021) Bio-organic fertilizers induce biochemical changes and affect seed oil fatty acids composition in black cumin (Nigella sativa Linn). Ind Crop Prod 164:113383. https://doi.org/10.1016/j.indcrop.2021.113383

Sammak AS, Anvari M, Matinizadeh M, Mirza M (2020) The synergistic effect of arbuscular mycorrhizal fungi and Pseudomonas fluorescens on growth and qualitative and quantitative yield of Thymus kotschyanus essential oil. J Essent Oil Bear Plants 23(3):532–547. https://doi.org/10.1080/0972060X.2020.1773586

Sangwan S, Prasanna R (2022) Mycorrhizae helper bacteria: unlocking their potential as bioenhancers of plant–arbuscular mycorrhizal fungal associations. Microb Ecol 84(1):1–10. https://doi.org/10.1007/s00248-021-01728-1

Sharma E, Anand G, Kapoor R (2017) Terpenoids in plant and arbuscular mycorrhiza-reinforced defense against herbivorous insects. Ann Bot 119:791–801. https://doi.org/10.1093/aob/mcw218

Silvia FA, Silva FSB, Maia LC (2014) Biotechnical application of arbuscular mycorrhizal fungi used in the production of foliar biomolecules in ironwood seedlings [Libidibia ferrea (Mart. Ex Tul.) L.P. Queiroz var. ferrea]. J Med Plant Res 8(20):814–819. https://doi.org/10.5897/JMPR2014.5483

Titanji VP, Zofou D, Ngemenya MN (2008) The antimalarial potential of medicinal plants used for the treatment of malaria in Cameroonian folk medicine. Afr J Tradit Complement Altern Med 5(3):302–321

Toussaint JP, Smith FA, Smith SE (2007) Arbuscular mycorrhizal fungi can induce the production of phytochemicals in sweet basil irrespective of phosphorus nutrition. Mycorrhiza 17(4):291–297. https://doi.org/10.1007/s00572-007-0123-1

Weisany W, Raei Y, Ghasemi-Golezani K (2016b) Funneliformis mosseae alters seed essential oil content and composition of dill in intercropping with common bean. Ind Crop Prod 79:29–38. https://doi.org/10.1016/j.indcrop.2015.10.041

Weisany W, Raei Y, Pertot I (2015) Changes in the essential oil yield and composition of dill (Anethum graveolens L.) as response to arbuscular mycorrhiza colonization and cropping system. Ind Crop Prod 77:295–306. https://doi.org/10.1016/j.indcrop.2015.09.027

Weisany W, Raei Y, Salmasi SZ, Sohrabi Y, Ghassemi-Golezani K (2016a) Arbuscular mycorrhizal fungi induced changes in rhizosphere, essential oil and mineral nutrients uptake in dill/common bean intercropping system. Ann Appl Biol 169(3):384–397. https://doi.org/10.1111/aab.12309

Weisany W, Tahir NAR, Schenk PM (2021) Coriander/soybean intercropping and mycorrhizae application lead to overyielding and changes in essential oil profiles. Eur J Agron 126:126283. https://doi.org/10.1016/j.eja.2021.126283

Yadav K, Aggarwal A, Singh N (2013) Arbuscular mycorrhizal fungi induced acclimatization and growth enhancement of Glycyrrhiza glabra L.: a potential medicinal plant. Agric Res 2:43–47. https://doi.org/10.1007/s40003-012-0026-8

Yadav K, Singh N, Aggarwal A (2012) Arbuscular mycorrhizal technology for the growth enhancement of micropropagated Spilanthes acmella Murr. Plant Prot Sci 48:31–36. https://doi.org/10.17221/60/2011-PPS

Yang Y, Ou XH, Yang G, Xia YS, Chen ML, Guo LP, Liu DH (2017) Arbuscular mycorrhizal fungi regulate the growth and phyto-active compound of salvia miltiorrhiza seedlings. Appl Sci 7(1):68. https://doi.org/10.3390/app7010068

Yilmaz A, Karik Ü (2022) AMF and PGPR enhance yield and secondary metabolite profile of basil (Ocimum basilicum L.). Ind Crop Prod 176:114327. https://doi.org/10.1016/j.indcrop.2021.114327

Zare AS, Ganjeali A, Kakhki MRV, Mashreghi M, Cheniany M (2023) Additive effects of arbuscular mycorrhizae and TiO2 nanoparticles on growth and essential oils enhancement of peppermint. Rhizosphere 25:100659. https://doi.org/10.1016/j.rhisph.2022.100659

Zeng Y, Guo LP, Chen BD, Hao ZP, Wang JY, Huang LQ, Yang G, Cui XM, Yang L, Wu ZX, Chen ML, Zhang Y (2013) Arbuscular mycorrhizal symbiosis and active ingredients of medicinal plants: current research status and perspectives. Mycorrhiza 23(4):253–265. https://doi.org/10.1007/s00572-013-0486-4

Zhi LY, Chuan CD, Lian QC (2007) Regulation and accumulation of secondary metabolites in plant-fungus symbiotic system. Afr J Biotechnol 6:1266–1271. https://doi.org/10.5897/AJB2007.000-2415

Zimare S, Borde M, Jite PK, Malpathak N (2013) Effect of AM fungi (gf, gm) on biomass and gymnemic acid content of Gymnema sylvestre (Retz.). Biol Fertil Soils 49(3):439–445. https://doi.org/10.1007/s00374-012-0734-x

Zitter-Eglseer K, Nell M, Lamien-Meda A, Steinkellner S, Wawrosch C, Kopp B, Novak J (2015) Effects of root colonization by symbiotic arbuscular mycorrhizal fungi on the yield of pharmacologically active compounds in Angelica archangelica L. Acta Physiol Plant 37:2–11. https://doi.org/10.1007/s11738-014-1695-5

Zolfaghari M, Nazeri V, Sefdkon F, Rejali F (2013) Effect of arbuscular mycorrhizal fungi on plant growth and essential oil content and composition of Ocimum basilicum L. Iranian J Plant Physiol 3:643–650

Zubek S, Stojakowska A, Anielska T, Turnau K (2010) Arbuscular mycorrhizal fungi alter thymol derivative contents of Inula ensifolia L. Mycorrhiza 20(7):497–504. https://doi.org/10.1007/s00572-010-0316-1

Effect of Environmental Factors on Essential Oil Biosynthesis, Chemical Stability, and Yields

10

Somenath Das and Bhanu Prakash

Abstract

Essential oils are major secondary metabolites of aromatic plants and composed of complex mixture of terpenoids and phenylpropanoids components. Complex metabolic cross talk between mevalonic acid, methylerythritol, malonic acid, and DOXP pathways are involved for synthesis of essential oils. The biosynthesis and yield of essential oils are regulated by various environmental factors like light, temperature, drought, salinity, heavy metals, seasonal variations, and mineral concentrations. These environmental factors also have significant impact on chemical nature and stability of essential oil components. The essential oil ingredients are converted into structurally related components by oxidation, reduction, polymerization, dehydrogenation, and cyclization. However, in changing environmental conditions plants develop various mechanisms related to the production of secondary messengers, and signal transduction through mitogen-activated protein kinase and calcium-dependent protein kinases for activation or repression of different transcription factors. The upstream and downstream regulation of genes regulates the metabolic responses of aromatic plants which ultimately serve a crucial role in essential oil biosynthesis. Based on the above background, the article comprises biosynthesis of essential oils through different pathways, cross talk of metabolic reactions, and effect of various environmental factors, namely, light, temperature, oxygen availability, seasonal variation, salinity, and toxic contaminants on chemical stability of essential oils.

S. Das (✉)
Department of Botany, Burdwan Raj College, Purba Bardhaman, West Bengal, India

B. Prakash
Centre of Advanced Study in Botany, Institute of Science, Banaras Hindu University, Varanasi, Uttar Pradesh, India

B. Prakash et al. (eds.), *Plant Essential Oils*,
https://doi.org/10.1007/978-981-99-4370-8_10

Keywords

Essential oil · Biosynthesis · Cross-talk · Transcription factors · Environmental conditions

10.1 Introduction

Essential oils are natural volatile liquids of aromatic plants isolated from varieties of plant parts like wood, seeds, bark, leaves, fruits, roots, rhizomes, and flower (Solórzano-Santos and Miranda-Novales 2012). They are complex mixture of low molecular weight organic volatile compounds which can partially stable in vapor pressure. The essential oils are mainly isolated from plant families like Asteraceae, Apiaceae, Cupressaceae, Lamiaceae, Myrtaceae, Oleaceae, Liliaceae, Rutaceae, and Rosaceae (Das et al. 2021a). India, Indonesia, China, and Brazil are recognized as major country for production of essential oil, whereas Japan, Europe, and the United States are the prime consumers of essential oils. The components of essential oils are categorized into different structural families like terpenoids and phenylpropanoids (Moghaddam and Mehdizadeh 2017). Terpenes are structurally and functionally different classes of components made up of five carbon isoprene units. The main terpenes are monoterpene, sesquiterpenes, diterpenes, triterpenes, and tetraterpenes (Ashour et al. 2010). Terpenes molecules containing oxygen atoms are called terpenoids. There are several types of terpenes like alcohol, aldehyde, ketone, esters, ethers, peroxides, and phenols. Terpenoids are mainly synthesized by two different pathways, namely, methylerythritol pathway in plastid and mevalonate pathway in cytosol-producing isopentenyl pyrophosphate and dimethylallyl pyrophosphate. These two molecules are condensed through geranyl pyrophosphate synthase producing ten carbon monoterpenes (Singh and Sharma 2015). These monoterpenes further develop di, tri, and tetra terpenes. The aromatic essential oil components are mainly synthesized through shikimate pathway (Gounaris 2010). They are composed of essence rich natural active ingredients especially internalized into specialized cells like glandular hairs, trichome, vacuoles, and pockets (Rehman et al. 2016). The global climate change significantly affects the quality of aromatic plants throughout the world. The environmental threats like cold stress, drought, salinity, heavy metal concentrations, and flooding severely affect the essential oil composition and yield in aromatic plants (Verma and Shukla 2015). Most notably, various environmental factors like light, temperature, oxygen availability, and water content significantly alter the biochemical and physiological equilibrium of plants (Chaudhry and Sidhu 2022). These environmental factors also influence the genetic makeup and plant developmental factors leading to changes in essential oil compositions. Different abiotic and biotic stimuli dramatically influence the secretion mechanism of terpenoids and phenolic components in specialized secretory structures like vacuoles, glands, and trichomes. The essential oils have various biological functions like antimicrobial, antifungal, anti-inflammatory, anti-carcinogenic, and green additive or preservative agent in perfumery, pharmaceuticals, and food industries (Nazzaro et al. 2017). Taken as a whole, all the compiled characteristics of essential oils completely depend on the stability and molecular orientation of components (Hyldgaard et al. 2012). The essential oil components are highly reactive in the presence of different environmental factors like heat, light, and air converting into structurally related components by oxidation, isomerization, polymerization, and dehydrogenation (Al-Harrasi et al. 2022). The knowledge of stability of biochemical components may be helpful in tracking the changes/alteration of components under various environmental factors.

On the basis of above background, this chapter highlights the effect of various environmental factors on chemical nature, and cross talk of metabolic reactions for the biosynthesis of essential oils. This chapter provides an insight on the effect of light, temperature, oxygen availability, and toxic contaminants on chemical stability and yield of essential oils.

10.2 Chemical Components of Essential Oils

Essential oils are complex mixture of 20–60 components at varying concentrations (León-Méndez et al. 2019). Two or three components contributing 5–30% of essential oils are called as major components, while the other components in trace amounts are termed as minor components. Essential oils are mainly constituted by terpenoids and phenylpropanoids. Basically, they are hydrocarbons and their oxygenated derivatives contain nitrogen or sulfur (Hüsnü et al. 2007). Essential oils are low molecular weight components with limited solubility in water (Barradas and de Holanda e Silva 2021). Terpenes are combinations of five carbon isoprene units and their modification by allylic prenyldiphosphate by terpene-specific synthetases develop the functional skeleton of different terpenes. The 10 carbon monoterpenes are representative of 90% of essential oils with variety of structures. The acyclic, monocyclic, and bicyclic carbures monoterpenes are p-cymene, phellandrene, camphene, sabinene, 3-carene, ocimene, and myrcene. Linalyl acetate, citronellyl acetate, pulegone, geranial, neral, 1,8-cineol, thymol, and carvacrol are grouped under aldehyde, ketone, ester, and ether monoterpenes (Bakkali et al. 2008). In addition to terpenoids, aromatic compounds are derived from phenylpropane. Common aldehyde, phenol, alcohol, and methoxy aromatic compounds are cinnamaldehyde, eugenol, chavicol, anethole, estragole, and safrole. Table 10.1 presents chemical components of different essential oils.

10.3 Routes and Cross Talk for Biosynthesis of Essential Oils

Essential oil components are mainly categorized under two different chemical classes, namely, terpenoids and phenylpropanoids. Terpenes are more frequently in occurrence as compared to phenylpropanoids which have significant contribution in odor and flavor of the essential oil (Stephane and Jules 2020). Terpenoids and phenylpropanoids are originated from different primary metabolites with variation in

Table 10.1 Chemical components of different essential oils

Terpenoids	Types	Components	Essential oils of specific plants	References
Monoterpenes	Hydrocarbons	Limonene	*Thymus serpyllum*, *Margotia gummifera*	Raal et al. (2004), Valente et al. (2013)
		Phellandrene	*Crithmum maritimum*, *Pinus contorta*, *Eucalyptus dives*, *Anethum graveolens*	Özcan et al. (2006), Sell (2010)
		Pinene	*Juniperus communis*, *Citrus Limon*, *salvia rosifolia*, *Sideritis erythrantha*	Gonny et al. (2006), Vekiari et al. (2002), Özek et al. (2010), Köse et al. (2010)
		Ocimene	*Lavandula multifida*	Zuzarte et al. (2012)
		Terpinene	*Melaleuca alternifolia*, *thymus* sp.	Rudbäck et al. (2012), Jamali et al. (2012)
		Sabinene	*Juniperus communis*, *Daucus carota*	Marzouki et al. (2010), Ottavioli et al. (2009)
	Oxygenated derivatives	Bornyl acetate	*Pinus* sp.	Wagner and Bladt (2009)
		Borneol	*Cinnamomum camphora*, *Abies nordmanniana*, *Artemisia* sp., *Boswellia carterii*	Bauer and Garbe (2001)
		Ascaridol	*Chenopodium ambrosoides*	Harborne and Baxter (2001)
		Carvone	*Mentha spicata*, *Citrus reticulata*, *Anethum graveolens*	Hunter (2009), Bauer and Garbe (2001), Das et al. (2021b)
		Carvacrol	*Origanum vulgare*, *Satureja montana*, *S. hortensis*, *Lavendula multifida*	Teixeira et al. (2013), Bauer and Garbe (2001), Zuzarte et al. (2012)
		Citronellal	*Backhousia citriodora*, *Cymbopogon* sp., *Eucalyptus citriodora*	Bauer and Garbe (2001)
		1,8 cineol	*Salvia officinalis*, *melaleuca alternifolia*	Hunter (2009)
		Geraniol	*Pelargonium graveolens*, *Citrus aurantium*, *Cymbopogon martini*	Bauer and Garbe (2001)
		Menthone	*Mentha spicata*, *M. piperata*	Bauer and Garbe (2001), Hunter (2009)
		Menthol	*Mentha arvensis*	Kamatou et al. (2013)
		Menthofuran	*Mentha aquatica*, *M. piperata*, *M. pulegium*	Derwich et al. (2010), Saharkhiz et al. (2012)
		Linalool	*Coriandrum sativum*, *Laurus nobilis*, *Ocimum basilicum*	Das et al. (2019), Saab et al. (2012)

Table 1.1 (continued)

Terpenoids	Types	Components	Essential oils of specific plants	References
		Pulegone	*Mentha pulegium*	Ysef (2013)
		Terpineol	*Thymus caespititius*, *Amomum subulatum*, *Pimenta dioca*	Pereira et al. (2000), Joshi et al. (2013), Chaudhari et al. (2020a)
		Terpinene-4-ol	*Pandamus odoratissimus*, *Juniperus wallichiana*, *Origanum majorana*, *Melaleuca alternifolia*	Carson et al. (2006), Lohani et al. (2013), Raina et al. (2004), Chaudhari et al. (2020b)
		Thymol	*Origanum* sp., *Thymus* sp.	Davazdahemami et al. (2011), Mota et al. (2012)
		Thujone	*Artemisia absinthium*, *Cedrus deodara*	Pinto et al. (2007), Hunter (2009)
Sequiterpenes	Hydrocarbons	Caryophyllene	*Cinnamomum zeylanicum*, *Piper nigrum*, *Cannabis sativa*	Jayaprakasha et al. (2003), Orav et al. (2004)
		Cadinene	*Juniperus oxycedrus*	Yannai (2004)
		Longifolene	*Pinus longifolia*	Yao et al. (2005)
		Zingiberene	*Zingiber officinale*	Baser and Demirci (2007)
		Elemene	*Curcuma aromatica*	Wang et al. (2012)
	Oxygenated derivatives	Cedrol	*Thuja* sp., *Cedrus* sp., *Juniperus* sp.	Sell (2010)
		Caryphyllene oxide	*Didymocarpus tomentosa*	Gowda et al. (2012)
		Carotol	*Daucus carota*	Bauer and Garbe (2001)
		Bisabolol	*Matricaria recutita* and *Myoporum crassifolium*	Dewick (2002), Sell (2010)
		Patchoulol	*Pogostemon cablin*	Jager (2010)
		Nerolidol	*Zanthoxylum gardneri*, *Citrus aurantium*	Baser and Demirci (2007), Craveiro et al. (1991)
Diterpene	–	Camphorene, citronellal, phytol		Baser and Demirci (2007)
Phenylpropanoid	–	Anethole	*Pimpinella anisum*, *Foeniculum vulgare*	Sharifi et al. (2008), Das et al. (2021c)
		Cinnamaldehyde	*Cinnamomum cassia*	Chang et al. (2013)
		Myristicine	*Myristica fragrans*	Das et al. (2020)
		Eugenol	*Cinnamomum zeylanicum*	Chericoni et al. (2005)
		Safrole	*Myristica fragrans*	Das et al. (2020)

biosynthetic routes. Shikimate pathway is responsible for synthesis of phenylpropanoid components along with the precursor amino acids tyrosine and phenyl alanine (Haslam 2014). Terpenoids are mainly synthesized from five carbon precursor isopentenyl pyrophosphate (IPP) and its isomer dimethylallyl pyrophosphate (DMAPP) (Böttger et al. 2018). The prenyl pyrophosphate enzyme helps in coupling of IPP units with DMAPP resulting into production of monoterpenoids (C_{10}), sesquiterpenoids (C_{15}), diterpenoids (C_{20}), etc. Different skeletal themes of terpenoids are developed by prenyl pyrophosphates causing generation of multiple parental terpenoids (Ashour et al. 2010). Stereochemical variation of parental terpenoids by oxidation, conjugation, reduction, derivatization, and isomerization develops myriad of terpenoids. IPP has been recognized as universal biological precursor of isoprenoids. The horizontal gene transfer event (HGTs) supported that mevalonate pathway is an ancestral metabolic route of isoprenoid biosynthesis (Lombard and Moreira 2011). In addition, the glyceraldehydes/pyruvate route of isoprenoid biosynthesis has been recently indicated which regulate the terpenoid metabolism in plants (Kiani 2017). Most notably, the plasticidal 2-C-methylerythritol-4-phosphate (MEP) pathway is also responsible for the synthesis of IPP and DMAPP from pyruvate and glyceraldehydes 3 phosphate (Kuzuyama and Seto 2012). Schwarz and Arigoni (1999) demonstrated that plastid-related isoprenoid compounds are developed by the involvement of MEP pathway. Rohmer et al. (2005) reported that 1-deoxy-D-xylulose-5-phosphate (DOXP) as the first stable product for the synthesis of isoprenoids. The DOXP pathway initiates by formation of 1-deoxy-D-xylulose-5-phosphate (DOXP) from pyruvate and glyceraldehydes 3 phosphate. DOXP further produces 2-C-methyl-D-erythritol 4-phosphate by DOXP reductoketoisomerase with operational biosynthesis of leucine, valine, and isoleucine. Thereafter, the 2-C-methyl-Derythritol 4-phosphate metabolized to produce the IPP (Wanke et al. 2001). Figure 10.1 represents the pathways for isoprenoids and terpenoid biosynthesis.

In addition to isoprenoids, phenylpropanoids are also common constituents of essential oils. Their abundance provides different sensory values of essential oils. Major phenylpropanoids which have been isolated from essential oils are eugenol, estragol, anethole, dillapiole, elemicine, methyl chavicol, and methyl cinnamate (Zuzarte and Salgueiro 2015). The phenylpropanoid components are developed from phenylalanine synthesized by Shikimate pathway. Phenylalanine ammonia lyase helps in conversion of phenylalanine to trans cinnamic acid. The cinnamic acid is converted to p-coumaric acid by cinnamate 4 hydroxylase. The tyrosine ammonia lyase synthesized coumarate directly from tyrosine (Hüsnü et al. 2007). In vivo modification of peroxidase developed the eugenol from coumare derivatives. The phenylpropane, elemicine, and myristicine are developed from methyleugenol (Zuzarte and Salgueiro 2015). The pathways for phenylpropanoid biosynthesis are presented in Fig. 10.2.

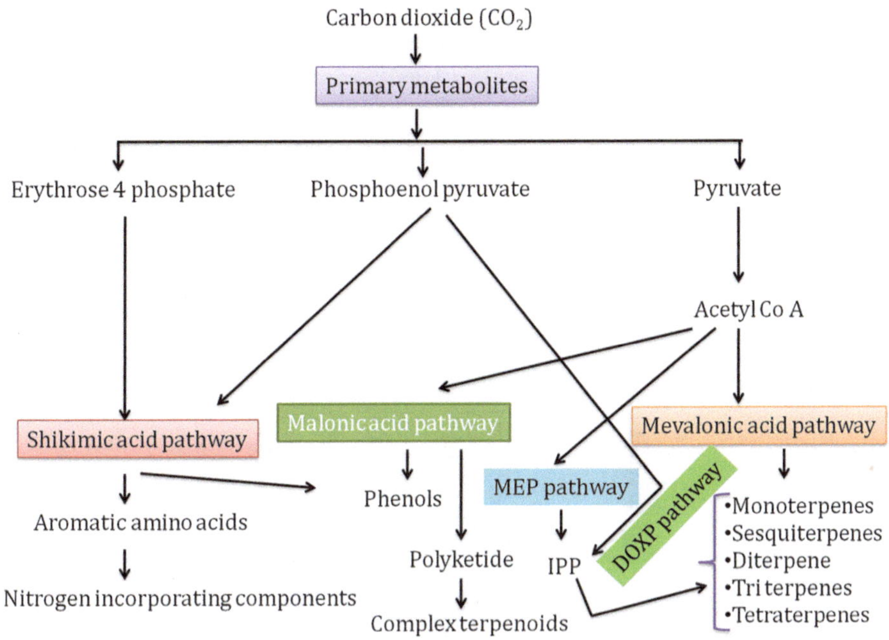

Fig. 10.1 Pathways for isoprenoids and terpenoid biosynthesis

10.4 Environmental Factors Affecting Essential Oil Biosynthesis

Environmental factors are key regulator for plant growth and biosynthesis of essential oils in specific plant organs (Li et al. 2020). Different abiotic factors play profound effect on essential oil compositions. Temperature, mineral nutrients, light quality, humidity, ozone, and oxygen are greatly involved in amelioration of essential oil production in plants.

10.4.1 Light

The light quality has special effect on essential oil biosynthesis. It has been observed that in *Pelargonium* spp. fraction of incident light at different wavelength modulated the biogenesis of essential oil in leaves. The red light induced the production of essential oil from exogenous primary precursors (Sangwan et al. 2001). Authors suggested that red-light-mediated essential oil production is associated with the phytochrome metabolism in plants. Yanhe et al. (1995) pointed the fact that quantum irradiation has profound impact on essential oil biosynthesis in thyme and sage plants. Higher amount of essential oil with thujone content (reduction in camphor occurrence) was recorded in thyme and sage plants at 45% sunlight. In case of

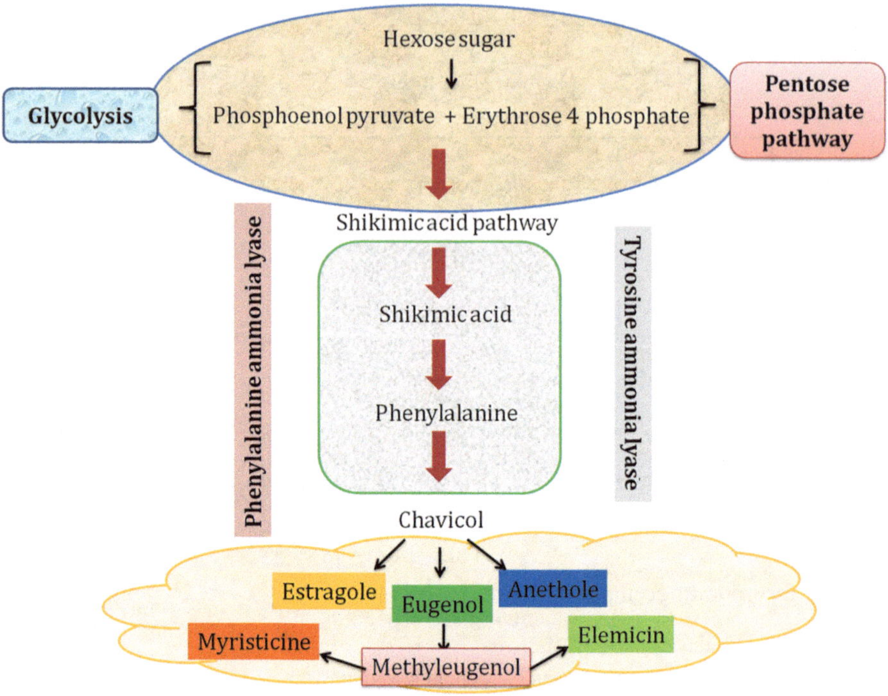

Fig. 10.2 Pathways for phenylpropanoid biosynthesis

thyme plant maximum amount of myrcene and thymol was found in full sunlight. Light has the capacity to induce phenyl propane and monoterpenes in *Ocimum basilicum* and *Satureja douglasii* (Barra 2009). Jaiswal et al. (2021) demonstrated UV-B-dependent regulation of secondary metabolism in plants which influence the biosynthetic reactions. Wang et al. (2020) reported increase in essential oil content of *Asarum heterotropoids* by light irradiation treatments of 12, 24, 50, and 100% sunlight. Cinnamic acid, phenylalanine, and p-coumarine level in *A. heterotropoids* were increased which directly induces the production of essential oil by shikimic acid pathway. Light affects the plastid differentiation and chlorophyll biosynthesis in *Rosamarinus officinalis* which in turn induce the production of essential oil. Batista et al. (2016) demonstrated the effect of spectral quality on leaf anatomy and expansion of leaf blade with increase in essential oil production in *Lippia alba*. Nishioka et al. (2008) reported increase in menthol content of *Mentha arvensis* in red light as compared to blue and green light.

10.4.2 Seasonal Variation

Seasonal variation is a combinatorial factor of radiation, precipitation, and temperature. Barroso et al. (1992) reported that sabinene was the major component in

Crithmum maritimum essential oil during the flowering periods while γ-terpinene was recorded as major ingredient rest of the season. Zouari-Bouassida et al. (2018) reported highest yield of essential oil in *Mentha longifolia*. In winter season 1,8 cineol, pulegone, and L-menthone were recognized as major component of *M. longifolia* essential oil in winter season. In the spring season, palmitic acid and pulegone were found as major ingredient. Cascade regulations of genes in different seasons are the prime reason for variation in chemical components. In *Achillea millefolium*, sesquiterpene was recorded as major ingredient of essential oil at vegetative period while monoterpene hydrocarbon was recognized as major components during the flowering periods (Figueiredo et al. 1992). Santos-Gomes and Fernandes-Ferreira (2001) reported that oxygenated monoterpene of *Salvia officinalis* essential oil decreased in winter, while increment in monoterpene hydrocarbon was recorded in summer. The essential oil of *Salvia libanotica* contains maximum amount of camphene, thujone, and camphor in winter time than spring time (Farhat et al. 2001). Rathore et al. (2022) demonstrated higher essential oil content in *Rosmarinus officinalis* during the autumn season as compared to rainy and summer seasons. Authors reported higher amount of β-pinene, 1,8-cineol, terpinen-4-ol, and cis sabinene hydrate as major component in rainy season, while terpinen-4-ol was found as prime ingredient in summer season. Similarly, β-caryophyllene was found in higher amount in the rainy season; however, it was not observed in autumn season. In *Ocimum basilicum*, oxygenated monoterpenes and sesquiterpenes were found higher in winter season, while lowest content was recorded in summer season (Hussain et al. 2008; Gazim et al. 2010). Seasonal temperature also has significant impact on essential oil yield. For instance, essential oil of *O. basilicum* was found highest at winter (0.8%) while reduced the oil content at summer (0.5%).

10.4.3 Temperature

Temperature plays prime role in release of terpenes especially monoterpenes and isoprene of essential oil in hot weather. Temperature has direct role to induce the enzyme catalyzing the synthesis of volatile components. Isoprene has better tolerance ability against high temperature and reactive oxygen species. Heydari et al. (2018) reported differentiation of essential oil composition of *Mentha piperita* and *Mentha arvensis* plants in heat stress. At higher temperature, the menthol level was reduced while menthyl acetate and pulegone were increased significantly. Ložienė et al. (2021) demonstrated the accumulation of geraniol in *Thymus pulegioides* at rising temperature condition. In addition to high temperature, cold stress (low temperature) plays significant role in variation of essential oil composition. Senji and Mandoulakani (2018) reported that higher level of cadinene and geraniol were observed at 4 °C for 24 h in *Ocimum basilicum*, while after continuation of same temperature for 48 h maximum concentration of 1,8-cineol was recorded. The experiment also extended up to the temperature level of 10 °C, where α-bergamotene and germacrene-D were recognized as major components of essential oil. Significant alteration in total amount of active ingredients of *Matricaria chamomilla* in cold stress has been reported by Ahmad et al. (2011).

10.4.4 Drought

Water plays prime role in growth and development of plants. Hence, scarcity or excessiveness of water may influence the essential oil contents of different plants. Under drought stress condition, increment in total sesquiterpene content and menthol was observed while the level of menthofuran, pulegone, and cineol were considerably decreased in peppermint essential oil (Charles et al. 1990). In contrast with this study, Idrees et al. (2010) reported decrement in essential oil yield in *Cymbopogon flexuosus* under drought stress conditions. Rahimi et al. (2017) threw light on the fact that increase in essential oil yield is related to upregulation of limonene synthase gene expression and downregulation of menthofuran synthase gene expression along with improved quality. Most importantly, the moderate and severe drought conditions reduced the essential oil content in aromatic plants. García-Caparrós et al. (2019) reported 53.8 and 36.6% reduction in essential oil content of *Lavendula latifolia* and *Salvia sclarea* essential oil under drought stress condition. Mandoulakani et al. (2017) demonstrated the escalation of methyl eugenol and methyl chavicol level of sweet basil essential oil in drought stress. It has also been revealed that extreme drought condition improved the essential oil contents of *Matricaria chamomilla* while the yield was considerably declined (Jeshni et al. 2017). Investigation of Chrysargyris et al. (2016) suggested that essential oil yield in Greek sage plant in moderate water-deficit condition was increased by 50%, while 83% increment in essential oil production was achieved in severe water-deficit conditions. Similar result was also reported by Bettaieb et al. (2009) for *Salvia officinalis* plant in water-deficit condition.

10.4.5 Plant Growth Regulators

Plant growth regulators or phytohormones have special impact on essential oil biosynthesis. The growth regulators also influence the genes regulating physiological and biochemical processes which could enhance the essential oil production in specific plant parts (Farooqi et al. 1996). The phytohormones induce the leaf and flower production with general increase in biomass causing higher essential oil yield. Povh and Ono (2007) demonstrated increase in essential oil of *Salvia officinalis* by applying 100 mg/L of gibberellic acid. Lactonic and ketonic spirostane (analogues of brassinosteroids) showed increased production of leaves which directly correlated with menthol synthesis in *Mentha arvensis* (Bovi et al. 2001). Essential oils are synthesized in special secretary structures which are the key factor for terpene biosynthesis. Fraternale et al. (2003) reported that at 0.1 mg/L of benziladenine *Thymus mastichina* plant produced maximum essential oil due to increased density of glandular hair at post secretary stage. Erbilgin et al. (2006) demonstrated that application of jasmonic acid induced the development of resiniferous ducts in *Picea abies* with threefold increase in concentration of limonene and β-piene. Application of 0.5 mM methyl jasmonate significantly increased the monoterpene quantity in *Ocimum basilicum*. As compared to control plant linalool and eugenol was increased by 43 and 56%. Kim et al. (2006) observed that increase in transcript number of phenylammonia lyase induced the production of eugenol.

10.4.6 Saline Conditions

Soil salinity is one of the prime regulators of plant growth and metabolism. Severe salinity reduces the growth of plants due to changes in root osmotic potential. Ahmad and Prasad (2011) reported accumulation of Na^+ and Cl^- ions in plant cells in salt stress resulting production of reactive oxygen species and inhibition of uptake of K^+ ions. Khorasaninejad et al. (2010) reported decrement in essential oil content of peppermint at salinity stress conditions. Menthone and pulegone level of *Mentha canadensis* was enhanced in saline conditions while menthol level was found to decrease significantly (Yu et al. 2015). Razmjoo et al. (2008) reported decline in essential oil content of *Matricaria chamomilla* at saline level of 336 mmol/L. Effects of salinity on peppermint essential oil has been carefully studied. It has been observed that reduction of pinene, menthol, sabinene, 1,8-cineol, and menthofuran at higher saline concentration whereas menthone concentration was enhanced promisingly (Ghorbani et al. 2018).

10.4.7 Toxicity of Heavy Metals

Heavy metals are major pollutant of soil and can be absorbed by different plant parts. After accumulation, the heavy metals carryforward to the food chain and influence varieties of physiological manifestations (Anjum et al. 2016). Lajayer et al. (2017) reported alteration in biosynthesis of essential oil of the aromatic plants grown in the metal-contaminated areas. Azimychetabi et al. (2021) demonstrated decreased menthol content from 54.1 to 40.8% and increased concentration of pulegone and menthofuran by exposure of 40 mg/L cadmium. It has been reported that cadmium concentration greater than 25 mg/L showed prominent decrease in essential oil yield of *Ocimum basilicum* (Youssef 2021). Sá et al. (2015) described the induction of carvone content of Lemongrass essential oil in presence of lead. Table 10.2. presents the effects of different environmental factors on essential oil composition and yield.

Table 10.2 Effect of different environmental factors on essential oil composition and yield

Environmental factors	Plant species	Effects	References
Drought	*Ocimum basilicum*	Increment in level of β-myrcene, methyleugenol, and methylchavicol	Mandoulakani et al. (2017)
	Thymus carmanicus	Reduce γ-terpinene and carvacrol content while increase the borneol and thymol content in essential oil	Bahreininejad et al. (2014)
Temperature	*Mentha piperata*	Escalation of production of menthofuran, menthone, and menthyl acetate while decrement in menthol concentration	Maffei and Scannerini (2000)
	Matricaria chamomilla	Low temperature increased the level of essential oil	Ahmad et al. (2011)
Saline conditions	*Rosmarinus officinalis*	Increased the level of essential oil	Dehghani Bidgoli et al. (2019)
	Mentha pulegium and *M. suaveolens*	Reduce the yield of essential oil	Aziz et al. (2008)
Heavy metals	*Melissa officinalis*	Reduce essential oil yield by 97%	Kilic and Kilic (2017)
	Mentha piperata	Decrement in menthol content while enhancement in pulegone and menthofuran content under cd stress	Azimychetabi et al. (2021)
	Mentha arvensis	Modest essential oil content at 20 mg/kg of arsenic (As) while at 100 mg/kg of As essential oil concentration was considerably reduced	Nabi et al. (2019)

10.5 Effect of Environmental Factors on Stability and Yield of Essential Oils

The essential oil components are structurally related to each other; therefore, they are easily converted into another forms by oxidation, cyclization, isomerization, and dehydrogenation reactions (Al-Harrasi et al. 2022). The chemical stability of the essential oil components is influenced by plant growth stages, edaphic factors, climate, and harvest time (Figueiredo et al. 2008). Terpenoids are volatile and thermolabile in nature. Hence, easily hydrolyzed or oxidized on the basis of their structures (Pavela and Benelli 2016). In the absence of protective compartmentation, the essential oils are prone to chemical transformations, oxidative damage, and polymerization reactions, affecting the stability of essential oil. For instance, p-cymene (monoterpene) has been identified in aged essential oils developed through acid-catalyzed depletion of citral or oxidative dehydrogenation of unsaturated monocyclic terpenes causing off-flavor of lemon oil (Nguyen et al. 2009). In addition, autooxidation has also been recorded for terpenoids deterioration. Hagvall et al. (2007) reported the development of hydroperoxide through oxidation of terpenoids which further decompose in the presence of heat and light. The chemical natures of secondary oxidation products remain unknown. Caryophyllene oxide is a secondary oxidation product of sesquiterpene caryophyllene which has been recognized in stored essential oils (Turek and Stintzing 2012).

The degradation of essential oils depends on several environmental factors like temperature, oxygen availability, light, water content, and metal contaminants which are described as follows.

10.5.1 Temperature

Temperature influences the stability of essential oil based on acceleration of chemical reactions according to the equation of Arrhenius (Atkins 2002). Van't Hohf's law also stated that an increase of 10 °C temperature doubled the chemical reaction rates. The autooxidation and hydroperoxide decomposition at high temperature develops free radicals. At low temperature, oxygen was soluble in liquid affecting the stability of essential oils (Bernhard and Marr 1960). Turek and Stintzing (2012) reported decrement in terpenic hydrocarbon, namely, γ-terpene, β-myrcene, β-caryophyllene, and β-pinene in pine, lavender, rosemary, clove bud, and cardamom essential oils upon increase in temperature. Terpenes and aldehydes are prone to rearrangement at higher temperatures. Nguyen et al. (2009) demonstrated the loss of neral, geraniol, and terpenic hydrocarbons of lemon essential oil when kept at 50 °C for 2 weeks. It has also been investigated that Rosemary oil was stable at low temperature in refrigerator over a period of 3 months. In contrast, the peroxide was developed in pine essential oil while keeping at 5 °C in refrigerator. Similar finding was reported for lavender essential oil with development of peroxide at 5 °C; however, at 38 °C peroxides were completely degraded. These data emphasized that stability of essential oils depends on chemical nature of the components (Turek et al. 2012).

10.5.2 Light

Visible and ultraviolet light plays an important role in autooxidation by accelerating the hydrocarbon abstraction leading to generation of alkyl radicals (Choe and Min 2006). Misharina et al. (2003) reported rapid degradation of monoterpenes in the presence of light. The rosemary essential oil was found susceptible at imitated daylight leading to considerable increase in camphor, p-cymene, and caryophyllene oxide. Nguyen et al. (2009) demonstrated the development of photoanethole and photocitral in stored anise and lemon essential oil by sunlight or UV light-mediated oxidation. Most importantly, in sweet fennel oil trans-anethole was isomerized into cis-anethole or oxidized to anisaldehyde in the presence of light within 2 months of storage. In the presence of UV light, trans-anethole was converted into cis-anethole which was found 10–12 times more toxic than cis isomers (Misharina and Polshkov 2005).

10.5.3 Availability of Oxygen

Oxygen concentration causes spoilage of essential oil and plays a decisive role in stability. Oxidation of essential oil is influenced by partial pressure in headspace and ambient temperature. At lower temperature, the essential oil oxidized to produce peroxide and hydroperoxide radicals. At higher temperature (limited oxygen concentration), the essential oils are oxidized to produce alkyl or hydroxyl radicals. Turek et al. (2012) reported peroxide radical formation of pine and lavender essential oil at 23 °C and 5 °C, respectively. Nguyen et al. (2009) demonstrated decrease in β-caryophyllene, limonene, α-terpinene, and β-phellandrene components of lemon essential oil when stored at 50 °C which was mainly associated with the oxidation of essential oil components. The essential oils may sometimes moderately oxidize to produce peroxide radicals at refrigerator (Tisserand and Young 2013). Hence, to avoid the oxidation process, essential oils should be treated with inert gas like argon to prevent the development of peroxide or hydroxyl radicals.

10.5.4 Contamination of Metals

Distillation unit made of primitive stills or preservation of essential oils in metallic containers may cause the oxidation of essential oil components by catalyzing reaction with the metallic impurities (Al-Harrasi et al. 2022). The hydroperoxide decomposition through Fe^{2+} to Fe^{3+} and Cu^{+} to Cu^{2+} mainly develops the peroxyl or alkoxyl radicals which in turn cause photooxidation of essential oil components (Choe and Min 2006).

10.6 Mechanism of Essential Oil Regulation by Environmental Factors

On the basis of physiological point of view, first of all the environmental factors are detected by plant plasma membrane, thereafter, the signal is transmitted toward the downstream regulatory signals through different secondary messengers like carbohydrates, cyclic adenosine monophosphate (cAMP), cyclic guanosine monophosphate (cGMP), inositol phosphate, calcium ions, and nitric oxide (Smékalová et al. 2014). The secondary messengers activated the Ca^{2+} sensors allowing the signal for further transduction (Zhou et al. 2019). Thereafter, the signal transduction pathway is regulated by protein kinase and phosphatases like calcium-dependent protein kinase (CDPKs) and mitogen-activated protein kinases (MAPKs) (Wurzinger et al. 2011). The protein kinase or phosphatases have specific role to activate or repress different transcriptional factors which regulate the production of secondary metabolites especially essential oils (Chan et al. 2005). The upstream components facilitate the regulation of transcription factors by various steps of posttranscriptional modifications namely, SUMOylation and ubiquitination (Debnath et al. 2011). The transcription factors are prime regulator of different terminal transducers by acting on cis-element in the promoter regions. They detect the environmental stress signals and direct the downstream expression of some defense genes (Jan et al. 2021). The stress gene products involving essential oils help in plant adaptation allowing the plants to grow in stressful conditions. The plant receptors and sensors facilitate in detection of environmental factors to stand the defensive reactions and expression of essential oil synthesizing genes in specialized organs of plants. Most importantly, the terpenoids and isoprenoids are not stored in permanent metabolic pools of plants. Their synthesis depends on the functional requirements and modulated by substrate concentration, light, temperature, and oxygen concentrations (Bustamante et al. 2020). Additionally, the critical environmental conditions trigger the emission of terpenoids in plants which has been suggested as a possible reason for compositional variation of essential oil (Ngumbi and Ugarte 2021). The carbon flux and allocation of resources are responsible for alteration of bioactive components and quality of essential oils in different environmental conditions. The mechanism of essential oil regulation by environmental factors is presented in Fig. 10.3.

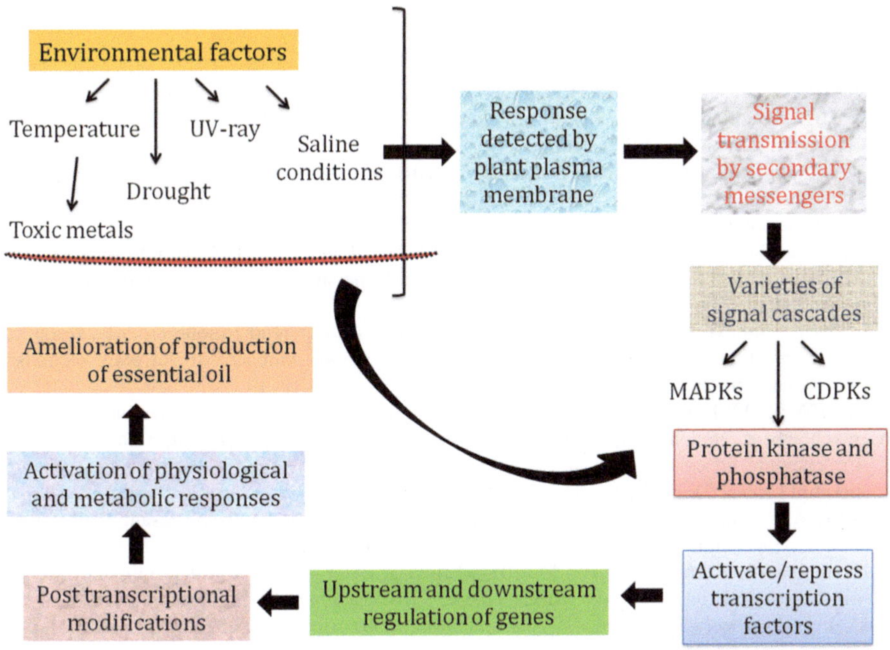

Fig. 10.3 Diagrammatic representation of mechanism of essential oil regulation by environmental factors

10.7 Conclusion

Essential oils are complex secondary metabolite of aromatic plants and have substantial interest for varieties of biological functions. Various environmental factors like temperature, radiation, seasonal variation, salinity, drought, and toxic metals play important role in the synthesis and regulation of essential oil production. The environmental factors could affect the biosynthesis of essential oils by regulating the mevalonic acid, methylerythritol-4-phosphate, and DOXP pathways. Additionally, the environmental factors have prominent impact on stability of essential oils. The structural similarities of essential oil components have special effect on oxidation, cyclization, polymerization, and dehydrogenation, leading to determination of the stability of components. The secondary messengers are prime factors which activate the protein kinase and phosphatase activities in genes leading to upstream regulation of transcription factors at the onset of various environmental factors. However, further investigation is required to unveil different analytic approaches for consideration of essential oil components analysis and stability.

Acknowledgments Somenath Das is thankful to Head, Department of Botany, and Principal Burdwan Raj College, Purba Bardhaman, West Bengal, India for necessary facilities. Bhanu Prakash is grateful to Head, Department of Botany, Banaras Hindu University, Varanasi, India for kind support.

References

Ahmad E, Payam M, Alireza TD, Hossein AF (2011) Effects of temperature and varieties on essential oil content and quantity features of chamomile. J Agric Ext Rural Dev 3:19–22

Ahmad P, Prasad MNV (2011) Abiotic stress responses in plants: metabolism, productivity and sustainability. Springer Science & Business Media, pp 1–465

Al-Harrasi A, Bhatia S, Behl T, Anwer K (2022) Essential oil stability. In: Role of essential oils in the management of COVID-19. CRC Press, pp 159–168

Anjum NA, Duarte B, Caçador I, Sleimi N, Duarte AC, Pereira E (2016) Biophysical and biochemical markers of metal/metalloid-impacts in salt marsh halophytes and their implications. Front Environ Sci 4:24–38

Ashour M, Wink M, Gershenzon J (2010) Biochemistry of terpenoids: monoterpenes, sesquiterpenes and diterpenes. In: Annual plant reviews volume 40: biochemistry of plant secondary metabolism, pp 258–303

Atkins PW (2002) Kurzlehrbuch Physikalische Chemie, 3rd edn. Wiley-VCH, Weinheim

Azimychetabi Z, Nodehi MS, Moghadam TK, Motesharezadeh B (2021) Cadmium stress alters the essential oil composition and the expression of genes involved in their synthesis in peppermint (*Mentha piperita* L.). Ind Crop Prod 168:113602

Aziz EE, Al-Amier H, Craker LE (2008) Influence of salt stress on growth and essential oil production in peppermint, pennyroyal, and apple mint. Int J Geogr Inf Syst 14:77–87

Bahreininejad B, Razmjoo J, Mirza M (2014) Effect of water stress on productivity and essential oil content and composition of *thymus carmanicus*. J Essent Oil-Bear Plants 17:717–725

Bakkali F, Averbeck S, Averbeck D, Idaomar M (2008) Biological effects of essential oils–a review. Food Chem Toxicol 46:446–475

Barra A (2009) Factors affecting chemical variability of essential oils: a review of recent developments. Nat Prod Comm 4:1934578X0900400827

Barradas TN, de Holanda e Silva KG (2021) Nanoemulsions of essential oils to improve solubility, stability and permeability: a review. Environ Chem Lett 19:1153–1171

Barroso JG, Pedro LG, Figueiredo AC, Pais MSS, Scheffer JJ (1992) Seasonal variation in the composition of the essential oil of *Crithmum maritimum* L. Flavour Fragr J 7:147–150

Baser KHC, Demirci F (2007) Chemistry of essential oils. In: Berger RG (ed) Flavours and fragrances—chemistry, bioprocessing and sustainability. Springer, Berlin, pp 43–86

Batista DS, de Castro KM, da Silva AR, Teixeira ML, Sales TA, Soares LI, Otoni WC (2016) Light quality affects in vitro growth and essential oil profile in *Lippia alba* (Verbenaceae). Vitro Cellular Develop Biol Plant 52:276–282

Bauer K, Garbe D (2001) Common fragrance and flavour materials. Preparation, properties and uses. VCH Verlagsgesellschaft, Weinheim

Bernhard RA, Marr AG (1960) The oxidation of terpenes. I. Mechanism and reaction products of D-limonene autoxidation. J Food Sci 25:517–530

Bettaieb I, Zakhama N, Wannes WA, Kchouk ME, Marzouk B (2009) Water deficit effects on *Salvia officinalis* fatty acids and essential oils composition. Sci Hortic 120:271–275

Böttger A, Vothknecht U, Bolle C, Wolf A (2018) Terpenes and terpenoids. In: Lessons on caffeine, cannabis & Co. Springer, Cham, pp 153–170

Bovi OA, Zullo MAT, Perecin MB, Granja NP, Carmello QAC, Robaina C, Maia NB (2001) Hydroponic cultivation of mint and vetiver with spirostane analogues of brassinosteroids. Int Sympos Growing Media Hydroponics 644:55–59

Bustamante MÁ, Michelozzi M, Barra Caracciolo A, Grenni P, Verbokkem J, Geerdink P, Nogues I (2020) Effects of soil fertilization on terpenoids and other carbon-based secondary metabolites in *Rosmarinus officinalis* plants: a comparative study. Plan Theory 9:830–851

Carson CF, Hammer KA, Riley TV (2006) *Melaleuca alternifolia* (tea tree) oil: a review of antimicrobial and other medicinal properties. Clin Microbiol Rev 19:50–62

Chan AS, Pang H, Yip EC, Tam YK, Wong YH (2005) Carvacrol and eugenol differentially stimulate intracellular Ca2+ mobilization and mitogen-activated protein kinases in Jurkat T-cells and monocytic THP-1 cells. Planta Med 71:634–639

Chang CT, Chang WL, Hsu JC, Shih Y, Chou ST (2013) Chemical composition and tyrosinase inhibitory activity of *Cinnamomum cassia* essential oil. Bot Stud 54:1–7

Charles DJ, Joly RJ, Simon JE (1990) Effects of osmotic stress on the essential oil content and composition of peppermint. Phytochemistry 29:2837–2840

Chaudhari AK, Singh VK, Das S, Prasad J, Dwivedy AK, Dubey NK (2020b) Improvement of in vitro and in situ antifungal, AFB1 inhibitory and antioxidant activity of *Origanum majorana* L. essential oil through nanoemulsion and recommending as novel food preservative. Food Chem Toxicol 143:111536

Chaudhari AK, Singh VK, Dwivedy AK, Das S, Upadhyay N, Singh A, Dubey NK (2020a) Chemically characterised *Pimenta dioica* (L.) Merr. Essential oil as a novel plant based antimicrobial against fungal and aflatoxin B1 contamination of stored maize and its possible mode of action. Nat Prod Res 34:745–749

Chaudhry S, Sidhu GPS (2022) Climate change regulated abiotic stress mechanisms in plants: a comprehensive review. Plant Cell Rep 41:1–31

Chericoni S, Prieto JM, Iacopini P, Cioni P, Morelli I (2005) In vitro activity of the essential oil of *Cinnamomum zeylanicum* and eugenol in peroxynitrite-induced oxidative processes. J Agric Food Chem 53:4762–4765

Choe E, Min DB (2006) Mechanisms and factors for edible oil oxidation. Compr Rev Food Sci Food Saf 5:169–186

Chrysargyris A, Laoutari S, Litskas VD, Stavrinides MC, Tzortzakis N (2016) Effects of water stress on lavender and sage biomass production, essential oil composition and biocidal properties against *Tetranychus urticae* (Koch). Sci Hortic 213:96–103

Craveiro AA, Alencar JW, Matos FJA, Machado MIL (1991) Essential oil of *Zanthoxylum gardneri* Engl. A new source of nerolidol. J Essential Oil Res 3:371–372

Das S, Singh VK, Dwivedy AK, Chaudhari AK, Dubey NK (2021a) Insecticidal and fungicidal efficacy of essential oils and nanoencapsulation approaches for the development of next generation ecofriendly green preservatives for management of stored food commodities: an overview. Int J Pest Manag:1–32

Das S, Singh VK, Dwivedy AK, Chaudhari AK, Dubey NK (2021b) *Anethum graveolens* essential oil encapsulation in chitosan nanomatrix: investigations on in vitro release behavior, organoleptic attributes, and efficacy as potential delivery vehicles against biodeterioration of rice (*Oryza sativa* L.). Food Bioprocess Technol 14:831–853

Das S, Singh VK, Dwivedy AK, Chaudhari AK, Dubey NK (2021c) Nanostructured Pimpinella anisum essential oil as novel green food preservative against fungal infestation, aflatoxin B1 contamination and deterioration of nutritional qualities. Food Chem 344:128574

Das S, Singh VK, Dwivedy AK, Chaudhari AK, Upadhyay N, Singh A, Dubey NK (2020) Assessment of chemically characterised *Myristica fragrans* essential oil against fungi contaminating stored scented rice and its mode of action as novel aflatoxin inhibitor. Nat Prod Res 34:1611–1615

Das S, Singh VK, Dwivedy AK, Chaudhari AK, Upadhyay N, Singh P, Dubey NK (2019) Encapsulation in chitosan-based nanomatrix as an efficient green technology to boost the antimicrobial, antioxidant and in situ efficacy of *Coriandrum sativum* essential oil. Int J Biol Macromol 133:294–305

Davazdahemami S, Sefidkon F, Jahansooz MR, Mazaheri D (2011) Chemical composition of the essential oils from Foliages and seeds of Ajowan (*Trachyspermum ammi* (L.) Sprague) in two planting dates (spring and summer). J Essential Oil Bear Plants 14:639–642

Debnath M, Pandey M, Bisen PS (2011) An omics approach to understand the plant abiotic stress. OMICS J Integr Biol 15:739–762

Dehghani Bidgoli R, Azarnezhad N, Akhbari M, Ghorbani M (2019) Salinity stress and PGPR effects on essential oil changes in *Rosmarinus officinalis* L. Agric Food Secur 8:1–7

Derwich E, Benziane Z, Taouil R (2010) GC/MS analysis of volatile compounds of the essential oil of the leaves of *Mentha pulegium* growing in Morocco. Chem Bull POLITEHNICA Univ 55:103–106

Dewick PM (2002) Chap. 5—the mevalonate and deoxyxylulose phosphate pathways: terpenoids and steroids. In: Medicinal natural products: a biosynthetic approach, 2nd edn. Wiley, pp 167–285

Erbilgin N, Krokene P, Christiansen E, Zeneli G, Gershenzon J (2006) Exogenous application of methyl jasmonate elicits defenses in Norway spruce (*Picea abies*) and reduces host colonization by the bark beetle Ips typographus. Oecologia 148:426–436

Farhat GN, Affara NI, Gali-Muhtasib HU (2001) Seasonal changes in the composition of the essential oil extract of East Mediterranean sage (*Salvia libanotica*) and its toxicity in mice. Toxicon 39:1601–1605

Farooqi AA, Shukla A, Sharma S, Khan A (1996) Effect of plant age and GA3 on artemisinin and essential oil yield in *Artemisia annua* L. Int J Geogr Inf Syst 4:73–80

Figueiredo AC, Barroso JG, Pais MSS, Scheffer JJ (1992) Composition of the essential oils from leaves and flowers of *Achillea millefolium* L. ssp. millefolium. Flavour Fragr J 7:219–222

Figueiredo AC, Barroso JG, Pedro LG, Scheffer JJ (2008) Factors affecting secondary metabolite production in plants: volatile components and essential oils. Flavour Fragr J 23:213–226

Fraternale D, Giamperi L, Ricci D, Rocchi MBL, Guidi L, Epifano F, Marcotullio MC (2003) The effect of triacontanol on micropropagation and on secretory system of *thymus mastichina*. Plant Cell Tissue Organ Cult 74:87–97

García-Caparrós P, Romero MJ, Llanderal A, Cermeño P, Lao MT, Segura ML (2019) Effects of drought stress on biomass, essential oil content, nutritional parameters, and costs of production in six Lamiaceae species. Water 11:573–594

Gazim ZC, Amorim ACL, Hovell AMC, Rezende CM, Nascimento IA, Ferreira GA, Cortez DAG (2010) Seasonal variation, chemical composition, and analgesic and antimicrobial activities of the essential oil from leaves of *Tetradenia riparia* (Hochst.) Codd in southern Brazil. Molecules 15:5509–5524

Ghorbani M, Movahedi Z, Kheiri A, Rostami M (2018) Effect of salinity stress on some morpho-physiological traits and quantity and quality of essential oils in peppermint (*Mentha piperita* L.). Environ Stress Crop Sci 11:413–420

Gonny M, Cavaleiro C, Salgueiro L, Casanova J (2006) Analysis of Juniperus communis subsp. alpina needle, berry, wood and root oils by combination of GC, GC/MS and 13C-NMR. Flavour Fragr J 21:99–106

Gounaris Y (2010) Biotechnology for the production of essential oils, flavours and volatile isolates. A review Flavour Fragr J 25:367–386

Gowda PJ, Ramakrishnaiah H, Krishna V, Narra S, Jagannath N (2012) Caryophyllene-rich essential oil of *Didymocarpus tomentosa*: chemical composition and cytotoxic activity. Nat Prod Commun 7:1934578X1200701131

Hagvall L, Bäcktorp C, Svensson S, Nyman G, Börje A, Karlberg AT (2007) Fragrance compound geraniol forms contact allergens on air exposure. Identification and quantification of oxidation products and effect on skin sensitization. Chem Res Toxicol 20:807–814

Harborne JB, Baxter H (2001) Chemical dictionary of economic plants: dictionary of useful plant products. Wiley, West Sussex

Haslam E (2014) The shikimate pathway: biosynthesis of natural products series. Elsevier

Heydari M, Zanfardino A, Taleei A, Shahnejat Bushehri AA, Hadian J, Maresca V, Rigano D (2018) Effect of heat stress on yield, monoterpene content and antibacterial activity of essential oils of *Mentha x piperita* var. Mitcham and *Mentha arvensis* var. piperascens. Molecules 23(8):1903–1914

Hunter M (2009) Essential oils: art, agriculture, science, industry and entrepreneurship (a focus on the Asia-Pacific region). In: Agriculture issues and policies series. Nova Science Publishers, Inc, New York

Hüsnü K, Başer C, Demirci F (2007) Chemistry of essential oils. In: Flavours and Fragrances. Springer, Berlin, pp 43–86

Hussain AI, Anwar F, Sherazi STH, Przybylski R (2008) Chemical composition, antioxidant and antimicrobial activities of basil (*Ocimum basilicum*) essential oils depends on seasonal variations. Food Chem 108:986–995

Hyldgaard M, Mygind T, Meyer RL (2012) Essential oils in food preservation: mode of action, synergies, and interactions with food matrix components. Front Microbiol 3:12–28

Idrees M, Khan MMA, Aftab T, Naeem M, Hashmi N (2010) Salicylic acid-induced physiological and biochemical changes in lemongrass varieties under water stress. J Plant Interact 5:293–303

Jager W (2010) Metabolism of terpenoids in animal models and humans. In: Baser KHC, Buchbauer G (eds) Handbook of essential oils, science, technology and applications. CRC Press, pp 209–234

Jaiswal D, Pandey-Rai S, Agrawal SB (2021) Untangling the UV-B radiation-induced transcriptional network regulating plant morphogenesis and secondary metabolite production. Environ Exp Bot 192:104655

Jamali CA, El Bouzidi L, Bekkouche K, Lahcen H, Markouk M, Wohlmuth H, Abbad A (2012) Chemical composition and antioxidant and anticandidal activities of essential oils from different wild Moroccan *thymus* species. Chem Biodivers 9:1188–1197

Jan R, Asaf S, Numan M, Kim KM (2021) Plant secondary metabolite biosynthesis and transcriptional regulation in response to biotic and abiotic stress conditions. Agronomy 11:968–985

Jayaprakasha GK, Rao JM, Sakariah KK (2003) Volatile constituents from *Cinnamomum zeylanicum* fruit stalks and their antioxidant activities. J Agric Food Chem 51:4344–4348

Jeshni MG, Mousavinik M, Khammari I, Rahimi M (2017) The changes of yield and essential oil components of German chamomile (*Matricaria recutita* L.) under application of phosphorus and zinc fertilizers and drought stress conditions. J Saudi Soc Agric Sci 16:60–65

Joshi R, Sharma P, Sharma V, Prasad R, Sud RK, Gulati A (2013) Analysis of the essential oil of large cardamom (*Amomum subulatum* Roxb.) growing in different agro-climatic zones of Himachal Pradesh, India. J Sci Food Agric 93:1303–1309

Kamatou GP, Vermaak I, Viljoen AM, Lawrence BM (2013) Menthol: a simple monoterpene with remarkable biological properties. Phytochemistry 96:15–25

Khorasaninejad S, Mousavi A, Soltanloo H, Hemmati K, Khalighi A (2010) The effect of salinity stress on growth parameters, essential oil yield and constituent of peppermint (*Mentha piperita* L.). World Appl Sci J 11:1403–1407

Kiani BH (2017) Biosynthetic pathway of artemisinin. In: Artemisia annua: Prospects, Applications and Therapeutic Uses, pp 15–129

Kilic S, Kilic M (2017) Effects of cadmium-induced stress on essential oil production, morphology and physiology of lemon balm (*Melissa officinalis* L., Lamiaceae). Appl Ecol Environ Res 15:1653–1669

Kim HJ, Chen F, Wang X, Rajapakse NC (2006) Effect of methyl jasmonate on secondary metabolites of sweet basil (*Ocimum basilicum* L.). J Agric Food Chem 54:2327–2332

Köse EO, Deniz IG, Sarıkürkçü C, Aktaş Ö, Yavuz M (2010) Chemical composition, antimicrobial and antioxidant activities of the essential oils of Sideritis erythrantha Boiss. And Heldr.(var. erythrantha and var. cedretorum PH Davis) endemic in Turkey. Food Chem Toxicol 48:2960–2965

Kuzuyama T, Seto H (2012) Two distinct pathways for essential metabolic precursors for isoprenoid biosynthesis. Proc Jpn Acad Ser B 88:41–52

Lajayer HA, Savaghebi G, Hadian J, Hatami M, Pezhmanmehr M (2017) Comparison of copper and zinc effects on growth, micro-and macronutrients status and essential oil constituents in pennyroyal (*Mentha pulegium* L.). Rev Bras Bot 40:379–388

León-Méndez G, Pájaro-Castro N, Pájaro-Castro E, Torrenegra-Alarcón M, Herrera-Barros A (2019) Essential oils as a source of bioactive molecules. Rev Colomb de Quimica 48:80–93

Li Y, Kong D, Fu Y, Sussman MR, Wu H (2020) The effect of developmental and environmental factors on secondary metabolites in medicinal plants. Plant Physiol Biochem 148:80–89

Lohani H, Haider SZ, Chauhan NK, Sah S, Andola HC (2013) Aroma profile of two Juniperus species from alpine region in Uttarakhand. J Nat Prod 6:38–43

Lombard J, Moreira D (2011) Origins and early evolution of the mevalonate pathway of isoprenoid biosynthesis in the three domains of life. Mol Biol Evol 28:87–99

Ložienė K, Vaičiulytė V, Maždžierienė R (2021) Influence of meteorological conditions on essential oil composition in geraniol-bearing *Thymus pulegioides* and *thymus hybrid*. Acta Physiol Plant 43:1–9

Maffei M, Scannerini S (2000) UV-B effect on photomorphogenesis and essential oil composition in peppermint (*Mentha piperita* L.). J Essential oil Res 12:523–529

Mandoulakani BA, Eyvazpour E, Ghadimzadeh M (2017) The effect of drought stress on the expression of key genes involved in the biosynthesis of phenylpropanoids and essential oil components in basil (*Ocimum basilicum* L.). Phytochemistry 139:1–7

Marzouki H, Khaldi A, Falconieri D, Piras A, Marongiu B, Molicotti P, Zanetti S (2010) Essential oils of *Daucus carota* subsp. carota of Tunisia obtained by supercritical carbon dioxide extraction. Nat Prod Commun 5:1934578X1000501226

Misharina TA, Polshkov AN (2005) Antioxidant properties of essential oils: autoxidation of essential oils from laurel and fennel and of their mixtures with essential oil from coriander. Appl Biochem Microbiol 41:610–618

Misharina TA, Polshkov AN, Ruchkina EL, Medvedeva IB (2003) Changes in the composition of the essential oil of marjoram during storage. Appl Biochem Microbiol 39:311–316

Moghaddam M, Mehdizadeh L (2017) Chemistry of essential oils and factors influencing their constituents. In: Soft chemistry and food fermentation. Academic Press, pp 379–419

Mota KSDL, Pereira FDO, De Oliveira WA, Lima IO, Lima EDO (2012) Antifungal activity of *Thymus vulgaris* L. essential oil and its constituent phytochemicals against *Rhizopus oryzae*: interaction with ergosterol. Molecules 17:14418–14433

Nabi A, Naeem M, Aftab T, Masroor M, Khan A (2019) Arsenic toxicity induced changes in growth, photosynthetic pigments, antioxidant machinery, essential oil, menthol and other active constituents of menthol mint (*Mentha arvensis* L.). J Essential Oil Bea Plant 22:1333–1348

Nazzaro F, Fratianni F, Coppola R, De Feo V (2017) Essential oils and antifungal activity. Pharmaceuticals 10:86–102

Ngumbi EN, Ugarte CM (2021) Flooding and herbivory interact to alter volatile organic compound emissions in two maize hybrids. J Chem Ecol 47:707–718

Nguyen H, Campi EM, Jackson WR, Patti AF (2009) Effect of oxidative deterioration on flavour and aroma components of lemon oil. Food Chem 112:388–393

Nishioka N, Nishimura T, Ohyama K, Sumino M, Malayeri SH, Goto E, Morota T (2008) Light quality affected growth and contents of essential oil components of Japanese mint plants. In: International Workshop on Greenhouse Environmental Control and Crop Production in Semi-Arid Regions, vol 797, pp 431–436

Orav A, Stulova I, Kailas T, Müürisepp M (2004) Effect of storage on the essential oil composition of *Piper nigrum* L. fruits of different ripening states. J Agric Food Chem 52:2582–2586

Ottavioli J, Gonny M, Casanova J, Bighelli A (2009) Chemical variability of the needle oil of Juniperus communis ssp. alpina from Corsica. Chem Biodivers 6:2192–2199

Özcan MM, Pedro LG, Figueiredo AC, Barroso JG (2006) Constituents of the essential oil of sea fennel (*Crithmum maritimum* L.) growing wild in Turkey. J Med Food 9:128–130

Özek G, Demirci F, Özek T, Tabanca N, Wedge DE, Khan SI, Hamzaoglu E (2010) Gas chromatographic–mass spectrometric analysis of volatiles obtained by four different techniques from *salvia rosifolia* Sm., and evaluation for biological activity. J Chromatogr A 1217:741–748

Pavela R, Benelli G (2016) Essential oils as ecofriendly biopesticides? Challenges and constraints. Trends Plant Sci 21:1000–1007

Pereira SI, Santos PAG, Barroso JG, Figueiredo AC, Pedro LG, Salgueiro LR, Scheffer JJC (2000) Chemical polymorphism of the essential oils from populations of *thymus caespititius* grown on the Island S. Jorge (Azores). Phytochemistry 55:241–246

Pinto E, Salgueiro LR, Cavaleiro C, Palmeira A, Gonçalves MJ (2007) *In vitro* susceptibility of some species of yeasts and filamentous fungi to essential oils of *Salvia officinalis*. Ind Crop Prod 26:135–141

Povh JA, Ono EO (2007) EFEITO DO ÁCIDO GIBERÉLICO NA COMPOSIÇÃO DO ÓLEO ESSENCIAL DE Salvia officinallis L. Publicatio UEPG: Ciências Biológicas e da Saúde 13:2–6

Raal A, Paaver U, Arak E, Orav A (2004) Content and composition of the essential oil of Thymus serpyllum L. growing wild in Estonia. Medicina (Kaunas) 40:795–800

Rahimi Y, Taleei A, Ranjbar M (2017) Changes in the expression of key genes involved in the biosynthesis of menthol and menthofuran in *Mentha piperita* L. under drought stress. Acta Physiol Plant 39:1–9

Raina VK, Kumar A, Srivastava SK, Syamsundar KV, Kahol AP (2004) Essential oil composition of 'kewda'(*Pandanus odoratissimus*) from India. Flavour Fragr J 19:434–436

Rathore S, Mukhia S, Kapoor S, Bhatt V, Kumar R, Kumar R (2022) Seasonal variability in essential oil composition and biological activity of *Rosmarinus officinalis* L. accessions in the western Himalaya. Sci Rep 12:1–13

Razmjoo K, Heydarizadeh P, Sabzalian MR (2008) Effect of salinity and drought stresses on growth parameters and essential oil content of *Matricaria chamomile*. Int J Agric Biol 10:451–454

Rehman R, Hanif MA, Mushtaq Z, Mochona B, Qi X (2016) Biosynthetic factories of essential oils: the aromatic plants. Nat Prod Chem Res 4:227–238

Rohmer M, Seemann M, Grosdemange-Billiard C (2005) Biosynthetic routes to the building blocks of isoprenoids. In: Biopolymers online. Wiley Online Library. Wiley, New York, p 49

Rudbäck J, Bergström MA, Börje A, Nilsson U, Karlberg AT (2012) α-Terpinene, an antioxidant in tea tree oil, autoxidizes rapidly to skin allergens on air exposure. Chem Res Toxicol 25:713–721

Sá RA, Sá RA, Alberton O, Gazim ZC, Laverde JA, Caetano J, Dragunski DC (2015) Phytoaccumulation and effect of lead on yield and chemical composition of *Mentha crispa* essential oil. Desalination Water Treat 53:3007–3017

Saab AM, Tundis R, Loizzo MR, Lampronti I, Borgatti M, Gambari R, Menichini F (2012) Antioxidant and antiproliferative activity of *Laurus nobilis* L. (Lauraceae) leaves and seeds essential oils against K562 human chronic myelogenous leukaemia cells. Nat Prod Res 26:1741–1745

Saharkhiz MJ, Motamedi M, Zomorodian K, Pakshir K, Miri R, Hemyari K (2012) Chemical composition, antifungal and antibiofilm activities of the essential oil of Mentha piperita L. Int Sch Res Notices 2012:1–7

Sangwan NS, Arora B, Sangwan RS (2001) Spectral modulation of essential oil synthesis in *Pelargonium graveolens* L. Int J Geogr Inf Syst 10:85–91

Santos-Gomes PC, Fernandes-Ferreira M (2001) Organ-and season-dependent variation in the essential oil composition of *Salvia officinalis* L. cultivated at two different sites. J Agric Food Chem 49:2908–2916

Schwarz M, Arigoni D (1999) Ginkgolide Biosynthesis. In: Cane DE (ed) Comprehensive natural product chemistry, Isoprenoids including carotenoids and steroids, vol 2. Pergamon, Oxford, pp 367–400

Sell C (2010) Chap. 5—the chemistry of essential oils. In: Baser CKH, Buchbauer G (eds) Handbook of essential oils: science, technology, and applications. CRC Press, Taylor & Francis Group, Boca Raton, pp 121–150

Senji BM, Mandoulakani BA (2018) The impact of cold stress on genes expression pattern of mono- and sesquiterpene biosynthesis and their contents in *Ocimum basilicum* L. Phytochemistry 156:250–256

Sharifi R, Kiani H, Farzaneh M, Ahmadzadeh M (2008) Chemical composition of essential oils of Iranian *Pimpinella anisum* L. and *Foeniculum vulgare* Miller and their antifungal activity against postharvest pathogens. J Essential Oil Bear Plants 11:514–522

Singh B, Sharma RA (2015) Plant terpenes: defense responses, phylogenetic analysis, regulation and clinical applications. 3. Biotech 5:129–151

Smékalová V, Doskočilová A, Komis G, Šamaj J (2014) Crosstalk between secondary messengers, hormones and MAPK modules during abiotic stress signalling in plants. Biotech Adv 32:2–11

Solórzano-Santos F, Miranda-Novales MG (2012) Essential oils from aromatic herbs as antimicrobial agents. Curr Opin Biotechnol 23:136–141

Stephane FFY, Jules BKJ (2020) Terpenoids as important bioactive constituents of essential oils. In: Essential oils-bioactive compounds, new perspectives and applications. IntechOpen, London

Teixeira B, Marques A, Ramos C, Serrano C, Matos O, Neng NR, Nunes ML (2013) Chemical composition and bioactivity of different oregano (*Origanum vulgare*) extracts and essential oil. J Sci Food Agric 93:2707–2714

Tisserand R, Young R (2013) Essential oil safety: a guide for health care professionals. Elsevier Health Sciences

Turek C, Kirschmann N, Stintzing FC (2012) Quality monitoring of selected essential oils upon storage at different temperature regimes. Zeitschrift für Arznei-& Gewürzpflanzen 17:73–79

Turek C, Stintzing FC (2012) Impact of different storage conditions on the quality of selected essential oils. Food Res Int 46:341–353

Valente J, Zuzarte M, Liberal J, Goncalves MJ, Lopes MC, Cavaleiro C, Salgueiro L (2013) *Margotia gummifera* essential oil as a source of anti-inflammatory drugs. Ind Crop Prod 47:86–91

Vekiari SA, Protopapadakis EE, Papadopoulou P, Papanicolaou D, Panou C, Vamvakias M (2002) Composition and seasonal variation of the essential oil from leaves and peel of a Cretan lemon variety. J Agric Food Chem 50:147–153

Verma N, Shukla S (2015) Impact of various factors responsible for fluctuation in plant secondary metabolites. J Appl Res Med Aromat Plants 2:105–113

Wagner H, Bladt S (2009) Plant drug analysis. A thin layer chromatography atlas, 2nd edn. Springer, Berlin

Wang B, Peng XX, Sun R, Li J, Zhan XR, Wu LJ, Xie T (2012) Systematic review of β-elemene injection as adjunctive treatment for lung cancer. Chin J Integr Med 18:813–823

Wang Z, Xiao S, Wang Y, Liu J, Ma H, Wang Y, Hou W (2020) Effects of light irradiation on essential oil biosynthesis in the medicinal plant *Asarum heterotropoides* Fr. Schmidt var. mandshuricum (maxim) Kitag. PLoS One 15:e0237952

Wanke M, Skorupinska-Tudek K, Swiezewska E (2001) Isoprenoid biosynthesis via 1-deoxy-D-xylulose 5-phosphate/2-C-methyl-D-erythritol 4-phosphate (DOXP/MEP) pathway. Acta Biochim Pol 48:663–672

Wurzinger B, Mair A, Pfister B, Teige M (2011) Cross-talk of calcium-dependent protein kinase and MAP kinase signaling. Plant Signal Behav 6:8–12

Yanhe L, Craker LE, Polter T (1995) Effect of light levels on essential oil production of sage (*S. officinalis*) and thyme (*T. vulgaris*). In: International Symps Med Arom Plants, Amherst, MA

Yannai S (2004) Dictionary of food compounds with CD-ROM: additives, flavors, and ingredients. Chapman & Hall/CRC, FL

Yao SS, Guo WF, Lu Y, Jiang YX (2005) Flavor characteristics of lapsang souchong and smoked lapsang souchong, a special Chinese black tea with pine smoking process. J Agric Food Chem 53:8688–8693

Youssef NA (2021) Changes in the morphological traits and the essential oil content of sweet basil (*Ocimum basilicum* L.) as induced by cadmium and lead treatments. Int J Phytoremediation 23:291–299

Ysef A (2013) Chemical composition of the essential oil of Mentha pulegium L. from Taftan area by means of gas chromatography/mass spectrometry (GC/MS). J Med Plant Res 7:3003–3007

Yu X, Liang C, Chen J, Qi X, Liu Y, Li W (2015) The effects of salinity stress on morphological characteristics, mineral nutrient accumulation and essential oil yield and composition in *Mentha canadensis* L. Sci Hortic 197:579–583

Zhou DR, Eid R, Miller KA, Boucher E, Mandato CA, Greenwood MT (2019) Intracellular second messengers mediate stress inducible hormesis and programmed cell death: a review. Biochim Biophys Acta 1866:773–792

Zouari-Bouassida K, Trigui M, Makni S, Jlaiel L, Tounsi S (2018) Seasonal variation in essential oils composition and the biological and pharmaceutical protective effects of *Mentha longifolia* leaves grown in Tunisia. Biomed Res Int 2018:7856517

Zuzarte M, Salgueiro L (2015) Essential oils chemistry. In: Bioactive essential oils and cancer. Springer, Cham, pp 19–61

Zuzarte M, Vale-Silva L, Gonçalves MJ, Cavaleiro C, Vaz S, Canhoto J, Salgueiro L (2012) Antifungal activity of phenolic-rich *Lavandula multifida* L. essential oil. Eur J Clin Microbiol Infect Dis 31:1359–1366

Role of Biotechnology and Combinatorial Chemistry Approaches in Molecular-Assisted Engineering of Plant Volatile Compounds

11

Ritu Singh and Prem Pratap Singh

Abstract

Plants produce a wide range of volatile compounds, which plays a crucial role in their growth, development, and interactions with the environment. In recent years, biotechnology and combinatorial chemistry approaches have enabled the efficient and precise manipulation of metabolic pathways, leading to the production of new or improved plant-based products with desired properties. Biotechnology approaches, such as metabolic engineering and synthetic biology, have emerged as powerful tool for genetic modification of plant metabolic pathways to produce targeted volatiles. Combinatorial chemistry approaches, on the other hand, involve the screening and optimization of large chemical libraries to identify novel compounds with desirable properties. The combination of these two approaches provides a powerful platform for the molecular engineering of plant volatile compounds, enabling the development of new fragrances, flavors, and functional ingredients for use in a wide range of applications, including food, cosmetic, and pharmaceutical industries. Additionally, the molecular engineering of plant volatile compounds has the potential to address global challenges, such as food security and sustainability, by providing new and improved plant-based products with enhanced nutritional and functional properties. In conclusion, the use of biotechnology and combinatorial chemistry approaches in the molecular engineering of plant volatile compounds is a rapidly growing area with significant potential to revolutionize the production of plant-based products and address global challenges.

R. Singh
National Institute of Plant Genome Research (NIPGR), Aruna Asaf Ali Marg, New Delhi, India

P. P. Singh (✉)
Centre of Advanced Study in Botany, Institute of Science, Banaras Hindu University, Varanasi, India

B. Prakash et al. (eds.), *Plant Essential Oils*,
https://doi.org/10.1007/978-981-99-4370-8_11

Keywords

Volatile compounds · Biosynthesis pathway · Biotechnology and combinatorial chemistry · Molecular-assisted engineering of volatile compounds

11.1 Introduction

Plants being sessile develop diverse strategies to compensate for this and protect themselves against challenges. One of these is the synthesis of several metabolic compounds that act as the language for plants to communicate with surrounding environments and with themselves. These compounds are grouped into two major categories: primary and secondary (Theis and Lerdau 2003). Only a small number of the incredible diversity of metabolites that plants create constitute the 'primary' metabolic pathways. These are commonly present in all species. Others, referred to as 'secondary' metabolites, are exclusive to smaller groups of plants (Pichersky and Gang 2000). These secondary metabolites are the results of many plant responses or certain demands. Volatile organic compounds (VOCs) have a starring role among these secondary metabolites (Dicke and Loreto 2010). VOCs are lipophilic low molecular weight compounds having high vapor pressure and a low boiling point at ambient temperatures. These properties make them amenable to cross the cellular membrane and being released into the environment (Pichersky et al. 2006). Plant VOCs can be released constitutively or in response to a range of stimuli and are produced by every type of tissue and vegetation as nitrogen-containing chemicals, green leaf volatiles, and aromatic compounds (Peñuelas and Llusià 2001; Dudareva et al. 2006; Holopainen and Gershenzon 2010; Holopainen et al. 2010). Intriguingly, VOC synthesis depends on the plant species as different plants utilize different metabolites to tackle similar problems. For instance, different flowers produce different odorous volatiles for attracting the same type of pollinator. The VOCs perform multifaceted functions in the plant system, namely, attract the pollinators, plant-to-plant communication, defense against predators and insects, and adaptability to environmental stress and heat (Dudareva and Pichersky 2000; Mumm et al. 2003; Baldwin et al. 2006; Heil and Karban 2010; War et al. 2012). They specifically contribute to food quality, biomass output, reproductive success, and many other significant agronomic aspects (Dudareva and Pichersky 2008).

The current breeding and crop improvement programs for enhanced growth and yield had a negative impact on secondary metabolism, leading to a deficiency in VOC production in many current crop cultivars. For instance, *Theobroma cacao* blooms are tiny, hardly scented that only draw opportunistic flies and ineffective pollinators. On the other hand, wild *Theobroma* sp. flowers are huge, aromatic, and attract bees for their pollination (Young & Severson, 1994). The same patterns were seen for vegetal VOCs. The corn domestication resulted in the loss of (E)-b-caryophyllene which attracts pest-eliminating parasitic wasps and entomopathogenic nematodes thus, makes the plants more susceptible to insect pests (Tamiru et al., 2011). Similarly, growth-defense trade-offs are also seen in even more recent cultivated crops. For instance, cranberries are derived from *Vaccinium* sp. wild progenitor. However, compared to their lower-yielding relatives, high-yielding cranberry genotypes exhibit suppressed herbivore-induced sesquiterpene emissions (Rodriguez-Saona et al., 2011). Together, these demonstrate how poorly crop VOCs are adapted to support beneficial insects in agricultural settings. Additionally, despite their increased size and intensity, agricultural monocultures are becoming a bigger part of modern agriculture. Because of this decline in plant variety, the ecosystem services supplied by the arthropod community such as pest control by beneficial insects and crop pollination are negatively impacted (Gardiner et al., 2010). Therefore, methods must be developed to address these issues and engineering VOCs is a viable technique for enhancing plant chemical variety as well as beneficial insect diversification.

These crucial functions of VOCs and their loss during crop domestication and improvement showed a great potential for biotechnological exploitation (Aharoni et al. 2005). Understanding the fundamental biochemical processes and identifying the genes and enzymes involved in the creation of volatile chemicals are prerequisites for engineering the VOCs using biotechnological approaches. Additionally, the significant rise in the discovery of the genes and enzymes of volatile biosynthesis have resulted from a resurgence of interest in these subjects during the past 10 years. To date, numerous attempts have been undertaken to modify the volatile profiles of plants – guarding against infections and abiotic stress, discouraging herbivores, affecting tritrophic relationships, and enhancing fruit and floral fragrances. The most evident biotechnological modulation application on these VOCs is modulation of their emissions in plants to enhance or repress their activities in order to accomplish specified outcomes. While some initiatives have succeeded in reaching their objectives, many others have only modestly improved volatiles or had unanticipated metabolic impacts, such as increased metabolism of the desired end products or harmful effects on plant growth and development.

Here, in this chapter, we will briefly go over the production of VOCs, talk about bioengineering strategies and targets, and use case studies to demonstrate recent developments in these fields. We also offer an outlook on where these innovative and intriguing technologies might take us in the future.

11.2 Classes and Biosynthetic Pathways of VOCs

The biosynthesis of VOCs depends on the amount of building blocks present in the cells that is, carbon, nitrogen, and sulfur. On the basis of biosynthetic origin and chemical constitution, VOCs are divided into several classes: isoprenoids and terpenoids. In some cases, oxygenated VOCs (OVOCs), such as acetaldehyde (C_2H_4O), acetone (C_3H_6O), methyl-ethyl-ketone (MEK, C_4H_8O), methanol (CH_4O), and methyl-vinyl-ketone (MVK, C_4H_6O), sulfur compounds (Brassicales), and furanocoumarins (Asterales, Apiales, Fabales, and Rosales), are also reported (Berenbaum and Zanger 2008; Agrawal 2011). Herein, we will discuss the classes of VOCs:

11.2.1 Terpenoids

Terpenoids constitute the largest and most diverse class of VOCs. In plants, terpenoids biosynthesis was conducted through two pathways: the mevalonic acid (MVA) and methylerythritol phosphate (MEP). Isopentenyl diphosphate (IPP) and dimethylallyl diphosphate (DMAPP) serve as the main precursor for this biosynthesis pathway (McGarvey and Croteau 1995). Further, the MVA produces precursor for volatile sesquiterpenes and MEP pathway for monoterpenes (C10), diterpenes (C20), and hemiterpenes (C5). The MEP pathway occurs in plastid (C15) (Hsieh et al. 2008). On the other side, site for MVA pathway is unclear as previously it was shown to happen in cytosol later evidences suggest its location in endoplasmic reticulum and peroxisome (Simkin et al. 2011; Pulido et al. 2012). According to Winterhalter and Rouseff (2001), plants also produce erratic volatile terpenoids with carbon skeletons ranging from C8 to C18 that are derived from carotenoids through a three-step modification process that starts with dioxygenase cleavage followed by enzymatic transformation, and is finally converted to volatile compounds by acid. In rare instances, such as in *Arabidopsis*, tomato, petunia, and melon, the dioxygenase cleavage process itself can produce a volatile product from a variety of carotenoid pigments, such as a- and b-ionone, geranylacetone, and pseudoionone. The acyclic C11- and C16-homoterpenes 4,8,12-trimethyltrideca-1,3,7,11-tetraene (TMTT) and 4,8,12-dimethylnona-1,3,7-triene (DMNT) are irregular and are produced from GGPP and FPP, respectively. A cytochrome P450 monooxygenase-catalyzed oxidative breakdown follows the creation of the tertiary C15- and C20-alcohol precursors (E)-nerolidol and (E, E)-geranyl linalool, respectively, in the first of two enzymatic steps that make up their biosynthesis.

11.2.2 Phenylpropanoid and Benzenoid Compounds

The second largest VOCs are phenylpropanoid and benzenoid. They are synthesized from the phenylalanine (Phe) (Knudsen et al. 2006). Central carbon metabolism and Phe are linked via seven enzyme processes of the shikimate pathway and three of the arogenate pathway (Maeda & Dudareva, 2012). Phosphoenolpyruvate (PEP) and D-erythrose 4-phosphate (E4P), which are the direct precursors of the shikimate pathway, are produced by glycolysis and the PPP, respectively. Since the MEP system depends on the same pathways for its precursors, it must compete with the shikimate/phenylpropanoid pathway for carbon allocation, especially since 30% of photosynthetically fixed carbon is diverted to Phe, mostly for the production of lignin (Razal et al., 1996). The 3-deoxy-D-arabino-heptulosonate 7-phosphate synthase (DAHP synthase), the first gene in the shikimate pathway, is crucial for regulating the carbon input into the process. The molecular processes behind this control in plants are, however, still largely unexplored.

11.2.3 Fatty Acid Derivatives

Next class of VOCs are fatty acid derivative compounds such as cis-3-hexenol, 1-hexanal, methyl jasmonate, and nonanal. These are synthesized from linoleic and linolenic, that is, the C_{18} unsaturated fatty acids. A plastidic pool of acetyl-CoA produced from pyruvate (Pyr), the end result of glycolysis, is necessary for the biosynthesis of these fatty acids. Unsaturated fatty acids undergo stereospecific oxygenation once they enter the lipoxygenase (LOX) route, resulting in the intermediates 9-hydroperoxy and 13-hydroperoxy, which are then further metabolized via the two branches of the LOX pathway, producing volatile chemicals. Only the 13-hydroperoxy intermediate is used by the allene oxide synthase branch, which results in the synthesis of jasmonic acid (JA), which is then changed into methyl jasmonate by JA carboxyl methyl transferase. The hydroperoxide lyase branch, on the other hand, transforms both varieties of hydroperoxide fatty acid derivatives into C6 and C9 aldehydes, which are frequently reduced to alcohols by alcohol dehydrogenases, and then further transformed into their esters. These saturated and unsaturated C6/C9 aldehydes and alcohols, often known as green leaf volatiles, are typically produced by plants' green organs in reaction to injury but also give fruits and vegetables their distinctive fresh green scent.

11.2.4 Branched Chain Amino Acid Derivative

Numerous amino acids, including alanine, valine, leucine, isoleucine, and methionine, as well as intermediates in their biosynthesis, are the source of many volatile compounds, particularly those that are highly abundant in floral scents and fruit aromas. These compounds also contain sulfur and nitrogen. These amino acid-derived volatiles production in plants is thought to follow a similar path to that in yeast or bacteria, where these pathways have been more thoroughly investigated. Aminotransferases, much like in microorganisms, catalyze the first deamination or transamination of the amino acids, which produces the α-ketoacid. Decarboxylation, reduction, oxidation, and/or esterification can be applied to these α-ketoacids in order to produce aldehydes, acids, alcohols, and esters. Further, alcohol acyltransferases (AATs) catalyze alcohol esterification processes, and amino acids can serve as the precursors of these acyl-CoAs.

11.3 VOCs Engineering: The Prerequisite Criteria

To modify the VOCs for beneficial purposes can be achieved by four ways: (1) extract the VOCs from naturally occurring species, (2) create it using synthetic chemistry, (3) breed or engineer desired features in the target organism, and (4) engineer excessive VOCs production in a model organism. For a particular need, one or all of these methods may be appropriate.

Where the desired VOC is abundant, extraction from natural sources may be feasible; nevertheless, natural sources infrequently provide enough for widespread industrial applications, and the volatility of these chemicals makes harvesting potentially impractical. Since almost two centuries, chemical synthesis has been used to create chemicals that are useful for agriculture; however, in the case of VOCs, these compounds are either overly complex or too expensive to synthesize at the required yields. Also, no synthetic pathway has been developed thus, making the production impossible. Furthermore, accurate chirality of these compounds is frequently needed for biological activity and the synthetic synthesis is frequently not enantiospecific. In these circumstances, utilizing living organisms for production might be the only practical option. Since harvesting, transport, and application are not required, the production of the desired VOCs in the host/target organism can provide the benefit in most effective way. This process involves breeding and genetic engineering approach. However, this approach is also very complicated and requires a detailed understanding of the biology and biochemistry in a specific system. If the desired feature is present in the germplasm of interest, desirable qualities can be introduced through selective breeding and backcrossing. Despite the fact that some crop species, like maize, have genotypes that can contribute desirable volatile components (Schnee et al. 2006), backcrossing to introduce these components into elite agricultural cultivars can be a time-consuming operation. Genetic manipulation can be employed to acquire genes from unrelated species when such germplasm is not present within the species. Even when germplasm is available, the latter method may be faster for cultivar development; however, the testing and regulatory requirements for genetically modified organisms (GMOs) may mean that the final product takes just as long as possible to reach the market.

Bioproduction in microbial platforms like *Saccharomyces cerevisiae* or *Escherichia coli* may offer a solution in cases where natural sources are unable to produce the desired traits or yield, chemical synthesis is impossible, and/or breeding and genetic engineering are impractical for biological, technological, or sociopolitical reasons. Nowadays, for a wide range of different biochemicals, production in tightly regulated bioprocesses using engineered microorganisms is emerging as the preferred synthesis route (Vickers et al. 2012).

11.4 Molecular-Assisted Engineering of VOCs

Today, the engineering of VOCs was performed in **(i)** plants and **(ii)** microbes. In plants, this involves breeding and genetic engineering approaches.

11.4.1 Engineering the Floral Volatiles

The impact of changes in floral aroma on insect attractiveness has not yet been addressed, in contrast to metabolic engineering of vegetative volatiles where the effect of changing emission profiles on insect behavior was investigated.

Furthermore, as humans have a much lower odor threshold than insects, perception tests have often only included sensory evaluations (Vosshall 2000; Stockhorst and Pietrowsky 2004). The metabolic engineering of floral volatiles in these trials was deemed successful when the alterations in scent profiles were detectable by humans. For instance, the introduction of three citrus monoterpene synthases to transgenic tobacco allowed for the olfactorily perceptible augmentation of volatiles released from flowers and leaves (El Tamer et al. 2003; Lücker et al. 2004). In a different experiment, transgenic carnations with metabolic flux redirected from the anthocyanin route to benzoic acid produced more methylbenzoate, which was enough for human olfactory detection (Zuker et al. 2002). However, many more attempts to alter the scent bouquet failed for a variety of reasons, such as the lack of appropriate substrates for the introduced reaction (Aranovich et al. 2007), transformation of the scent compound into a nonvolatile form (Lücker et al. 2001), insufficient levels of emitted volatiles for human olfactory detection (Lavy et al. 2002), or masking of the introduced compound(s) by other volatiles.

Another strategy that has lately been employed for smell modification is the removal of some volatile components from the floral bouquet. Using RNA interference (RNAi)-mediated posttranscriptional gene silencing, it was possible to create transgenic petunias devoid of methylbenzoate (Underwood et al. 2005), phenylacetaldehyde (Kaminaga et al. 2006), benzylbenzoate and phenylethylbenzoate (Orlova et al. 2006), and isoeugenol (Dexter et al. 2007). With the exception of the plants that emit less methylbenzoate, the impact of these modifications on human perception has not yet been investigated. The panelists have unfavorable response in this cases as according to them the modified flowers were not able to produce desired level fragrance (Dexter et al. 2007).

11.4.2 Improvement of Aroma Quality of Fruits, Vegetables, and Herbs

The olfactory characteristics and flavor of fruits are significantly influenced by volatiles (Goff and Klee 2006). In the natural world, volatiles make fruits more alluring thus helping in the spreading of the seeds. Several efforts have been made to enhance the aroma of fruits by utilizing breeding strategies. For instance, in tomato, crossing with its relative *L. peruvianum* reintroduced the desired volatile compounds thus, enhancing the aroma (Kamal et al. 2001). However, the monitoring of the complicated aroma trait is necessary for this process which is time-consuming and tedious process. Since it is still not possible to forecast how people would react to a certain mixture of volatile substances and volatile collections an analysis must first be carried out with pricey GC-MS instruments, and human evaluations must then additionally be carried out by subjective test panels. Further, due to interspecific diversity in the capacity to identify particular molecules and the lack of a common vocabulary to characterize particular odors, human perception of smells is very subjective. In fact, these issues have made it less common for breeding initiatives to prioritize the aroma trait.

Some of the shortcomings of traditional plant breeding can be abridged by genetic engineering and aroma trait of the fruits can be improved. Genetic engineering has the benefit of being less complicated because it introduces one attribute at a time (McCaskill and Croteau 1998; Pichersky and Dudareva 2007). The insertion of genes whose coding information may not be present in the crop is another benefit. However, the general issues and traps of genetic engineering for biochemical properties are as follows: (1) If the generation of the desired volatile is the end result of a lengthy metabolic pathway, the addition of a single gene is unlikely to result in a significant production of the desired volatile. (2) A single new volatile is often improbable to modify the consumers' perceptions about the aroma and flavor quality of the produce.

Early attempts to solve these issues involved overexpression of the yeast D9-desaturase (Wang et al. 1996) and a nonspecific alcohol dehydrogenase (ADH) (Speirs et al. 1998; Prestage et al. 1999) in tomato fruit. The amounts of different aroma compounds such as (Z)-3- hexenol and (Z)-3- hexenal as well as the proportions of aldehydes to alcohols were altered in these transgenics. As per taste panelists, the higher alcohol content in these transgenic fruits provided more potent ripe flavors (Speirs et al. 1998); however, none of these changes added any novel scent components.

The first attempt to introduce a new compound to fruit flavor was the introduction of the *Clarkia breweri* linalool synthase (LIS) gene into tomato under the control of the fruit specific E8 promoter. As a result, small amount of linalool and its oxidation product, 8-hydroxylinalool, were accumulated in the fruit, and both substances could be detected by GC-MS as well as human nose (Lewinsohn et al. 2001). This metabolic modification was successful because linalool is produced by LIS from geranyl diphosphate (GPP) in a single step and GPP is a precursor in the synthesis of carotenoids, a pathway that is very active in ripening tomato fruits.

Davidovich-Rikanati et al. (2007) expressed geraniol synthase (GES) in tomato under the promoter of polygalacturonase, another fruit ripening-specific promoter, to produce a considerably higher influence on flavor perception. Geraniol is another acyclic monoterpene alcohol that can be made from GPP and can easily be converted to geranial by nonspecific alcohol dehydrogenases, in contrast to linalool, which is a tertiary alcohol and its hydroxyl group cannot be further oxidized (Davidovich-Rikanati et al. 2007). Large levels of geraniol were produced by GES-transgenic tomato fruit, which significantly reduced pigmentation. Additionally, geraniol was further metabolized by transgenic fruits into geranial, which naturally tautomerized into neral. Citral, a concoction made from neral and geranial, has a powerful lemon flavor. Additionally, geranial and neral underwent metabolism to form geranic and neric acids, respectively. Nerol, citronellol, citronellal, citronelic acid, citronellyl acetate, and rose oxide were created by further modifications of geranial and neral (Davidovich-Rikanati et al. 2007). A test panel of 34 persons evaluated these transgenic fruits, and the majority of the participants (80%) said that the fruits had a stronger aroma and that they preferred the transgenic fruits over the non-transgenic ones.

The development of volatile plant production in vegetative plants for human consumption has lagged behind that of tomato fruit. However, recent efforts to increase the production of terpenes that humans prefer (such as menthol) and decrease the synthesis of unfavorable substances (such as menthofuran) in peppermint have only partially been effective (Mahmoud and Croteau 2001, 2003; Mahmoud et al. 2004). While other transgenic plants showed a slight increase in the amounts of limonene, a cyclic monoterpene, antisense technology was used to create transgenic plants with a 50% drop in the concentration of menthofuran. However, a review of consumer reactions to these transgenic mint plants fragrance qualities has not yet been conducted.

11.4.3 Engineering the VOCs for Plant Defense

It has been widely established over the past decades that plants release more than 200 distinct volatile chemical compounds in response to herbivore attacks (Dicke and Van Loon 2000). These released volatiles may directly inebriate, repel, or discourage herbivorous insects (Bernasconi et al. 1998; Kessler and Baldwin 2001; Vancanneyt et al. 2001; Aharoni et al. 2003; Seybold et al. 2006) or they may draw predators and parasitoids that would ordinarily prey on the offending herbivores, thereby indirectly protecting the signaling plant from further harm (such as through tritrophic interactions) (Arimura et al. 2004; Degen et al. 2004; Mercke et al. 2004). The progression in biotechnology approaches and growing research indicating the role of VOCs in plant defense attracts scientists to modulate the volatiles in order to improve defense of agricultural and forest ecosystem. This provides an alternate biological pest management strategy. However, in order to successfully strengthen plant defense, a number of conditions must be fulfilled (Degenhardt et al. 2003). First, the area where the crop is planted must have herbivore enemies that can effectively manage the herbivore population. Second, the newly introduced or improved plant volatile blends must include significant herbivore adversary's principal attractants, and volatile emission must coincide with herbivore activity. Finally, the volatiles that have been produced should not make the plant attractive to other non-herbivores.

To date, engineering in several plant species enhance the plant herbivore defense strategies via emitting appropriate VOCs. For instance, the overexpression of strawberry nerolidol/linalool synthase gene (*FaNES1*) in *Arabidopsis* and potato enhances the production of linalool, thus preventing the plants from aphids. However, the protecting strategy differs in these plants – in *Arabidopsis* the *FaNES1* transgenics deter the aphid *Mysus persicae* while in potato it enhances tritrophic interactions by attracting the predatory mites. In another study in *Arabidopsis,* the *FaNES1* was overexpressed in mitochondria where sesquiterpene precursor farnesyl diphosphate (FPP) is present. This results in the production of (3S) (E) nerolidol and C11 homoterpene 4, 8 dimethyl 1, 3 (E), 7 nonatriene [(E) DMNT] compounds which enhances the attractiveness of plants toward *Phytoseiulus persimilis*, a predatory mite that are natural enemy of spider mites (Kappers et al. 2005). Another successful instance was the overexpression of the maize terpene synthase gene, that is, *TPS10* in *Arabidopsis*. The generated transgenic plants strongly emit a number of sesquiterpenes that are generally released in maize during herbivory by lepidopteran larvae. The female parasitic wasp *Cotesia marginiventris*, which had previously experienced oviposition on larvae of the potential host, found these transgenic plants to be more alluring (Schnee et al. 2006). Furthermore, transgenics of tobacco overexpressing patchoulol synthase (PTS) causes production of the volatile patchoulol and 13 additional sesquiterpene which deter the tobacco hornworms. It was also observed that the majority of tobacco hornworms had moved from the leaves of the transgenic plants to the leaves of wild-type plants and consumed 20–50% more of the wild-type plants in 6 h (Wu et al. 2006).

Terpenoids have not been the only volatile signals that have been subject to metabolic engineering in direct and indirect defenses. The fatty acid derived VOCs were also modified by the researchers. Transgenic tobacco plants overexpressing either the yeast acyl-CoAD9 desaturase or the insect acyl-CoA D11 desaturase produced more (Z)-3-hexenal, a significant green leaf volatile. The expression of these transgenes causes increased level of 13-lipoxygenase activity and 16:1 fatty acid, which catalyzes the initial step in the formation of hexenal from α-linolenic acid (Hong et al. 2004). Although the impact of increased levels of (Z)-3-hexenal on insect behavior was not examined in this study, the detrimental impact of this substance on aphid performance was shown in transgenic potato plants with lower levels of the hydroperoxide lyase enzyme, which is in charge of cleaving fatty acid hydroperoxides to C6 aldehydes (Vancanneyt et al. 2001).

11.5 Limitations of Engineering the VOCs in Plants: The Unwanted Side

Plant metabolic engineering is still in its infancy compared to microbials as their metabolic processes are substantially less complex. Though, significant advancement in biotechnology approaches provides us a significantly better understanding of biosynthetic pathways, regulation, compartmentalization, as well as the general biology of plant volatile-mediated compounds and their interactions, however, still our limited knowledge of the plant systems and the unique difficulties posed by engineering volatiles continue to be some of our limitations. The major limitations are:

11.5.1 Detrimental Effect on Growth and Development

The researches have shown the potential of genetic engineering to strengthen plant defense as well as the part that particular volatile molecules play in interactions between plants and insects. Despite these advancements, they have also shown how genetic changes affect plant growth and development and identified several obstacles to obtaining effective production of the necessary volatiles. While the levels of isoprenoids derived from plastids, such as chlorophylls, lutein, and β carotene, were not affected by the divergence of carbon to linalool production in *Arabisopsis* via *FaNES1* overexpression, it did result in a growth-retardation phenotype that was passed down through multiple generations of transgenic plants (Aharoni et al. 2003). A more severe phenotype was produced by the linalool-emitting transgenic potato; plants had bleached leaves after being moved from the in vitro to the greenhouse in addition to growth retardation (Aharoni et al. 2006). Due to the expression of *PTS* combined with *FPP* synthase, both of which are targeted to the plastids, transgenic tobacco that produces high levels of patchoulol also exhibits leaf chlorosis, vein clearing, and shortened stature (Wu et al. 2006). These reported phenotypes might result from the loss of isoprenoid precursors for other metabolites necessary for plant growth and development, or they might be the result of the harmful effects of the newly introduced terpenoids on plant cells.

11.5.2 Product Alteration

For volatile spectrum engineering to succeed it is crucial to create and emit enough of the appropriate molecules. The full biochemical toolkit of the plant employed will determine the metabolic destiny of newly produced molecules. Predicting how much of the intended molecule will actually survive in the desired form is difficult since the biochemical repertoire of each plant species is not completely understood. For instance, enzymes with broad substrate specificity that are typically found in cells, such as glutathione transferases, hydroxylases, dehydrogenases, glycosyl transferases, and others, may have an impact on freshly generated molecules (Dudareva et al. 2004). Currently, little is known about these enzymes in general and their precise distribution in various plant species. Research suggests that these broad substrate specificity enzymes work as competing enzymes and drain the intermediate/product of the pathway, often generating the unwanted products. This issue is more problematic when pathway involve multiple steps. In *FaNES1 Arabidopsis* transgenic plant, the generated linalool was exposed to hydroxylation and glycosylation by endogenous enzymes (and some of the nerolidol generated was degraded to the C11 homoterpene (E)-DMNT). Thus decreasing the overall concentrations of linalool produced (Aharoni et al. 2003).

11.5.3 Nonspecificity of the Enzymes

Nonspecific catalytic activity of the enzymes makes the outcome of metabolic engineering unpredictable. For instance, peppermints (E)-β-farnesene synthase enzyme can produce terpinolene, myrcene, and limonene from GPP; however, it can also produce δ-cadinene and (E)- and (Z)-β farnesene from FPP (Crock et al. 1997). This can be overcome by targeting specific subcellular organelle; however, it may affect the catalytic efficiency of the enzyme. Targeting of FaNES to the mitochondria mainly produces nerolidol while the plastid resulted in linalool production (Aharoni et al. 2003; Kappers et al. 2005). Moreover, the enzymes may also behave differently in vitro and in vivo due to difference in biochemical conditions and cofactors (Ginglinger et al. 2013).

11.5.4 Complex Interspecies Interaction

Plants frequently emit volatiles to draw in advantageous interactions. The equilibrium is delicate and these interactions are often distraught. For instance, in the *Cucurbita pepo* var. *texana* (wild Texas gourd), boosting volatile production increased the attraction of florivores and decreased the production of seeds. However, it did not improve the attraction of pollinators (Theis & Adler 2012). (E)-caryophyllene an antibacterial and mutualist attractant volatile component in Huang et al. (2012) is currently the focus of various engineering initiatives for its advantageous properties. But it competes with the aphid alarm pheromone (E)-farnesene (Vet & Dicke 1992). Tritrophic interactions can also be hampered by the production of isoprene (Loivamäki et al. 2008) and even (E)-farnesene at large concentrations. With these examples, we can say that appropriate gene expression control may be able to reduce potential cost-benefit scenarios.

11.6 Generating Desired VOCs Via Microbial Engineering

Although engineering plants is undoubtedly the most direct way to change how plants and the environment interact, but doing so comes with a number of drawbacks. An alternate strategy is to engineer microbes to produce certain products, which can then be applied as needed. For instance, a volatile attractant or repellant may be released from a reservoir into a field of crops at the right moment. The relatively straightforward metabolism, ease of genetic modification, well-developed engineering tools, and speed of engineering in microorganisms over plants are important advantages of engineering. Furthermore, it is much more practical to harvest volatiles in a controlled fermentation setting than it is to harvest them from plants for large-scale industrial production. Additionally, because microbial bioprocesses based on fermentation are less susceptible to the whims of seasonality, climate, and pest or disease attack, the circumstances of the bioprocess, including production yields (and thus, market pricing) may be more constant.

The first stage in creating a microbial cell factory to create volatile compounds is to select an appropriate production host. Here, it is important to take into account factors (1) native ability to produce VOCs, (2) genetic engineerability, (3) competition with essential requirements, (4) potential toxicity of the end product, and (5) potential to carry out desired biotransformation. The wide range of microbes may be utilized to produce VOCs; however, up to now majority of engineering has been done in the widely used model microorganisms, that is, *E. coli* and *S. cerevisiae*. One should also note that the production of VOCs also varies upon the strain of the host used. Next the genetic capability must be imported after the host organism has been chosen. Plant genes are frequently employed for this. The introns need to be removed in yeast and *E. coli* to maintain a continuous protein coding sequence. The signal peptides or other targeting sequences (e.g., chloroplast targeting region in case of plant chloroplast protein) must also be eliminated. As a result, eliminating the targeting peptide causes microorganisms to produce more of the desired chemical (Burke & Croteau 2002; Vickers et al. 2011; Bott et al. 2012). Also, the precise truncation site is crucial since it might have a big impact. Sequence alignment with microbial homologs that lack targeting sequences and closely related genes that have already been characterized can be used to help choose truncation sites in these cases; however, for a completely uncharacterized gene, it is wise to test several truncations. Codon-optimizing plant isoprenoid genes to alter codon usage toward the host's preference can also improve translation (Anthony et al. 2009; Bott et al. 2012; Calabria et al. 2013). Finally, selection of an appropriate expression system [promoter and terminator sequences, plasmid expression systems or chromosomal integration, variable copy number (plasmid/chromosome), etc.] is required. Balancing all of these elements appropriately may have a substantial effect on the overall success of the project. For biotechnological applications, strict product and enantiomeric specificity is often required. As discussed above, many isoprenoid synthases have relaxed substrate/product specificities, and catalytic activities of different terpene synthases can vary widely (Schomburg et al. 2013).

11.7 Conclusion and Future Direction

The cultivation of many plant species is significantly influenced by the volatile compounds found in fruits, vegetables, and herbs. However, very less focus is given on boosting or even maintaining volatile production as significant breeding programs are undertaken to improve certain food qualities such as overall yield, total solids, sugar content, or pigmentation. Therefore, compared to their wild cultivars, many modern cultivars of domesticated plant species produce fewer volatiles (Gutterson 1993). Progression in breeding programs can enhance this loss of genetic diversity in volatile compounds. Plants release volatile substances into the rhizosphere from their roots, a fact that has only lately been recognized (Chen et al. 2004; Rasmann et al. 2005; Bouwmeester et al. 2007). These volatiles may support the plant's ability to both draw in and repel helpful microbes. They might also be helpful in interspecies plant competition (Horiuchi et al. 2007). However, it has been demonstrated that some parasitic plants detect their hosts by using underground volatile chemicals (Bouwmeester et al. 2007). There have not been any reports of attempts at genetic engineering to alter root volatile emission in order to increase plant fitness, but this is obviously a very promising field for future research.

References

Agrawal AA (2011) Current trends in the evolutionary ecology of plant defence. Funct Ecol 25:420. https://doi.org/10.1111/j.1365-2435.2010.01796.x

Aharoni A, Giri AP, Deuerlein S et al (2003) Terpenoid metabolism in wild-type and transgenic Arabidopsis plants. Plant Cell 15:2866. https://doi.org/10.1105/tpc.016253

Aharoni A, Jongsma MA, Bouwmeester HJ (2005) Volatile science? Metabolic engineering of terpenoids in plants. Trends Plant Sci 10(12):594–602

Aharoni A, Jongsma MA, Kim TY et al (2006) Metabolic engineering of terpenoid biosynthesis in plants. Phytochem Rev 5:49

Anthony JR, Anthony LC, Nowroozi F, Kwon G, Newman JD, Keasling JD (2009) Optimization of the mevalonate-based isoprenoid biosynthetic pathway in Escherichia coli for production of the anti-malarial drug precursor amorpha-4,11-diene. Metab Eng 11:13–19

Aranovich D, Lewinsohn E, Zaccai M (2007) Post-harvest enhancement of aroma in transgenic lisianthus (Eustoma grandiflorum) using the Clarkia breweri benzyl alcohol acetyltransferase (BEAT) gene. Postharvest Biol Technol 43:255. https://doi.org/10.1016/j.postharvbio.2006.09.001

Arimura GI, Ozawa R, Kugimiya S et al (2004) Herbivore-induced defense response in a model legume. Two-spotted spider mites induce emission of (E)-β-ocimene and transcript accumulation of (E)-β-ocimene synthase in Lotus japonicus. Plant Physiol 135:1976. https://doi.org/10.1104/pp.104.042929

Baldwin IT, Halitschke R, Paschold A et al (2006) Volatile signaling in plant-plant interactions: “talking trees” in the genomics era. Science (80-)

Berenbaum MR, Zanger AR (2008) Facing the future of plant-insect interaction research: Le retour à la “raison d’être”. Plant Physiol 146:804

Bernasconi ML, Turlings TCJ, Ambrosetti L et al (1998) Herbivore-induced emissions of maize volatiles repel the corn leaf aphid. Entomol Exp Appl, Rhopalosiphum maidis. https://doi.org/10.1046/j.1570-7458.1998.00315.x

Bouwmeester HJ, Roux C, Lopez-Raez JA, Bécard G (2007) Rhizosphere communication of plants, parasitic plants and AM fungi. Trends Plant Sci 12:224

Bott RR, Cervin MA, Kellis JT Jr, McAuliffe JC, Miasnikov A, Peres CM, Whited GM (2012) Isoprene synthase variants for improved microbial production of isoprene. USPTO Patent number US8173410 B2

Burke C, Croteau R (2002) Geranyl diphosphate synthase from Abies grandis: cDNA isolation, functional expression, and characterization. Arch Biochem Biophys 405(1):130–136

Calabria AR, Cervin MA, Chotani GK, La Duca R, McAuliffe JC, Miller MC, Whited GM (2013) Compositions and methods for producing isoprene free of C5 hydrocarbons under decoupling conditions and/or safe operating ranges. USPTO Patent number US 8,420,360 B2

Chen F, Ro DK, Petri J et al (2004) Characterization of a root-specific arabidopsis terpene synthase responsible for the formation of the volatile monoterpene 1,8-cineole. Plant Physiol 135:1956. https://doi.org/10.1104/pp.104.044388

Crock J, Wildung M, Croteau R (1997) Isolation and bacterial expression of a sesquiterpene synthase cDNA clone from peppermint (Mentha x piperita, L.) that produces the aphid alarm pheromone (E)-β-farnesene. Proc Natl Acad Sci 94(24):12833–12838

Davidovich-Rikanati R, Sitrit Y, Tadmor Y et al (2007) Enrichment of tomato flavor by diversion of the early plastidial terpenoid pathway. Nat Biotechnol 25:899. https://doi.org/10.1038/nbt1312

Degen T, Dillmann C, Marion-Poll F, Turlings TCJ (2004) High genetic variability of herbivore-induced volatile emission within a broad range of maize inbred lines. Plant Physiol 135:1928. https://doi.org/10.1104/pp.104.039891

Degenhardt J, Gershenzon J, Baldwin IT, Kessler A (2003) Attracting friends to feast on foes: engineering terpene emission to make crop plants more attractive to herbivore enemies. Curr Opin Biotechnol 14:169

Dexter R, Qualley A, Kish CM et al (2007) Characterization of a petunia acetyltransferase involved in the biosynthesis of the floral volatile isoeugenol. Plant J 49:265. https://doi.org/10.1111/j.1365-313X.2006.02954.x

Dicke M, Loreto F (2010) Induced plant volatiles: from genes to climate change. Trends Plant Sci 15:115

Dicke M, Van Loon JJA (2000) Multitrophic effects of herbivore-induced plant volatiles in an evolutionary context. Entomol Exp Appl 97:237

Dudareva N, Negre F, Nagegowda DA, Orlova I (2006) Plant volatiles: Recent advances and future perspectives. CRC. Crit Rev Plant Sci 25:417

Dudareva N, Pichersky E (2000) Biochemical and molecular genetic aspects of floral scents. Plant Physiol 122:627

Dudareva N, Pichersky E (2008) Metabolic engineering of plant volatiles. Curr Opin Biotechnol 19(2):181–189

Dudareva N, Pichersky E, Gershenzon J (2004) Biochemistry of plant volatiles. Plant Physiol 135:1893

El Tamer MK, Smeets M, Holthuysen N et al (2003) The influence of monoterpene synthase transformation on the odour of tobacco. J Biotechnol 106:15. https://doi.org/10.1016/j.jbiotec.2003.09.003

Gardiner MA, Tuell JK, Isaacs R, Gibbs J, Ascher JS, Landis DA (2010) Implications of three biofuel crops for beneficial arthropods in agricultural landscapes. BioEnergy Res 3:6–19

Ginglinger JF, Boachon B, Höfer R, Paetz C, Köllner TG, Miesch L, Lugan R, Baltenweck R, Mutterer J, Ullmann P, Beran F (2013) Gene coexpression analysis reveals complex metabolism of the monoterpene alcohol linalool in Arabidopsis flowers. Plant Cell 25(11):4640–4657

Goff SA, Klee HJ (2006) Plant volatile compounds: sensory cues for health and nutritional value? Science (80-.)

Gutterson NC (1993) Molecular breeding for color, flavor and fragrance. Sci Hortic (Amsterdam) 55:141. https://doi.org/10.1016/0304-4238(93)90029-P

Heil M, Karban R (2010) Explaining evolution of plant communication by airborne signals. Trends Ecol Evol 25:137. https://doi.org/10.1016/j.tree.2009.09.010

Holopainen JK, Gershenzon J (2010) Multiple stress factors and the emission of plant VOCs. Trends Plant Sci 15:176

Holopainen JK, Heijari J, Oksanen E, Alessio GA (2010) Leaf volatile emissions of Betula pendula during autumn coloration and leaf fall. J Chem Ecol 36:1068. https://doi.org/10.1007/s10886-010-9857-4

Hong M, Zilinskas BA, Knipple DC, Chin CK (2004) Cis-3-Hexenal production in tobacco is stimulated by 16-carbon monounsaturated fatty acids. Phytochemistry 65:159. https://doi.org/10.1016/j.phytochem.2003.10.013

Horiuchi JI, Badri DV, Kimball BA et al (2007) The floral volatile, methyl benzoate, from snapdragon (Antirrhinum majus) triggers phytotoxic effects in Arabidopsis thaliana. Planta 226:1. https://doi.org/10.1007/s00425-006-0464-0

Hsieh MH, Chang CY, Hsu SJ, Chen JJ (2008) Chloroplast localization of methylerythritol 4-phosphate pathway enzymes and regulation of mitochondrial genes in ispD and ispE albino mutants in Arabidopsis. Plant Mol Biol 66:663. https://doi.org/10.1007/s11103-008-9297-5

Huang M, Sanchez-Moreiras AM, Abel C, Sohrabi R, Lee S, Gershenzon J, Tholl D (2012) The major volatile organic compound emitted from Arabidopsis thaliana flowers, the sesquiterpene (E)-β-caryophyllene, is a defense against a bacterial pathogen. New Phytol 193(4):997–1008

Kamal, Takashina, Egashira, et al (2001) Introduction of aromatic fragrance into cultivated tomato from the "peruvianum complex." Plant Breed https://doi.org/10.1046/j.1439-0523.2001.00581.x

Kaminaga Y, Schnepp J, Peel G et al (2006) Plant phenylacetaldehyde synthase is a bifunctional homotetrameric enzyme that catalyzes phenylalanine decarboxylation and oxidation. J Biol Chem 281:23357. https://doi.org/10.1074/jbc.M602708200

Kappers IF, Aharoni A, Van Herpen TWJM et al (2005) Plant science: genetic engineering of terpenoid metabolism attracts bodyguards to Arabidopsis. Science (80-). https://doi.org/10.1126/science.1116232

Kessler A, Baldwin IT (2001) Defensive function of herbivore-induced plant volatile emissions in nature. Science (80-). https://doi.org/10.1126/science.291.5511.2141

Knudsen JT, Eriksson R, Gershenzon J, Ståhl B (2006) Diversity and distribution of floral scent, vol 72. Bot, Rev, p 1

Lavy M, Zuker A, Lewinsohn E et al (2002) Linalool and linalool oxide production in transgenic carnation flowers expressing the Clarkia breweri linalool synthase gene. Mol Breed. https://doi.org/10.1023/A:1026755414773

Lewinsohn E, Schalechet F, Wilkinson J et al (2001) Enhanced levels of the aroma and flavor compound S-linalool by metabolic engineering of the terpenoid pathway in tomato fruits. Plant Physiol 127:1256. https://doi.org/10.1104/pp.010293

Loivamäki M, Mumm R, Dicke M, Schnitzler JP (2008) Isoprene interferes with the attraction of bodyguards by herbaceous plants. Proc Natl Acad Sci 105(45):17430–17435

Lücker J, Bouwmeester HJ, Schwab W et al (2001) Expression of clarkia S-linalool synthase in transgenic petunia plants results in the accumulation of S-linalyl-β-D-glucopyranoside. Plant J 27:315. https://doi.org/10.1046/j.1365-313X.2001.01097.x

Lücker J, Schwab W, Van Hautum B et al (2004) Increased and altered fragrance of tobacco plants after metabolic engineering using three monoterpene synthases from lemon. Plant Physiol 134:510. https://doi.org/10.1104/pp.103.030189

Maeda H, Dudareva N (2012) The shikimate pathway and aromatic amino acid biosynthesis in plants. Annu Rev Plant Biol 63:73–105

Mahmoud SS, Croteau RB (2001) Metabolic engineering of essential oil yield and composition in mint by altering expression of deoxyxylulose phosphate reductoisomerase and menthofuran synthase. Proc Natl Acad Sci U S A 98:8915. https://doi.org/10.1073/pnas.141237298

Mahmoud SS, Croteau RB (2003) Menthofuran regulates essential oil biosynthesis in peppermint by controlling a downstream monoterpene reductase. Proc Natl Acad Sci U S A 100:14481. https://doi.org/10.1073/pnas.2436325100

Mahmoud SS, Williams M, Croteau R (2004) Cosuppression of limonene-3-hydroxylase in peppermint promotes accumulation of limonene in the essential oil. Phytochemistry 65:547. https://doi.org/10.1016/j.phytochem.2004.01.005

McCaskill D, Croteau R (1998) Some caveats for bioengineering terpenoid metabolism in plants. Trends Biotechnol 16:349

McGarvey DJ, Croteau R (1995) Terpenoid metabolism. Plant Cell 7:1015–1026

Mercke P, Kappers IF, Verstappen FWA et al (2004) Combined transcript and metabolite analysis reveals genes involved in spider mite induced volatile formation in cucumber plants. Plant Physiol 135:2012. https://doi.org/10.1104/pp.104.048116

Mumm R, Schrank K, Wegener R et al (2003) Chemical analysis of volatiles emitted by Pinus sylvestris after induction by insect oviposition. J Chem Ecol. https://doi.org/10.1023/A:1023841909199

Orlova I, Marshall-Colón A, Schnepp J et al (2006) Reduction of benzenoid synthesis in petunia flowers reveals multiple pathways to benzoic acid and enhancement in auxin transport. Plant Cell 18:3458. https://doi.org/10.1105/tpc.106.046227

Peñuelas J, Llusià J (2001) The complexity of factors driving volatile organic compound emissions by plants. Biol Plant 44:481

Pichersky E, Dudareva N (2007) Scent engineering: toward the goal of controlling how flowers smell. Trends Biotechnol 25:105

Pichersky E, Gang DR (2000) Genetics and biochemistry of secondary metabolites in plants: an evolutionary perspective, vol 5. Trends Plant Sci, p 439

Pichersky E, Noel JP, Dudareva N (2006) Biosynthesis of plant volatiles: Nature's diversity and ingenuity. Science (80-.):808

Prestage S, Linforth RST, Taylor AJ et al (1999) Volatile production in tomato fruit with modified alcohol dehydrogenase activity. J Sci Food Agric 79:131. https://doi.org/10.1002/(SICI)1097-0010(199901)79:1<131::AID-JSFA196>3.0.CO;2-Z

Pulido P, Perello C, Rodriguez-Concepcion M (2012) New insights into plant isoprenoid metabolism. Mol Plant 5:964. https://doi.org/10.1093/mp/sss088

Rasmann S, Köllner TG, Degenhardt J et al (2005) Recruitment of entomopathogenic nematodes by insect-damaged maize roots. Nature 434:732. https://doi.org/10.1038/nature03451

Razal RA, Ellis S, Singh S, Lewis NG, Towers GN (1996) Nitrogen recycling in phenylpropanoid metabolism. Phytochemistry 41(1):31–35

Rodriguez-Saona C, Kaplan I, Braasch J, Chinnasamy D, Williams L (2011) Field responses of predaceous arthropods to methyl salicylate: a meta-analysis and case study in cranberries. Biol Control 59(2):294–303

Schnee C, Köllner TG, Held M et al (2006) The products of a single maize sesquiterpene synthase form a volatile defense signal that attracts natural enemies of maize herbivores. Proc Natl Acad Sci U S A 103:1129. https://doi.org/10.1073/pnas.0508027103

Schomburg I, Chang A, Placzek S, Söhngen C, Rother M, Lang M, Schomburg D (2013) BRENDA in 2013: integrated reactions, kinetic data, enzyme function data, improved disease classification: new options and contents in BRENDA. Nucleic Acids Res 41:D764–D772

Seybold SJ, Huber DPW, Lee JC et al (2006) Pine monoterpenes and pine bark beetles: a marriage of convenience for defense and chemical communication, vol 5. Phytochem, Rev, p 143

Simkin AJ, Guirimand G, Papon N et al (2011) Peroxisomal localisation of the final steps of the mevalonic acid pathway in planta. Planta 234:903. https://doi.org/10.1007/s00425-011-1444-6

Speirs J, Lee E, Holt K et al (1998) Genetic manipulation of alcohol dehydrogenase levels in ripening tomato fruit affects the balance of some flavor aldehydes and alcohols. Plant Physiol 117:1047. https://doi.org/10.1104/pp.117.3.1047

Stockhorst U, Pietrowsky R (2004) Olfactory perception, communication, and the nose-to-brain pathway. Physiol Behav 83:3. https://doi.org/10.1016/j.physbeh.2004.07.018

Tamiru A, Bruce TJ, Woodcock CM, Caulfield JC, Midega CA, Ogol CK, Mayon P, Birkett MA, Pickett JA, Khan ZR (2011) Maize landraces recruit egg and larval parasitoids in response to egg deposition by a herbivore. Ecol Lett 14(11):1075–1083

Theis N, Adler LS (2012) Advertising to the enemy: enhanced floral fragrance increases beetle attraction and reduces plant reproduction. Ecology 93(2):430–435

Theis N, Lerdau M (2003) The evolution of function in plant secondary metabolites. Int J Plant Sci 164:S93

Underwood BA, Tieman DM, Shibuya K et al (2005) Ethylene-regulated floral volatile synthesis in petunia corollas. Plant Physiol 138:255. https://doi.org/10.1104/pp.104.051144

Vancanneyt G, Sanz C, Farmaki T et al (2001) Hydroperoxide lyase depletion in transgenic potato plants leads to an increase in aphid performance. Proc Natl Acad Sci U S A 98:8139. https://doi.org/10.1073/pnas.141079498

Vet LEM, Dicke M (1992) Ecology of infochemical use by natural enemies in a tritrophic context. Annu Rev Entomol 37:141–172

Vickers CE, Klein-Marcuschamer D, Krömer JO (2012) Examining the feasibility of bulk commodity production in Escherichia coli. Biotechnol Lett 34:585

Vickers CE, Possell M, Laothawornkitkul J, Ryan AC, Hewitt CN, Mullineaux PM (2011) Isoprene synthesis in plants: lessons from a transgenic tobacco model. Plant Cell Environ 34:1043–1053

Vosshall LB (2000) Olfaction in drosophila. Curr Opin Neurobiol 10:498

Wang C, Chin CK, Ho CT et al (1996) Changes of fatty acids and fatty acid-derived flavor compounds by expressing the yeast Δ-9 desaturase gene in tomato. J Agric Food Chem 44:3399. https://doi.org/10.1021/jf960174t

War AR, Paulraj MG, Ahmad T et al (2012) Mechanisms of plant defense against insect herbivores. Plant Signal Behav 7:1306

Winterhalter P, Rouseff R (2001) Carotenoid-Derived Aroma Compounds: An Introduction

Wu S, Schalk M, Clark A et al (2006) Redirection of cytosolic or plastidic isoprenoid precursors elevates terpene production in plants. Nat Biotechnol 24:1441. https://doi.org/10.1038/nbt1251

Young AM, Severson DW (1994) Comparative analysis of steam distilled floral oils of cacao cultivars (Theobroma cacao L., Sterculiaceae) and attraction of flying insects: implications for a Theobroma pollination syndrome. J Chem Ecol 20:2687–2703

Zuker A, Tzfira T, Ben-Meir H et al (2002) Modification of flower color and fragrance by antisense suppression of the flavanone 3-hydroxylase gene. Mol Breed. https://doi.org/10.1023/A:1019204531262

Application of Mathematical Modelling and Statistical Approaches to Boost the Industrial Application of Plant Volatiles

12

Surya Pratap and Horesh Kumar

Abstract

Plant volatiles possess inherent biological properties with potent industrial application in diverse field such as agriculture, medicine, and food sector. However, the complex chemical profile and biological activities of plant volatile are often influenced by several intrinsic and extrinsic parameter such as extraction methods, temperature, moisture, relative humidity, pH, water activity (aw), and oxidation reduction potential of substrates. Hence, the exploration of optimized conditions to enhance the effectiveness of volatile oils could boost the industrial application of plant volatiles. This chapter deals with the use of mathematical modelling approaches in combination with the experimental design strategies to optimize the intrinsic and extrinsic environmental factors to boost biological activities of plant volatiles.

Keywords

Response surface methodology · Biological application · Central composite rotatable design · Experiment design · Mathematical modelling

S. Pratap
Department of Physics, Institute of Science, Banaras Hindu University, Varanasi, India

Department of Physics, Harish Chandra Post Graduate College, Varanasi, India

H. Kumar (✉)
Department of Physics, Institute of Science, Banaras Hindu University, Varanasi, India

B. Prakash et al. (eds.), *Plant Essential Oils*,
https://doi.org/10.1007/978-981-99-4370-8_12

12.1 Introduction

Essential oils are cocktail mixture of biologically active compounds often extracted from different parts (root, leaves, stem, bark, and fruits) of aromatic plants. In general, essential oils harbor tremendous biological properties against a range of microbes, oxidative deterioration, and insect pests due to its inherent wide range of chemical profiles such as terpene, terpenoids, and phenylpropanoids (Prakash et al. 2015). The chemical profile of essential oils was often affected by different plant species, plant parts, geographical location, and storage conditions, which influence their biological properties. Hence, the extraction process and biological properties need to be optimized to obtain the desirable results, hence, the use of mathematical modelling, statistical approaches in combination with the experimental design strategies could be used to optimize the intrinsic and extrinsic environmental factors and to boost biological activities of plant volatiles.

Mathematical modelling is algorithms which represent a real-world system or process based on the developed mathematical equations. The classical method of the optimization process was based on changing the selected independent variable with the fixing of the other variables. However, the major obstacle of conventional methods is its time consuming, quite laborious and often not suitable to consider the interaction among the selected variables. Hence, the response surface methodology and central composite rotatable design (CCRD) that deal with the methodical and statistical approaches have been widely used to determine the corelation between selected independent variables and their responses (Kılıç et al. 2013, 2014). The simplex centroid design method is one of the currently used mathematical optimization techniques to find the global minimum or maximum of a function with multiple variables (X1, X2, and X3 as independent variables, while (Y1), (Y2), and Y3 as the response variables). Hence, these methods could be used to optimize the different factors, such as temperature, pH, and substrate composition, often associated with the yield and biological properties of essential oils to develop cost effective products. The chapter provides an overview of mathematical modelling, experimental design approaches, recent report on their effective utilization in different field, and potential application to boost the industrial application of plant volatiles.

12.2 Experimental Designs

To develop a hypothesis and the study of the relationship between independent and response variables, in this chapter, mathematical modelling approaches such as the simplex-centroid design method and central composite rotatable design in combination with the full-factorial design, fractional factorial design, Plackett–Burman design, and Box–Behnken design were explored.

12.3 Simplex-Centroid Mixture Design

Simplex-centroid mixture design is an augmented assay which have often been used to optimize the desired biological properties of plant volatiles and other biologically active compounds which can significantly contribute to the development of synergistic cost-effective products effective at low doses with desirable effects. In general, a total of 10 different formulation could be designed based on the selected adjustment parameters (X_1, X_2, and X_3) in a triangle. The triangle vertices points (1, 2, 3) correspond to the selected adjustment parameters, that is, X_1, X_2 and X_3; points 4, 5, and 6 (binary mixture) and point 7 represent the equal proportion of compounds; and points (8, 9, 10) are augmented. The best-fitted combination could be explored using mathematical model Eq. (12.1):

$$Y = \alpha_1 X_1 + \alpha_2 X_2 + \alpha_3 X_3 + \alpha_{12} X_1 X_2 + \alpha_{13} X_1 X_3 + \alpha_{23} X_2 X_3 + \alpha_{123} X_1 X_2 X_3 \quad (12.1)$$

Y: Response of optimized test biological properties as dependent variable; (α_1, α_2, α_3), (α_{12}, α_{13}, α_{23}), and α_{123}, linear, binary and ternary terms coefficients; X_1, X_2, and X_3 denote the selected adjustment parameter and their proportions in the mixture.

Kumar et al. (2020) developed antifungal formulation based on combination of trans-cinnamaldehyde (C), methyl eugenol (M), and estragole (E) using Simplex-centroid mixture design. They have reported the effective combination ratio (4:1:1 of trans-cinnamaldehyde (C), methyl eugenol (M), and estragole (E)) exhibited synergistic antifungal activity against the toxigenic species of *Aspergillus flavus* and aflatoxin B_1.

Kumar et al. (2019a, b) reported the mixture formulation of selected volatile compounds (thymol (T), methyl cinnamate (M), and linalool (L)) exhibited enhanced antifungal activity against the fungal species associated with the food spoilage and aflatoxin B_1 contamination. Further, authors have reported the mechanism of action which was found to be related with membrane perturbance due to decrease in ergosterol contents, elevation in membrane ion leakage, and utilization of different carbon-source required for the normal physiological functioning, mitochondrial dysfunction, change in anti-oxidative defence (SOD, CAT, and GR), and genes involved in aflatoxin B_1 biosynthesis process.

In a similar observation Kumar et al. (2019a, b) reported that the nanoencapsulation of optimized antifungal formulation based on mixture of eugenol, t-anethole, and menthol at 1:1:4 exhibited potent antimicrobial effects against food-selected fungi and aflatoxin B_1 producing toxigenic species of *Aspergillus flavus* (EC-03). The authors also elucidated the detailed mechanism of toxicity and without any significant effects on organoleptic profile of *Eleusine coracana* millet grain and recommended the use of mixture design for the development of green chemical for food safety.

Kumar et al. (2022) developed the antifungal formulation based on different ration of citral (C), geraniol (G), and terpineol (T) using the augmented simplex-centroid design and reported that the nanoencapsulated formulation exhibited enhanced antifungal property in vitro and in vivo condition. They have also explored the mechanism of toxicity and safety assessment based on in silico and quantitative structure-activity relationship (QSAR) approach.

Yadav et al. (2021) reported the antifungal formulation by the combination of different ration of traditionally used spices essential oils *Myristica fragrans* (M), *Bunium persicum* (B), and *Zanthoxylum alatum* (Z) (1:1:1 v/v/v) using mixture design assay. The authors also reported the pesticidal mode of action and safety limit profile using in silico approaches. The developed formulation showed strong pesticidal effects against the toxigenic species of *Aspergillus flavus*, aflatoxin B_1 production, and pulse beetle insect pest *Callosobruchus chinensis*.

Gupta et al. (2023) reported that adjuvant antifungal formulation based on combination of *Carum carvi*, and *Illicium verum* essential oils and methyl anthranilate compound using mathematical modelling approach. Further, the developed formulation was encapsulated inside the chitosan, and it was reported that the nanoencapsulated formulation shows enhanced antifungal, aflatoxin B_1inhibition, and also protects the biodeterioration active ingredients of selected herbal raw material of *Piper nigrum, P. longum, Andrographis paniculata, Silybum marianum,* and *Withania somnifera*, often used in Ayurvedic formulation. They recommend the developed antifungal adjuvant formulation as green chemical agent to enhance shelf life of herbal raw materials.

Crespo et al. (2019) reported the synergistic formulation possess enhanced antioxidant activity based on different ratio mixture of traditionally used aromatic plant essential oils such as *Apium graveolens* L., *Thymus vulgaris* L., and *Coriandrum sativum* L., using the simplex mixture design. The antioxidant effects were found to be superior for the ratio (66.7% of *T. vulgaris*, 16.7% of *C. sativum*, and 16.7% of *A. graveolens*) of EOs determined by ferric-reducing antioxidant power (FRAP) and 2,2′-azino-bis-3-ethylbenzthiazoline-6-sulphonic acid (ABTS) techniques.

Chraibi et al. (2023) studied the antimicrobial effects of synergistic composition prepared based on combination of essential oils of *Thymus satureioides, Myrtus communis,* and *Artemisia herba alba* using simplex-centroid design and optimized the antimicrobial effect against pathogenic microbial species such as *Escherichia coli, Staphylococcus aureus*, and *Candida tropicalis.* They have reported that the best mixture composition (*T. satureioides* (60%) and *M. communis* (40%), *T. satureioides* (72%) and *A. herba alba* (28%), and *thyme* (75%) and *mugwort* (25%)) of essential oils exhibited highest antimicrobial effect against test microbes.

Soulaimani et al. (2022) reported the optimized antibacterial effects of mixture of *Ammodaucus leucotrichus* Cosson, *Thymus vulgaris* L., and *Lavandula maroccana* (Murb.) essential oils against three multidrug-resistant bacteria using augmented simplex-centroid design. They have reported the optimal EOs mixture (55% *T. vulgaris*, 41% *L. maroccana*, and 4% *A. leucotrichus*), which significantly lowers the minimum inhibitory concentration required for the growth inhibition of *Staphylococcus aureus, Escherichia coli*, and *Pseudomonas aeruginosa.*

Jafarizadeh-Malmiri et al. (2022) prepared the nanoemulsion of essential oil mixture composition containing three different spices plant such as ginger, cinnamon, and cardamom. They reported the best essential oil-based formulation contained the 10% ginger, 68% cinnamon, and 22% cardamom and recommended it as a potential source of natural food preservatives with highest zeta potential and antioxidant and antibacterial activity.

12.4 Checkerboard Method

In order to assess the combined effects of selected biologically active compounds to enhance their industrial application in different sectors, checkerboard design is of the most widely used method which provides accuracy between in vitro and in vivo observation. This method can be used to explore the possible synergism, additive, and antagonistic effects between the different combination of selected compounds. The obtained data were interpretated based on fractional inhibitory concentration index (FICI) and ΔE model (response surface approach, Segatore et al. 2012).

Bhattacharya et al. (2023) prepared the synergistic antifungal formulation using checkerboard method based on mixture of *Aegle marmelos* (L.) *Correa* essential oil and commercial antifungal drug (amphotericin B and fluconazole) and reported the effective combination (2MIC of oil (0.625%) and ½ MIC of fluconazole (0.312 mg/ml) and ½ MIC of oil (0.156%) and MIC of amphotericin B (1.25 mg/ml)) as the best combination against human pathogen *Candida albicans* ATCC90028 with 34- and 65-fold increases of individual fluconazole and amphotericin B.

Sharma et al. (2023) developed an essential oil-based gel formulation to enhance the shelf life of apples against the postharvest rot of apples caused by *Aspergillus foetidus*. They have explored the combinatorial treatment of *Monarda citriodora* essential oil (MEO144.23 μL L^{-1} air) and hexanal (96.15 μL L^{-1} air) exhibited synergy (FICI 0.375). The phytotoxicity effects of combinatorial treatment of *Monarda citriodora* essential oil and hexanal was successfully overcome by the gel formulations based on their different combination with a synergistic effect and increment in antioxidant activity and total phenolic content in the stored apples.

Magi et al. (2015) investigated the antibacterial efficacy of selected essential oils such as *Origanum vulgare, Thymus vulgaris, Lavandula angustifolia, Mentha piperita*, and *Melaleuca alternifolia* against 32 erythromycin-resistant bacterial species. They have reported the synergistic antibacterial effect for the combination of carvacrol and erythromycin by the checkerboard assay based on the calculation of the Fractional Inhibitory Concentration (FIC) Index. Results suggested that a 2- to 2048-fold reduction of the erythromycin minimum inhibitory concentration. The synergy (FIC Index ≤0.5) was reported in 17/23 strains using 24-h time-kill curves in combination based on carvacrol and erythromycin.

da Silva et al. (2023) reported the possible synergism between the essential oils and commercially available antibiotics, resulting in the reduction of the required dosage of synthetic antibiotics against the resistant pathogens. The synergy was explored based on minimum inhibitory concentration (MIC) in microplates by a checkerboard assay. The synergistic interaction (IFIC <0.5) was observed for *Poiretia latifolia* essential oil, reduced the three times reduction in MIC of gentamicin against all microorganisms except for *E. coli* (FICI = 3) indicated indifference in their interaction. Similarly, the combination of the antibiotic and *Lithraea molleoides* essential oil exhibited synergism against *B. cereus* (IFIC = 0.4 and 0.5), but was indifferent for the other microorganisms.

Jafri and Ahmad (2020) explored the in vitro synergy between essential oils and antifungal drugs in sessile and planktonic mode of growth using a checkerboard microtiter assay. The possible interaction between the *thyme oil, thymol* with commercial antifungal drug fluconazole, and amphotericin B were evaluated against *Candida* strains. The result indicated that the thyme oil and thymol exhibited synergy with fluconazole in planktonic as well as biofilm mode against the growth of *C. albicans* and *C. tropicalis*. Similar synergy was observed for amphotericin B combination effective against only for planktonic *Candida* cells.

12.5 Response Surface Methodology (RSM)

RSM is one of the widely used statistical technique to optimize the developed model and analyze the all-possible relationship between a response variable and numerous independent variables. These techniques could be used to optimize the conditions for a process or experiment, such as essential oil extraction (Pereira et al. 2021).

12.5.1 Central Composite Rotational Design

The central composite rotational design is one of the most common methods currently being used in response surface modelling for the optimization of independent factor on mixture combination to get the desired results (Pereira et al. 2021). This method design explores the first-order and second-order (quadratic) term. This kind of design happens to be effective when both axial points play the role of factorial points in central composite design. More often the central points are augmented with the axial points. These points are arranged in such a manner from the center at the same distance, ensuring that all places on a centered sphere at the origin have equal probability of receiving the available answer. One of important aspect of CCD is rotatability, which implies that the prediction error is same from all the points to the central points from the same distance. The second-order mathematical model under CCRD is shown in the form of Eq. (12.2):

$$Z = \alpha_0 + \sum_{n}^{i=1} \alpha_i x_i + \sum_{n}^{i=1} \alpha_{ii} x_i^2 + \sum_{n}^{i=1} \sum_{n}^{j=1} \alpha_{ij} x_{ij} + \varepsilon. \tag{12.2}$$

where Z is the response, $\alpha_0$is the intercept, $\alpha_i$values represent the coefficient of the main effects, α_{ij} are the coefficients of interaction effects, α_{ii} are the second-order terms, and ε is the random error component that is determined by fitting the model to the data. The star point lies outside the desired range and the center point, representing the experimentally desired range, helping in determination of response surface plot. The precise calculation of surface responses helps in finding the appropriate value for α. This quadratic design consists of factorial (−1, +1) and axial (−α, + α) points and of repetitions at the central point, which provides properties such as orthogonality and rotativity to the adjustment of quadratic polynomials. The axial points (2 k) offer the detailed analysis of curvature under evaluation and estimation of new extremes for all the study factors in high and low configurations. This design is useful because it provides full knowledge of responses with the smallest number of experiments [36, 44]. The number of experiments for this design can be obtained from Eq. (12.3):

$$N = n^2 + 2n + \mu \tag{12.3}$$

where n represents the number of factors and μ represents the number of replicates of the central point. The value of α can be determined according to the calculation possibilities and offer precision, which can be traced from surface responses. The positioning of α values enhances the quality of the design or estimation under observation. The rate offered by design is identified by determining the proper positioning of the points. The evaluation of the estimation many a times influenced by the number of trials at the center of the domine to estimate precise the coefficients' variability and responses, quality by design approach is fundamental need. One of important aspect of CCD is rotatability, which implies that the prediction error is same from all the points to the central points from the same distance.

Azarifar et al. (2019) reported the antibacterial effects optimized formulation (gelatin-based nanocomposite films) based on varied concentration of selected compounds such as carboxymethyl cellulose, chitin nanofiber concentrations, and *Trachyspermum ammi* essential oil by response surface methodology (RSM). The optimization was done based on selected and varied parameter such as contact angle, tensile strength, strain at break, and lightness, and antibacterial activity, minimizing water vapour permeability, solubility, swelling, yellowness index, and total color difference. They have reported the optimal values of carboxymethyl cellulose (15.83 wt%), chitin nanofiber concentrations (3 wt%), and *Trachyspermum ammi* essential oil (0.74 (v/v%),) for composition of the gelatine film.

Rezzoug et al. (2005) studied the effects of different optimized parameter such as processing pressure (410 kPa), time (15.8 min), and the initial moisture content of leaves (0.40 gH2O/g) on global extraction yield and extraction yield of oil components. Based on RSM results, authors reported that the processing pressure and processing time as the foremost parameters which influence the yield and the chemical constituent of essential oils.

Sutaphanit and Chitprasert (2014) explored the optimum encapsulating conditions for holy basil oil using RSM considering two independent variables (X1: gelatin concentrations (4–16% w/v) and X2: holy basil essential oil amounts (7.5–37.5 ml)) and three response variables yield (Y1), oil content (Y2), and encapsulation efficiency (Y3). They have reported the gelatin concentration (11.75% (w/v)), and an oil (31 ml) as an optimized condition for enhanced yield, oil content, and encapsulation efficiency.

Zermane et al. (2014) explored the potential application of RSM methodology on optimization of *Myrtus communis* L. leaves essential oil using supercritical fluid extraction process. The optimization conditions selected for the extraction recovery study were range of pressure (10–30 MPa), temperature (308–323 K), a solvent flow rate (fixed at 0.42 kg h^{-1}) and a mean particle diameter (<0.5 mm).

Benmoussa et al. (2018) reported the optimized condition for the microwave hydro-diffusion and gravity (MHG) process used for the efficient extraction of the *Cuminum cyminum* seeds essential oils. The selected parameters for the optimization protocol used for the rapid extraction were time (8, 12, and 16 min), power (150, 200, and 250 W) for the microwave irradiation, and moisture content (40, 50, and 60) based on central composite design (CCD) assay. Authors reported that the combined effect of the microwave irradiation, the gravity, and the in situ seed water content played an important role overall quantity and the quality of the essential oil extracted aromatic plants.

Yingngam et al. (2021) optimized the two variables (microwave power and irradiation time), which significantly affected the essential oil yield using RSM method and reported that 700 W power and 25 min time are effective for the maximal yield of *Limnophila aromatica* essential oil.

Azevedo et al. (2014) reported the optimized concentration of cassava starch, chitosan and *Lippia gracilis* (LGRA) essential oil which exhibited potent antimicrobial activity against foodborne bacteria species such as *Staphylococcus aureus, Bacillus subtilis, Serratia marcescens*, and *Salmonella Enteritidis* using response surface methodology. Further, they reported the mixture ration based on combination of cassava starch (1.6%), chitosan (0.6%), and *Lippia gracilis* (2.4%), best for the edible coating as its MIC is remained below than the recommended doses for psychrophilic aerobic bacteria, yeast, and mold during the in situ application as an edible coating for the strawberries.

Yang et al. (2022) reported the effects of ultrasonic power, time, and types of emulsification chemical on antimicrobial effects of oil using response surface methodology. Based on single-factor experiments, authors reported that the ultrasonic power (527.45 W), time 9.97 min, and tween 80 (5.94 mg/mL) killed *E. coli* through extracellular release and exhibited good stability. While power (350 W), time (14.44 min), and cetylpyridinium chloride (0.15 mg/mL) exhibited excellent antibacterial activity (through electrostatic interaction).

Teshale et al. (2022) explored the potential use of RSM to obtain the maximum yield, that is, 2.8% (by wt.) of rosemary leaves essential oil by optimizing the extraction temperature (103–118 °C), time (2–6 h), and feed mass (200–600 g).

Zermane et al. (2016) reported the optimized condition of extraction temperature (X1) and pressure (X2) for high yield of Algerian rosemary essential oil using supercritical CO_2 based on response surface methodology. Authors reported that the maximum yield of essential oil 3.52 wt% (relative to the initial mass of leaf powder) at 313 K and 22 MPa.

Ebadollahi and Taghinezhad (2020) reported the optimized insecticidal effects of *Teucrium polium* essential oil against *Tribolium castaneum* Herbst using response surface methodology and reported that the effective dose of essential oil 20 μl/l and 72 min exposure time as the optimum conditions with 97.97% mortality and 87.8% desirability.

12.5.2 Full-Factorial Design with Level 2

Practically in every field of science, mathematical modelling has very fundamental importantce. This is so that we may create, evaluate new controllers, and assess original ones by better evaluation of system through simulation and prediction of its behavior. In general, 2 k full-factorial design is the common first-order designs, where k represents the number of control variables (Pereira et al. 2021). In a 2 k full-factorial design, variables under observations are examined at two levels (values +1, −1) that correspond to the high and low levels for the selected variables. The combination of the elements called as factor (k) and their levels results in a design that contains every potential combination. It provides the desirable information about the effects and interactions of the selected variables such as temperature, pH, pressure, and other selected parameters as per the requirement. The full-fFactorial design may be symmetric, in which case all of the factors under study have the same number of levels, or asymmetric, in which case each component has a different number of levels. Possibility of high expense and complex calculations offered for establishing several experimental conditions in case of larger value of factor k, this approach may not be practical for many factors. To avoid such circumstances, the researcher has the option of switching to another form of mathematical modelling called as fractional factorial design.

Machado et al. (2023) reported optimized nanoemulsion of *Ocotea indecora* (Shott) Mez essential oil using a 2^3 factorial design suggesting the variables conditions of essential oil (5% w/w) and surfactants at 1:1. The smallest droplets size and polydispersity index were used as selection criteria for the optimization of nanoemulsion. The stability of optimized nanoemulsion stored at room temperature and refrigerated was observed for up to 1 year.

Shehata et al. (2021) optimized the prepared nanoemulsion based on a two-factor, three-level, full-factorial design assay and reported that developed formulations were stable for 3 months with enhanced functional potential such as systemic bioavailability, synergistic anti-obesity effect. The formulations fullfill all the desirable criteria of drug-excipient compatibility such as, physical appearance, rheological behavior, drug release, and in vivo anti-obesity potential with nano-size of droplet.

12.5.3 Fractional Factorial Design

In scientific research and industrial applications, fractional factorial designs are very useful and often employed technique (Pereira et al. 2021). In the fields of engineering and physical and chemical sciences, several successful applications have been recorded. Fractional factorial designs become effective than full-factorial designs especially when number of factors are large in number as it allows to construct statistical models with a limited number of runs as it drop insignificant factors. We can forecast the most effective experimental combinations using the models. The two-level fractional factorial design is denoted as 2 K − p, where k is the number of factors utilized and p is the size of the percentage of the applied complete factorial analysis. In general, it is denoted as Zk − p, where Z is level.

Souiy et al. (2023) reported the optimized factors (time, temperature, plant/water mass ratio, and sampling period) that significantly affected the extraction yield, chemical composition, antioxidant, and antimicrobial activities of *Thymus algeriensis* Boiss. and Reut essential oil using the full-factorial design 2^4.

12.5.4 Plackett–Burman Design

The Plackett–Burman design (PBD) was proposed by R.L. Plackett and J.P. Burman in 1946, in order to find out the most important factors in design experiments. It allows the quality control procedure in a more effective manner so that it could be used to examine the experiment (Pereira et al. 2021). This approach needs the minimum of four (4n) experiment (sample size should be multiple of four rather than a power of two (4 k observations with k = 1, 2, ... n)) to get the desired observation and wise judgments, the consequences of design factors on the system state. PB designs are used to evaluate $n - 1$ variables in each (n) experiments. The primary benefit of such a design is the low observational required to determine an impact for a given component in experiment.

Spadi et al. (2021) studied the optimal significant factors to maximize the yield of rosemary essential oil using Plackett–Burman design. Authors have considered the several factors such as extraction method (hydro-distillation and steam), time (extraction 30 min and 120 min), cooling (water flow rate 1 L/m and 5 L/m), power (heating 600 W and 2000 W), and ratio of rosemary leaves to deionized water (solid/liquid ratio 1:2 and 1:6). The results based on Plackett–Burman design (PBD) revealed extraction time and the extraction process as the two foremost significant factors.

Boateng and Yang (2021) used Plackett–Burman design assay for the screening of drying parameter (infrared drying) for dried seeds to improve the paramount of bioactive components to preserved or protect the desired functionality. The results reported that the following factors, drying time, temperature, and distance between emitter of infrared and samples for green-red, yellowness, ascorbic acid, and rehydration capacity responses, have significant effects on functionality of dried seed samples.

12.5.5 Box–Behnken Design

The Box–Behnken designs (BBD) modelling proposed by G.E.P. Box and D. Behnken 1960 is another form of rotatable or close-to-rotatable quadratic designs based on three-level partial factorial designs. This design approach generates high-order response surface with the minimum run compared with the other factorial approach (Pereira et al. 2021). The number of experiments (N) required for the development of BBD is defined by the following equation: $N = 2n(n - 1) + \lambda o$ ....(2), where n is number of factors and λo is the number of central points. The basic difference between this design and CCRD lies in the fact that it uses three levels for the study of observable factors (−1, 0, +1), while CCRD uses five levels (−α, −1, 0, +1, +α). This method expresses the second-order mathematical model, and estimates parameters which limits underlying experiment and eliminates unnecessary combinations of treatments.

Pongsumpun et al. (2020) explored the effects of selected parameters such as sonication time, temperature, and Tween® 80 concentration on the droplet size, PDI, and viscosity of essential oil-based antifungal formulation using Box–Behnken design (BBD). The results revealed that sonication time (266 s), temperature (4.82 °C), and Tween® 80 (3% w/w) produced the optimum (droplet size 65.98 nm, PDI of 0.15, and viscosity 0.67 mPa.s.) cinnamon essential oil-based antifungal formulation effective against *Aspergillus niger, Rhizopus arrhizus, Penicillium* sp., and *Colletotrichum gloeosporioides.*

Wei et al. (2023) studied the effects of optimized parameter on quantity and chemical profile of essential oils extracted using solvent-free microwave using single-factor and Box–Behnken design. The analyzed factors were microwave irradiation time (20–40 min), microwave irradiation power (280–560 W), and homogenization time (4–8 min). Based on three-factor Box-Behnken design, the optimized solvent-free microwave extraction of fresh peel *of Citrus medica* L. var. arcodactylis Swingle was reported which fulfills the need of green chemical approaches with negligible use of solvent.

Tunç and Odabaş (2021) explored the potential uses of Box–Behnken design to optimize the ohmic heating-assisted extraction/hydro-distillation (OHAE/H), to enhance the yield of pectin and essential oils from lemon waste. The optimized parameter was reported as liquid to solid ratio (8.7:1), extraction/hydro-distillation time (58.4 min), and voltage gradient (14.2 V/cm) with the predicted pectin (16.58) and essential oil yields (3.62 g/100 g).

Labri et al. (2022) reported that optimize hydro-distillation parameters to get the high yield of selected essential oils extracted from aerial parts aromatic plants such as *Lavandula stoechas, Eucalyptus camaldulensis*, and *Carum carvivapor*. Based on the observation, vapor volume, distillate flow, particle size, and interaction between volume and flow rate were reported as the most significant factors affecting the EOs yield.

12.6 Conclusion

This chapter explored the role of mathematical modelling and statistical approaches, such as the simplex-centroid mixture design, central composite rotatable design, and factorial design, in optimizing the parameters which significantly affects the biological properties of plant volatiles. The literature review suggested that the application of statistical approaches could provide valuable information for the development of efficient, optimize, and cost-effective biologically active formulation based on plant volatiles for various industrial application in food, pharmaceutical, cosmetic, and agriculture industries.

References

Azarifar M, Ghanbarzadeh B, Khiabani MS, Basti AA, Abdulkhani A, Noshirvani N, Hosseini M (2019) The optimization of gelatin-CMC based active films containing chitin nanofiber and Trachyspermumammi essential oil by response surface methodology. Carbohydr Polym 208:457–468

Azevedo AN, Buarque PR, Cruz EMO, Blank AF, Alves PB, Nunes ML, de Aquino Santana LCL (2014) Response surface methodology for optimisation of edible chitosan coating formulations incorporating essential oil against several foodborne pathogenic bacteria. Food Control 43:1–9

Benmoussa H, Elfalleh W, He S, Romdhane M, Benhamou A, Chawech R (2018) Microwave hydrodiffusion and gravity for rapid extraction of essential oil from Tunisian cumin (Cuminum cyminum L.) seeds: optimization by response surface methodology. Ind Crop Prod 124:633–642

Bhattacharya R, Sourirajan A, Sharma P, Kumar A, Upadhyay NK, Shukla RK et al (2023) Bioenhancer potential of Aegle marmelos (L.) Corrêa essential oil with antifungal drugs and its mode of action against Candida albicans. Biocatalysis Agric Biotechnol:102647

Boateng ID, Yang XM (2021) Process optimization of intermediate-wave infrared drying: Screening by Plackett–Burman; comparison of Box-Behnken and central composite design and evaluation: A case study. Ind Crop Prod 162:113287

Chraibi M, Fadil M, Farah A, Benkhaira N, Lebrazi S, Fikri-Benbrahim K (2023) Simplex-centroid design as innovative approach in the optimization of antimicrobial effect of thymus satureioides, Myrtus communis and Artemisia herba alba essential oils against Escherichia coli, Staphylococcus aureus and Candida tropicalis. Exp Parasitol 247:108472

Crespo YA, Sánchez LRB, Quintana YG, Cabrera AST, Del Sol AB, Mayancha DMG (2019) Evaluation of the synergistic effects of antioxidant activity on mixtures of the essential oil from Apium graveolens L., Thymus vulgaris L and Coriandrum sativum L using simplex-lattice design. Heliyon 5(6):e01942

Ebadollahi A, Taghinezhad E (2020) Modeling and optimization of the insecticidal effects of Teucrium polium L. essential oil against red flour beetle (Triboliumcastaneum Herbst) using response surface methodology. Information processing. Agriculture 7(2):286–293

Gupta V, Singh PP, Prakash B (2023) Synthesis, characterization, and assessment of chitosan-nanomatrix enriched with antifungal formulation against biodeterioration of active ingredients of selected herbal raw materials. Int J Biol Macromol 123684:123684

Jafarizadeh-Malmiri H, Anarjan N, Berenjian A (2022) Developing three-component ginger-cinnamon-cardamom composite essential oil nanoemulsion as natural food preservatives. Environ Res 204:112133

Jafri H, Ahmad I (2020) Thymus vulgaris essential oil and thymol inhibit biofilms and interact synergistically with antifungal drugs against drug resistant strains of Candida albicans and Candida tropicalis. Journal de mycologieMedicale 30(1):100911

Kılıç M, Pütün E, Pütün AE (2014) Optimization of Euphorbia rigida fast pyrolysis conditions by using response surface methodology. J Anal Appl Pyrolysis 110:163–171

Kılıç M, Uzun BB, Pütün E, Pütün AE (2013) Optimization of biodiesel production from castor oil using factorial design. Fuel Process Technol 111:105–110

Kumar A, Gupta V, Singh PP, Kujur A, Prakash B (2020) Fabrication of volatile compounds loaded-chitosan biopolymer nanoparticles: optimization, characterization and assessment against aspergillus flavus and aflatoxin B1 contamination. Int J Biol Macromol 165:1507–1518

Kumar A, Kujur A, Singh PP, Prakash B (2019a) Nanoencapsulated plant-based bioactive formulation against food-borne molds and aflatoxin B1 contamination: preparation, characterization and stability evaluation in the food system. Food Chem 287:139–150

Kumar A, Kujur A, Yadav A, Pratap S, Prakash B (2019b) Optimization and mechanistic investigations on antifungal and aflatoxin B1 inhibitory potential of nanoencapsulated plant-based bioactive compounds. Ind Crop Prod 131:213–223

Kumar A, Singh PP, Kumar M, Prakash B (2022) Nanoencapsulated plant-based antifungal formulation against the aspergillus flavus and aflatoxin B1 contamination: unraveling the biochemical and molecular mechanism of action. Int J Food Microbiol 372:109681

Labri K, Moghrani H, Ihadadene R, Hamitouche H, Maachi R (2022) The application of the full factorial design and response surface methodology in optimization conditions for essential oils extraction from Lavandula stoechas, Carum carvi and Eucalyptus camaldulensis: effect of plants particle size on the extraction of essential oils. Sustain Chem Pharm 30:100830

Machado FP, Folly D, Enriquez JJS, Mello CB, Esteves R, Araújo RS et al (2023) Nanoemulsion of Ocotea indecora (Shott) Mez essential oil: Larvicidal effects against Aedes aegypti. Ind Crop Prod 192:116031

Magi G, Marini E, Facinelli B (2015) Antimicrobial activity of essential oils and carvacrol, and synergy of carvacrol and erythromycin, against clinical, erythromycin-resistant group a streptococci. Front Microbiol 6:165

Pereira LMS, Milan TM, Tapia-Blácido DR (2021) Using response surface methodology (RSM) to optimize 2G bioethanol production: a review. Biomass Bioenergy 151:106166

Pongsumpun P, Iwamoto S, Siripatrawan U (2020) Response surface methodology for optimization of cinnamon essential oil nanoemulsion with improved stability and antifungal activity. Ultrason Sonochem 60:104604

Prakash B, Kedia A, Mishra PK, Dubey NK (2015) Plant essential oils as food preservatives to control moulds, mycotoxin contamination and oxidative deterioration of Agri-food commodities–potentials and challenges. Food Control 47:381–391

Rezzoug SA, Boutekedjiret C, Allaf K (2005) Optimization of operating conditions of rosemary essential oil extraction by a fast controlled pressure drop process using response surface methodology. J Food Eng 71(1):9–17

Segatore B, Bellio P, Setacci D, Brisdelli F, Piovano M, Garbarino JA et al (2012) In vitro interaction of usnic acid in combination with antimicrobial agents against methicillin-resistant Staphylococcus aureus clinical isolates determined by FICI and ΔE model methods. Phytomedicine 19(3–4):341–347

Sharma M, Rani P, Nidhin PT, Katoch M (2023) A polymer-based gel fumigator, having a combination of Monarda citriodora essential oil (MEO) and hexanal: a solution for the postharvest rot and phytotoxicity of apple during storage. Postharvest Biol Technol 112248:112248

Shehata TM, Khalil HE, Elsewedy HS, Soliman WE (2021) Myrrh essential oil-based nanolipid formulation for enhancement of the antihyperlipidemic effect of atorvastatin. J Drug Deliv Sci Technol 61:102277

da Silva LDFF, Pallaoro RB, de Freitas EM, Hoehne L, Heidrich D, Ethur EM (2023) Antibacterial activity of Lithraeamolleoides hook et Arn. And Poiretia latifolia Vogel essential oils combined with gentamicin on foodborne disease-causing bacteria. Biocatalysis and agricultural. Biotechnology 48:102620

Souiy Z, Al-Shaikh TM, Alghamdi OA, Krifi B (2023) Optimization of extraction yield, chemical composition, antioxidant, and antimicrobial activities of Thymus algeriensisBoiss. & Reut essential oils. Biocatalysis Agric Biotechnol 47:102593

Soulaimani B, Abbad I, Varoni E, Iriti M, Mezrioui NE, Hassani L, Abbad A (2022) Optimization of antibacterial activity of essential oil mixture obtained from three medicinal plants: evaluation of synergism with conventional antibiotics and nanoemulsion effectiveness. S Afr J Bot 151:900–908

Spadi A, Angeloni G, Guerrini L, Corti F, Michelozzi M, Cencetti G et al (2021) Using a Plackett–Burman design to maximise yield of rosemary essential oil by distillation. Ind Crop Prod 166:113488

Sutaphanit P, Chitprasert P (2014) Optimisation of microencapsulation of holy basil essential oil in gelatin by response surface methodology. Food Chem 150:313–320

Teshale F, Narendiran K, Beyan SM, Srinivasan NR (2022) Extraction of essential oil from rosemary leaves: optimization by response surface methodology and mathematical modeling. Appl Food Res 2(2):100133

Tunç MT, Odabaş Hİ (2021) Single-step recovery of pectin and essential oil from lemon waste by ohmic heating assisted extraction/hydrodistillation: a multi-response optimization study. Innovative Food Sci Emerg Technol 74:102850

Wei L, Yu X, Li H, Zhu M, Pu D, Lu Q et al (2023) Optimization of solvent-free microwave extraction of essential oil from the fresh peel of Citrus medica L. var. arcodactylis Swingle by response surface methodology, chemical composition and activity characterization. Sci Hortic 309:111663

Yadav A, Kumar A, Singh PP, Prakash B (2021) Pesticidal efficacy, mode of action and safety limits profile of essential oils based nanoformulation against Callosobruchus chinensis and aspergillus flavus. Pestic Biochem Physiol 175:104813

Yang Z, He Q, Ismail BB, Hu Y, Guo M (2022) Ultrasonication induced nano-emulsification of thyme essential oil: optimization and antibacterial mechanism against Escherichia coli. Food Control 133:108609

Yingngam B, Brantner A, Treichler M, Brugger N, Navabhatra A, Nakonrat P (2021) Optimization of the eco-friendly solvent-free microwave extraction of Limnophilaaromatica essential oil. Ind Crop Prod 165:113443

Zermane A, Larkeche O, Meniai AH, Crampon C, Badens E (2014) Optimization of essential oil supercritical extraction from Algerian Myrtus communis L. leaves using response surface methodology. J Supercrit Fluids 85:89–94

Zermane A, Larkeche O, Meniai AH, Crampon C, Badens E (2016) Optimization of Algerian rosemary essential oil extraction yield by supercritical CO2 using response surface methodology. ComptesRendusChimie 19(4):538–543

Prospects of Bioinformatics and Data Acquirement Tools in Boosting the Application of Phytochemicals in Food Sciences

13

Atul Kumar Jaiswal, Prem Pratap Singh, and Bhanu Prakash

Abstract

Bioinformatics has been established as a multidisciplinary research area bringing about a paradigm shift in the discipline of biological researches. In silico resources made biologists to treat the huge and heterogeneous amount of bio-digital data in rapid mode to predict models for real-life applications. Its application in the biomedical field has been well explored; however, the significance of computational studies has remained underappreciated in food and nutritional sciences, despite their data-rich attributes. In the present study, a broader framework of bioinformatics in food sciences was discussed, especially plant-based databases, third-generation sequencing, genome assembly, and annotation. In addition, its role in the elucidation of the probable mechanism of action against foodborne microbes, recent advances, associated challenges, and future perspectives have also been discussed.

Keywords

Bioinformatics · Databases · Genome assembly · Third-generation sequencing · Food Sciences

---

Atul Kumar Jaiswal and Prem Pratap Singh contributed equally with all other contributors.

A. K. Jaiswal
Department of Computer Science and Engineering, School of Engineering Sciences and Technology, Jamia Hamdard University, New Delhi, India

P. P. Singh · B. Prakash (✉)
Centre of Advanced Study in Botany, Institute of Science, Banaras Hindu University, Varanasi, India
e-mail: bprakash@bhu.ac.in

B. Prakash et al. (eds.), *Plant Essential Oils*,
https://doi.org/10.1007/978-981-99-4370-8_13

13.1 Introduction

Bioinformatics is an interdisciplinary scientific area of research, which deals with the management and interpretation of complex biological data by computational approaches. The operational idea of bioinformatics exemplifies the umbrella concept by integrating the computer science tools, mathematical algorithms, and biostatistics to the single domain of biological sciences (Rhee et al. 2006; Bellgard and Bellgard 2011). In recent years, advancements in plant bioinformatics have made significant progress in the food industries with specific applicability for phytochemicals. The data generated via computational analysis can be used to explore and characterize the role of phytochemicals as dietary components (preservatives and nutritional agents) at the molecular level, as well as their interactions with the targeted sites of action. These massive and diverse sets of bio-digital data can act as a significant resource for identifying pathogenic patterns, allowing the development of legitimate prediction models for industrial applications in a fast and reliable way (Bellgard and Bellgard 2011). In addition, in silico research has made substantial efforts to promote the nutrition and quality of food sources. For instance, major elements related to the food industry, namely, allergenicity, flavor, functioning, safety issues, probiotics, and prebiotics, have received a lot of attention from applied bioinformatics (Holton et al. 2013).

The combinatorial aspect of food research and bioinformatics will be driven by the integration of heritable agronomic features, including essential metabolic profiles, with new omics data. Genome-sequencing techniques are being used with genome assembly and annotation procedures to analyze and integrate the large sets of molecular data, in preparation for elucidating the genetic basis of phytochemical biosynthesis (Saito 2013). Whereas databases serve as a vital component in not only speeding up the pace of research but also enabling investigations that would otherwise be difficult to perform. Furthermore, the existing literature and resources on phytochemicals and their incorporation into dietary supplements are often scattered, making it difficult to exploit the diversity of compounds for the betterment of mankind (Sharma and Sarkar 2013). In that view, bioinformatics not only provides the set of tools for identifying and comprehending genes and pathways linked to major cellular processes, but its scope has expanded to include everything from the genome to the phenome.

Despite having a definite and concerted perspective, bioinformatics role in the food industry remains underappreciated. There are numerous challenges of data management, accurate interpretation, and biological optimization associated with the in silico studies (Holton et al. 2013). In this review, we highlighted the examples of bioinformatic approaches that could be used successfully to enhance and extend the research on phytochemicals in the food and nutrition sciences, as well as their current limits and future research directions.

13.2 Bioinformatic Resources

Food science is an information-based science that includes the triangular relationship of food, human health, and environmental factors. Bioinformatics resources are required to systematize and integrate the knowledge of food biomaterials with human nutritional requirements, as well as nutritional and trophic variables. In silico tools evaluate the biological effects of phytochemicals based on specific condition markers (deficiencies in nutrition and health problems) as well as novel health improvement markers with metabolic consistency (Desiere et al. 2001). Computational approaches are co-emerging with the integration of genomics to unravel the molecular mechanism of biosynthesis, regulation, function, and evolution of phytochemicals. Approaches to integrating the gene functions with metabolite biosynthesis require the development of a wide platform storing and analyzing large-scale biological datasets (Saito 2013). In light of the aforementioned requirement, databases are built for extracting, storing, and presenting these massive observations in a useful format for researchers to mine subsequently (Sharma and Sarkar 2013).

13.2.1 Databases

A database is a centralized collection of theoretically linked data records that encompasses one or more topic areas with their integrated information. Any consolidated electronic data record with a combination of text, numbers, and pictures in various formats can be utilized as a database in today's informatics era, with the condition that users can retrieve and edit the data (Scalbert et al. 2011). Based on the checkpoints of evaluation and curation, databases can be classified into two broad categories of archival and curated databases. Archival databases (such as PubChem, GenBank, Gene Expression Omnibus, and Protein Data Bank) are designed to capture data of a specific format despite its quality or authenticity, whereas curated databases (such as MassBank, KEGG, and UniProt) are maintained with high-quality data based on expert evaluations (Scalbert et al. 2011; Sharma and Sarkar 2013).

There are several databases developed for cataloguing the information available on phytochemicals, from their chemistry and occurrence in foods to biology and health impacts. To start with, some databases provide reference nuclear magnetic resonance (NMR) or mass spectrometry (MS) spectra of pure and authentic phytochemicals for their routine and rapid identification. HMDB (http://www.hmdb.ca), NMRShiftDB (https://nmrshiftdb.nmr.uni-koeln.de/), NAPROC-13 (http://c13.usal.es/), BMRB-Metabolomics (http://www.bmrb.wisc.edu/metabolomics/), SDBS (http://riodb01.ibase.aist.go.jp/sdbs), and Golm Metabolome Database (http://gmd.mpimp-golm.mpg.de/) are some of the freely accessible and query-able databases of the NMR and MS spectra. Furthermore, PubChem (https://pubchem.ncbi.nlm.nih.gov/), KEGG (https://www.genome.jp/kegg/compound/), KNApSAcK (http://www.knapsackfamily.com/KNApSAcK/), and Dictionary of Food Compounds (https://dfc.chemnetbase.com) provide the taxonomic and compositional report of phytochemicals with their biological properties. Along with the taxonomic and compositional data, databases like KEGG, Plant Metabolic Network (https://plantcyc.org/), and Gramene Pathway (https://www.gramene.org/pathway/) also provide biosynthetic routes of phytochemicals.

The broad and rising intake of phytochemicals in food warrants a rigorous examination of the phytochemicals consumed. The purpose of the research is to regulate and improve the nutritional quality of foods in terms of their therapeutic and health-promoting effects. The USDA database (https://fdc.nal.usda.gov/), Phenol-Explorer (http://phenol-explorer.eu/), and Dr. Duke's Phytochemical database (https://phytochem.nal.usda.gov) provide concentration values for several phytochemicals found in various types of food. Also, the Dietary Supplement Label Database (DSLD) of the National Institutes of Health, indexes all printed information on dietary supplement products sold in the United States (Scalbert et al. 2011; Sharma and Sarkar 2013).

13.2.2 Genome Sequencing

Today, next-generation sequencing does not only serve as a research tool but also applied to various fields including diagnostics, pandemic investigations, food authenticity, food adulteration, and microbiome analysis (Goodwin et al. 2016; Quainoo et al. 2017; Allard et al. 2018). Whole-genome sequencing (WGS) and annotation of microorganisms used in the food industry (such as the various variety of yeast) (Piškur et al. 2012) and their closely related strains are of paramount importance, as this will help in understanding key differences and similarities at the DNA level. In the past few decades, a significant advancement in the sequencing technologies and related analysis tools have been achieved, which enlighten the understanding of various plants genomes such as rice (McNally et al. 2006), potato (Diambra 2011), tomato (Sahu and Chattopadhyay 2017), and their cultivars. As far as food safety is concerned, the present-day WGS technologies and tools enable us with fast identification and characterization of microorganisms with high precision along with antimicrobial resistance (FAO 2016). As the cost of WGS has rapidly gone down in the past decade, we can apply them in a much more cost-effective manner in diverse areas of interest which were relatively less feasible in the past. In the below sections, we will discuss genome sequencing, assembly, and annotation in detail.

13.2.2.1 First-Generation Sequencing

DNA sequencing by Sanger and Maxam-Gilbert method is classified as first-generation sequencing (FGS) technique. In the Sanger method, dideoxynucleotides (dNTP) are used for chain termination of elongation reaction and as a result, different sized fragments ending with dNTP (ddA, ddT, ddC, ddG) are obtained (Sanger et al. 1977; França et al. 2002). They are further separated on gel slab based on their size, and visualized by imaging techniques (such as UV or X-ray) (Masoudi-Nejad et al. 2013; El-Metwally et al. 2014). The first few genomes sequenced by the Sanger method are phiX174 and bacteriophage λ with genome sizes of 5374 bp and 48,501 bp, respectively (Sanger and Coulson 1975; Sanger et al. 1980). The other FGS method, Maxam-Gilbert, is a chemical degradation-based method where chemicals were used to cleave nucleotides (Maxam and Gilbert 1977). Both of these methods (especially Sanger) were in active use for the past three decades and are still used for sequencing short DNA fragments. These methods are very expensive as well as time consuming which limits it for sequencing different plant species.

13.2.2.2 Second- and Third-Generation Sequencing

Advancement of sequencing technologies in the past decade gave rise to second-generation sequencing (SGS) which has many advantages over FGS such as (a) massive parallelization resulting in millions of short reads, (b) accelerated sequencing rate compared to FGS, (c) no need of electrophoresis and visualization rather getting sequence information in real-time and low cost of sequencing (Kchouk et al. 2017). The SGS is used for short-read sequencing (maximum up to 1000 bp but most commonly 200–300 bp) and are mainly of two types: (a) sequencing by synthesis and (b) sequencing by ligation (Myllykangas et al. 2012; Van Dijk et al. 2014). The main disadvantages of these sequencing methods are low-read length and cost. There are several platforms for SGS as shown in Table 13.1.

SGS has revolutionized sequencing techniques which allows from understanding the genetic basis of cancer treatment to improving crop yield. However, SGS still has some disadvantages such as (a) requirement of PCR amplification step, (b) difficulty for sequencing high repetitive DNA sequences, (c) short-read length causes difficulties in genome assembly, and (d) high cost of sequencing (Kchouk et al. 2017). These limitations of SGS gave rise to the development of third-generation sequencing (TGS). The TGS technologies are cost-effective and rapid compared to SGS. Notably, they do not require PCR amplification and the read length is much longer (several kilobases) as compared to SGS, which makes the genome assembly process easy even for the genomes abundant in repetitive sequences. TGS can be characterized via two approaches (Van Dijk et al. 2014): The single-molecule real-time sequencing (SMRT) (Bentley et al. 2008) developed by Quake laboratory and synthetic approach used by Illumina and 10xGenomics (Braslavsky et al. 2003; Harris et al. 2008; Eid et al. 2009). At present, Pacific Biosciences (PacBio) and Oxford Nanopore Technology (ONT) are the two most abundantly used TGS technologies (Lu et al. 2016). These sequencers can generate reads of length 10 kb–15 kb and in some cases up to 100 Kbp (Del Angel et al. 2018). However, TGS mainly suffers from a high error rate (10% to 15%). The errors can be corrected from Illumina reads, large consensus data, and bioinformatics tools. The reads were also further refined by combining optical genome mapping and chromosome conformation capture sequencing data, thus generating chromosome-level assembly. The TGS has been successfully used to fill the gaps in the human genome, reconstructing highly contiguous regions of complex eukaryotic genomes and de novo microbial genome assembly (Chen et al. 2014; Loman et al. 2015; Rhoads and Au 2015). Comprehensive guidelines (including pros and cons) for selecting the correct sequencing technology have been extensively described in previous works (Sohn and Nam 2018; Jung et al. 2019; Jayakumar and Sakakibara 2019; Wee et al. 2019).

Table 13.1 Second-generation sequencing

Company	Platform	Mechanism	Read length (bp)	Read types	Data generated per run (GB)	Error type (error rate %)
Roche/454	GS FLX	Pyrosequencing	Up to 1000	SE, PE	0.02 to 0.7	Indel (1)
	Ion S5/ S5XL 530	Detection of hydrogen ion	400	SE	3 to 5	Indel (1)
Ion torrent	PGM 316		200		0.6 to 1	
	Ion proton		200		10	
	MiSeq	Reversible terminator	300	SE, PE	15	Mismatch (0.1)
	HiSeq		150		1.5 TB	Mismatch (0.1)
Illumina	miniSeq		150		7.5	Mismatch (1)
	NextSeq		150		120	Mismatch (1)
	HiSeq X		150		1.8 TB	Mismatch (0.1)

Note: *SE* Single-read End, *PE* Paired End

13.2.3 Genome Assembly

Plant genome assembly is a complicated and multistep process. Many factors such as polyploidy, presence of highly repetitive sequence, and genome size make the genome assembly process complicated (Mishra et al. 2017). There are mainly three types of genome assemblies: (1) reference-based genome assembly preferred when the reference genome of desired species or very closely related species is available, (2) de novo assembly preferred when a reference genome is not present, and (3) hybrid assembly approach.

13.2.3.1 Reference-Based Assembly

Firstly, reads are aligned to a reference genome and chains of mutually consistent matches are obtained with the help of the longest increasing subsequence (LIS) algorithm. The layout of overlapping reads is thus generated. The layout formed by reading alignment has several shortcomings (such as indels, and rearrangement of target and reference genome) for which it requires further refinement. After the layout refinement is achieved consensus sequences are generated with the help of multiple sequence alignment (MSA). Here, MSA is used to generate overlapping reads in an interactive manner which on convergence produces consensus sequence or contigs. Finally, scaffolds are generated by placing contigs in correct order and orientation which require information on mate pairs, genetic or physical maps, and so on.

13.2.3.2 De Novo Assembly

The de novo genome assembly is done from scratch when no reference genome is available (Chaisson et al. 2004; Chaisson et al. 2009). There are two main approaches of de novo assembly: (1) de Bruijn graph approach and (2) overlap layout consensus (OLC) approach. The de Bruijn approach is k-mer-based approach, where k-mers are generated from the reads and a directed graph is generated (Chaisson et al. 2004). In this graph, the edges represent overlap and nodes represent k-mers. The Euclidean path from this directed graph gives contigs that are further extended via contig extension algorithms. De Bruijn approach is preferred for short-read sequences. Some of the commonly used de Bruijn assemblers are ALLPATHS (Butler et al. 2008), SOAPdenovo (Li et al. 2010), and AbySS (Simpson et al. 2009), and some of the contig extension-based assemblers are PE-Assembler (Ariyaratne and Sung 2011) and SSAKE (Warren et al. 2007). On the other hand, OLC-based approach is best suited for long reads (generated via TGS) as it faces difficulties when utilized for large number of sequences. MIRA is OLC-based assembler used mainly for complex plant genome assembly. Few OLC-based assemblers use the MinHash Alignment algorithm which is very efficient in detecting overlap regions. OLC-based assemblers are particularly good for plant genome assembly as they are good in resolving highly repetitive regions of the genome. The basic workflow for de novo genome assembly is removal of contaminant DNA --> read quality control --> contig assembly --> scaffold generation --> chromosomal assembly.

13.2.3.3 Hybrid Assembly

It is proposed to use a combinatorial approach (SGS as well as TGS reads) to improve the quality of plant genome assembly, as SGS produces short reads (e.g., Illumina) with a very low error rate ($\leq 1\%$), but it has trouble assembling repetitive sections. TGS, on the other hand, produces lengthy readings with a high error rate (PacBio) (up to 15%). For instance, a number of error correction and assembly methods/pipelines have been developed in the past (readers are advised to read these references for detailed information (Koren et al. 2012; Wang et al. 2012; Utturkar et al. 2014)). Jabba is one of the widely used hybrid assemblers in which SGS reads are used to generate contigs via de Bruijn graph approach and errors in TGS reads are corrected by mapping them onto these contigs (Miclotte et al. 2016). DBG2OLC is also a widely used and efficient plant genome assembler (Ye et al. 2016). Commonly used plant genome assemblers are given in Table 13.2.

The quality of genome assembly depends on the choice of assembler, read length, sequencing error rate, complexity of the genome, etc. The obtained assembled genome should be subjected to statistical and biological validation. The present-day bioinformatics tools are very efficient in this regard and they do take care of these in an automatic fashion (Hunt et al. 2013; Yang et al. 2019; Matthews and Vosshall 2020).

13.2.4 Annotation

Genome annotation provides meaningful biological information such as gene structure and function of the presented genome (Cook et al. 2019) and can be broadly classified into the structural and functional annotation. The structural annotation involves predicting the gene locus and structure (e.g., intron, exon, and UTR). It captures the cellular and physiological activities of these genes, their functions, and interactions with other genes or nucleic acids, for example, transcription factors. Various annotation servers, tools, and pipelines are provided in Table 13.3.

13.2.4.1 Structural Annotation

Before making any gene prediction it is important to predict/mask repeat elements (including low complexity region and transposable elements) in the genome (Smith et al. 2007) because these sequences can cause a false homology in the gene prediction process. For this purpose, RepeatMasker (Nishimura 2000) can be used which utilizes repeat library from repbase database. RepeatRunner (Smith et al. 2007) is another tool used for the same purpose but RepeatMasker is widely accepted in the scientific community for repeat sequence prediction.

Identifying gene structure involves the prediction of the intron, exon, coding sequences (CDS) starting codon, and end codon. There are mainly two approaches of gene prediction: (1) ab initio gene structure prediction and (2) evidence-based gene structure prediction. In ab initio gene structure prediction, certain pre-calculated parameters are used, for example, codon bias, intron-exon length distribution which is obtained from gene models of closely related model organisms (e.g., *Arabidopsis thaliana* and *Drosophila melanogaster*). When no closely related gene structure models are available then the predictor is required to be trained on the genome under study. Augustus and SNAP are the most widely used ab initio gene structure prediction tools (Korf 2004; Stanke et al. 2006). The ab initio predictors show good performance in identifying CDS but they lack in predicting UTRs and alternately spliced transcripts. These limitations led to the development of many recent methods where information from an external source is used. In these methods, expressed sequence tags (ESTs), RNA-Seq data, protein sequence, etc. are

Table 13.2 Genome assemblers for plant genomes

Assembler type	Assembler	Read type
Reference based	Phrap	Genomic reads
	TIGR assembler	–
De-novo	SOAPdenovo	–
	Velvet	–
	ALLPATHS-LG	–
	ABySS	–
Hybrid	LORDEC	–
	DBG2OLC	–
	Jabba	–

Table 13.3 List of genome annotation tools

Mode	Tools (weblink)
Online	Ensembl (http://ensemblgenomes.org/info/data/annotation)
	GenSAS (https://www.gensas.org)
	Blast2GO (https://www.blast2go.com)
	AmiGO (http://amigo.geneontology.org/amigo)
	KAAS (https://www.genome.jp/tools/kaas/)
	Augustus (http://bioinf.uni-greifswald.de/augustus/)
	GAAP (http://GAAP.hallym.ac.kr)
Command based	BRAKER (https://github.com/Gaius-Augustus/BRAKER)
	MAKER (https://www.yandell-lab.org/software/maker. html)
	GSNAP (http://research-pub.gene.com/gmap)
	TopHat (https://ccb.jhu.edu/software/tophat/index. shtml)
	Evigan (http://www.seas.upenn.edu/~strctlrn/evigan/ evigan.html)
	Cufflinks (http://cole-trapnell-lab.github.io/cufflinks/)
	StringTie (https://ccb.jhu.edu/software/stringtie/)
	BLAST (https://blast.ncbi.nlm.nih.gov)

used as external evidence. These predictors/methods should be used in combination with ab initio gene predictors to improve the quality of gene structure prediction. MAKER, BAKER (includes ab initio prediction), and StringTie (Cantarel et al. 2008; Jung et al. 2020) (only for evidence-based prediction) are a few of such gene structure predictors. Since many methods/tools are used for gene structure prediction of a single genome, it is always advisable to generate a consensus gene structure (Cruz et al. 2019; Hosmani et al. 2019; Wilbrandt et al. 2019) with the following tools: Evidence Modeler, GLEAN, Evigan, GAAP, etc. These tools generate a consensus gene structure by minimizing the error from each prediction source and utilize a combination of evidence to do so (Haas et al. 2008; Liu et al. 2008).

13.2.4.2 Functional Annotation

Functional annotation of genes or protein sequences refers to associating biological information to these sequences. There are two types of functional annotation: (1) blast-based homology search and (2) GO term mapping based on gene ontology (GO) searches. To investigate the function of an unknown gene in a newly assembled genome or to predict its evolutionary relationship with other related sequences, blast is performed with the interest of finding a high homology. There are several tools for homology search such as blast, TopHat, Cufflinks (Trapnell et al. 2009; Ghosh and Chan 2016), and GSNAP (Wu et al. 2016). Some tools/pipelines, such as Blast2GO, allow you to execute blast on both local and cloud-based servers (Conesa et al. 2005). Depending upon the requirement, one can opt for Localblast or Cloudblast though later is preferred when opting for mass sequence alignment. GAAP (Kong et al. 2019) is another pipeline that provides both the options of Localblast as well as Cloudblast. Blas2GO result provides numerous information related to the query sequence such as number of hits mapped to query sequence, their e-values, mean similarity of mapped sequence, number of mapped terms in GO, and number of enzymes searched by these GO terms.

GO-term mapping is the process of obtaining GO terms associated with mapped sequences or hits via Blast search. GO stores information on relations between genes and gene-related terms. It is classified into three categories: (1) biological process, (2) molecular function, and (3) cellular component. The information regarding gene ontology, gene products, their functions, and activity sites are stored in the GO database, a relational database, and can be downloaded from AmiGO. There are various analysis tools in Blast2GO that can be used for predicting interaction between gene products with the help of the KEGG database (Kanehisa 2002).

13.3 Bioinformatic Application and its Challenges

There have been many biological implications of bioinformatics tools in plant and food sciences. They intend to augment the knowledge of various biological processes, namely, pathogenic machinery, food poisoning, and crop improvement, by elucidating the underlying mode of action. Further, bioinformatics approaches such as docking methods and pharmacophore searching are proving to be potential tools for deciphering molecular relationships between phytochemicals, food, and human health. Using in silico methods, Aziz et al. (2021) investigated the antibacterial, antioxidant, and urease inhibitory potential of different solvent extracts of *Strobilanthes glutinosus*. Similarly, Elgamal et al. (2021) reported the significant inhibitory activity of isolated phytochemicals of *Euphorbia retusa* against collagenase, elastase, hyaluronidase, and tyrosinase enzyme involved in skin remodelling and aging processes. Further, Khatun et al. (2021) showed the potential antimicrobial, antioxidative, and anti-cancerous potential of isolated phytochemical constituents from *Stevia rebaudiana* (Bert.) by targeting glutathione reductase, urase oxidase, and dihydrofolate reductase (DHFR), respectively, via bioinformatic approaches. Table 13.4 has shown some significant prospects of computational tools for phytochemicals in food research.

For availing the importance of the docking approaches in unravelling the therapeutic activities of phytochemicals, an example of evaluating their antibacterial property has been conducted using Protein*Plus* web service of the University of Hamburg, Germany. The bacterial DNA polymerase III α subunit has been chosen as the target protein as it is an essential enzyme for DNA replication and a potential target for antimicrobials (Van Eijk et al. 2017). Five phytochemicals of flavonoid and alkaloid families were selected from the NAPROC-13 database, namely 5,7,2′-trihydroxyflavone, 7,3’-Dihydroxy-5,4′,5′-trimethoxyisoflavone, Kushenol S, Isoflavanone, and 4-Hydroxy-3-methoxyphenyl ferulate. DoGSiteScorer, a grid-based tool has been used to predict binding pockets in DNA polymerase III (PDB ID: 5FKV) based on Gaussian filter differences of the protein 3D structure (Volkamer et al. 2012). Subsequently, the binding pockets along with the ligand molecules were transferred to the JAMDA module of Protein*Plus* for docking procedures and

Table 13.4 Significant prospects of computational tools for phytochemicals in human health development

Sr. No.	Phytochemical source	Phytochemicals	Bioinformatic technique	Target (Disease if any)	Reference
1.	*Nigella sativa*	α-Hederin, dithymoquinone, nigellicine, and nigellidine were potential molecules for drug development strategies against SARS-CoV-2	Molecular docking (AutoDock Vina software)	Inhibit RNA-dependent RNA polymerase (RdRp) (SARS-CoV-2)	Mir et al. (2021)
2.	*Moringa oleifera*	Anthraquinone, 2-phenylchromenylium (anthocyanins), hemlock tannin, sitogluside (glycoside), and A-phenolic steroid	In silico molecular docking (patch dock server), toxicity assessment, and pharmacophoric studies	Mutated insulin receptor kinase (diabetes mellitus)	Zainab et al. (2020)
3.	Aloe, black cohosh, cascara, chaparral, comfrey, ephedra, Fo-Ti, garcinia, germander, greater celandine, green tea, horse chestnut, kava, kratom, noni, pennyroyal, senna, usnea, valerian	Anthraquinones, glycosides (triterpene, anthracene, stilbene), polyphenols, nordihydroguaiaretic acid, alkaloids (pyrrolizidine, Isoquinoline, indole) ephedrine, hydroxycitric acid, Neoclerodane, monoterpenoids, diterpenoids, catechins, Aescin, kavalactones, flavonoids, sennosides, flavones, coumarins, Usnic acid, valepotriates	In silico quantitative structure–activity relationship (QSAR) for ADMET and hepatotoxicity	Alanine transaminase, aspartate transaminase, and lactate dehydrogenase (liver failure)	Liu (2018)
4	*Cinnamomum zeylanicum, Cinnamomum tamala, Amomum subulatum, Trigonella foenumgraecum, Mentha piperita, Coriandrum sativum, Lactuca sativa, and brassica oleraceae var. italica*	Rutin, catechin, quercetin, kaempferol, vanillic acid and ferulic acid (polyphenols)	In silico molecular docking by Pyrx-virtual screening tool	Inhibition of fungal enzymes (14-alpha demethylase (CYP51) and nucleoside diphosphokinase (NDK)) by the polyphenols	Khanzada et al. (2021)

Table 13.4 (continued)

Sr. No.	Phytochemical source	Phytochemicals	Bioinformatic technique	Target (Disease if any)	Reference
5	*Clerodendrum colebrookianum*	Acteoside, martinoside, and osmanthuside β6	Molecular docking by Glide extra precision (XP) method, GROningen MAchine for chemical simulations (GROMACS v5.0.4)	Interacted with the anti-hypertensive drug targets (ROCK I, ROCK II, and PDE5)	Arya et al. (2018)
6	*Semecarpus anacardium*	Catechol alkenyls	GOLD 3.1 software	Inhibit AChE from *Torpedo californica* (TcAChE) and from *Electrophorus electricus* (EeAChE)	Adhami et al. (2012)
7	*Emblica officinalis*	β-Glucogallin	Computational molecular docking by discovery studio software v2.5.5 (Accelrys) and ligand scout v2.3 (MacResearch)	β-Glucogallin interacts and inhibits AKR1B1 (diabetic cataracts)	Puppala et al. (2012)
8	*Withania somnifera*	Withaferin A (Wi-A) and Withanone (Wi-N)	Molecular docking by AutoDock 4.2 and analyzed using the graphical user interface of ADT	Mortalin, p53, p21, Nrf2 (anticancerous activity)	Vaishnavi et al. (2012)
9	*Parthenium hysterophorus*	Saponins	Docking using Cerius2 LigandFit and Schrodinger Glide modules	TNF-α inhibitory activity (anti nflamatory)	Shah et al. (2009)

Sr. No.	Phytochemical source	Phytochemicals	Bioinformatic technique	Target (Disease if any)	Reference
10	Vitis vinifera	ε-Viniferin	In silico approach using SELNERGY, an inverse docking computer software	ε-Viniferin possesses anti-inflammatory properties by inhibiting PDE4 subtype	Do et al. (2005)
11	*Dacryodes edulis*	DES4	MOE software running on a Linux workstation	Interactions of DES4 with artemether and quinine	Zofou et al. (2013)
12	*Mikania cordifolia and Artemisia dracunculus*	Limonene	Molegro virtual docker with Moldock score and Moldock SE	Inhibition of the MepA efflux pump by limonene	Freitas et al. (2021)
13	*Lygodium microphyllum*	Isoeleutherol, loperamide, quercetin, Isoquercetin and Stigmast-4-en-3-one	UCSF chimera applied for the docking study and an online tool was used to explore the ADME/T properties of selected bioactive compounds	Cyclooxygenase-1 and cyclooxygenase-2(anti-inflammatory), 5HT3 receptor inhibitor (antidiarrheal), tubulin-colchicine(anthelmintic)	Alam et al. (2021)
14	Vernonia cinerea.	Rutin, Methanone, phenol, 2,4-bis(1-phenylethyl)- and 1,2-Benzenedicarboxylic acid, diisooctyl ester	PyRx open-source software (GUI version 0.8 of AutodockVina)	D-alanine ligase (Ddl) and peptide deformylase (PDF)	Joshi et al. (2021)
15	*Plumbago zeylanica* (Chitrak), *Piper nigrum* (black pepper)	Plumbagin and Piperine	PyRX, Autodock vina virtual screening tool	Inhibit the Alpha-glucosidase receptor (in diabetes mellitus)	Tuli et al. (2021)

(continued)

Table 13.4 (continued)

Sr. No.	Phytochemical source	Phytochemicals	Bioinformatic technique	Target (Disease if any)	Reference
16	*Caesalpinia digyna*	Intricatinol, Bonducellin, Isobonducellin, Caesalpinine A, Caesalpinine C	PyRx Autodock Vina with a semi-flexible docking system, online tool SwissADME	COX-1, COX-2, and mPGES-1 receptors (antipyretic effect)	Emon et al. (2021)
17	*Azadirachta indica, Mangifera indica, Crinum glaucum*	Nimbin, nimbindiol, nimbinene, and quercetin, Crinamine and lycorine, catechin, ellagic acid, epicatechin, gallic acid, and kaempferol	Python prescription 0.8, a suite comprising automated molecular docking tools (auto dock Vina)	Nitric oxide synthase and arginase (in asthma)	Umar et al. (2021)
18	*Plumbago zeylanica*	Plumbagin	AutoDock 4.2, for conformation Lamarckian Genetic algorithm	Inhibition of CDK6 active site.	Roy and Bhatia (2021)
19	Tea, Grapes, berries, *Kochia scoparia,*	Gallic acid, caffeic acid, thymoquinone, thymol, betaine, alkannin, ellagic acid, betanin, Shikonin, betanidin, and Momordin IC	AutoDock 4 and molecular dynamics (MD) simulation with GROMACS 4.6.5	SENP1 protease inhibition (anticancer)	Taghvaei et al. (2021)
20	*Ocimum basilicum, Spinacia oleracea Capsicum annuum*, and *citrus reticulate and Prunus cerasus*	Luteolin and abyssinone II	NAMD 2.14 (CUDA version) graphical interface module with CHARMM36 as a force filed integrated with VMD 1.9.3 program	Mpro/3CLpro, PLpro and ACE2,	Shawan et al. (2021)

Sr. No.	Phytochemical source	Phytochemicals	Bioinformatic technique	Target (Disease if any)	Reference
21	*Calendula officinals*	Rutin, isorhamnetin-3-O-β-D, calendoflaside, narcissin, calendulaglycoside B, calenduloside, calendoflavoside	Docking performed by dock prep tool of UCSF chimera	Inhibiting M^{pro}, the main protease for SARS-CoV-2 (M^{pro} responsible for the proteolytic mutation of this virus and important for their life cycle)	Das et al. (2021)
22	*Torreya nucifera, Chamaecyparis obtusa, Ginkgo biloba*, and Pichia pastoris	Amentoflavone and Gallocatechin gallate	Docking by Autodock Vina	Potential inhibitors of 3-chymotrypsin-like (3CLpro) and papain-like proteases (PLpro) of SARS-CoV2	Swargiary et al. (2020)
23	Gallnuts, *toona Sinensis, Terminalia belerica camellia sinensis* Blueberries, *Vitis vinifera silymarin marimun* artichoke, Basil, celery	Gallic acid, catechin, resveratrol, apigenin, Silibinin, dasabuvir	Docking by AutoDock 4.2 tool	Bind with active site of NS5B polymerase of hepatitis C virus (HCV)	Shakya (2019)
24	*Verbena officinalis* green tea Citrus aurantium Cinnamomum camphora chamomile plants	Verbenalin, epigallocatechin, nobiletin, Sesamin, Swertisin	Docking algorithm of MOE software	Interact with RNA-dependent RNA polymerase (nsp-12) (major constituent of viral replication and RNA synthesis machinery)	Mahrosh and Mustafa (2021)
25	*Camellia sinensis*	Theaflavin, 1-O-caffeoylquinic acid, genistein, epigallocatechin 3-gallate, ethyl trans-caffeate	GLIDE docking module of Schrodinger suite software	Inhibit matrix metalloproteinase (MMPs) against SARS-CoV-2 main protease	Kanbarkar and Mishra (2021)

Table 13.5 Binding details of selected phytochemicals against the bacterial DNA polymerase III α subunit

Sr. No.	Name	Formula	SMILES	M. W. (g/mol)	Docking score
1	5,7,2′-Trihydroxyflavone	$C_{15}H_{10}O_5$	O = c2cc(c1ccccc1O)oc3cc(O)cc(O)c23	270.24	−2.03883
2	7,3'-Dihydroxy-5,4′,5′-trimethoxyisoflavone	$C_{18}H_{16}O_7$	COc3cc(c2coc1cc(O)cc(OC)c1c2 = O)cc(O)c3OC	344.19	−2.25947
3	(2S)-5,7-Dihydroxy-2-(2-hydroxyphenyl)-8-(3-methylbut-2-enyl)-2,3-dihydrochromen-4-one (Kushenol S)	$C_{20}H_{20}O_5$	C/C(C) = C\Cc2c(O)cc(O)c3C(=O)C[C@@H](c1ccccc1O)Oc23	340.21	−2.30394
4	(3R)-3-phenyl-2,3-dihydrochromen-4-one (Isoflavanone)	$C_{15}H_{12}O_2$	O=C1c2ccccc2OC-[C@H]1c1ccccc1	224.26	−1.92734
5	4-Hydroxy-3-methoxyphenyl ferulate	$C_{17}H_{16}O_6$	COc2cc(/C=C/C(=O)Oc1ccc(O)c(OC)c1)ccc2O	316.18	−2.49985

scoring function. JAMDA is a totally automated docking tool for proteins and ligands. The structures are preprocessed and docked automatically; neither coordinates nor protonation states are required (Flachsenberg et al. 2020). The docking parameters were kept at default. As per Table 13.5, significant inhibition of the DNA polymerase III α subunit by the selected phytochemicals has been shown based on the JAMDA docking score. Docking poses were generated by the PoseView section (Stierand et al. 2006) and represented in Fig. 13.1.

In addition to the promising domains of bioinformatics research covered in this review, there are many obstacles and opportunities that must be addressed to fully realize the potential of bioinformatics tools. There is a vast amount of information available on phytochemicals including their taxonomic and compositional data, but it is usually theoretical since it is scattered across various literature sources in unstructured form (natural language). There is an urgent need for developing a community-based-cum-automated framework for compiling the diverse set of biological data at a commonplace. It will provide a strong motive for busy researchers to join and contribute, as well as automated sorting of submitted descriptors (ontological terms) to their respective shells (Scalbert et al. 2011). Considering the huge variation in the research quality, testing parameters, and analytical interpretations, mean content values should be used to reflect the average content of the phytochemicals. Furthermore, data obtained in interventional clinical trials, as well as analytics available on experimental animals, should be included in the databases along with the taxonomic and compositional data. The systemic integration, collaborative effort, and "expertise" curation will provide the reasons for establishing and developing innovative success.

Genome assembly may seem a simple process but there are various limitations and challenges such as computational complexity, biological complexity, and

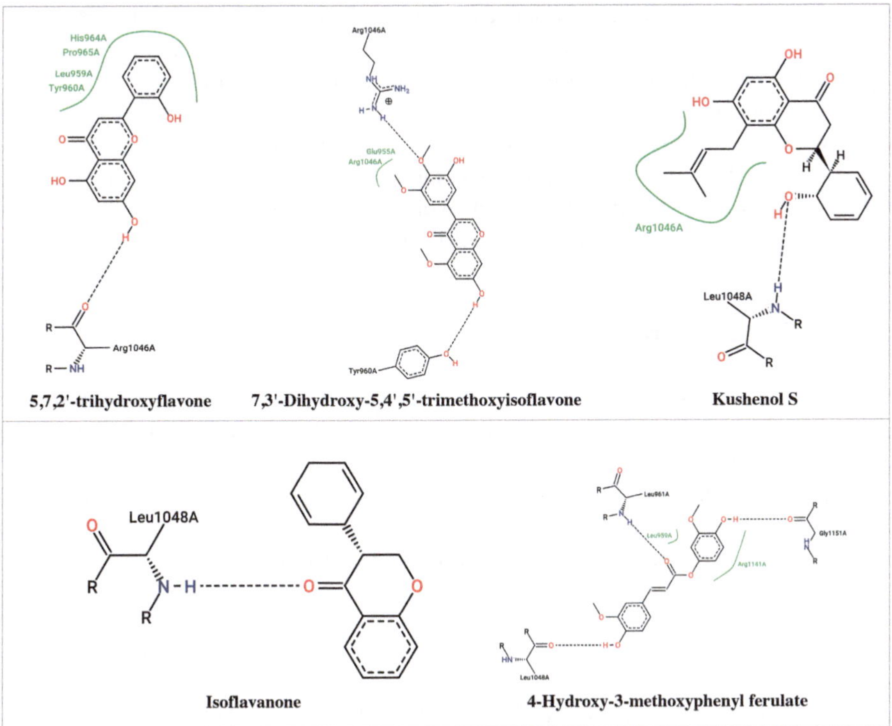

Fig. 13.1 Docking poses of selected phytochemicals against the bacterial DNA polymerase III α subunit

quality of the finished genome. The most popular methods for genome assembly, that is, OLC and de Bruijn approaches also suffer from time as well as space complexity. In the OLC approach, the limiting step is finding the overlaps which require pair-wise alignment of the reads which has O(n2) level of complexity for n number of reads and O(ij) level of complexity for alignment of reads of length i and j. On the contrary, the de Bruijn graph approach is much faster than OLC-based methods but suffers from space complexity. The most computationally intensive step in de Bruijn graph-based methods is the construction of the graph. Several assemblers utilize algorithms based on these methods (OLC and de Bruijn) but with some solution to time and space complexity but again they compromise on the accuracy.

Biological complexity comprises of several factors such as the size of the genome, the read numbers, and large repetitive/duplicated regions, and these factors also increase computational complexity. The presence of repetitive sequences can complicate the assembly process as these sequences may not differ from one another but can belong to different regions of the genome and since the reads are put together for contig formation by searching overlapping reads hence repeat sequences can cause an error-prone assembly. Another biological complexity arises especially in the plant genome because of polymorphism. The plant genome sequencing is further complicated because sequences from step-loop structures of centromere and telomeric regions are ignored which may cause problems in plant genome assembly. As genome is assembled by different assemblers hence it should be reexamined and validated concerning low coverage reads, poor handling of repeated regions, and quality of reads. This process is also known as genome finishing and it can be divided into three categories, namely, gap closure, assembly validation, and genome refinement. This is done in automatic as well as the manual mode to make sure of a robust genome assembly (Mishra et al. 2017).

13.4 Conclusion

This chapter provides an overview on the emerging bioinformatic approaches, currently being used in plant science to elucidate the structural and functional attributes of phytochemicals. It also explored the effect of in silico studies in food sciences to integrate the traditional ethnobotanical knowledge with the health promising outcomes. The chapter outlined the major challenges associated with the bioinformatics framework that need to be addressed in the near future.

References

Adhami HR, Linder T, Kaehlig H, Schuster D, Zehl M, Krenn L (2012) Catechol alkenyls from *Semecarpus anacardium*: acetylcholinesterase inhibition and binding mode predictions. J Ethnopharmacol 139:142–148

Alam MM, Emon NU, Alam S, Rudra S, Akhter N, Mamun MMR et al (2021) Assessment of pharmacological activities of *Lygodium microphyllum* Cav. Leaves in the management of pain, inflammation, pyrexia, diarrhea, and helminths: in vivo, in vitro and in silico approaches. Biomed Pharmacother 139:111644

Allard MW, Bell R, Ferreira CM, Gonzalez-Escalona N, Hoffmann M, Muruvanda T et al (2018) Genomics of foodborne pathogens for microbial food safety. Curr Opin Biotechnol 49:224–229

Ariyaratne PN, Sung WK (2011) PE-assembler: de novo assembler using short paired-end reads. Bioinformatics 27:167–174

Arya H, Syed SB, Singh SS, Ampasala DR, Coumar MS (2018) In silico investigations of chemical constituents of *Clerodendrum colebrookianum* in the anti-hypertensive drug targets: ROCK, ACE, and PDE5. Interdiscip Sci Comput Life Sci 10:792–804

Aziz M, Ahmad S, Iqbal MN, Khurshid U, Saleem H, Alamri A et al (2021) Phytochemical, pharmacological, and in-silico molecular docking studies of Strobilanthes glutinosus Nees: an unexplored source of bioactive compounds. S Afr J Bot 147:618

Bellgard MI, Bellgard SE (2011) A bioinformatics framework for plant pathologists to deliver global food security outcomes. Australas Plant Pathol 41:113–124

Bentley DR, Balasubramanian S, Swerdlow HP, Smith GP, Milton J, Brown CG et al (2008) Accurate whole human genome sequencing using reversible terminator chemistry. Nature 456(7218):53–59

Braslavsky I, Hebert B, Kartalov E, Quake SR (2003) Sequence information can be obtained from single DNA molecules. Proc Natl Acad Sci 100:3960–3964

Butler J, MacCallum I, Kleber M, Shlyakhter IA, Belmonte MK, Lander ES et al (2008) ALLPATHS: de novo assembly of whole-genome shotgun microreads. Genome Res 18:810–820

Cantarel BL, Korf I, Robb SM, Parra G, Ross E, Moore B et al (2008) MAKER: an easy-to-use annotation pipeline designed for emerging model organism genomes. Genome Res 18:188–196

Chaisson MJ, Brinza D, Pevzner PA (2009) De novo fragment assembly with short mate-paired reads: does the read length matter? Genome Res 19:336–346

Chaisson M, Pevzner P, Tang H (2004) Fragment assembly with short reads. Bioinformatics 20:2067–2074

Chen X, Bracht JR, Goldman AD, Dolzhenko E, Clay DM, Swart EC et al (2014) The architecture of a scrambled genome reveals massive levels of genomic rearrangement during development. Cell 158:1187–1198

Conesa A, Götz S, García-Gómez JM, Terol J, Talón M, Robles M (2005) Blast2GO: a universal tool for annotation, visualization and analysis in functional genomics research. Bioinformatics 21:3674–3676

Cook DE, Valle-Inclan JE, Pajoro A, Rovenich H, Thomma BP, Faino L (2019) Long-read annotation: automated eukaryotic genome annotation based on long-read cDNA sequencing. Plant Physiol 179:38–54

Cruz F, Lagoa D, Mendes J, Rocha I, Ferreira EC, Rocha M et al (2019) SamPler—a novel method for selecting parameters for gene functional annotation routines. BMC Bioinformatics 20:454

Das P, Majumder R, Mandal M, Basak P (2021) In-silico approach for identification of effective and stable inhibitors for COVID-19 main protease (Mpro) from flavonoid based phytochemical constituents of Calendula officinalis. J Biomol Struct Dyn 39:6265–6280

Del Angel VD, Hjerde E, Sterck L, Capella-Gutierrez S, Notredame C, Pettersson OV et al (2018) Ten steps to get started in genome assembly and annotation. F1000 Res 7

Desiere F, German B, Watzke H, Pfeifer A, Saguy S (2001) Bioinformatics and data knowledge: the new frontiers for nutrition and foods. Trends Food Sci Technol 12:215–229

Diambra LA (2011) Genome sequence and analysis of the tuber crop potato. Nature 475

Do QT, Renimel I, Andre P, Lugnier C, Muller CD, Bernard P (2005) Reverse pharmacognosy: application of Selnergy, a new tool for lead discovery. The example of ε-viniferin. Curr Drug Discov Technol 2:161–167

Eid J, Fehr A, Gray J, Luong K, Lyle J, Otto G et al (2009) Real-time DNA sequencing from single polymerase molecules. Science 323:133–138

El-Metwally S, Ouda OM, Helmy M (2014) New horizons in next-generation sequencing. In: Next generation sequencing technologies and challenges in sequence assembly. Springer, New York, NY, pp 51–59

Elgamal AM, El Raey MA, Gaara A, Abdelfattah MA, Sobeh M (2021) Phytochemical profiling and anti-aging activities of Euphorbia retusa extract: in silico and in vitro studies. Arab J Chem 14:103159

Emon NU, Alam S, Rudra S, Al Haidar IK, Farhad M, Rana MEH (2021) Antipyretic activity of the leaves extract of *Caesalpinia digyna* Rottl along with phytoconstituent's binding affinity to COX-1, COX-2 and mPGES-1 receptors: an in vivo and in silico approaches (antipyretic activity of *Caesalpinia digyna* Rottl). *Saudi.* J Biol Sci 28:5302

Flachsenberg F, Meyder A, Sommer K, Penner P, Rarey M (2020) A Consistent Scheme for Gradient-Based Optimization of Protein–Ligand Poses. J Chem Inf Model 60:6502–6522

Food and Agriculture Organization of the United Nations (2016) Applications of whole genome sequencing (WGS) in food safety management. https://www.fao.org/documents/card/en/c/61e44b34-b328-4239-b59c-a9e926e327b4/

França LT, Carrilho E, Kist TB (2002) A review of DNA sequencing techniques. Q Rev Biophys 35:169–200

Freitas PR, de Araújo ACJ, dos Santos Barbosa CR, Muniz DF, de Almeida RS, de Menezes IRA et al (2021) Inhibition of the MepA efflux pump by limonene demonstrated by in vitro and in silico methods. Folia Microbiol 1-6:15

Ghosh S, Chan CKK (2016) Analysis of RNA-Seq data using TopHat and cufflinks. In: Plant bioinformatics. Humana Press, New York, NY, pp 339–361

Goodwin S, McPherson JD, McCombie WR (2016) Coming of age: ten years of next-generation sequencing technologies. Nat Rev Genet 17:333–351

Haas BJ, Salzberg SL, Zhu W et al (2008) Automated eukaryotic gene structure annotation using EVidenceModeler and the program to assemble spliced alignments. Genome Biol 9:R7

Harris TD, Buzby PR, Babcock H, Beer E, Bowers J, Braslavsky I et al (2008) Single-molecule DNA sequencing of a viral genome. Science 320:106–109

Holton TA, Vijayakumar V, Khaldi N (2013) Bioinformatics: current perspectives and future directions for food and nutritional research facilitated by a food-wiki database. Trends Food Sci Technol 34:5–17

Hosmani PS, Shippy T, Miller S, Benoit JB, Munoz-Torres M, Flores-Gonzalez M et al (2019) A quick guide for student-driven community genome annotation. PLoS Comput Biol 15:1006682

Hunt M, Kikuchi T, Sanders M, Newbold C, Berriman M, Otto TD (2013) REAPR: a universal tool for genome assembly evaluation. Genome Biol 14:1–10

Jayakumar V, Sakakibara Y (2019) Comprehensive evaluation of non-hybrid genome assembly tools for third-generation PacBio long-read sequence data. Brief Bioinform 20:866–876

Joshi T, Pandey SC, Maiti P, Tripathi M, Paliwal A, Nand M et al (2021) Antimicrobial activity of methanolic extracts of *Vernonia cinerea* against *Xanthomonas oryzae* and identification of their compounds using in silico techniques. PLoS One 16:0252759

Jung H, Ventura T, Chung JS, Kim WJ, Nam BH, Kong HJ et al (2020) Twelve quick steps for genome assembly and annotation in the classroom. PLoS Comput Biol 16:1008325

Jung H, Winefield C, Bombarely A, Prentis P, Waterhouse P (2019) Tools and strategies for long-read sequencing and de novo assembly of plant genomes. Trends Plant Sci 24:700–724

Kanbarkar N, Mishra S (2021) Matrix metalloproteinase inhibitors identified from *Camellia sinensis* for COVID-19 prophylaxis: an *in-silico* approach. Adv Tradit Med 21:173–188

Kanehisa M (2002) The KEGG database. In: Novartis found symp, vol 247, pp 91–103

Kchouk M, Gibrat JF, Elloumi M (2017) Generations of sequencing technologies: from first to next generation. Biol Med 09:9

Khanzada B, Akhtar N, Okla MK, Alamri SA, Al-Hashimi A, Baig. et al (2021) Profiling of antifungal activities and in silico studies of natural polyphenols from some plants. Molecules 26:7164

Khatun MCS, Muhit MA, Hossain MJ, Al-Mansur MA, Rahman SA (2021) Isolation of phytochemical constituents from *Stevia rebaudiana* (Bert.) and evaluation of their anticancer, antimicrobial and antioxidant properties via in vitro and in silico approaches. Heliyon 7:08475

Kong J, Huh S, Won JI, Yoon J, Kim B, Kim K (2019) GAAP: a genome assembly + annotation pipeline. Biomed Res Int 4767354:1

Koren S, Schatz MC, Walenz BP, Martin J, Howard JT, Ganapathy G et al (2012) Hybrid error correction and de novo assembly of single-molecule sequencing reads. Nat Biotechnol 30:693–700

Korf I (2004) Gene fnding in novel genomes. BMC Bioinform 5:59

Li R, Zhu H, Ruan J, Qian W, Fang X, Shi Z et al (2010) De novo assembly of human genomes with massively parallel short read sequencing. Genome Res 20:265–272

Liu Q, Mackey AJ, Roos DS, Pereira FCN (2008) Evigan: a hidden variable model for integrating gene evidence for eukaryotic gene prediction. Bioinformatics 24:597–605

Liu Y (2018) Incorporation of absorption and metabolism into liver toxicity prediction for phytochemicals: a tiered in silico QSAR approach. Food Chem Toxicol 118:409–415

Loman NJ, Quick J, Simpson JT (2015) A complete bacterial genome assembled de novo using only nanopore sequencing data. Nat Methods 12:733–735

Lu H, Giordano F, Ning Z (2016) Oxford Nanopore MinION sequencing and genome assembly. Genom Proteom Bioinform 14:265–279

Mahrosh HS, Mustafa G (2021) An *in-silico* approach to target RNA-dependent RNA polymerase of COVID-19 with naturally occurring phytochemicals. Environ Dev Sustain:1–14

Masoudi-Nejad A, Narimani Z, Hosseinkhan N (2013) Next generation sequencing and sequence assembly: methodologies and algorithms, vol 4. Springer Science and Business Media

Matthews BJ, Vosshall LB (2020) How to turn an organism into a model organism in 10 'easy' steps. J Exp Biol 223:218198

Maxam AM, Gilbert W (1977) A new method for sequencing DNA. Proc Natl Acad Sci 74:560–564

McNally KL, Bruskiewich R, Mackill D, Buell CR, Leach JE, Leung H (2006) Sequencing multiple and diverse rice varieties. Connecting whole-genome variation with phenotypes. Plant Physiol 141:26–31

Miclotte G, Heydari M, Demeester P, Rombauts S, Van de Peer Y, Audenaert P et al (2016) Jabba: hybrid error correction for long sequencing reads. Algorithms Mol Biol 11:1–12

Mir SA, Firoz A, Alaidarous M, Alshehri B, Dukhyil AAB, Banawas S (2021) Identification of SARS-CoV-2 RNA-dependent RNA polymerase inhibitors from the major phytochemicals of *Nigella sativa*: an in-silico approach. Saudi J Biol Sci 29:394

Mishra DC, Lal SB, Sharma A, Kumar S, Budhlakoti N, Rai A (2017) Strategies and tools for sequencing and assembly of plant genomes. In: The potato genome. Springer, Cham, pp 81–93

Myllykangas S, Buenrostro J, Ji HP (2012) Overview of sequencing technology platforms. In: Bioinformatics for high throughput sequencing. Springer, New York, NY

Nishimura D (2000) Repeat masker. Biotech Software and Internet Report 1, 36–39

Piškur J, Ling Z, Marcet-Houben M, Ishchuk OP, Aerts A, LaButti K et al (2012) The genome of wine yeast Dekkera bruxellensis provides a tool to explore its food-related properties. Int J Food Microbiol 157:202–209

Puppala M, Ponder J, Suryanarayana P, Reddy GB, Petrash JM, LaBarbera DV (2012) The isolation and characterization of β-glucogallin as a novel aldose reductase inhibitor from *Emblica officinalis*. PLoS One 7:31399

Quainoo S, Coolen JP, van Hijum SA, Huynen MA, Melchers WJ, van Schaik W et al (2017) Whole-genome sequencing of bacterial pathogens: the future of nosocomial outbreak analysis. Clin Microbiol Rev 30:1015–1063

Rhee SY, Dickerson J, Xu D (2006) Bioinformatics and its applications in plant biology. Annu Rev Plant Biol 57:335–360

Rhoads A, Au KF (2015) PacBio sequencing and its applications. Genom Proteom Bioinform 13:278–289

Roy A, Bhatia KS (2021) *In-silico* analysis of plumbagin against cyclin-dependent kinases receptor. Vegetos 34:50–56

Sahu KK, Chattopadhyay D (2017) Genome-wide sequence variations between wild and cultivated tomato species revisited by whole genome sequence mapping. BMC Genomics 18:1–10

Saito K (2013) Phytochemical genomics-a new trend. Curr Opin Plant Biol 16:373–380

Sanger F, Coulson AR (1975) A rapid method for determining sequences in DNA by primed synthesis with DNA polymerase. J Mol Biol 94:441–448

Sanger F, Coulson A, Barrell BG, Smith AJH, Roe BA (1980) Cloning in single-stranded bacteriophage as an aid to rapid DNA sequencing. J Mol Biol 143:161–178

Sanger F, Nicklen S, Coulson AR (1977) DNA sequencing with chain-terminating inhibitors. Proc Natl Acad Sci 74:5463–5467

Scalbert A, Andres-Lacueva C, Arita M, Kroon P, Manach C, Urpi-Sarda M (2011) Databases on food phytochemicals and their health-promoting effects. J Agric Food Chem 59:4331–4348

Shah BA, Chib R, Gupta P, Sethi VK, Koul S, Andotra SS (2009) Saponins as novel TNF-α inhibitors: isolation of saponins and a nor-pseudoguaianolide from *Parthenium hysterophorus*. Org Biomol Chem 7:3230–3235

Shakya AK (2019) Natural phytochemicals: potential anti-HCV targets in silico approach. J App Pharm Sci 9:094–100

Sharma V, Sarkar IN (2013) Bioinformatics opportunities for identification and study of medicinal plants. Brief Bioinform 14:238–250

Shawan MMAK, Halder SK, Hasan MA (2021) Luteolin and abyssinone II as potential inhibitors of SARS-CoV-2: an in silico molecular modeling approach in battling the COVID-19 outbreak. Bull Natl Res Cent 45:1–21

Simpson JT, Wong K, Jackman SD, Schein JE, Jones SJ, Birol I (2009) ABySS: a parallel assembler for short read sequence data. Genome Res 19:1117–1123

Smith CD, Edgar RC, Yandell MD et al (2007) Improved repeat identification and masking in dipterans. Gene 38:1–9

Sohn JI, Nam JW (2018) The present and future of de novo whole-genome assembly. Brief Bioinformatics 19:23–40

Stanke M, Tzvetkova A, Morgenstern B (2006) AUGUSTUS at EGASP: using EST, protein and genomic alignments for improved gene prediction in the human genome. Genome Biol 7:1–8

Stierand K, Maaß P, Rarey M (2006) Molecular complexes at a glance: automated generation of two-dimensional complex diagrams. Bioinformatics 22:1710–1716

Swargiary A, Mahmud S, Saleh MA (2020) Screening of phytochemicals as potent inhibitor of 3-chymotrypsin and papain-like proteases of SARS-CoV2: an *in-silico* approach to combat COVID-19. J Biomol Struct Dyn:1–15

Taghvaei S, Sabouni F, Minuchehr Z, Taghvaei A (2021) Identification of novel anti-cancer agents, applying in silico method for SENP1 protease inhibition. J Biomol Struct Dyn:1–15

Trapnell C, Pachter L, Salzberg SL (2009) TopHat: discovering splice junctions with RNA-Seq. Bioinformatics 25:1105–1111

Tuli HS, Sood S, Bhatia GK, Debnath P, Aggarwal D, Upadhyay SK (2021) In silico analysis and molecular docking studies of plumbagin and piperine ligands as potential inhibitors of alpha-glucosidase receptor. Biointerface Res Appl Chem 11:9629–9637

Umar HI, Saliu TP, Josiah SS, Ajayi A, Danjuma JB (2021) In silico studies of bioactive compounds from selected African plants with inhibitory activity against nitric oxide synthase and arginase implicated in asthma. Egypt J Med Hum Genet 22:1–16

Utturkar SM, Klingeman DM, Land ML, Schadt CW, Doktycz MJ, Pelletier DA et al (2014) Evaluation and validation of de novo and hybrid assembly techniques to derive high-quality genome sequences. Bioinformatics 30:2709–2716

Vaishnavi K, Saxena N, Shah N, Singh R, Manjunath K, Uthayakumar M et al (2012) Differential activities of the two closely related withanolides, Withaferin a and Withanone: bioinformatics and experimental evidences. PLoS One 7:e44419

Van Dijk EL, Auger H, Jaszczyszyn Y, Thermes C (2014) Ten years of next-generation sequencing technology. Trends Genet 30:418–426

Van Eijk E, Wittekoek B, Kuijper EJ, Smits WK (2017) DNA replication proteins as potential targets for antimicrobials in drug-resistant bacterial pathogens. J Antimicrob Chemother 72:1275–1284

Volkamer A, Kuhn D, Grombacher T, Rippmann F, Rarey M (2012) Combining global and local measures for structure-based druggability predictions. J Chem Inf Model 52:360–372

Wang Y, Yu Y, Pan B, Hao P, Li Y, Shao Z et al (2012) Optimizing hybrid assembly of next-generation sequence data from enterococcus faecium: a microbe with highly divergent genome. BMC Syst Biol 6(3):1–13

Warren RL, Sutton GG, Jones SJ, Holt RA (2007) Assembling millions of short DNA sequences using SSAKE. Bioinformatics 23:500–501

Wee Y, Bhyan SB, Liu Y, Lu J, Li X, Zhao M (2019) The bioinformatics tools for the genome assembly and analysis based on third-generation sequencing. Brief Bioinform 18:1–12

Wilbrandt J, Misof B, Panfilio KA, Niehuis O (2019) Repertoire-wide gene structure analyses: a case study comparing automatically predicted and manually annotated gene models. BMC Genomics 20:753

Wu TD, Reeder J, Lawrence M, Becker G, Brauer MJ (2016) GMAP and GSNAP for genomic sequence alignment: enhancements to speed, accuracy, and functionality. In: Statistical genomics. Humana Press, New York NY, pp 283–334

Yang LA, Chang YJ, Chen SH, Lin CY, Ho JM (2019) SQUAT: a sequencing quality assessment tool for data quality assessments of genome assemblies. BMC Genomics 19:1–12

Ye C, Hill CM, Wu S, Ruan J, Ma ZS (2016) DBG2OLC: efficient assembly of large genomes using long erroneous reads of the third-generation sequencing technologies. Sci Rep 6:1–9

Zainab B, Ayaz Z, Alwahibi MS, Khan S, Rizwana H, Soliman DW et al (2020) In-silico elucidation of *Moringa oleifera* phytochemicals against diabetes mellitus. Saudi J Biol Sci 27:2299–2307

Zofou D, Tematio EL, Ntie-Kang F, Tene M, Ngemenya MN, Tane P (2013) New antimalarial hits from Dacryodes edulis (Burseraceae)-part I: isolation, in vitro activity, in silico "drug-likeness" and pharmacokinetic profiles. PLoS One 8:79544

Recent Advances in Nanotechnological Approaches to Enhance the Industrial Application of Essential Oils and Their Application in Food Packaging

14

Tarsila Rodrigues Arruda, Clara Suprani Marques, Patrícia Fontes Pinheiro, Taíla Veloso de Oliveira, Bruno Ricardo de Castro Leite Júnior, Patrícia Campos Bernardes, Allan Robledo Fialho e Moraes, and Nilda de Fátima Ferreira Soares

Abstract

Essential oils (EOs) are lipophilic volatile compounds synthesized by plants and used for centuries in several areas, including medicine and cosmetic. Currently, they have gained great visibility in the food industry, including their application in food packaging development. However, EOs present inherent characteristics that may limit their application in food products/packaging, especially triggered by the nature of their constituents (e.g., strong sensorial aspects, high sensibility to external factors: temperature, and oxygen). Thus, efforts have been made in the search for technologies to help circumvent these drawbacks. Among these alternatives, nanotechnology has been highlighted as an important one. Studies have shown that there are some promising nanotechnological approaches to

T. R. Arruda (✉) · T. V. de Oliveira · B. R. de C. Leite Júnior · N. de F. F. Soares
Department of Food Technology, Federal University of Viçosa, Viçosa, Minas Gerais, Brazil
e-mail: tarsila.arruda@ufv.br

C. S. Marques
Federal Institute of Minas Gerais, Campus Bambuí, Bambuí, Minas Gerais, Brazil

P. F. Pinheiro
Department of Chemistry, Federal University of Viçosa, Viçosa, Minas Gerais, Brazil

P. C. Bernardes
Department of Food Engineering, Center of Agrarian Sciences and Engineering, Federal University of Espírito Santo, Alegre, Espírito Santo, Brazil

A. R. F. e Moraes
Agricultural Science Institute, Federal University of Viçosa, Rio Paranaiba, Minas Gerais, Brazil

B. Prakash et al. (eds.), *Plant Essential Oils*,
https://doi.org/10.1007/978-981-99-4370-8_14

improve the application of EOs in food and food packaging. This chapter describes the main aspects circumventing EOs' hole in the food industry, especially when used as an active additive in food packaging, and how nanotechnology can be an important tool to increase their application in this area. Some examples of nanotechniques are presented and discussed in order to present an overview of this important and current topic.

Keywords

Essential oils · Volatile compounds · Nanotechnology · Food packaging · Food industry

14.1 Introduction

Essential oils (EOs) are secondary metabolites synthesized by plants, mainly composed of lipophilic volatile compounds. They have been employed over centuries mainly for pharmacological, medicinal, aromatic, or cosmetic purposes (Carpena et al. 2021). Nevertheless, since the nineteenth century, their application has been extended, attending great areas in the nutrition field. Several studies have already covered the bioactive properties of EOs, such as antibacterial, antifungal, antioxidant, antiviral, antiparasitic, and insecticidal, broadening the application spectrum of these complex mixtures (Viuda-Martos et al. 2011; Galié et al. 2018; Marsin et al. 2020; Sharma et al. 2022).

Triggered by the current consume prospects that seek healthier diets and natural consumption, EOs have been widely applied by the food industry not only as a flavoring ingredient (i.e., due to their strong sensorial characteristics) but also as a food preservative (Pisoschi et al. 2018; Zanetti et al. 2018; El-Sayed and El-Sayed 2021). Furthermore, EOs have evidenced great results in active food packaging (e.g., antioxidant and antimicrobial activities along with the enhancement of water vapor barrier of the active films) (de Andrade et al. 2020; Scaffaro et al. 2020).

However, some limitations of EOs, discussed in the following sections of this chapter, may restrict their application in food packaging systems, especially considering the large-scale produce. Hence, some strategies have been adopted to improve the applicability of EOs by circumventing these drawbacks and, in some cases, enhancing their properties and bioactivity. One great example of this strategy is nanotechnology. Some EO nano-systems are already well established, while others still need further investigation (Zanetti et al. 2018; Das et al. 2021).

This chapter aims to provide a comprehensive approach to EOs background in the food industry and present the aspects surrounding how nanotechnology has been applied to EOs to improve their applicability in food packaging. The main nanosystems (nanoemulsions, nanocapsules, nanoliposomes, nanocubosomes, and others) are listed and discussed along with examples of successful application in food packaging materials. Safety and toxicological concerns are also assessed in order to provide an entire overview of the topic.

14.2 Essential Oils

Records have revealed that EOs play a remarkable role in traditional medicine (Prakash et al. 2018a). In food industry, as a food additive, the application of EOs as a natural, safe, and effective alternative to the use of commercial synthetic chemicals has increased (Reyes-Jurado et al. 2019, Ni et al. 2021; Rehman et al. 2020; El Khetabi et al. 2022), due to their antimicrobial, coloring, seasoning, and antioxidant properties (Calo et al. 2015; Prakash et al. 2018a). Furthermore, EOs reduce spoilage of crops and foods with shelf-life extension, as well as control weed growth (Abd-ElGawad et al. 2020; Senthil-Nathan 2020; Walia and Kumar 2021; El Khetabi et al. 2022). Figure 14.1 illustrates the main applications of EOs in food processing and packaging.

In general, EOs are natural volatile liquids composed of a complex mixture of various secondary metabolites such as terpenes, terpenoids, phenylpropenes, aldehydes, ketones, ethers, alcohols, and phenolic compounds (Prakash et al. 2018a, b; Ni et al. 2021; Vianna et al. 2021). Most of these substances are simple lipophilic compounds characterized by a strong aroma that can be obtained from plant materials such as flowers, buds, leaves, seeds, bark, and stems (Prakash et al. 2018a; Reyes-Jurado et al. 2019; Doost et al. 2020; Ni et al. 2021; Arruda et al. 2022; Teixeira et al. 2022).

As a complex mixture, these metabolites have been explored concerning their biological properties such as antimicrobial and antioxidant activity. In this context, EOs can inactivate pathogenic microorganisms without promoting the acquisition of resistance (Prakash et al. 2014; Prakash et al. 2018a). In addition, most EOs have low mammalian toxicity, which makes them relatively safe and due to their relatively low cost, they play an important role in the food, pharmaceutical, and cosmetic industries (Calo et al. 2015; Prakash and Kiran 2016; Prakash et al. 2018a; Vianna et al. 2021).

Some examples of EOs categorized as Generally Recognised as Safe (GRAS) by the U.S. Food and Drug Administration (U.S. Code of Federal Regulations 2016) include jasmine (*Jasminum officinale* L.), cumin (*Cuminum cyminum* L.), coriander (*Coriandrum sativum* L.), ylang-ylang (*Cananga odorata* Hook. f. and Thoms.), basil (*Ocimum basilicum* L.), origanum (*Origanum* spp.), winter savory (*Satureia montana* L.), chamomile flowers (*Matricaria chamomilla* L.), peppermint (*Mentha piperita* L.), and pimenta leaf (*Pimenta officinalis* Lindl.); in which as main active components are highlighted linalool, benzyl acetate, linalyl acetate, thymol, germacrene, carvone, cinnamaldehyde, eugenol, menthol, vanillin, citral, limonene, carvacrol, among others (Prakash et al. 2018a; Vianna et al. 2021; El Khetabi et al. 2022).

EOs can be extracted by hydrodistillation, steam and water distillation, solvent extraction, or supercritical fluid extraction (Shukla et al. 2019; Kaya et al. 2020; Ni et al. 2021; Teixeira et al. 2022). The composition and quality of EOs can be affected by numerous factors inherent to the plant, such as botanical source, geographical location, stage of development, climatic conditions, growing season, stage of ripeness, and harvest (Sabo and Knezevic 2019; Arruda et al. 2022). In addition, the

Fig. 14.1 Applications of essential oils in food processing and packaging

extraction method and storage conditions can strongly affect the stability of EOs (Vianna et al. 2021).

In this sense, the variability in the composition of EOs can confer different biological properties and directly affect their antimicrobial and antioxidant potential (Arruda et al. 2022; Teixeira et al. 2022). Therefore, knowing the mechanism of action of each EO is essential to define which compound will be used depending on the target microorganism and the food matrix used (Arruda et al. 2022; Teixeira et al. 2022).

14.2.1 Bioactivity of EOs: Mechanism of Antimicrobial and Antioxidant Activities

The main compounds associated with EOs' antimicrobial activity are described as low molecular weight substances, most of them are aldehydes, terpenes, and terpenoids, with aromatic or aliphatic carbon chains (Ju et al. 2019; Arruda et al. 2022; Teixeira et al. 2022). The antimicrobial activity of these complex mixtures has already been verified in a wide spectrum of microorganisms, including Gram-positive and Gram-negative bacteria, yeasts, and molds (Nionelli et al. 2018; Arruda et al. 2022).

The mechanisms of action of EOs are directly associated with the chemical structure of their components (Ju et al. 2019). In general, these compounds can damage the structure and affect the function of the bacterial cell membrane (Ni et al. 2021). In addition, EOs may have other pathways of action, such as cell wall damage, inhibition of peptidoglycan and ATP synthesis, mitochondrial disruption, ribosomal dysfunction, DNA damage, and inhibition of gene expression (Ju et al. 2019; Ni et al. 2021; Arruda et al. 2022). Furthermore, EOs may exert interesting effects on virulence, motility, sporulation, and biofilm formation factors of microorganisms (Sabo and Knezevic 2019; Các̃eres et al. 2020; Moradi et al. 2020; Arruda et al. 2022).

In parallel, EOs are excellent sources of natural antioxidants (Simionato et al. 2019; Bastos et al. 2020). The antioxidant mechanism of each fraction strongly depends on its chemical composition (Arruda et al. 2022). Mainly, the antioxidant activity of EOs is associated with biologically active compounds such as terpenoids and phenolic acids (Thokchom et al. 2020), acting by donating hydrogen to highly reactive compounds, preventing the formation of radicals (Arruda et al. 2022).

14.2.2 Essential Oils and their Prospects in Food Packaging

EOs present some challenges, such as low water solubility, strong lipophilicity, and high volatility (Prakash and Kiran 2016; Prakash et al. 2018a; Ju et al. 2019). Furthermore, they lack photothermal resistance and are prone to oxidative decomposition, which limits their direct application in food systems (Ju et al. 2019). Furthermore, EOs have several intrinsic challenges that make their application difficult as food preservatives. For example, scarcity of raw materials, chemotypic variation, low solubility in water, low stability during storage, and the threat of biodiversity loss are some of the main challenges to producing EOs on an industrial scale (Prakash and Kiran 2016; Prakash et al. 2018a). In parallel, the characteristics of the food matrix together with extrinsic factors such as temperature, gaseous composition, and microbial species can reduce the bioactivity of EOs (Calo et al. 2015; Prakash et al. 2018a). Therefore, improving the stability and durability of EOs has become essential to maintain biological activity, reduce volatility, and minimize adverse effects on aroma and flavor (Ju et al. 2019).

For this, EOs are being incorporated into food packaging to overcome these barriers and enhance their performance by increasing the contact area and dispersibility of these components on the food surface (where there is a higher incidence of microorganisms) (Donsì and Ferrari 2016; Ni et al. 2021). As a result, there is an increase in the shelf life of perishable foods, protecting against oxidation, vitamin loss, enzymatic browning, and microbial growth (Mir et al. 2018; Vianna et al. 2021; Arruda et al. 2022).

In recent years, there have been reports of food preservation based on EOs (directly applied into food matrix and as active additive in food packaging) obtained from fennel (*Pimpinella anisum*), cloves (*Syzygium aromaticum*), peppermint (*Mentha piperita*), oregano (*Origanum vulgare*), cinnamon (*Cinnamomum zeylanicum*), lemongrass (*Cymbopogon citratus*), rosemary (*Salvia rosmarinus*), thyme (*Thymus vulgaris*), ginger (*Zingiber officinale Roscoe*), and other plants (Tu et al. 2018; Ju et al. 2019; Satyal and Setzer 2020).

Strategically, the incorporation of EOs in packages modifies the characteristics of the packaging material and can increase its durability along with the improvement of the packaged food's shelf life (Mir et al. 2018; Alizadeh-Sani et al. 2020; Arruda et al. 2022). In addition, the combination of different EOs is an interesting strategy to obtain a synergistic effect (Arruda et al. 2022).

However, several aspects must be considered to guarantee the desired effects, including the polymer used, the composition and concentration of the EO, the desired activity, and the characteristics of the food to be packaged (Arruda et al. 2022). In specific situations, the sensorial aspects of EOs can carry a positive interpretation for consumers due to their harmony with packaged food (Moraczewski et al. 2020; Jafarzadeh et al. 2021). Nevertheless, in other cases, the presence of EOs directly incorporated into films can negatively affect consumer perception (Noori et al. 2018; Arruda et al. 2022), requiring other incorporation strategies.

In this sense, nanotechnology has been extensively studied to improve the application of EOs in food packaging to overcome limitations such as low solubility, high instability, and degradability (Bora et al. 2020; Alfei et al. 2020). In the following section, this topic is discussed in more detail.

14.3 Carrier Systems: Nanotechnological Approaches Versus Industrial Application of Essential Oils

The field of nanochemistry research has shown great progress in the development of novel nanocarriers as potential active compound delivery systems. As mentioned, nanomaterials have at least one dimension in the range between about 0.1 to 100 nm, which differs from the bulk material, resulting in novel characteristics. Nanotechnological materials have been prepared by the top-down approach, which involves the breakdown of materials, or through "bottom-up" forms in which nanostructures and nanomaterials are synthetized through the utilization of supramolecular and biomimetic materials. Both approaches have interfaces with biology and biomimetic chemistry, engendering the field of nanobiology (Steed et al. 2007).

The nanocarrier materials offer many advantages such as (1) improving pharmacokinetics and biodistribution of active agents due to a higher ratio of surface area to volume; (2) diminishing toxicity by their preferential accumulation at the target site, facilitating intracellular delivery; and (3) improving compound stabilization (Steed et al. 2007; Pardakhty and Moazeni 2013; Moghassemi and Hadjizadeh

2014). Therefore, the use of nanotechnology to develop EOs-based preservatives may increase their effectiveness in food packaging (Abdalkarim et al. 2021; Ahmed et al. 2022; Arruda et al. 2022).

There are distinct systems that can be used as carriers and their forms are important for the application as followed by the discussion.

14.3.1 Nanoemulsions

Despite the huge interest in the use of EOs by the food industry, some issues still need to be solved. To overcome those disadvantages mentioned in previous sections, the nanoemulsions (NEs) can be explored. NEs are emulsions with droplet sizes around 100 nm, in which the dispersed phase is oil, and the continuous phase is water (O/W). These two immiscible phases are kinetically stabilized by a surfactant. Some characteristics of NEs, such as high surface area per unit volume, higher kinetic stability, optically transparency, and increase of bioavailability of EOs (Safaya and Rotliwala 2020) make these nanomaterials a great choice for a lot of applications in food production, especially in food packing.

NEs can be prepared by high-energy methods, including high-pressure homogenization, ultrasound emulsification and microfluidization, or low-energy methods, such as spontaneous emulsification, phase inversion temperature, phase inversion composition, and emulsion inversion point. Each method has advantages and disadvantages. In general, low-energy methods are easy to scale up to industrial production, however, requiring high concentrations of surfactants. On the other hand, the high-energy methods are less suitable for industrial scale compared to low-energy methods and, due to their higher energy requirements, are more expensive (Singh and Pulikkal 2022).

Although some issues still need to be improved, several studies have shown great potential for the application of NEs in food packaging. Functional gelatin-based composite films incorporated with oil-in-water lavender essential oil nanoemulsion were effective for cherry tomatoes preservation. The films exhibited strong antibacterial activity against *Staphylococcus aureus*, *Escherichia coli*, and *Listeria monogytogenes*. Additionally, the films present good UV light barrier and heat-sealing properties (Sun et al. 2021).

Another study, with *Trachyspermum ammi* essential oil nanoemulsion (TOEN) loaded in alginate-based edible coating, evidenced that TOEN prevented the growth of *L. monocytogenes* in turkey fillets even after 12 days of cold storage. The study comparatively investigated TOEN in the forms of emulsion and NE against inoculated *L. monocytogenes* in turkey fillets for 12 days of storage at 4 ± 1 °C, with NE being very effective (Kazemeini et al. 2021).

Studies also have proved that NEs can be successfully applied in food coating. Carnauba NE-based coatings improved shelf life and maintained papaya fruit quality during storage by retarding firmness loss, and color changes, also reducing respiration rates, resulting in delayed ripening. Furthermore, the incorporation of ginger essential oil in the coating presented some positive effects on fungal disease control (Miranda et al. 2022).

The influence of cumin EO NE (COEN) used as a brined solution on the quality of white soft cheese was reported (El-Sayed and El-Sayed 2021). *S. aureus, Bacillus cereus, Pseudomonas aeruginosa, E. coli, L. monocytogenes, Salmonella* Typhimirum, *Yersinia enterocolitica, Aspergillus niger,* and *Aspergillus flavus* were significantly inhibited by COEN. In cheese preserved with 1% COEN, the counts of psychrotrophic, yeasts and molds were not detected after 60 days of storage.

The antimicrobial mechanism of action of EOs NE is related to the cell membrane. The high surface area of NE facilitates the interaction of cell membranes and EOs. Thyme EO NE (TEON) prepared by ultrasonication promoted membrane disintegration with massive leakage of cytoplasmatic inclusion on *E. coli* 0157:H7 and *S. aureus*. The bacterial death was associated with modifications in fatty acid composition after TEON exposure (He et al. 2022).

Frank et al. (2018) developed alginate-based films incorporated with cinnamon EO NE (CEO-NE) to assess antimicrobial potential. These films based on CEO-NE/alginate showed antibacterial activity against Gram-positive and Gram-negative bacteria and may be useful in obtaining active packaging against the proliferation of bacteria and preservation of fresh foods.

In another example, Souza et al. (2021) developed thermoplastic starch (TPS) films containing Pickering emulsions stabilized with nanocellulose of different EOs (ho wood, cardamom, and cinnamon). All obtained films showed lower crystallinity than pure TPS, high thermal stability, and lower water vapor transmission rate.

14.3.2 Polymeric Nanomaterials

14.3.2.1 Nanoparticles

Polymeric nanoparticles (PNPs) are formed by polymers, which can be of natural, semisynthetic, or synthetic origin, nontoxic, biodegradable, biocompatible, and responsible for carrying bioactive compounds. PNPs are particles with a size ranging from 1 to 1000 nm. Some of these PNPs can adsorb bioadditive compounds on the surface of the polymeric core or can carry these compounds through entrapment. The term "nanoparticle" refers to nanocapsules and nanospheres, which differ in structural organization and composition, as shown in Fig. 14.2 (Zielińska et al. 2020; Lima et al. 2022).

Nanocapsules are reservoir-type systems that have a polymeric shell arranged around an oil core. The EO (or another bioactive compound) may be contained in the core and/or adsorbed to the polymeric wall. The nanospheres are of the matrix type and do not have oil in their formulations. Despite being formed by a polymeric matrix, it is not possible to identify a differentiated nucleus. In this case, the bioactive constituents can be homogeneously dispersed or solubilized within the polymeric matrix (Singh and Lillard 2009; Jawahar and Meyyanathan 2012; Ivanova et al. 2020).

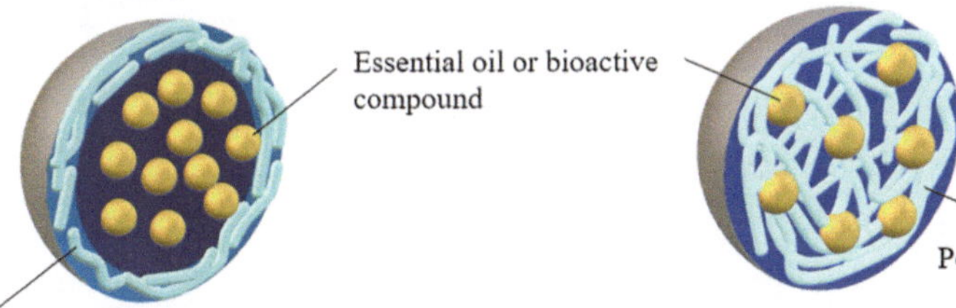

Fig. 14.2 The two main types of polymeric nanoparticles: nanocapsule and nanosphere

PNPs are used as carriers of proteins, antigens, and drugs for topical, oral, injectable administration, and bioadhesive systems, being the main component of controlled release systems in the form of implants (Pudlarz and Szemraj 2018; Lima et al. 2022). One of the main advantages of PNPs encompasses the carrying of bioactive compounds that can have controlled release. In addition, PNPs can protect bioactive constituents from the environment, resulting in better biological activity, therapeutic index, and bioavailability (Begines et al. 2020; Zielińska et al. 2020).

PNPs can be used in the area of pollution control and environmental technology, in conductive materials, sensors, biotechnology, medicine, the food industry, and others. One of the highlights of the use of PNPs is the application in active materials for food packaging, which can provide a barrier to gases, humidity, temperature, and stability (Kumar et al. 2021a; Gupta et al. 2022).

14.3.2.2 Nanofibers

Along with PNPs, nanofibers (i.e., fibers with nanometric diameter) have been widely used to provide a suitable vehicle system to deliver EOs. As well as PNPs, the nanofibers can supply naturally derived chemical compounds, such as EOs, in a controlled manner and prevent their degradation. However, the main characteristic of these 1-D structured nanomaterials is their large surface-to-volume ratio compared to other nanostructures (Partheniadis et al. 2022).

Despite the large number of procedures that can be applied to nanofibers manufacturing (e.g., freeze-drying, self-assembly, phase separation, template synthesis, and spinneret-based engineered parameters), so far, the most widely applied to bioactives' delivery is electrospinning. This technique is based on the application of a strong electric field, with high voltages being used during the nanofibers produce (c.a.,1–30 kV) (Hu et al. 2014). The core material (EO) is located inner the nanofiber, covered by the shell material (polymer), but the dispersion of EO on the nanofiber depends on which generation method is employed during the fibers' manufacturing: nanoemulsion, conventional, and coaxial electrospinning (Partheniadis et al. 2022) (Fig. 14.3).

Furthermore, considering nanofibers produced from electrospinning, some advantages may be acquired, such as (1) it can produce very thin fibers (of few nanometers in diameter) with large specific surface areas (high surface-to-volume

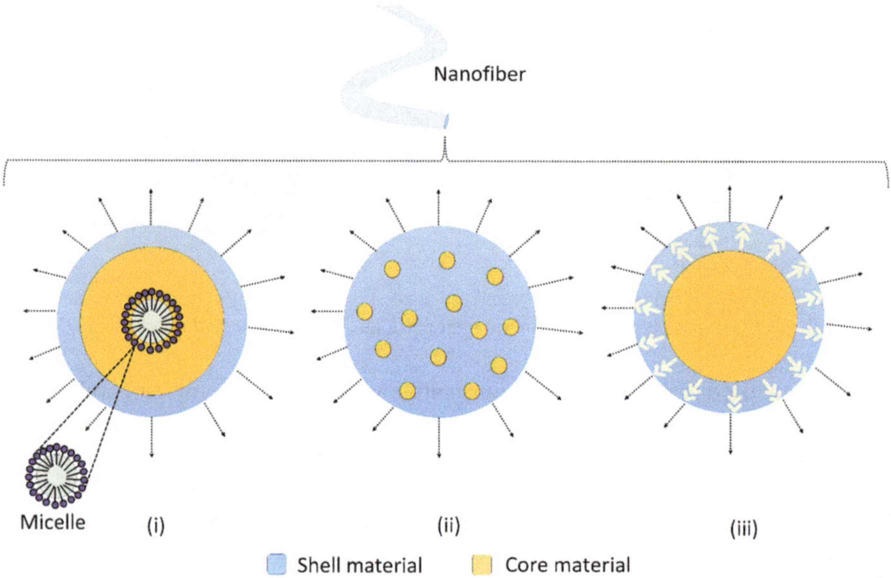

Fig. 14.3 Schematic illustration of shell and core materials' disposition into nanofibers, depending on electrospinning procedure: (1) nanoemulsion, (2) conventional and (3) coaxial

ratio) and large interstitial spaces (voids), (2) it can provide ease of functionalization, (3) the produced electrospun nanofibers exhibit superior mechanical properties, (4) it is a versatile process, and (5) it is amenable for scale-up (Partheniadis et al. 2020); thus, several studies have been covering the EOs encapsulation by electrospun nanofibers (Rather et al. 2021).

14.3.2.3 Natural Biopolymers Used in Polymeric Nanomaterials Development

Polimeric materials can be employed as the package material perself or as the suport material in nanocarriers. The use of synthetic polymers in food packaging is very practical and cost-effective. The most used are polyethylene, polypropylene, polyvinyl chloride, polyethylene terephthalate, polyurethane, polyamides, etc. These materials have excellent properties concerning chemical and physical resistance, sealing qualities, mechanical properties, lightness, and durability. However, most of the plastic is discarded and can already be found in a widespread way: in the soil, in the oceans, in drinking water, in the air, and in the bodies of animals and human beings (Nielsen et al. 2020). Worldwide, only 18% of synthetic polymers are recycled, 24% are incinerated and the rest, 58%, enter the natural environment or are landfilled, potentially accumulating with long persistence (Geyer et al. 2017; Chamas et al. 2020). Thus, employing natural materials in the development of polymeric nanocarrier systems may act as a great strategy, especially concerning food packaging development.

The main biodegradable polymers of natural origin that are currently applied at nanozised carriers to EOs are cellulose, chitosan, alginate, carrageenans, agar, starch, proteins, and pectins. We can also cite other biopolymers that have been currently studied on the production of PNPs, such as PLA (polylactic acid), PHA (polyhydroxyalkanoate), PCL (poly-ε-caprolactone), PGA (polyglycolide), PVA (poly vinyl alcohol), and PBS (polybutylene succinate). The advantages of these polymers are that they are easily hydrolyzed by enzymes and are environmentally friendly (Teixeira-Costa and Andrade 2022; Zheng et al. 2022).

Cellulose and Derivatives

Cellulose is the most abundant and renewable biopolymer found in nature, which plays a structural role in the cell walls of higher plants, aquatic species, and microorganisms (e.g., bacteria, fungi, and algae) (Favier et al. 1995; Andriani et al. 2020). Cellulose can be extracted from materials such as wood, cotton, straw, and hemp. This polymer is a polysaccharide formed by linear D-glucose chains joined by β(1–4) bonds, which do not exist in the form of molecules in nature but crystallize due to intermolecular hydrogen bonds (Klemm et al. 2005).

Cellulose-based packaging presents a barrier to water vapor, air, and oxygen. Hydrogen bonds within and between cellulose fibers are considered to be an important factor in determining their chemical and physical properties. Furthermore, some derivatives with greater solubility and processability can be obtained from cellulose (Liu et al. 2021). For example, cellulose acetate is one of the most widely used cellulose derivatives, being obtained from the reaction of cellulose with acetic acid or acetic anhydride, catalyzed by sulfuric or perchloric acid, resulting in a molecule with one or more -OH groups of glucose replaced by an acetate group (Fig. 14.4).

Cellulose acetate can be dissolved in acetone and has been used as a food packaging material, although it has a relatively high permeability to water and oxygen. In addition to forming films, cellulose can also be processed into nanocrystals and nanofibrils, which are widely used as fillers in other polymers. Cellulose nanocrystals are obtained by acid or enzymatic hydrolysis and cellulose nanofibers are obtained by mechanical homogenization of pure cellulose fibers, which have previously been chemically or enzymatically treated. Cellulose nanofibers have properties of some fluids and gels, being considered a pseudo-plastic (Romão et al. 2022).

Bacterial nanocellulose is a β-D-glucopyranose-type biopolymer produced by specific types of bacteria *(Agrobacterium, Rhodobacter, Komagataeibacter, Alcaligenes, Achromobacter, Aerobacter, Pseudomonas, Rhizobium, Sarcina*, and

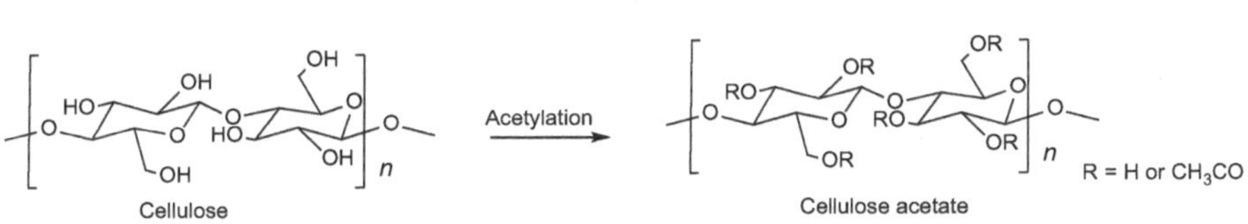

Fig. 14.4 Scheme of the cellulose acetylation reaction to obtain cellulose acetate

Dickeya). The material is secreted extracellularly as a network of randomly distributed nanofibers, linked in a highly cross-linked three-dimensional fashion with a unique, porous matrix. Compared to plant cellulose, the advantages of bacterial nanocellulose are: they are physically and mechanically more resistant, purity, they are produced free of hemicelluloses and lignin, they can be shaped during cultivation into three-dimensional structures, they have a high degree of crystallinity, high porosity, excellent water retention capacity and biocompatibility. Thus, it is a material of interest not only to the food industry but also to several other sectors (e.g., electronics, pharmacy, medicine, cosmetics, and textiles) (Sharma and Bhardwaj 2019; Ludwicka et al. 2020).

Cellulose has great sensorial advantages (e.g., non-pronounced sensorial aspects, such as aroma and color, good appearance) that reflect on their applicability to food packaging development, being also a great option to carry antimicrobials and antioxidants, for example, by nanoencapsulating bioactive compounds for their controlled release in the matrix (Gupta et al. 2022; Silva et al. 2020).

Razavi et al. (2022) used cinnamon EO (CEO) nanocapsules at concentrations of 0.03–0.48% (v/w), using fish gelatin as the main polymeric phase (3% w/w), stabilized using bacterial cellulose nanocrystals at 0.06% (w/w). The obtained film effectively provided a great barrier against water vapor, oxygen, and carbon dioxide, also showing interesting potential for increasing the shelf life of mollusks, crustaceans, vegetables, and fruits.

Chitosan

Chitosan is one of the most studied biopolymers with antimicrobial properties for food packaging, due to its ability to form films and coatings. Chitosan exhibits antimicrobial activity against foodborne bacteria, yeasts, and a wide variety of filamentous fungi, being more active against yeasts (Díaz-Montes and Castro-Muñoz 2021).

Chitosan is obtained from chitin by a partial deacetylation reaction (Fig. 14.5). Chitin, the second most abundant polysaccharide in nature after cellulose, is found mainly in the exoskeleton of crustaceans, crabs, shrimp shells, and insects. The structure of chitin is similar to that of cellulose, with the difference being the presence of the acetamide group ($-NHCOCH_3$), located on carbon 2, meanwhile, in cellulose, there is a hydroxyl (-OH) in this position. Chitin is composed of (1,4)-β-2-acetamido-2-deoxy-D-glucopyranose units and small amounts of (1,4)-β-2-amino-2-deoxy-D-glucopyranose units, and chitosan is composed of

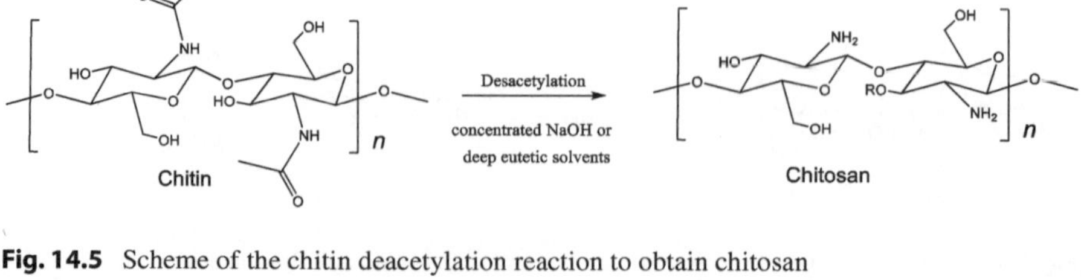

Fig. 14.5 Scheme of the chitin deacetylation reaction to obtain chitosan

N-acetyl-D-glucosamine and deacetylated D-glucosamine units (Jin et al. 2021; Kozma et al. 2022).

Chitin is insoluble in many organic and inorganic solvents, while chitosan is soluble in slightly acidic solutions, such as formic, acetic, malic, citric, tartaric, or lactic acid, among others, due to the presence of the amino group (-NH_2) located in the C-2 from D-glucosamine (Melro et al. 2020). This NH_2 group can be protonated in an acid medium, promoting the polycationic behavior of chitosan chains, which can influence its physical-chemical, mechanical, and biological properties. Furthermore, the polycationic chain of chitosan can electrostatically interact with polyanionic biopolymers, such as sodium alginate, and amphiphilic proteins. The protonation of amino groups provides the interaction of chitosan with microorganisms' cell membranes that are mainly negatively charged, altering their permeability and metabolism, chelating trace metals and spores of microorganisms, thus leading to cell death and inhibiting microbial growth (Li et al. 2020).

Chitosan can be chemically modified, due to the presence of -OH and -NH2 groups, with the introduction of other functional groups that aim to enhance its solubility and antimicrobial activity (Aljawish et al. 2015).

Gasti et al. (2022) developed chitosan/pullulan (CS/PL) nanocomposite films loaded with clove EO (CEO):chitosan-ZnO hybrid PNPs. Incorporation of the active PNPs led to higher film hydrophobicity, higher oxygen barrier, higher UV light blocking ability, higher tensile strength, and higher water vapor barrier compared to pure CS/PL film by the inclusion of PNPs. There was an increase in antioxidant activity and relevant antibacterial activity against *P. aeruginosa*, *S. aureus* and *E. coli*. Additionally, the obtained films showed potential to poultry meat's shelf life extension by up to 5 days when stored at 8 ± 2 °C.

Ferreira et al. (2021) prepared thermoplastic starch (TPS) films filled with chitosan PNPs containing ho wood and cinnamon EOs. The films containing 1 and 3% w/w of active PNPs showed lower permeability to water. At a concentration of 3% w/w of PNPs, there was inhibition of the bacterial growth of *E. coli* and *Bacillus subtillis*. Applying the active films on the shelf life extension of strawberries, films containing 1 or 3% w/w loaded-PNPs were responsible to inhibit fungal growth, also preventing fruits' weight loss.

Alginates

Alginate is a biopolymer extracted from brown algae such as *Ascophyllum nodosum*, *Laminaria digitata*, and *Macrocystis pyrifera*. In the food industry, alginate is often used as a gelling agent, thickener, and stabilizer (Abka-Khajouei et al. 2022). Alginate is a polysaccharide composed of molecules of β-D-mannuronic acid and α-L-guluronic acid linked by 1,4-bonds, which can be in the form of β-D-mannuronate (M) and α-L- guluronate (G) (Abasalizadeh et al. 2020). These units are randomly distributed in a linear chain, formed by links between M-G, M-M and G-G (Fig. 14.6).

The properties of alginate are influenced by changes in pH, due to the presence of carboxylic groups in the structure at pHs lower than 3.4, that is, below the pKas of M (3.38) and G (3.65) the carboxylic groups become non-ionized (-COOH), so

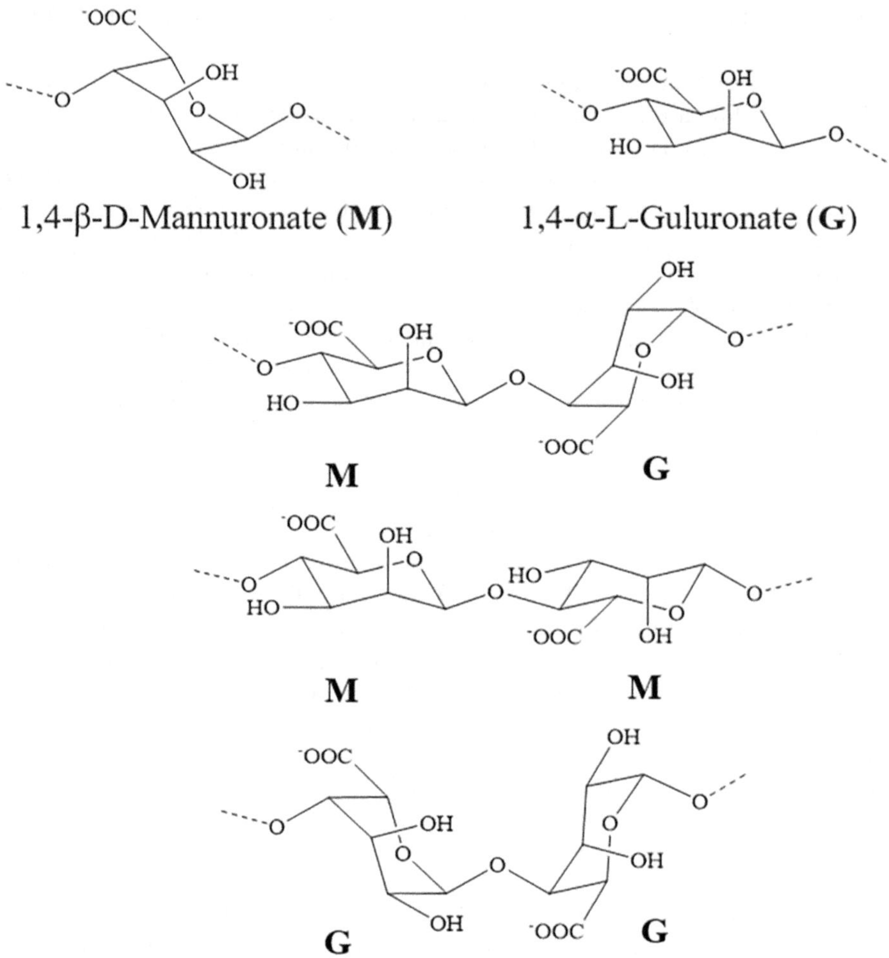

Fig. 14.6 Structures of alginates monomers: 1,4-β-D-Mannuronate (M) and 1,4-α-L-Guluronate (G) and the possible bonds between them forming M-G, M-M and G-G

the alginate is insoluble. In pH conditions greater than 4.4, the carboxylic groups become ionized (-COO$^-$) and the expansion of the alginate chain occurs due to the electrostatic repulsion of negative charges, with the apex at higher pH values, in this case, pH above 7.4 (Agüero et al. 2017; Teixeira-Costa and Andrade 2022).

Alginate can absorb cationic ions, due to its anionic nature, and can therefore be used to remove pollutants in an aqueous solution. In addition, alginate can absorb dyes and can be useful in the food packaging industry, which can prevent food contact with dyes (Zhang et al. 2022b; ALSamman and Sánchez 2022). This interesting property also allows alginate to be a potential material in the development of PNPs, especially along with other polymeric material. Some examples of alginate-based PNPs are alginate/cashew gum nanoparticles for *Lippia sidoides* EO encapsulation (de Oliveira et al. 2014) and chitosan/alginate nanocapsules as carrier materials for turmeric (*Curcuma longa*) and lemongrass (*Cymbopogon citratus*) EOs (Natrajan et al. 2015).

Carrageenans and Agar

Carrageenans are anionic sulfated linear polysaccharides obtained from red marine algae of the Rhodophyceae family. Carrageenans contain alternating units of 1,3-β-D-galactopyranose and 1,4-α-D-galactopyranose (Chauhan and Saxena 2016; Lee 2020). The most used types in the industry are the κ-(kappa), ι-(iota), and λ-(lambda) carrageenans (Fig. 14.7), this classification is according to their molecular structures, such as the sulphation arrangements, as well as the presence or absence of 3,6-anhydro-α-galactopyranose into the 1,4-linked α-galactopyranose unit (Lin et al. 2022a; Rupert et al. 2022).

The κ-(kappa) and ι-(iota) carrageenans present 25–30% and 28–30% of sulfate content, respectively. These carrageenans can be used as a design for edible films and coatings, due to their high hydrophilicity, water solubility, and structural ability to form a helical assembly (Jancikova et al. 2020; Teixeira-Costa and Andrade 2022), but also present great prospects on the development of natural PNPs.

Fani et al. (2022) designed κ-carrageenan (κC) PNPs containing the main constituent of citrus EO, D-limonene, performing one-step electrospray synthesis. The D-limonene-κC PNPs of spherical morphology showed excellent levels of thermostability and photostability. Toxicological studies of the material must be carried out

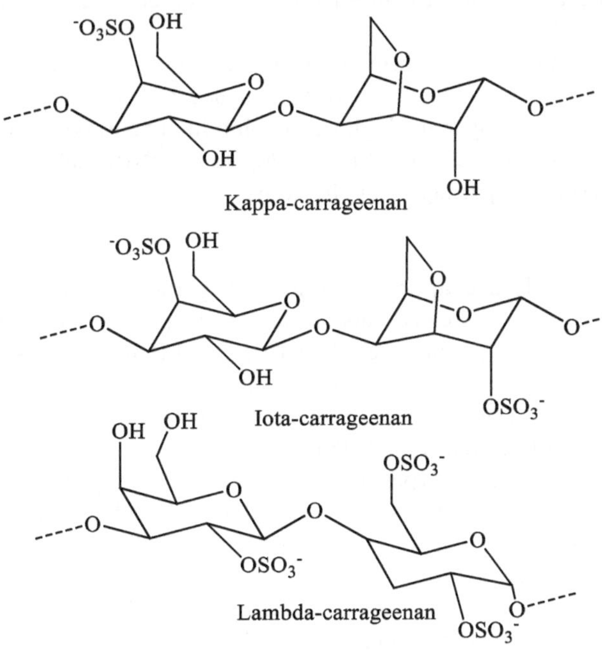

Fig. 14.7 Structures of κ-(kappa), ι-(iota) and λ-(lambda) carrageenans

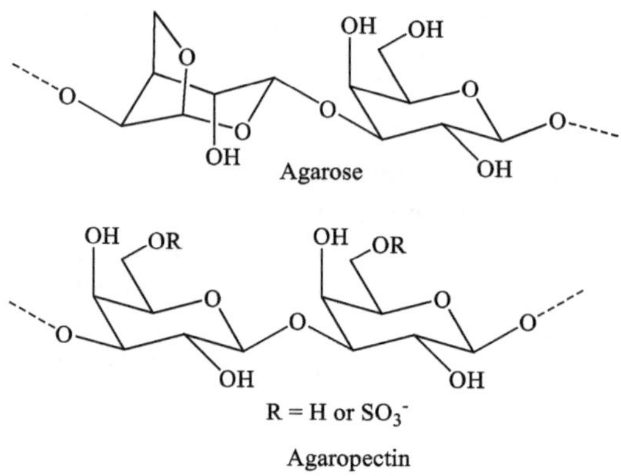

Fig. 14.8 Structures of agar components

to verify its potential application in the release of bioactive molecules for drugs or foods.

On the other hand, agar is a gelatinous polysaccharide derived from red algae, of the class Rhodophyceae, the main sources of this biopolymer are *Gracilaria* sp. and *Gelidium* sp. Agar has two components in its composition: agarose and agaropectin, which structures are shown in Fig. 14.8.

Agarose is the gelling portion of agar, which is neutral and is made up of disaccharide units formed by galactose and 3,6-anhydro-α-L-galactopyranose. Agaropectin is the non-gelling portion, which contains a sulfate group, D-glucuronic acid and pyruvic acid in its structure (Martínez-Sanz et al. 2019; Mostafavi and Zaeim 2020).

Carrageenans and agar have been used to make biodegradable packaging films (i.e., together, these materials can improve their negative aspects, such as the low mechanical properties of agar and high water solubility of carrageenan). Their properties can be also successfully used to develop EOs' PNPs, helping to overcome EOs limitations on food packaging applications.

Starch

Starch is a natural polysaccharide used as a carbohydrate reserve in various plants, such as rice, potatoes, soybeans, oats, corn, sweet potatoes, yams, barley, sago, wheat, and vegetables (Horstmann et al. 2017). The different origins of starch and environmental factors during raw material cultivation can lead to differences in its crystallinity, which will significantly influence its processing properties (Dos Santos et al. 2016; Cornejo-Ramírez et al. 2018).

Starch is composed of two types of polysaccharides: amylopectin (75–80%) and amylose (20–25%) (Fig. 14.9). Amylopectin has α-1,4-glycosidic linkages, and α-1,6-glycosidic linkages, which form highly branched polymers with 24 to 30 glucose residues. Amylose is composed of end-to-end connected α-1,4-glycosidic bonds, which is an unbranched helical structure (Alcázar-Alay and Meireles 2015). This biopolymer is of great importance in the biotechnological and food industries, having as its great advantages: great abundance, low cost, worldwide availability,

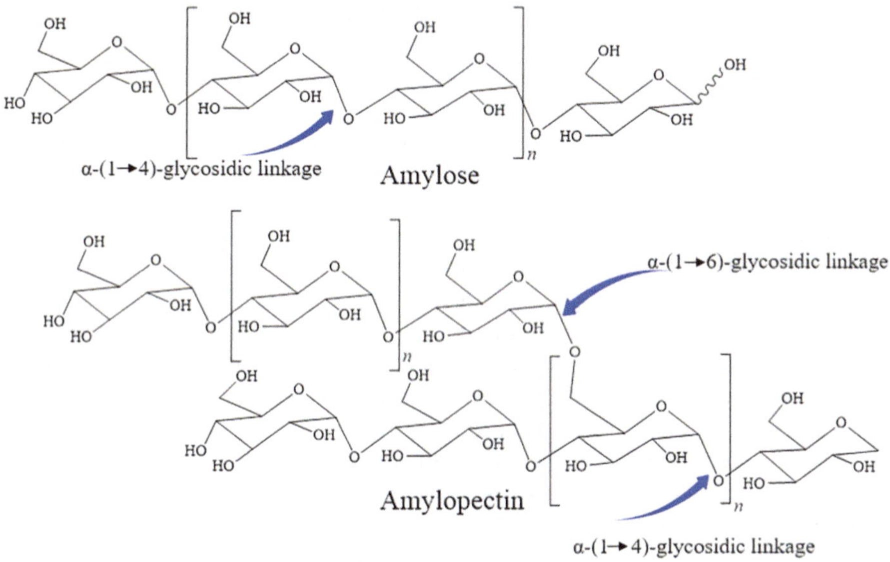

Fig. 14.9 Structures of starch components

biodegradability, good processability in conventional plastic processing equipment, and edibility (García-Guzmán et al. 2022).

The high amylose content attributes better properties to starches in terms of strength, elongation capacity, and flow properties. One of the disadvantages of films made from starch is that they are generally brittle due to retrogradation, a phenomenon that occurs when amylose in the solution can crystallize forming two left-handed helices, which are packaged into amorphous types A and B, in this case, on cooling the starch reorganizes into a more crystalline structure (Liu et al. 2022).

To circumvent the problem of retrogradation, mixing starch with water-soluble polymers such as PVA, or modifying and cross-linking the –OH groups can effectively suppress retrogradation, resulting in starch-based polymers that are useful for packaging applications (García-Guzmán et al. 2022).

Starch molecules are highly hydrophilic as they contain many hydroxyl groups, which results in poor water resistance and poor mechanical properties in humid environments. In this case, it is necessary to strengthen the mechanical and barrier properties of starch-based packaging materials (Liu et al. 2022).

One of the strategies is to convert natural starch into thermoplastic starch (TPS) by treatment with plasticizers (glucose, amino acids, urea, glycerol, etc.) under heat and/or shear. Still, TPS has lower mechanical and barrier properties than conventional petroleum-based plastics (Lopez-Gil et al. 2014). The TPS has to be reinforced to obtain homogeneous properties for packaging, which can be done by mixing the surface, modifying the presentation, cross-linking, acetylation, organic and inorganic fillers, and mixing with other polymers. (PVA, PLA, PE, PP, PS, and others) (Mohammadi Nafchi et al. 2013; Tarique et al. 2021).

Wang et al. (2022) developed a film obtained from starch/polyvinyl alcohol (STA/PVA) loaded with oregano essential oil (OEO) adsorbed on microporous starch (MS) to be used in sea bass packaging. The films with the incorporation of OEO-MS obtained better tensile strength, lower permeability to water vapor and oxygen, and exhibited antimicrobial and antioxidant activities. Using such film extended the shelf life and better preserved the freshness of the sea bass.

Proteins

Natural proteins can be used in food packaging development. Protein-based biopolymers have low oxygen permeability and good mechanical properties, which may be superior to those presented by biopolymers based on polysaccharides and lipid compounds (Hammam 2019; Kumar et al. 2021a). Along with this usage as the polymer matrix in food packaging manufacturing, proteins also have been studied as shell material in PNPs carriers for EOs.

Proteins naturally exist as fibrous or globular, the former serves as a structural material for animal tissues and are insoluble in water. Globular proteins are present in living systems, being soluble in water or aqueous solutions of acids, bases, and salts. Fibrous proteins are shown to be extended and with chains associated with each other by hydrogen bonding to form fibers. Globular proteins fold into complex spherical structures held together by hydrogen bonds, ionic, hydrophobic, and covalent forces (disulfide bonds) (Basiak et al. 2017; Senthilkumaran et al. 2022)

The physical and chemical properties of these proteins depend on the location and amount of amino acid residues along the polypeptide chain. Collagen is a fibrous protein that can be used in the production of edible films. Among the globular proteins, soy protein, whey protein, mung bean protein, wheat gluten, and corn zein have been studied for use in food packaging (Kumar et al. 2022).

Proteins are not completely hydrophobic, due to the predominantly hydrophilic amino acid residues, there is a limitation in their moisture barrier property. For example, to obtain edible films of proteins with low permeability to water vapor, it is necessary to add hydrophobic constituents (Huo et al. 2018; Mihalca et al. 2021; Chaudhary et al. 2022).

Gelatin is a high molecular weight polypeptide (composed of 19 amino acids) obtained from the hydrolysis of collagen. Gelatin has a high content of proline, hydroxyproline, and glycine, having a mixture of unfolded single and double chains with a hydrophilic character (Fig. 14.10). Gelatin has excellent film-forming properties, reasonable barrier and mechanical properties, and high flexibility. One of the great advantages of film gelatine is that it intrinsically has antimicrobial and antioxidant activity (Ramos et al. 2016; Ma et al. 2018; Lu et al. 2022).

The limitation of the use of gelatin as a film is the issue of its high sensitivity to humidity, in contact with water, they swell, crack and dissolve. To improve the water barrier of gelatin-based materials, it is common to mix them with other materials, such as fibers and other polymers (Said and Sarbon 2022).

For example, Angourani et al. (2022) successfully developed plant-protein-based PNPs containing Rosemary (Rosmarinus officinalis) EO, which can be applied not only in food packaging but also in other industry sectors, such as the

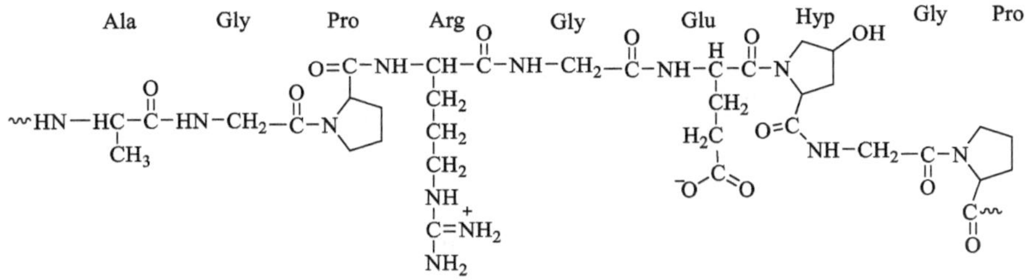

Fig. 14.10 Gelatin structure

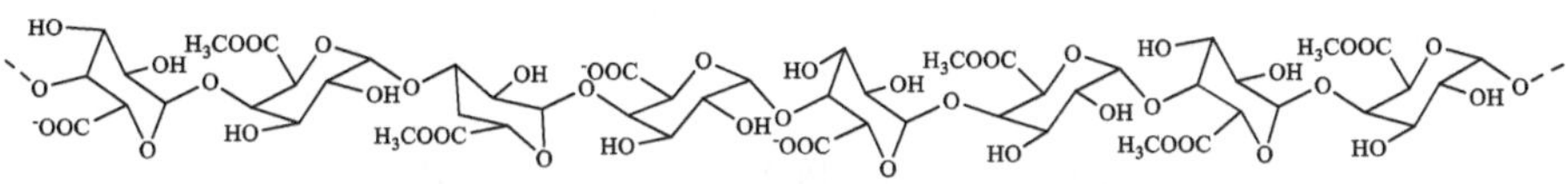

Fig. 14.11 Structure of a pectin molecule. Residues of rhamnose, galactose, arabinose, and xylose are not included. Adapted from Alkorta et al. (1998)

pharmaceutical. Furthermore, proteins can also be applied in nanofibers. Using the electrospinning technique, a study was carried out with gelatin nanofibers with encapsulated angelica essential oil (AEO) in order to obtain a material with the potential to be used in active packaging. With the addition of AEO, there was an improvement in hydrophobicity, an increase in antioxidant activity and an antibacterial effect against Gram-positive and Gram-negative bacteria. In addition, there was no effect of cytotoxicity (Zhou et al. 2020).

Pectin

Pectin is a natural and abundant heteropolysaccharide, present in plant cell walls, composed of β-(1–4)-D-galacturonic acid, galactose, arabinose, and rhamnose (Fig. 14.11). The pectin chain can have different degrees of methyl esterification, an important factor in its film-forming and gelling properties (Ravishankar et al. 2012; Chandel et al. 2022).

Pectin has excellent gelling ability in acidic solutions, which is why it is widely used as an additive in the beverage and food industry. Pectin is used in ice cream, dairy drinks, and jellies as a thickening agent or colloidal stabilizer. Pectin can be used in the production of films for food packaging, as it is edible, can form gels and films, and is biocompatible and biodegradable (Jiang et al. 2021; Freitas et al. 2021).

The use of pectin in packaging has major limitations, as this biopolymer is highly hydrophilic, fragile, and has low tensile strength. To overcome these obstacles, other biopolymers must be added to the pectin matrix to improve its mechanical properties and reduce its flexibility (Almasi et al. 2020; Huang et al. 2021). For example, D-limonene (i.e., an important EO's component) was successfully nanoencapsulated by pectin and whey protein concentrate (Ghasemi et al. 2018; Rehman et al. 2019). The authors investigated different pectin (0.5, 0.75, 1%) and whey protein concentration (4, 6, 8%), in 3 distinct pH values (3, 6, 9), determining the

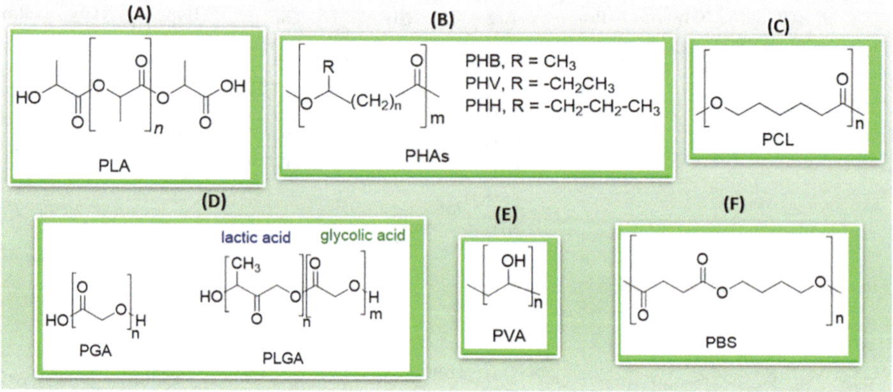

Fig. 14.12 Structures of biopolymers. (**a**): PLA (polylactic acid); (**b**): PHAs (polyhydroxyalkanoates) - poly(3-hydroxybutyrate) (PHB), polyhydroxyvalerate (PHV) and polyhydroxyhexanoate (PHH); (**c**): PCL (polycaprolactone); (**d**): polyglycolic acid (PGA) and poly(lactic acid-co-glycolic acid) (PLGA); (**e**): polyvinyl alcohol (PVA); and (**f**): poly(butylene succinate) (PBS)

stability of the product. The results indicated that optimum formulation regarding stability, color, and viscosity was obtained with 1% pectin and 4% WPC, with pH 3.

Other Biopolimers

Some relatively novel bioplastics have been used in food packaging development. These materials have been developed to obtain sustainable packages and replace synthetic polymers. Examples of these polymers include polylactic acid (PLA), polyhydroxyalkanoate (PHA), polycaprolactone (PCL), polyglycolic acid (PGA), polyvinyl alcohol (PVA), poly lactic acid-co-glycolic acid (PLGA), and poly butylene succinate (PBS). The structures of these polymers are shown in Fig. 14.12.

PLA

Polylactic acid (PLA) is an aliphatic polyester synthesized from lactic acid monomers, which can be obtained from the fermentation of renewable materials such as corn and waste cellulosic materials. PLA has important properties that make it attractive to be used as packaging, such as low permeability to gases and water, high transparency, and moderate tensile strength (Balla et al. 2021).

Some studies have already investiaged PLA as nanocapsules material for EO. Antonioli et al. (2020) nanoencapsulated lemongrass EO in PLA nanocapsules and evaluated their antifungal activity in vitro and in vivo. They have shown in vitro antifungal activity against *Colletotrichum acutatum* and *Colletotrichum gloeosporioides* with a MIC dosage of 0.1% (v/v) for both phytopathogens. The in vivo were performed with postharvest apples, and the ones treated with nanoencapsulated EO showed bitter rot lesions three times smaller than the ones treated with non-encapsulated EO, or in comparison to the apples in positive control.

PHAs

Polyhydroxyalkanoates (PHAs) are thermoplastic, biodegradable polymers produced by microorganisms. Currently, more than 100 types of PHA are known, the most widespread being polyhydroxybutyrate (PHB), followed by poly(3-hydroxybutyrate) (PHB), polyhydroxyvalerate (PHV), and polyhydroxyhexanoate (PHH). The use of encapsulation in PNPs based on polyhydroxyalkanoate (PHA) is being explored in obtaining biodegradable material for obtaining active packaging (Gadgil et al. 2017; Kumar et al. 2021b).

Zheng et al. (2022) encapsulated Mexican oregano EO in polyhydroxybutyrate (PHB) and poly-3-hydroxybutyrate-co-hydroxyhexanoate (PHB-HHx) and studied in vitro release in simulated food media. Both nanosystems containing EO showed antimicrobial activity against *Micrococcus luteus*, PNPs based on PHB-HHx were more efficient than pure EO. The use of PLA/PHA containing oregano EO was efficient in active packaging for puffer fish fillets.

PCL.

PCL (poli-ε-caprolactona) is obtained by ring-opening polymerization of ε-caprolactone or polycondensation of 6-hydroxyhexanoic acid. PCL has lower tensile strength than PLA and has greater permeability to water vapor and oxygen than other biopolymers. To be used in active packaging, PCL is usually blended with other biopolymers (Shaikh et al. 2021).

Concerning its application as a nanocarrier material, PCL was successfully used for the development of nanofibers aiming active packaging (Ferreira et al. 2021). *Lavandula luisieri* EO was successfuly incorporated in the form of a nonwoven substrate for museological packaging, but we can state that it can be potentially extended for further studies as food packaging.

PGA and PLGA

PGA (polyglycolide acid) is obtained by direct polycondensation of glycolic acid, ring-opening polymerization of glycolide (ROP), and solid-state polycondensation of halogen acetates. This polymer has high biodegradability, similar to cellulose. The greatest utility of PGA is as copolymers, such as poly(lactic-co-glycolic acid) (PLGA) obtained in 88:12, which means that the polymer is made of 88% lactic acid and 12% glycolic acid (Ayyoob et al. 2017).

Zhu et al. (2019) encapsulated thymol (a constituent of oregano and rosemary essential oil) using PLGA microparticles, obtaining spherical and smooth microparticles, with optimal efficiency when encapsulating 20% (w/w) of thymol. These thymol-loaded microparticles showed relevant antibacterial activities against *S. aureus* and *E. coli*. Thus, PLGA microparticles loaded with thymol have the potential for use as preservation additives in foods.

PVA

PVA (polyvinyl alcohol) is a synthetic biopolymer. It is synthesized by the polymerization of polyvinyl acetate and then the subsequent hydrolysis of the acetate group. PVA has high polarity and hydrophilicity, thus, it is usually used in mixtures with more hydrophobic constituents to obtain material to be used in active packaging (Moulay 2015).

Lamarra et al. (2020) produced PVA electrospun nanofibers and chitosan-based emulsions functionalized with cabreuva EO extracted from the wood of *Myrocarpus fastigiatus.* To overcome the low solubility of the EO and to protect it, the authors proposed a two-step process, emulsion formation compound by chitosan and PVA, and subsequent ionic cross-linking with sodium citrate. Nanofibers were also manufactured through the electrospinning method. The electrospun nanofibers exhibited the ability to be an effective carrier of the cabreuva EO and the capacity of controlling the compound release that proved an effective activity against broad spectra of microorganisms (*Candida albicans*, *E. coli*, *S. aureus*, and *S. epidermidis*).

PBS

PBS is a biodegradable polymer that can be obtained by the polycondensation of succinic acid (or dimethyl succinate) and 1,4-butanediol, these monomers can be obtained from renewable or fossil resources (Rafiqah et al. 2021).

Using the PBS/geraniol mixture, solid and porous plaques with antimicrobial activity were obtained, which were inserted into bread packages. The bread preservation period was extended by 5 and 10 days with the insertion of these solid and porous antimicrobial plates, containing 8% by weight of geraniol in the PBS matrix. (Petchwattana et al. 2021).

There is a long way to go in terms of spreading the use of bioplastics. Globally, less than 1% of plastics used are of biological origin. These materials are hydrobiodegradable and despite some of them are still poorly adressed as nanocarriers for essential oils aiming food packaging application, there is a remarkable increase on the interest on this topic (Atta et al. 2022).

14.3.3 Inclusion Complexation: Cyclodextrins Cases

Within the area of supramolecular chemistry, molecular inclusion or inclusion complexes (ICs) are studied, which refer to the confinement of a guest molecule within the cavity of a host molecule, also known as a cage compound, molecular capsule, or molecular container. The interactions between the trusting molecule and the host molecule are noncovalent, they occur through intermolecular forces (Huang and Anslyn 2015). This way, the interaction between the guest and host molecules differ to the encapsulation descibed before.

Cyclodextrins (CDs) have been widely used in supramolecular encapsulation, mainly in the formation of inclusion complexes with volatile compounds, such as EOs (Kfoury et al. 2018). CDs are cyclic oligosaccharides composed of at least six D-glucose units linked by α-1,4 bonds. The production of CDs occurs from starch, by the action of cyclodextrin-α-glucosyltransferase, a transglycosidase responsible for cleaving an α-1,4 bond in amylose and forming the cycle (Wüpper et al. 2021).

The most studied CDs are called α-, β-, and γ-cyclodextrins, which are composed of six, seven, and eight D-glucose residues, respectively. The shape of CDs molecules is similar to a truncated cone, where the outer surface is hydrophilic due to the presence of hydroxyl groups (-OH) and the inner cavity has lipophilic/hydrophobic characteristics. In this way, compounds with hydrophobic characteristics can remain inside the internal hydrophobic cavity of CDs, leading to the formation of inclusion complexes with several hydrophobic substances (Fig. 14.13).

One of the applications of the entrapment of EOs in CDs is to enable and modulate the controlled release of these volatile constituents when incorporated into active food packaging. ICs with basil EO (BEO) and *Pimenta dioica* (PDEO) in β-cyclodextrin (β-CD) were prepared in order to obtain sachets with antimicrobial activity for food preservation. The prepared ICs showed great potential for antimicrobial application and the sachets can be used in packaging for food preservation (Marques et al. 2019b).

Marques et al. (2022b) used garlic essential oil (GEO) complexed or not with β-cyclodextrin, incorporated into a blend formed by cellulose acetate and zein, and added by glycerol or tributyrin as plasticizers, aiming to obtain films for active packaging. The IC addition into films, however, did not ensure antibacterial action, albeit that the IC with GEO, when tested alone, showed activity against both bacteria.

14.3.4 Nanoliposomes

The Greek roots of the word liposome mean "fat body"; however, we can describe them as hollow structures made of phospholipids (Lasič 1992). Liposomes are well-defined by Laouini et al. (2012) as microscopic spherical-shaped vesicles, which consist of an internal aqueous compartment, entrapped by one or multiple

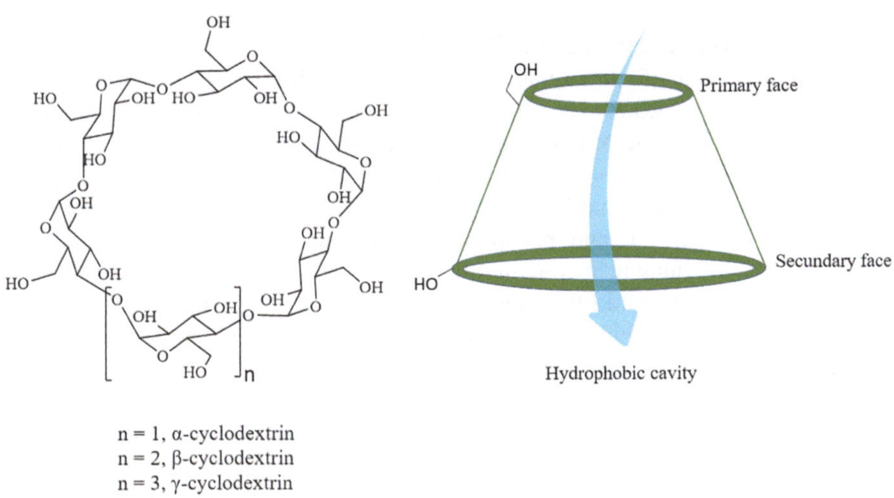

Fig. 14.13 Structures of α-, β-, and γ-cyclodextrins and the truncated cone shape of CDs molecules

concentric lipid bilayers. Because of their characteristics, hydrophilic substances can be encapsulated in the interior aqueous compartments, while lipophilic compounds can be mainly entrapped within lipid bilayers. The similarity to the structure of biological membranes allows liposomes to mingle with the cells and tissues of human bodies (Lasič 1992).

Liposomes were discovered in the early 1960s by the British scientist Alec Bangham, who soon recognized the encapsulated capacity of phospholipids with spherical forms (Lasič 1992). From 1970 to 1980, the potential applications of liposomes in the pharmaceutical and medical industries produced an increasing interest as potential drug carriers (Lasič 1992, Laouini et al. 2012). The versatility of liposomes leads them to a large number of applications including pharmaceutical, cosmetics, and food industrial fields (Lin et al. 2022b; Muñoz-Shugulí et al. 2021; Ohishi et al. 2022).

The phospholipid reorganization is mainly driven by the hydrophobic effect to minimize entropically unfavorable interactions between hydrophobic acyl-chains and surrounding aqueous medium (Lasič 1998, Lasič and Papadjopoulos 1995, Laouini et al. 2012). This effect is further settled by various intermolecular forces such as electrostatic interactions, hydrogen bonding, Van der Waals, and dispersion forces (Israelachvili et al. 1980; Laouini et al. 2012).

The liposomes enhance the compounds' performance through the increase of compounds' solubility and stability, the delivery of encapsulated substances to specific target sites, the sustained compounds release, the high encapsulation efficiency, due to the low toxicity of the system, the substance protection against degradations factors such as pH and light, and the reduction of tissue irritation (Laouini et al. 2012; Guimarães et al. 2021).

To ensure proper liposome performance, batch-to-batch reproducibility and stability of the liposome dispersions have to be established (Laouini et al. 2012).

14.3.4.1 Classical Liposomes Procedures

The liposome is manufactured mainly with a phospholipid molecule, consisting of a polar hydrophilic "head," typically a phosphate group, attached to a long nonpolar hydrophobic "tail" made of two long hydrocarbon chains (Lasič 1992). Generally, liposome composition includes natural and, or synthetic phospholipids (phosphatidylserine, phosphaticylethanolamine, phosphatidylinositol, phosphatidylcholine, also known as lecithin, and phosphatidylethanolamine) (Laouini et al. 2012). Liposome bilayers may also contain other constituents such as cholesterol, hydrophilic polymer conjugated lipids, and water to improve the membrane fluidity and bilayer stability, and reduces the permeability of water-soluble molecules through the membrane (Mozafari 2005; Laouini et al. 2012).

Three attributes interest a scientist preparing a batch of liposomes, which are (1) the chemical composition of the lipid bilayer, (2) the size distribution of the liposomes, and (3) the number of layers in each liposome (Lasič 1992).

There are four classical methods of liposome manufacture. The difference between those methods is the step of drying the liposomes in the organic solvents, followed by their redispersion in aqueous media (Mozafari 2005; Laouini et al. 2012).

Briefly, the classical methods are (1) the Bangham method or hydration of a thin lipid film, in which the phospholipid and cholesterol were dispersed in an organic solvent, removed by evaporation, and then, hydrated with water under agitation (Laouini et al. 2012; Bangham et al. 1962); (2) the reverse-phase evaporation technique, a lipid film is prepared by evaporating organic solvent under reduced pressure with nitrogen and the lipids are redissolved in a second organic phase (diethyl ether and/or isopropyl ether); (3) the solvent (ether or ethanol) injection technique involves the dissolution of the lipid into an organic phase (ethanol or ether), followed by the injection of it into warmed aqueous media; (4) the detergent dialysis technique that produces liposome size ranging of 40–180 nm through lipids solubilization with detergent which is removed by controlled dialysis, forming homogeneous unilamellar vesicles (Laouini et al. 2012).

Bangham method is used to produce multilamellar heterogeneous liposomes with different sizes and shapes such as single bilayer shells around 10 nm diameter (Johnson et al. 1971), or multivesicular liposomal, when incorporated with lavender essential oil, sizing around 0.4–1.3 μm (Varoma et al. 2011). The reverse-phase evaporation technique is used to produce large unilamellar and oligolamellar vesicles (Laouini et al. 2012). While the solvent injection technique forms small liposomes, with narrow distribution, without extrusion or sonification. The ether is removed from the liposomal product through heating while ethanol remains in the product (Laouini et al. 2012, Kryeziu et al. 2022).

Classical techniques require large amounts of organic solvents, which are harmful to the environment and human health, and consume a large amount of energy (Laouini et al. 2012). Therefore, new methods have been developed.

14.3.4.2 New Large-Scale Liposome Technique

An important point in the development of a suitable liposomal system is whether it is possible to prepare, isolate, and characterize a stable particular liposomal formulation on an industrial scale, with clearly defined and reproducible properties, and accessible cost. The manufacturing of liposomes for human and animal use needs a sterilization step such as filtration; however, this technique cannot remove viruses (Brandl et al. 1993, Mozafari 2005).

Heating is a method for the fast production of liposomes that involves the hydration of liposome components and glycerol (3 wt.) in an aqueous medium, followed by heating up to 120 °C (Kikuchi et al. 1991). Glycerol is a water-soluble and physiologically acceptable chemical with the ability to increase the stability of lipid vesicles and does not need to be removed from the product (Laouini et al. 2012). No degradation of the lipids occurred at the above-mentioned temperatures, and no further sterilization procedures are necessary, reducing the time and cost of liposome production (Mozafari et al. 2004).

Another technique is the spray-drying process which is considered to be a fast single-step procedure applied in the nanoparticles' formulation (Chougule et al. 2008). Hence, liposomes were prepared by suspending lecithin and mannitol in chloroform, sonicated, and subjected to spray-drying equipment. The conditions established were inlet and outlet temperatures, airflow rate, and flow rate (Laouini et al. 2012). Liposomal powder formulations are more stable and more appropriate for long-term storage compared with liposomal dispersions (Ingvarsson et al. 2011). Spray drying is less expensive, and less time and energy-consuming compared with freeze drying, making it a suitable strategy to dry liposomal dispersions (Sepúlveda et al. 2021). However, one of the challenges of using liposome spray drying is to avoid membrane bilayer instability caused by heat-induced phase transitions (Van Den Hoven et al. 2012). Furthermore, both heat and high shear forces involved in the process may also induce degradation of the liposomal bilayer structure (Ingvarsson et al. 2011).

Freeze drying process converts a liquid into a solid through three primary steps: (1) freezing, (2) primary drying (sublimation of ice), (3) and secondary drying (desorption of the remaining water under vacuum) (Franks 1998). This process, alone, generates various stresses on liposomes, induced by the dehydration and crystallization of ice. The stresses often lead to phase separation, aggregation, and drug leakage due to liposome disruption (Habib et al. 2022). The lyophilized product spontaneously forms homogenous liposome preparation on the readdition of water. According to Laouini et al. (2012), the lipid/carrier ratio is the key factor that affects the size and the polydispersity of the liposome preparation.

Alternatively, supercritical reverse-phase evaporation allowed aqueous dispersions of liposomes to be obtained through emulsion formation. A given amount of water is introduced into a homogeneous mixture of supercritical carbon dioxide and phospholipids, with sufficient stirring and subsequent pressure reduction (Otake et al. 2001). The vesicles produced by this method are large unilamellar, with diameters ranging from 0.1 μm to 1.2 μm (Laouini et al. 2012). Varoma et al. (2011) produced a gas-saturated solution with particle size between 1.4 μm and 24.8 μm, and poor incorporation efficiency of EOs (3–14.5%). However, Otake et al. (2001) indicate the method as an excellent technique that permits the one-step preparation of large unilamellar liposomes with a high trapping efficiency for both water-soluble and oil-soluble substances.

Another technique is the crossflow injection which has been recently used as a promising novel scalable approach. The ethanol injection method is one of the preferred reported techniques used to produce small unilamellar liposomes, simply and rapidly. The method involves the injection of an ethanolic solution of lipids into a large volume of an aqueous phase, which leads to the rapid formation of vesicles without passing through any intermediate process (Gouda et al. 2021). Upon injection, the unfavorable exposure of lipids to the aqueous medium results in the arrangement and precipitation of lipids at the boundary phase between ethanol and water in the form of bilayer phospholipid fragments (Lasič 1982, 1995). The main parameters influencing the process are injection velocity, stirring rate, injection hole diameter, lipid concentration, osmolality/pH/viscosity of the aqueous phase, the volume of utilized ethanol and its residual amount, dimensions of the reaction vessel, and type of drug to be loaded (Gouda et al. 2021).

Other methods to produce liposomes are the extrusion method, high shear homogenization, sonication, dual asymmetric centrifugation, supercritical anti-solvent, rapid expansion of the supercritical solution, calcium-induced fusion, nanoprecipitation, and emulsion techniques. All methods can be checked in Andra et al. (2022)) review and the works of Papahadjopoulos et al. (1990), Holland et al. (1996), Cauchetier et al. (1999), Piacentini et al. (2022), and Sun et al. (2022).

14.3.4.3 Classification

Liposomes can be classified in terms of composition and mechanism of intercellular delivery into five types (Sharma and Sharma 1997; Laouini et al. 2012) such as (1) conventional liposomes, (2) pH-sensitive liposomes; (3) cationic liposomes; (4) immunoliposomes; and (5) long-circulating liposomes.

Otherwise, liposomes were typically classified based on their size and number of bilayers into (1) small unilamellar vesicles or nanovesicles, 20–100 nm; (2) large unilamellar vesicles, >100 nm; (3) giant unilamellar vesicles, >1000 nm; (4) oligolamellar vesicle: 100-–00 nm; and (5) multilamellar vesicles, > 500 nm (Laouini et al. 2012)

New developed types of liposomes, designated as double liposomes and multissecular vesicles were reported (Laouini et al. 2012).

14.3.4.4 Nanosized Liposomes

Nanoliposomes (NLs) are a category of liposomes whose structures form vesicles composed of one or more lipid bilayers and the size does not exceed 100 nm. NLs are stabilized by Brownian motion, which explains the lack of need for surfactants to stabilize these preparations. The methods to produce NLs are almost the same used to produce liposomes; however, the parameters should be well-established to obtain the system at the size required, nanosized (Bondu and Yen 2022).

In reality, NLs suspensions are not perfectly stable and they are susceptible to self-aggregation or fusing into larger vesicles. This so-called coalescence phenomenon should be avoided to preserve the encapsulated active substances and the functional size of the vesicles (Bondu and Yen 2022). Factors can be controlled favoring the stability of NLs such as temperature, size, and the superficial charge of the system.

Transition temperature (TT) is the temperature at which a lipid bilayer loses its organization, becoming initially more fluid and then unstructured (Mozafari et al. 2008). The control of the temperature occurs at two levels. First, during the preparation of the NLs, the temperature must be higher than the TT in order to mix with the active substances to be encapsulated. Secondly, to maintain the integrity of the structure of NLs, the temperature of conservation must be lower than the TT (Bondu and Yen 2022). The TT depends on the type of lipids of the bilayer, including the polarity of the phospholipid heads, the fatty acid chain lengths, the degree of unsaturation, and the ionic strength of the suspension medium (Szoka Jr and Papahadjopoulos 1980; Leserman et al. 1994; Maherani et al. 2011).

Through electrophoretic mobility measures, the surface charge of the lipid droplets can be determined which considerably influences NLs' stability and their ability to repel each other well enough to avoid precipitation during storage. The charge density that determines this electrophoretic mobility and the binding affinity of the different ions is provided by the zeta potential (Bondu and Yen 2022). The composition of phospholipids is therefore important since they are not all charged in the same way. Some phospholipids are neutral such as phosphatidylcholine or phosphatidylethanolamine, and others negative such as phosphatidylserine, phosphatidylglycerol, phosphatidic acid, or diacetylphosphate. Other molecules besides phospholipids can be added such as dioleoyl trimethyl ammonium propane or stearylamine, which are positively charged (Maherani et al. 2011). The combinations in different proportions of these phospholipids can, therefore, significantly change the value of the NLs' zeta potential, influenced by the pH of the system.

Stable NLs have become a subject of growing interest in the field of nanotechnology, particularly given their ability to encapsulate bioactive substances, which makes them interesting as vectors or carriers. Because of their biocompatibility and biodegradability, along with their nano size, nanoliposomes have potential applications in a vast range of fields, including nanotherapy (e.g., diagnosis, cancer therapy, and gene delivery), cosmetics, food technology, and agriculture (Mozafari et al. 2008).

14.3.5 Nanoniosomes

The difference between liposomes and niosomes is slight. The liposomes were very well elucidated until here. Otherwise, niosomes are vesicles composed of non-ionic surfactants, amphipathic compounds with an overall neutral charge, in many cases, cholesterol, or its derivatives. These non-ionic surfactants are cheap and safe for use in biomedicine, for example, as niosomal drug carriers for both hydrophilic and hydrophobic drugs, similar to liposomes. They could be coated by various types of agents such as polyethylene glycol (Van den Bogaart et al. 2007), hyaluronic acid (Lee 2003), antibodies (Allen and Cleland 1980), for specific applications.

While most niosomes are in the nano or sub-micron (colloidal) size range, not many authors used the "nano-niosome" or "nanovesicle" term in their published articles.

Bartelds et al. (2018) approach a complete characterization of the functional properties of niosomes composed of ternary surfactant mixtures compared to liposomes made of saturated or unsaturated phosphatidylcholine and phosphatidylethanolamine-based lipids plus cholesterol. Similar to liposomes, niosomes composed of saturated amphiphiles compounds are more stable and less leaky when tested below the phase TT than vesicles composed of unsaturated components. Taken together, niosomes behave in many aspects similar to phospholipid-based vesicles but they may not provide the right lipid head-group composition for functional membrane transport, that is, the requirement of many transporters for a fraction of anionic headgroups (Lee 2003).

The membrane packing of niosomes is somewhat less dense than that of liposomes. The authors concluded that niosomes exhibited physical chemical properties similar to those of liposomes, albeit that permeability for small ions/solutes is higher, which makes them potentially attractive as drug carriers or delivery systems for all sorts of molecules (Bartelds et al. 2018).

Drug incorporation into niosomes has been accomplished and the authors show that for retention of cargo, the vesicles should not be frozen and thawed. The same factors e.g., size, charge, and stability, should be controlled to use niosomes as encapsulated systems or compound delivery (Bartelds et al. 2018). The niosomes system can be produced through the same techniques than liposomes.

14.3.6 Nano-Cubosomes and Others

Depending on the physicochemical properties of the compounds, different characteristics of the carriers are required. Therefore, there is no perfect delivery system that meets the demands of all kinds of compounds. A wide range of carriers is available, from liquid NE, polymer conjugates, mixed micelles, liposomes, niosomes, and cubosomes (Weiss et al. 2012). The most complex of the liquid crystalline phases formed in lipid/aqueous systems is the cubic phase. One general type of cubic phase is both lipid- and aqueous-continuous, where the lipids form curved, non-intersecting bilayers. The degree of swelling and the curvature of the interfaces as well as the type of cubic phase will be important factors in the entrapment of compounds (Nylander et al. 1996)

Recently, nano-structured cubosomes (NCs) have been used as novel drug nanocarriers owing to their great potential as a substitute delivery system for liposomes (Farag et al. 2022). NCs, particularly composed of binary systems of water and monoolein, are the most investigated systems (Larsson 1983). They can be considered hydrophilic surfactant systems that can self-assemble in the form of a bicontinuous cubic liquid crystalline phase (Bei et al. 2009). They are characterized by their viscous isotropic nature and their large internal surface area (Nylander et al. 1996).

NCs can incorporate lipophilic, hydrophilic, and amphiphilic drugs. Additionally, their lipidic contents are biocompatible, bioadhesive, and digestible (Barauskas et al. 2005a, b). Also, nano-structured cubosomal systems have been explored for diverse pharmaceutical purposes (delivery of enzymes, peptides, analgesics, and antibiotics) (Drummond and Fong 1999; Garg et al. 2007; Almoshari et al. 2022; Farag et al. 2022).

In this context, the aqueous phase behavior of unsaturated monoglycerides, such as glycerol monooleate and glycerol monolinoleate, has been thoroughly investigated due to their extensive polymorphism and widespread use in industrial products, e.g., food (Clogston et al. 2000). Depending on various molecular and ambient conditions, unsaturated monoglycerides can form four liquid crystalline mesophases, lamellar, reversed hexagonal, and two reversed bicontinuous cubic phases (Lutton 1965; Hyde et al. 1984; Qiu and Caffrey 2000).

Unsaturated monoglycerides have a unique property to form the cubic phase in equilibrium with the excess aqueous solution, expanding the application. Moreover, the unsaturated monoglycerides cubic phases have also been suggested for use in practical/technical applications such as bioelectrodes (Rowinski et al. 2004), biosensor construction (Barauskas et al. 2003), and protein (Katona et al. 2003) and metal nanoparticle (Norling et al. 1992) crystallization.

The simplest method of producing cubic phase particles is through the agitation of the glycerol monooleate with a magnetic stirrer in the presence of poly(ethylene oxide)-based stabilizers, resulting in a coarse dispersion with a particle size in the range of 1–100 μm. Further size reduction can be achieved by the use of high-shear energy input techniques such as ultrasonication, homogenization, and emulsification of the coarse dispersion (Farag et al. 2022).

Another method for producing cubic phase dispersions is based on mixing the glycerol monooleate with ethanol in miscible proportions and diluting the system with an aqueous solution containing a stabilizer. If the dilution trajectory falls into a cubic phase region, the cubic phase particles are formed spontaneously due to molecular diffusion difference at the liquid/liquid interface. However, also with this preparation method, a substantial amount of vesicular material was obtained and typically the particle size distributions were rather broad (Barauskas et al. 2005a, b).

Although several methods have been developed for the manufacturing of cubic phase dispersion, there is still no established process available for controlling the properties and quality of cubic phase particle dispersions. Importantly the particle dispersions of the method exhibit excellent colloidal stability during storage and dilution.

According to Barauskas et al. (2005a), a combination of high-shear energy homogenization and heat treatment provides a powerful and scalable way of producing glycerol monooleate-based cubic phase nanoparticles. The heat treatment of the homogenized dispersions consisting of predominantly vesicular-like particles results in their effective and reproducible conversion to cubic phase particles with narrow particle size distribution and well-defined inner morphology. The process is simple and results in cubosome nanoparticles containing minimal amounts of lamellar aggregates and possessing good colloidal stability. In addition, the amphiphile concentration, the amount of charged species, and salt content provide an elegant way of further controlling dispersion particle size and nanostructure (Almoshari et al. 2022).

14.4 Nanotechnological Approaches in Essential Oil-Based Food Packaging Systems

The natural appeal of EOs, combined with their bioactive properties and GRAS status, greatly interests the food industry. Nowadays, it is no secret that consumers' eating habits are changing, and a growing number of individuals are seeking more natural and health diets, as well as showing concerns for the environment. As a result, recent trends such as "clean label" products, sustainable packaging, and nanotechnology are thriving (Delgado-Pando et al. 2021; Ibrahim et al. 2022; Nile et al. 2020). The incorporation of EOs into the packaging to manufacture active packaging for the preservation of different food systems is another trend that deserves to be highlighted (Marques et al. 2022a; Sayadi et al. 2022; Tavares et al. 2021). An advantage of active packaging over conventional ones is the inclusion of the preservative agent in the packaging material, which ensures longer shelf life for the packaged products and promotes greater microbiological safety.

However, we must not ignore that there are several characteristics of the EOs that hinders their application as a preservative for foods, as well as potential additives for active packaging. They are prone to oxidative damage, possibly will degrade when exposed to light, and are extremely volatile and thermolabile. They are also hydrophobic compounds, and this feature may result in incompatibility with the polymer matrix, especially when working with bio-based polymers since the majority of them have a hydrophilic nature. In addition, their strong odor may compromise the food sensory attributes, such as aroma and flavor, and therefore affect negatively consumer acceptance, as reported by Ghabraie et al. (2016) and Marques et al. (2019a). In this sense, nanotechnology may be the key to bypassing these obstacles and ensuring the successful application of these compounds into active packaging (Arruda et al. 2022; Nile et al. 2020). Several approaches have been studied over the years, and a few recent examples are summed up in Table 14.1.

Nano-encapsulation of EOs, or their major compounds, is one of the most studied strategies to overcome EOs' drawbacks due to the great versatility regarding techniques, materials, and advantages provided, such as thermal protection, higher stability, a more sustained release, which ensure a longer bioactive of the compound, and more bioavailability and solubility (Delshadi et al. 2020; Liao et al. 2021; Surendhiran et al. 2022; Tavares et al. 2021). Moreover, the wall materials used to compose the nanocapsules varies widely according to the desired encapsulation system: cholesterol and phospholipids to produce nanoliposomes (Fattahian et al. 2022; Tavares et al. 2021); several oils, proteins (whey protein, for example) and carbohydrates (alginates, for example) for the obtainment of nanoemulsions (Almasi et al. 2021; Sun et al. 2021); distinct biopolymers (casein, zein, gelatin, chitosan, alginate, among others) to produce nanoparticles and nanofibers (Liao et al. 2021). The necessary equipment will also depend on the desired particles: ultrasonicators, homogenizers, spray dryers, and/or freeze dryers are commonly used for the manufacturing of liposomes, emulsions, and nanoparticles; while electrospinning and electrospray are required for the elaboration of nanofibers.

Tavares et al. (2021) used cholesterol and lecithin to produce liposomes for carvacrol nanoencapsulation through the lipid film hydration technique. The resulting vesicles provided effective thermal protection to the volatile compound and showed activity against *E. coli* and *S. aureus*. Later, the authors incorporated the vesicles into poly(vinyl) alcohol films to produce active packaging.

Ghoshal and Shivani (2022), in turn, used an ultrasonicator to incorporate thyme EO nanoemulsion into edible films composed of tamarind starch and whey protein. The films were evaluated on tomatoes and displayed the potential to be used as active edible packaging. At last, Zhang et al. (2022a) manufactured a multifunctional cellulose acetate membrane by electrospinning to act as both intelligent and active packaging. The authors incorporated a natural pigment (anthocyanin) as the freshness indicator and chamomile EO as the bioactive agent. The resulting nanofiber membrane was able to monitor freshness in pork, changing color as a pH

Table 14.1 Nanotechnological strategies to allow the use of EOs as preservative agents in active packaging aiming food preservation

Essential oil	Packaging material	Nanotechnological approach	Outcomes	Reference
Allyl isothiocyanate	Cellulose acetate	Nano-composite with carbon nanotubes and β-cyclodextrin	Controlled release of antimicrobials in the food model	Dias et al. (2018)
Rosemary and ginger	Chitosan	Development of montmorillonite nanobiocomposites	Extension of poultry shelf life	Pires et al. (2018)
Pimenta dioica and basil	Cellulose sachets	Complexation with β-cyclodextrin	Greater thermal stability; In vitro antimicrobial activity against *listeria monocytogenes* and spoilage mold *Byssochlamys nivea*	Marques et al. (2019b)
Peppermint and chamomile	Gelatin	Nanofibers manufactured by electrospinning	In vitro antimicrobial activity against *Escherichia coli* and *Staphylococcus aureus*; Antioxidant activity; UV barrier property; Not cytotoxic	Tang et al. (2019)
Coriander	–	Cyclodextrin nanosponges	In vitro antimicrobial activity against *L. monocytogenes*, verotoxigenic *E. coli*, *campylobacter* strains, and other pathogenic bacteria	Silva et al. (2019)
Carvacrol	Poly(vinyl) alcohol	Nanoencapsulation of carvacrol into liposomes	Greater thermal stability	Tavares et al. (2021)
Lavander	Gelatin	Nanoemulsion	Antimicrobial and antioxidant activity UV barrier; Heat-sealing property; Extension of cherry tomato shelf life	Sun et al. (2021)
Garlic	Chitosan	Nano-liposome	Extension of chicken fillet shelf life	Kamkar et al. (2021)
Cuminum cyminum	Chitosan	Nano-encapsulation of the EO into liposomes	Reduce microbial counts in meat fillets; Delay lipid oxidation in samples	Fattahian et al. (2022)

(continued)

Table 14.1 (continued)

Essential oil	Packaging material	Nanotechnological approach	Outcomes	Reference
Green tea extract and ginger EO	Starch	Development of nanocomposites with nanofibrillated cellulose	Extension of strawberry shelf life	Rodrigues et al. (2021)
Cumin	Gelatin	Development of nanocomposites with TiO_2	Delay lipid oxidation in fresh chicken meat	Sayadi et al. (2022)
Garlic	Cellulose acetate and zein blend	Complexation with β-cyclodextrin	Greater thermal stability	Marques et al. (2022b)
Thyme	Tamarind starch/whey protein	Nano-emulsion	Antimicrobial activity against *E. coli* and *S. aureus*; Extension of tomato shelf life	Ghoshal and Shivani (2022)
Clove	Poly(lactic acid)	Development of nanocomposite films with alkali-treated halloysite nanotubes	Better water vapor barrier, surface hydrophobicity, mechanical properties, and thermal stability; Delay weight loss of fresh-cut apples	Boro et al. (2022)
Turmeric	Chitosan	Development of magnetic-silica nanocomposite	Sustained release of the EO; Extension of surimi shelf life	Surendhiran et al. (2022)
Chamomile	Cellulose acetate	Fabrication of nanofiber membrane by electrospinning	Extension of pork shelf life	Zhang et al. (2022a)

response. Also, the membrane was able to extend the food shelf life by displaying both antimicrobial and antioxidant properties.

In another work, the cinnamon EO was also used in the preparation of new biodegradable material based on poly(butylene adipate-co-terephthalate) (PBAT) containing cellulose nanofibers (CNF) for application in food packaging. To prepare the films, CNFs were impregnated with cinnamon essential oil in a 2:1 ratio. The modified CNF (M-CNF) was then mixed with the PBAT solution at different concentrations of 0.5%, 1%, and 3%. The results were promising, as the films showed good thermal stability, evidencing a decrease in water vapor permeability values and relevant antimicrobial activity against *Salmonella* and *Listeria monocytogenes*. The films were tested as packaging for strawberries, the results indicated that in films with 0.5% by weight of modified CNF, the fruits had no fungal attack, less mass loss in 15 days of storage and greater preservation of freshness (De Matos Costa et al. 2020).

Roy and Rhim (2021) obtained a carrageenan/agar-based film containing Pickering tea tree oil emulsion (PET) and zinc sulfide nanoparticles (ZnSNP). PET was obtained with tea tree essential oil stabilized with nanocellulose fibers. PET and ZnSNPs were uniformly dispersed in the binary polymeric matrix and formed compatible films. There was an improvement in thermal stability, water vapor barrier, water resistance, and the composite membrane based on carrageenan/agar showed antibacterial and antioxidant activity. Thus, the developed film could be useful to be applied in active packaging for food.

The complexation of EOs with CDs is also an interesting nanotechnological strategy to enhance the EO stability and enable their application in packaging. The obtained EO/cyclodextrin inclusion complex usually is available in solid form, making it easier to handle and allowing its use in different ways. For example, it could be contained in sachets, as a powder (Marques et al. 2019a, 2019b) or dispersed in a polymer matrix (Dias et al. 2018; Marques et al. 2022b). The EOs could also be loaded in cyclodextrin-based nanosponges, a highly cross-linked polymeric network (Simionato et al. 2019; Silva et al. 2019), or nanofibers films prepared by electrospinning (Qin et al. 2022; Shi et al. 2022).

Regardless of the strategy chosen, EO/cyclodextrin inclusion complexes are known to enhance EO stability, mainly concerning thermal stability. Since the methods used to produce packaging usually involve high temperatures, it is of utmost importance to ensure thermal protection for the extremely thermolabile EO compounds, which can be successfully achieved by complexation with cyclodextrins (Arruda et al. 2021, 2022).

However, the properties of both CDs and the main polymer must be carefully studied before choosing the working materials. There are several cyclodextrins available on the market, usually classified into two groups: native cyclodextrins (α-, β-, and γ-cyclodextrins) and derivative cyclodextrins (e.g., hydroxyl-propyl-β-cyclodextrin) (Arruda et al. 2021). Each CD has particular features, such as higher or lower solubility in water and/or different organic solvents. β-cyclodextrin, for example, is not soluble in acetone, the main solvent used for cellulose acetate films manufactured by the casting technique. This characteristic may have been the main issue faced by Marques et al. (2021) during the elaboration of cellulose acetate films incorporated with β-cyclodextrin and allyl isothiocyanate, the major component of mustard essential oil. The authors verified that the films incorporated with β-cyclodextrin were more heterogeneous, brittle, and fragile than films without the compound. A similar result was verified by Dias et al. (2018), who described the films produced with cellulose acetate and β-cyclodextrin as less homogeneous, porous, and cracked when compared to the other elaborated films.

The choice of the technique could be a strategy to overcome drawbacks like that. In this sense, electrospinning has been successfully used to manufacture cyclodextrin nanofibers with different polymers, cellulose acetate among them. For example, Nthunya et al. (2017) synthesized β-cyclodextrin/cellulose acetate nanofibers via electrospinning to produce an active material aiming for bacteria removal from water. Although the antimicrobial agent of choice is not an EO, but silver/iron nanoparticles, the technique allowed the manufacture of nanocomposite fibers with strong antibacterial activity. Ghorani et al. (2019) also used the electrospinning technique to fabricate β-cyclodextrin/cellulose acetate nanofibrous webs aiming adsorption of undesirable volatile molecules. The authors used binary solvent systems composed of acetone and N,N-dimethylacetamide or acetone and dimethylformamide to synthesize uniform webs with adsorbent properties.

In fact, the electrospinning technique can be employed to produce EO/cyclodextrin-loaded nanofibers with several polymers. Figueroa-Lopez et al. (2020) used the technology to incorporate oregano EO/γ- or α-cyclodextrin inclusion complex into poly(3-hydroxybutyrate-*co*-3-hydroxyvalerate) fibers to act as active packaging. The films elaborated with 25% (w/w) of γ-CD inclusion complex displayed a homogeneous and continuous surface, and transparency, ensuring thermal protection for the EO, as well as antimicrobial and antioxidant activity when evaluated in vitro. Along the same line, Shi et al. (2022) developed electrospun nanofibers with polylactic acid (PLA) and polycaprolactone (PCL) incorporated with oregano EO loaded in β-cyclodextrin. The material acted as an active packaging when tested on blackberry, extending the shelf life of the fruit and maintaining its postharvest quality.

At last, the development of nanocomposites or nanobiocomposites formed by a blend of EO-polymer-nanostructure is also worth mentioning. Nanocomposites are materials that have at least one component at nano dimensions and are widely studied as potential active packaging. They could involve the incorporation of pure or encapsulated EOs in the polymer matrix aiming an antimicrobial and/or antioxidant activity, and a nanostructure as a filler to enhance certain features of the packaging. Carbon nanotubes, nanocrystalline cellulose and inorganic materials (such as nanoclay) are examples of nanoparticles studied as nanocomposites components to improve mechanical, optical, thermal, and/or barrier properties of the developed packaging, or even contribute to the active property, stability, or the controlled release of the EOs (Pola et al. 2016; Pires et al. 2018; Marques et al. 2021; Boro et al. 2022; Mahmud et al. 2022).

A potential concern that could guide future research in the active packaging field is the possibility of the acquisition of resistance by the bacteria in contact with the antimicrobial films. As discussed by Marques et al. (2022a), several studies showed that subinhibitory concentrations of EOs could lead to more resistant bacterial strains, which could, in turn, become a serious public health issue. In this sense, it is also necessary to evaluate the risk of the developed active packaging to induce microbial resistance acquisition.

14.5 Safety Issues and Toxicological Aspects

In general, EOs are considered safer than synthetic preservatives because they are naturally derived from plants; however, being aware of the complexity of the components of these aromatic substances and the conclusions of toxicological assessments, comprehensive safety concerns and regulatory issues must be considered before applying to food/packaging systems, especially when it comes through nanotechnology (Prakash et al. 2018b). Unfortunately, until the moment, there is still a lack of regulation of EOs and nanotechnological-based EOs in food matrices and packaging.

In a broad aspect, Regulation 450/2009/EC established the regulatory aspects surrounding the application of active packaging for food contact in Europe, claiming other guidelines such as that nanotechnology could not be used without further evaluation, even when direct contact with packaged food is not considered because of a functional barrier (European Commission 2009; Arruda et al. 2022). Also, the released substance must be listed on the European country's Positive List of Additives, and its use must characterize a technological need (Regulation 1333/2008/EC, European Parliament and the Concil of the European Union 2008).

The use of EOs can have negative health consequences (i.e., especially considering oils containing phenol and aldehyde groups), including the irritation of some tissues such as mucous membranes, eyes, and skin (Sharma et al. 2022). The Federal Food Drugs and Cosmetics Act (FFDCA) describes that a different safety standard has to be considered for naturally occurring substances in foods, other than that applied to intentionally added food ingredients. However, as EOs are mostly employed as flavoring agents, and many of which are purposefully added to foods as separate chemicals, a current standard cannot be easily applied to the safety evaluation of these complex substances; they are neither a direct food additive nor food themselves, falling somewhere in the middle (Reis et al. 2022).

EU No. 10/2011 establish guidelines concerning plastic materials and articles intended to come into contact with food, specifying that nanosized particles must be assessed case by case before their incorporation into active food packaging materials (European Union 2011).

The interaction of nanosized EOs substances with the food system raises a concern about human and animal health. In other words, these components on the surface of the packaging material are not detrimental to human health, but their translocation and integration into food may be. Their toxicity is stimulated by dynamic, kinetic, and catalytic properties and functionalization, net particle reactivity, agglomeration, and functional environment (Mahmud et al. 2022). Toxicologic issues caused by the nanotechnological EOs are mainly because of their persistent, non-dissolvable, and nondegradable behavior (Nile et al. 2020).

During in vitro and (the few conducted) in vivo studies, the main toxicity mechanisms that have been investigated by observing changes in diverse biomarkers are, for example, levels of glutathione (GSH), inflammation response, DNA damage, cell death, and ROS generation, with special attention to the last one, being one of the most significant factors (Nile et al. 2020; Mahmud et al. 2022). However, despite the advances already evidenced, so far, there is no standard protocol for the toxicity testing of nanomaterials that assesses their excretion mechanisms after digestion (Guidotti-Takeuchi et al. 2022).

It is important to state that, although these systems have the potential to be implemented in the recent future by the food industry, some nanostructures, such as carbon nanotubes, raise concerns due to safety hazards and toxic effects caused by their possible migration/release from the packaging to the food. The impact on the environment due to the disposal of the nanocomposite packaging is also a concern (Kotsilkov et al. 2018; Emamhadi et al. 2020). The lack of information regarding exposure, availability, and toxicity to humans can be an important drawback to the industrial application of nano-EOs in food packaging. Thus, continuous research and developments along with comprehensive government regulations are required, especially on the migration/toxicity of these nanosubstances from the nanocomposite to packaged food and their potential impacts on human health and to the environment.

References

Abasalizadeh F, Moghaddam SV, Alizadeh E et al (2020) Alginate-based hydrogels as drug delivery vehicles in cancer treatment and their applications in wound dressing and 3D bioprinting. J Med Biol Eng 14(1):1–22

Abdalkarim SYH, Chen LM, Yu HY et al (2021) Versatile nanocellulose-based nanohybrids: a promising-new class for active packaging applications. Int J Biol Macromol 182. https://doi.org/10.1016/j.ijbiomac.2021.05.169

Abd-ElGawad AM, El Gendy AENG, Assaeed AM et al (2020) Essential oil enriched with oxygenated constituents from invasive plant *Argemone ochroleuca* exhibited potent phytotoxic effects. Plan Theory 9(8):998. https://doi.org/10.3390/plants9080998

Abka-Khajouei R, Tounsi L, Shahabi N et al (2022) Structures, properties and applications of alginates. Mar Drugs 20(6):18–21

Agüero L, Zaldivar-Silva D, Peña L, Dias M (2017) Alginate microparticles as oral colon drug delivery device: a review. Carbohydr Polym 168:32–43

Ahmed MW, Haque MA, Mohibbullah M et al (2022) A review on active packaging for quality and safety of foods: current trends, applications, prospects and challenges. Food Packag Shelf Life 33:100913

Alcázar-Alay SC, Meireles MAA (2015) Physicochemical properties, modifications and applications of starches from different botanical sources. Food Sci Technol 35:215–236

Alfei S, Marengo B, Zuccari G (2020) Nanotechnology application in food packaging: a plethora of opportunities *versus* pending risks assessment and public concerns. Food Res Int 137:109664. https://doi.org/10.1016/j.foodres.2020.109664

Alizadeh-Sani M, Mohammadian E, McClements DJ (2020) Eco-friendly active packaging consisting of nanostructured biopolymer matrix reinforced with TiO_2 and essential oil: application for preservation of refrigerated meat. Food Chem 322:126782. https://doi.org/10.1016/j.foodchem.2020.126782

Aljawish A, Chevalot I, Jasniewski J et al (2015) Enzymatic synthesis of chitosan derivatives and their potential applications. J Mol Catal B Enzym 112:25–39

Alkorta I, Garbisu C, Llama MJ, Serra JL (1998) Industrial applications of pectic enzymes: a review. Process Biochem 33(1):21–28

Allen TM, Cleland LG (1980) Serum-induced leakage of liposome contents. Biochim Biophys Acta Biomembr 597(2):418–426. https://doi.org/10.1016/0005-2736(80)90118-2

Almasi L, Radi M, Amiri S, McClements DJ (2021) Fabrication and characterization of antimicrobial biopolymer films containing essential oil-loaded microemulsions or nanoemulsions. Food Hydrocoll 117:106733. https://doi.org/10.1016/j.foodhyd.2021.106733

Almasi H, Azizi S, Amjadi S (2020) Development and characterization of pectin films activated by nanoemulsion and Pickering emulsion stabilized marjoram (*Origanum majorana* L.) essential oil. Food Hydrocoll 99:105338

Almoshari Y, Iqbal H, Razzaq A et al (2022) Development of nanocubosomes co-loaded with dual anticancer agents curcumin and temozolomide for effective colon cancer therapy. Drug Deliv 29(1):2633–2643. https://doi.org/10.1080/10717544.2022.2108938

ALSamman MT, Sánchez J (2022) Chitosan- and alginate-based hydrogels for the adsorption of anionic and cationic dyes from water. Polymers 14(8):1498–1521

Andra VVSN, Pammi SVN, Bhatraju LVKP, Ruddaraju LK (2022) A comprehensive review on novel liposomal methodologies, commercial formulations, clinical trials and patents. BioNanoScience 12:274–291. https://doi.org/10.1007/s12668-022-00941-x

de Andrade MF, de Silva IDL, da Silva GA et al (2020) A study of poly (butylene adipate-co-terephthalate)/orange essential oil films for application in active antimicrobial packaging. LWT - Food Sci Technol 125:109148. https://doi.org/10.1016/j.lwt.2020.109148

Andriani D, Apriyana AY, Karina M (2020) The optimization of bacterial cellulose production and its applications: a review. Cellulose 27:6747–6766

Angourani HR, Heydari M, Yousefi AR et al (2022) Nanoparticles based-plant protein containing *Rosmarinus officinalis* essential oil; fabrication, characterization, and evaluation. Appl Sci 12(19):9968. https://doi.org/10.3390/app12199968

Antonioli G, Fontanella G, Echeverrigaray S et al (2020) Poly(lactic acid) nanocapsules containing lemongrass essential oil for postharvest decay control: *in vitro* and *in vivo* evaluation against phytopathogenic fungi. Food Chem 326:126997. https://doi.org/10.1016/j.foodchem.2020.126997

Arruda TR, Bernardes PC, Moraes ARF, de Soares NFF (2022) Natural bioactives in perspective: the future of active packaging based on essential oils and plant extracts themselves and those complexed by cyclodextrins. Food Res Int 156:111160. https://doi.org/10.1016/j.foodres.2022.111160

Arruda TR, Marques CS, Soares NFF (2021) Native cyclodextrins and their derivatives as potential additives for food packaging: a review. Polysaccharides 2:825–842. https://doi.org/10.3390/polysaccharides2040050

Atta OM, Manan S, Shahzad A, Ul-Islam M et al (2022) Biobased materials for active food packaging: a review. Food Hydrocoll 125:107419

Ayyoob M, Lee DH, Kim JH, Nam SW, Kim YJ (2017) Synthesis of poly (glycolic acids) via solution polycondensation and investigation of their thermal degradation behaviors. Fibers Polym 18(3):407–415

Balla E, Daniilidis V, Karlioti G et al (2021) Poly(lactic acid): a versatile biobased polymer for the future with multifunctional properties - from monomer synthesis, polymerization techniques and molecular weight increase to PLA applications. Polymers 13(11):1822–1872

Bangham AD, Glover JC, Hollingshead S, Pethica BA (1962) The surface properties of some neoplastic cells. Nature 84(3):513–517. https://doi.org/10.1042/bj0840513

Barauskas J, Johnsson M, Joabsson F, Tiberg F (2005a) Cubic phase nanoparticles (cubosome): principles for controlling size, structure, and stability. Langmuir 21(6):2569–2577. https://doi.org/10.1021/la047590p

Barauskas J, Johnsson M, Tiberg F (2005b) Self-assembled lipid superstructures: beyond vesicles and liposomes. Nano Lett 5(8):1615–1619. https://doi.org/10.1021/nl050678i

Barauskas J, Razumas V, Talaikyteù Z et al (2003) Towards redox active liquid crystalline phases of lipids: a monoolein/water system with entrapped derivatives of ferrocene. Chem Phys Lipids 123(1):87–97. https://doi.org/10.1016/S0009-3084(02)00170-6

Bartelds R, Nematollahi MH, Tjeerd Pols T, Stuart MCA, Pardakhty A, Asadikaram, G, Poolman B (2018) Niosomes an alternative for liposomal delivery PLOS ONE 13(4): e0194179-10. https://doi.org/1371/journal.pone.0194179

Basiak E, Lenart A, Debeaufort F (2017) Effects of carbohydrate/protein ratio on the microstructure and the barrier and sorption properties of wheat starch-whey protein blend edible films. J Sci Food Agric 97(3):858–867

Bastos LPH, dos Santos CHC, de Carvalho MG, Garcia-Rojas EE (2020) Encapsulation of the black pepper (*Piper nigrum* L.) essential oil by lactoferrin-sodium alginate complex coacervates: structural characterization and simulated gastrointestinal conditions. Food Chem 316:126345

Begines B, Ortiz T, Pérez-Aranda M, Martínez G et al (2020) Polymeric nanoparticles for drug delivery: recent developments and future prospects. Nano 10(7):1403–1441

Bei D, Marszalek J, Youan BBC (2009) Formulation of dacarbazine-loaded cubosomes—part I: influence of formulation variables. AAPS PharmSciTech 10:1032. https://doi.org/10.1208/s12249-009-9293-3

Bondu C, Yen FT (2022) Nanoliposomes, from food industry to nutraceuticals: Interests and uses. Innov Food Sci Emerg Technol (81): 103140. https://doi.org/10.1016/j.ifset.2022.103140

Bora H, Kamle M, Mahato DK, Tiwari P, Kumar P (2020) Citrus essential oils (ceos) and their applications in food: an overview. Plan Theory 9(3):357

Boro U, Priyadarsini A, Moholkar VS (2022) Synthesis and characterization of poly(lactic acid)/clove essential oil/alkali-treated halloysite nanotubes composite films for food packaging applications. Int J Biol Macromol 216:927–939. https://doi.org/10.1016/j.ijbiomac.2022.07.209

Brandl MM, Bachman D, Drechsler M, Bauer KH (1993) Liposome preparation using high-pressure homogenizers. In: Gregoriadis G (ed) Liposome technology, 2nd edn. CRC Press, Boca Raton, FL, pp 49–65

Cáceres, M, Hidalgo, W, Stashenko et al (2020) Essential oils of aromatic plants with antibacterial, anti-biofilm and anti-*quorum sensing* activities against pathogenic bacteria. Antibiotics 9(4). https://doi.org/10.3390/antibiotics9040147

Calo JR, Crandall PG, O'Bryan CA, Ricke SC (2015) Essential oils as antimicrobials in food systems: a review. Food Control 54:111–119

Carpena M, Nuñez-Estevez B, Soria-Lopez A et al (2021) Essential oils and their application on active packaging systems: a review. Resources 10:1–20. https://doi.org/10.3390/resources10010007

Cauchetier E, Fessi H, Boulard Y et al (1999) Preparation and physicochemical characterization of atovaquone-containing liposomes. Drug Dev Res 47(4):155–161. https://doi.org/10.1002/(sici)1098-2299

Chamas A, Moon H, Zheng J et al (2020) Degradation rates of plastics in the environment. ACS Sustain Chem Eng 8(9):3494–3511

Chandel V, Biswas D, Roy S et al (2022) Current advancements in pectin: extraction, properties and multifunctional applications. Foods 11(17):2683–2703

Chaudhary V, Kajla P, Kumari P et al (2022) Milk protein-based active edible packaging for food applications: an eco-friendly approach. Front Nutr 9:942524

Chauhan PS, Saxena A (2016) Bacterial carrageenases: an overview of production and biotechnological applications. Biotech 6(2) 146:1–18

Chougule M, Padhi B, Misra A (2008) Development of spray dried liposomal dry powder inhaler of dapsone. AAPS PharmSciTech 9:47–53. https://doi.org/10.1208/s12249-007-9024-6

Clogston J, Rathman J, Tomasko D et al (2000) Phase behavior of a monoacylglycerol:(myverol 18-99K)/water system. Chem Phys Lipids 107(2):191–220. https://doi.org/10.1016/S0009-3084(00)00182-1

Cornejo-Ramírez YI, Martínez-Cruz O, Del Toro-Sánchez CL et al (2018) The structural characteristics of starches and their functional properties. CyTA J Food 16(1):1003–1017

Das S, Ghosh A, Mukherjee A (2021) Nanoencapsulation-based edible coating of essential oils as a novel green strategy against fungal spoilage, mycotoxin contamination, and quality deterioration of stored fruits: an overview. Front Microbiol 12:1–14. https://doi.org/10.3389/fmicb.2021.768414

De Matos Costa AR, Crocitti A, Hecker de Carvalho LH et al (2020) Properties of biodegradable films based on poly(butylene succinate) (PBS) and poly(butylene adipate-*co*-terephthalate) (PBAT) blends. Polymers 12(10):2317–2334

de Oliveira EF, Paula HCB, Paula RCM, d. (2014) Alginate/cashew gum nanoparticles for essential oil encapsulation. Colloids Surfaces B Biointerfaces 113:146–151. https://doi.org/10.1016/j.colsurfb.2013.08.038

Delgado-Pando G, Ekonomou SI, Stratakos AC, Pintado T (2021) Clean labels alternatives in meat products. Foods 10(7):1615. https://doi.org/10.3390/foods10071615

Delshadi R, Bahrami A, Tafti AG, Barba FJ, Williams LL (2020) Micro and nano-encapsulation of vegetable and essential oils to develop functional food products with improved nutritional profiles. Trends in Food Sci Technol 104:72–83. https://doi.org/10.1016/j.tifs.2020.07.004

Dias MV, Sousa MM, Lara BRB, de Azevedo VM, Soares NFF, Borges SV, Queiroz F (2018) Thermal and morphological properties and kinetics of diffusion of antimicrobial films on food and a simulant, Food Packag. Shelf Life (16):15–22. https://doi.org/10.1016/j.fpsl.2018.01.007

Díaz-Montes E, Castro-Muñoz R (2021) Trends in chitosan as a primary biopolymer for functional films and coatings manufacture for food and natural products. Polymers 13(5):767–791

Donsì F, Ferrari G (2016) Essential oil nanoemulsions as antimicrobial agents in food. J Biotechnol 10(233):106–120. https://doi.org/10.1016/j.jbiotec.2016.07.005

Doost AS, Nasrabadi MN, Kassozi V et al (2020) Recent advances in food colloidal delivery systems for essential oils and their main components. Trends in Food Sci Technol 99:474–486

Dos Santos TPR, Leonel M, Garcia ÉL et al (2016) Crystallinity, thermal and pasting properties of starches from different potato cultivars grown in Brazil. Int J Biol Macromol 82:144–149

Drummond CJ, Fong C (1999) Surfactant self-assembly objects as novel drug Deliv vehicles. Curr Opin Colloid Interface Sci 4(6):449–456. https://doi.org/10.1016/S1359-0294(00)00020-0

El Khetabi A, Lahlali R, Ezrari S et al (2022) Role of plant extracts and essential oils in fighting against postharvest fruit pathogens and extending fruit shelf life: a review. Trends in Food Sci Technol 120:402–417

El-Sayed HS, El-Sayed SM (2021) A modern trend to preserve white soft cheese using nano-emulsified solutions containing cumin essential oil. Environ Nanotechnol Monit Manag 16:100499. https://doi.org/10.1016/j.enmm.2021.100499

Emamhadi MA, Sarafraz M, Akbari M et al (2020) Nanomaterials for food packaging applications: a systematic review. Food Chem Toxicol 146:111825. https://doi.org/10.1016/j.fct.2020.111825

European Commission (2009) Commission regulation (EC) no. 450/2009 on active and intelligent materials and articles intended to come into contact with food. Off J Eur Union

European Parliament and the Concil of the European Union (2008) Regulation (EC) no 1333/2008 of the European Parliament and of the council of 16 December 2998 on food additives. Off J Eur Union:16–33

European Union (2011) Commission regulation (EU) no 10/2011 of 14 January 2011. Off J Eur Union:1–89

Fani N, Enayati MH, Rostamabadi H, Falsafi SR (2022) Encapsulation of bioactives within electrosprayed κ-carrageenan nanoparticles. Carbohydr Polym 294:119761

Farag DBE, Yousry C, Al-Mahallawi AM et al (2022) The efficacy of *Origanum majorana* nanocubosomal systems in ameliorating submandibular salivary gland alterations in streptozotocin-induced diabetic rats. Drug Deliv 29(1):62–74. https://doi.org/10.1080/10717544.2021.2018522

Fattahian A, Fazlara A, Maktabi S, Pourmahdi M, Bavarsad N (2022) The effects of chitosan containing nano-capsulated *Cuminum cyminum* essential oil on the shelf-life of veal in modified atmosphere packaging. Food Measure 16:920–933. https://doi.org/10.1007/s11694-021-01213-0

Favier V, Chanzy H, Cavaille JY (1995) Polymer nanocomposites reinforced by cellulose whiskers. Macromolecules 28:6365–6367

Ferreira RR, Souza AG, Quispe YM, Rosa DS (2021) Essential oils loaded-chitosan nanocapsules incorporation in biodegradable starch films: a strategy to improve fruits shelf life. Int J Biol Macromol 188:628–638

Figueroa-Lopez KJ, Enescu D, Torres-Giner S et al (2020) Development of electrospun active films of poly(3-hydroxybutyrate-co-3-hydroxyvalerate) by the incorporation of cyclodextrin inclusion complexes containing oregano essential oil. Food Hydrocoll 108:106013. https://doi.org/10.1016/j.foodhyd.2020.106013

Franks F (1998) Freeze-drying of bioproducts: putting principles into practice European Journal of Pharmaceutics and Biopharmaceutics 45(3):221–229. https://doi.org/10.1016/S0939-6411(98)00004-6

Frank K, Garcia CV, Shin GH, Kim JT (2018) Alginate biocomposite films incorporated with cinnamon essential oil nanoemulsions: physical, mechanical, and antibacterial properties. Int J Polym Sci:1–8

Freitas CMP, Coimbra JSR, Souza VGL, Sousa RCS (2021) Structure and applications of pectin in food, biomedical, and pharmaceutical industry: a review. Coatings 11:922–944

Gadgil BST, Killi N, Rathna GV (2017) Polyhydroxyalkanoates as biomaterials. MedChemComm 8(9):1774–1787

Galié S, García-Gutiérrez C, Miguélez EM et al (2018) Biofilms in the food industry: health aspects and control methods. Front Microbiol 9:1–18. https://doi.org/10.3389/fmicb.2018.00898

García-Guzmán L, Cabrera-Barjas G, Soria-Hernández CG et al (2022) Progress in starch-based materials for food packaging applications. Polysaccharides 3(1):136–177

Garg G, Saraf S, Saraf S (2007) Cubosomes: an overview. Biol Pharmaceut Bull 30(2):350–353. https://doi.org/10.1248/bpb.30.350

Gasti T, Dixit S, Hiremani VD et al (2022) Chitosan/pullulan-based films incorporated with clove essential oil loaded chitosan-ZnO hybrid nanoparticles for active food packaging. Carbohydr Polym 277:118866

Geyer R, Jambeck JR, Law KL (2017) Production, use, and fate of all plastics ever made. Sci Adv 3(7):e1700782

Ghabraie M, Vu KD, Tata L, Salmieri S, Lacroix M (2016) Antimicrobial effect of essential oils in combinations against five bacteria and their effect on sensorial quality of ground meat. LWT - Food Sci Technol 66:332–339. https://doi.org/10.1016/j.lwt.2015.10.055

Ghasemi S, Jafari SM, Assadpour E, Khomeiri M (2018) Nanoencapsulation of D-limonene within nanocarriers produced by pectin-whey protein complexes. Food Hydrocoll 77:152–162. https://doi.org/10.1016/j.foodhyd.2017.09.030

Ghorani B, Kadkhodaee R, Rajabzadeh G, Tucker N (2019) Assembly of odour adsorbent nanofilters by incorporating cyclodextrin molecules into electrospun cellulose acetate webs. React Funct Polym 134:121–132. https://doi.org/10.1016/j.reactfunctpolym.2018.11.014

Ghoshal G, Shivani (2022) Thyme essential oil nano-emulsion/tamarind starch/whey protein concentrate novel edible films for tomato packaging. Food Control 138:108990. https://doi.org/10.1016/j.foodcont.2022.108990

Gouda A, Sakr OS, Nasr M, Sammour O (2021) Ethanol injection technique for liposomes formulation: an insight into development, influencing factors, challenges and applications. J Drug Deliv Sci Technol 61:102174. https://doi.org/10.1016/j.jddst.2020.102174

Guidotti-Takeuchi M, de Ribeiro LNM, dos Santos FAL et al (2022) Essential oil-based nanoparticles as antimicrobial agents in the food industry. Microorganisms 10. https://doi.org/10.3390/microorganisms10081504

Guimarães D, Cavaco-Paulo A, Nogueira E (2021) Design of liposomes as drug Deliv system for therapeutic applications. Pharmaceutics 60:120571. https://doi.org/10.1016/j.ijpharm.2021.120571

Gupta RK, Guha P, Srivastav PP (2022) Natural polymers in bio-degradable/edible film: a review on environmental concerns, cold plasma technology and nanotechnology application on food packaging- a recent trends. Food Chem Adv 1:1–15

Habib L, Alyan M, Ghantous,Y et al (2022) A mucoadhesive patch loaded with freeze-dried liposomes for the local treatment of oral tumors. Drug Deliv Transl Res https://doi.org/10.1007/s13346-022-01224-4

Hammam AR (2019) Technological, applications, and characteristics of edible films and coatings: a review. SN Appl Sci 1(6):1–11

He Q, Zhang L, Yang Z et al (2022) Antibacterial mechanisms of thyme essential oil nanoemulsions against *Escherichia coli* O157:H7 and *Staphylococcus aureus*: alterations in membrane compositions and characteristics. Innov Food Sci Emerg Technol 75:102902. https://doi.org/10.1016/j.ifset.2021.102902

Holland JW, Hui C, Cullis PR, Madden TD (1996) Madden poly(ethylene glycol)−lipid conjugates regulate the calcium-induced fusion of liposomes composed of phosphatidylethanolamine and phosphatidylserine. Biochemistry 35(8):2618–2624. https://doi.org/10.1021/bi952000v

Horstmann SW, Lynch KM, Arendt EK (2017) Starch characteristics linked to gluten-free products. Foods 6(4):29–50

Hu X, Liu S, Zhou G et al (2014) Electrospinning of polymeric nanofibers for drug deliv applications. J Control Release 185:12–21. https://doi.org/10.1016/j.jconrel.2014.04.018
Huang F, Anslyn EV (2015) Introduction: supramolecular chemistry. Chem Rev 115(15):6999–7000
Huang J, Hu Z, Hu L, Li G, Yao Q, Hu Y (2021) Pectin-based active packaging: a critical review on preparation, physical properties and novel application in food preservation. Trends in Food Sci Technol 118:167–178
Huo W, Wei D, Zhu W, Li Z, Jiang Y (2018) High-elongation zein films for flexible packaging by synergistic plasticization: preparation, structure and properties. J Cereal Sci 79:354–361
Hyde S, Andersson S, Eriksson B et al (1984) A cubic structure consisting of a lipid bilayer forming an infinite periodic minimum surface of the gyroid type in the glycerolmonooleat-water system. Z Kristallog 168:213–219. https://doi.org/10.1524/zkri.1984.168.14.213
Ibrahim ID, Hamam Y, Sadiku ER et al (2022) Need for sustainable packaging: an overview. Polymers 14(20):4430. https://doi.org/10.3390/polym14204430
Ingvarsson PT, Yang M, Nielsen HM, Rantanen J, Foged C (2011) Stabilization of liposomes during drying. Expert Opinion on Drug Deliv 8(3):375–388. https://doi.org/10.1517/17425247.2011.553219
Israelachvili J, Marcelja S, Horn R (1980) Physical principles of membrane organization. Q Rev Biophys 13(2):121–200. https://doi.org/10.1017/S0033583500001645
Ivanova K, Ivanova A, Ramon E, Hoyo J et al (2020) Antibody-enabled antimicrobial nanocapsules for selective elimination of *Staphylococcus aureus*. ACS App Mat Interfaces 12:35918–35927
Jafarzadeh S, Salehabadi A, Mohammadi Nafchi A et al (2021) Cheese packaging by edible coatings and biodegradable nanocomposites; improvement in shelf life, physicochemical and sensory properties. Trends in food. Sci Technol 116:218–231. https://doi.org/10.1016/j.tifs.2021.07.021
Jancikova S, Dordevic D, Jamroz E, Behalova H, Tremlova B (2020) Chemical and physical characteristics of edible films, based on κ- and ι-carrageenans with the addition of lapacho tea extract. Foods 9:357–372
Jawahar N, Meyyanathan S (2012) Polymeric nanoparticles for drug deliv and targeting: a comprehensive review. Int J Health Allied Sci 1:217–223
Jiang W-x, Qi J-r, Liao J-s, Yang X-q (2021) Acid/ethanol induced pectin gelling and its application in emulsion gel. Food Hydrocoll 118:106774
Jin T, Liu T, Lam E, Moores A (2021) Chitin and chitosan on the nanoscale. Nanoscale Horizons 6(7):505–542
Johnson SM, Bangham AD, Hill MW, Korn ED (1971) Single bilayer liposomes. Biochim Bioophys ACTA 233:820–826. https://doi.org/10.1016/0005-2736(71)90184-2
Ju J, Chen X, Xie Y, Yu H et al (2019) Application of essential oil as a sustained release preparation in food packaging. Trends in Food Sci Technol 92:22–32
Kamkar A, Molaee-aghaee E, Khanjari A et al (2021) Nanocomposite active packaging based on chitosan biopolymer loaded with nano-liposomal essential oil: its characterizations and effects on microbial, and chemical properties of refrigerated chicken breast fillet. Int J Food Microbiol 342:109071. https://doi.org/10.1016/j.ijfoodmicro.2021.109071
Katona G, Andreasson U, Landau EM, Andreasson LE, Neutze RJ (2003) Lipidic cubic phase crystal structure of the photosynthetic reaction Centre from *Rhodobacter sphaeroides* at 2.35 Å resolution. J Mol Biol 331(3):681–692. https://doi.org/10.1016/S0022-2836(03)00751-4
Kaya DA, Ghica MV, Danila E et al (2020) Selection of optimal operating conditions for extraction of *Myrtus communis* L. essential oil by the steam distillation method. Molecules 25(10):2399
Kazemeini H, Azizian A, Adib H (2021) Inhibition of *listeria monocytogenes* growth in Turkey fillets by alginate edible coating with *Trachyspermum ammi* essential oil nano-emulsion. Int J Food Microbiol 344:109104. https://doi.org/10.1016/j.ijfoodmicro.2021.109104
Kfoury M, Landy D, Fourmentin S (2018) Characterization of cyclodextrin/volatile inclusion complexes: a review. Molecules 23(5):1204–1227
Kikuchi H, Carlsson A, Yachi K, Hirota S (1991) Possibility of heat sterilisation of liposomes. Chem Pharm Bull 39(4):1018–1022. https://doi.org/10.1248/cpb.39.1018

Klemm D, Heublein B, Fink HP, Bohn A (2005) Cellulose: fascinating biopolymer and sustainable raw material. Angew Chem Int Ed 44:3358–3393

Kotsilkov S, Ivanov E, Vitanov NK (2018) Release of graphene and carbon nanotubes from biodegradable poly(lactic acid) films during degradation and combustion: risk associated with the end-of-life of nanocomposite food packaging materials. Materials 11(12):2346. https://doi.org/10.3390/ma11122346

Kozma M, Acharya B, Bissessur R (2022) Chitin, chitosan, and nanochitin: extraction, synthesis, and applications. Polymers 14(19):3989–4017

Kryeziu TL, Halogi E, Loshaj-Shala A et al (2022) Nanoencapsulation of *Origanum vulgare* essential oil into liposomes with anticancer potential. Pharmazie 77:172–178. https://doi.org/10.1691/ph.2022.1230

Kumar VA, Hasan M, Mangaraj S et al (2022) Trends in edible packaging films and its prospective future in food: a review. Appl Food Res 2(1):100118

Kumar L, Ramakanth D, Akhila K, Gaikwad KK (2021a) Edible films and coatings for food packaging applications: a review. Environ Chem Lett:1–26

Kumar V, Sehgal R, Gupta R (2021b) Blends and composites of polyhydroxyalkanoates (PHAs) and their applications. Eur Polym J 161:110824

Lamarra J, Calienni MN, Rivero S, Pinotti A (2020) Electrospun nanofibers of poly(vinyl alcohol) and chitosan-based emulsions functionalized with cabreuva essential oil. Int J Biol Macromol 160:307–318. https://doi.org/10.1016/j.ijbiomac.2020.05.096

Laouini A, Jaafar-Maalej C, Limayerm-Blouza I, Sfar S, Charcosset C, Fessi H (2012) Characterization and applications of liposomes: state of the art. J Colloid Sci Biotechnol 1(2):147–168. https://doi.org/10.1166/jcsb.2012.1020

Larsson K (1983) Two cubic phases in monoolein–water system. Nature 304:664

Lasič D (1992) Liposomes. Am Sci 80(1):20–31

Lasič DD (1982) A molecular model for vesicle formation. Biochim Biophys Acta Biomembr 692:501–502. https://doi.org/10.1016/0005-2736(82)90404-7

Lasič DD (1995) Mechanisms of liposome formation. J Liposome Res 5:431–441. https://doi.org/10.3109/08982109509010233

Lasič DD (1998) Novel applications of liposomes. Trends Biotechnol 16(7):307–321. https://doi.org/10.1016/S0167-7799(98)01220-7

Lasič DD, Papadjopoulos D (1995) Liposomes revisited. Science 267(5202):1275–1276. https://doi.org/10.1126/science.7871422

Lee AG (2003) Lipid–protein interactions in biological membranes: a structural perspective. Biochim Biophys Acta Biomembr 1612(1):1–40. https://doi.org/10.1016/S0005-2736(03)00056-7

Lee C (2020) Carrageenans as broad-spectrum microbicides: current status and challenges. Mar Drugs 18(9):435–462

Leserman L, Machy P, Zelphati O (1994) Immunoliposome-mediated delivery of nucleic acids: a review of our laboratory's experience. J Liposome Res 4:107–119. https://doi.org/10.3109/08982109409037032

Li B, Elango J, Wu W (2020) Recent advancement of molecular structure and biomaterial function of chitosan from marine organisms for pharmaceutical and nutraceutical application. Appl Sci 10(14):4719–4753

Liao W, Badri W, Dumas E, Ghnimi S, Elaissari A, Saurel R, Gharsallaoui A (2021) Nanoencapsulation of essential oils as natural food antimicrobial agents: an overview. Appl Sci 11(13):5778. https://doi.org/10.3390/app11135778

Lima AL, Gratieri T, Cunha-Filho M, Gelfuso GM (2022) Polymeric nanocapsules: a review on design and production methods for pharmaceutical purpose. Methods 199:54–66

Lin J, Jiao G, Kermanshahi-pour A (2022a) Algal polysaccharides-based hydrogels: extraction, synthesis, characterization, and applications. Mar Drugs 20(5):306

Lin W, Goldberg R, Klein J (2022b) Poly-phosphocholination of liposomes leads to highly-extended retention time in mice joints. J Pharm Sci 10:2820–2827. https://doi.org/10.1039/D1TB02346B

Liu D, Zhao P, Chen J, Yan Y, Wu Z (2022) Recent advances and applications in starch for intelligent active food packaging: a review. Foods 11(18):2879–2890

Liu Y, Ahmed S, Sameen DE et al (2021) A review of cellulose and its derivatives in biopolymer-based for food packaging application. Trends Food Sci Technol 112:532–546

Lopez-Gil A, Rodriguez-Perez MA, De Saja JA (2014) Strategies to improve the mechanical properties of starch-based materials: plasticization and natural fibers reinforcement. Polímeros 24:36–42

Lu Y, Luo Q, Chu Y, Tao N, Deng S, Wang L, Li L (2022) Application of gelatin in food packaging: a review. Polymers 14(3):436–455

Ludwicka K, Kaczmarek M, Białkowsk A (2020) Bacterial nanocellulose - a biobased polymer for active and intelligent food packaging applications: recent advances and developments. Polymers 12(10):1–23

Lutton ES (1965) Phase behavior of aqueous systems of monoglycerides. J Am Oil Chem Soc 42:1068–1070. https://doi.org/10.1007/BF02636909

Ma L, Yang H, Ma M, Zhang X, Zhang Y (2018) Mechanical and structural properties of rabbit skin gelatin films. Int J Food Prop 21(1):1203–1218

Maherani B, Arab-Arab-Tehrany E, Mozafari MR et al (2011) Liposomes: a review of manufacturing techniques and targeting strategies. Curr Nanosci 7:436–452. https://doi.org/10.2174/157341311795542453

Mahmud J, Sarmast E, Shankar S, Lacroix M (2022) Advantages of nanotechnology developments in active food packaging. Food Res Int 154:111023. https://doi.org/10.1016/j.foodres.2022.111023

Marques CS, Arruda TR, Silva RRA et al (2022a) Exposure to cellulose acetate films incorporated with garlic essential oil does not lead to homologous resistance in *listeria innocua* ATCC 33090. Food Res Int 160:111676. https://doi.org/10.1016/j.foodres.2022.111676

Marques CS, Silva RRA, Arruda TR et al (2022b) Development and investigation of zein and cellulose acetate polymer blends incorporated with garlic essential oil and β-cyclodextrin for potential food packaging application. Polysaccharides 3:277–291. https://doi.org/10.3390/polysaccharides3010016

Marques CS, Dias MV, Soares NFF et al (2021) Ultrastructural and antimicrobial impacts of allyl isothiocyanate incorporated in cellulose, β-cyclodextrin, and carbon nanotubes nanocomposites. J Vinyl Addit Technol 1:1–11. https://doi.org/10.1002/vnl.21850

Marques CS, Grillo RP, Bravim DG et al (2019a) Preservation of ready-to-eat salad: a study with combination of sanitizers, ultrasound, and essential oil-containing β-cyclodextrin inclusion complex. LWT - Food Sci Technol 115:108433. https://doi.org/10.1016/j.lwt.2019.108433

Marques CS, Carvalho SG, Bertoli LD et al (2019b) β-Cyclodextrin inclusion complexes with essential oils: obtention, characterization, antimicrobial activity and potential application for food preservative sachets. Food Res Int 119:499–509. https://doi.org/10.1016/j.foodres.2019.01.016

Marsin AM, Muhamad II, Anis SNS et al (2020) Essential oils as insect repellent agents in food packaging: a review. Eur Food Res Technol 246:1519–1532. https://doi.org/10.1007/s00217-020-03511-1

Martínez-Sanz M, Gómez-Mascaraque LG, Ballester AR et al (2019) Production of unpurified agar-based extracts from red seaweed *Gelidium sesquipedale* by means of simplified extraction protocols. Algal Res 38:101420

Melro E, Antunes FE, da Silva GJ et al (2020) Chitosan films in food applications. Tuning film properties by changing acidic dissolution conditions. Polymers 13(1):1–12

Mihalca V, Kerezsi AD, Weber A et al (2021) Protein-based films and coatings for food industry applications. Polymers 13:769–792

Mir SA, Dar BN, Wani AA, Shah MA (2018) Effect of plant extracts on the techno-functional properties of biodegradable packaging films. Trends in Food Sci Technol 80:141–154. https://doi.org/10.1016/j.tifs.2018.08.004

Miranda M, Sun X, Marín A et al (2022) Nano- and micro-sized carnauba wax emulsions-based coatings incorporated with ginger essential oil and hydroxypropyl methylcellulose on papaya:

preservation of quality and delay of post-harvest fruit decay. Food Chem X 13:100249. https://doi.org/10.1016/j.fochx.2022.100249

Moghassemi S, Hadjizadeh A (2014) Nano-niosomes as nanoscale drug Deliv systems: an illustrated review. J Control Release 185:22–36. https://doi.org/10.1016/j.jconrel.2014.04.015

Mohammadi Nafchi A, Moradpour M, Saeidi M, Alias AK (2013) Thermoplastic starches: properties, challenges, and prospects. Starch-Stärke 65(1–2):61–72

Moraczewski K, Pawłowska A, Stepczýnska M et al (2020) Plant extracts as natural additives for environmentally friendly polylactide films. Food Packaging Shelf Life 26. https://doi.org/10.1016/j.fpsl.2020.100593

Moradi F, Hadi N, Bazargani A (2020) Evaluation of quorum-sensing inhibitory effects of extracts of three traditional medicine plants with known antibacterial properties. New Microbes New Infect 38:100769. https://doi.org/10.1016/j.nmni.2020.100769

Mostafavi FS, Zaeim D (2020) Agar-based edible films for food packaging applications - a review. Int J Biol Macromol 159:1165–1176

Moulay S (2015) Poly (vinyl alcohol) functionalizations and applications. Polym Plast Technol Eng 54(12):1289–1319

Mozafari M, Johnson C, Hatziantoniou S, Demetzos C (2008) Nanoliposomes and their applications in food nanotechnology. J Liposome Res 18:309–327. https://doi.org/10.1080/08982100802465941

Mozafari MR (2005) Liposomes: an overview of manufacturing techniques. Cell Mol Biol Lett 10:711–719

Mozafari MR, Reed CJ, Rostron C, Martin DS (2004) Transfection of human airway epithelial cells using a lipid-based vector prepared by the heating method. J Aerosol Med 17(14):54–58

Muñoz-Shugulí C, Vidal CP, Cantero-López P, Lopez-Polo J (2021) Encapsulation of plant extract compounds using cyclodextrin inclusion complexes, liposomes, electrospinning and their combinations for food purposes. Trends in Food Sci Technol 108:177–186. https://doi.org/10.1016/j.tifs.2020.12.020

Natrajan D, Srinivasan S, Sundar K, Ravindran A (2015) Formulation of essential oil-loaded chitosan-alginate nanocapsules. J Food Drug Anal 23:560–568. https://doi.org/10.1016/j.jfda.2015.01.001

Ni ZJ, Wang X, Shen Y et al (2021) Recent updates on the chemistry, bioactivities, mode of action, and industrial applications of plant essential oils. Trends Food Sci Technol 110:78–89

Nielsen TD, Hasselbalch J, Holmberg K, Stripple J (2020) Politics and the plastic crisis: a review throughout the plastic life cycle. Wiley Interdiscipl Rev Energy Environ 9(1):e360

Nile SH, Baskar V, Selvaraj D et al (2020) Nanotechnologies in food science: applications, recent trends, and future perspectives. Nano-Micro Letters 12:45. https://doi.org/10.1007/s40820-020-0383-9

Nionelli L, Pontonio E, Gobbetti M, Rizzello CG (2018) Use of hop extract as antifungal ingredient for bread making and selection of autochthonous resistant starters for sourdough fermentation. Int J Food Microbiol 266:173–182. https://doi.org/10.1016/j.ijfoodmicro.2017.12.002

Noori S, Zeynali F, Almasi H (2018) Antimicrobial and antioxidant efficiency of nanoemulsion-based edible coating containing ginger (*Zingiber officinale*) essential oil and its effect on safety and quality attributes of chicken breast fillets. Food Control 84:312–320

Norling T, Lading P, Engstrom S, Larsson K, Krog N, Nissen SS (1992) Formulation of a drug Deliv system based on a mixture of monoglycerides and triglycerides for use in the treatment of periodontal disease. J Clin Periodontol 19(9):687–692. https://doi.org/10.1111/j.1600-051X.1992.tb02529.x

Nthunya LN, Masheane ML, Malinga SP et al (2017) Greener approach to prepare electrospun antibacterial β-cyclodextrin/cellulose acetate nanofibers for removal of bacteria from water. ACS Sustain Chem Eng 5(1):153–160. https://doi.org/10.1021/acssuschemeng.6b01089

Nylander T, Mattisson C, Razumas V, Miezis Y, Håkansson B (1996) A study of entrapped enzyme stability and substrate diffusion in a monoglyceride-based cubic liquid crystalline phase. Colloids Surf A Physicochem Eng Asp 114:311–320. https://doi.org/10.1016/0927-7757(96)03563-7

Ohishi K, Tsuchiya K, Ogura T et al (2022) Characterization of lecithin liposomes prepared by polyol dilution method using 1,3-butylene glycol. Colloids Surf A Physicochem Eng Asp 650:129592. https://doi.org/10.1016/j.colsurfa.2022.129592

Otake K, Imura T, Sakai H, Abe M (2001) Development of a new preparation method of liposomes using supercritical carbon dioxide. Langmuir 17(13):3898–3901. https://doi.org/10.1021/la010122k

Papahadjopoulos D, Nir S, Duzgunes N (1990) Molecular mechanisms of calcium-induced membrane fusion. J Bioenerg Biomembr 22(2):157–179. https://doi.org/10.1007/BF00762944

Pardakhty A, Moazeni E (2013) Nano-niosomes in drug, vaccine and gene delivery: a rapid overview. Nanomed J 1(1):1–12

Partheniadis I, Nikolakakis I, Laidmäe I, Heinämäki J (2020) A mini-review: needleless electrospinning of nanofibers for pharmaceutical and biomedical applications. PRO 8. https://doi.org/10.3390/PR8060673

Partheniadis I, Stathakis G, Tsalavouti D et al (2022) Essential oil—loaded nanofibers for pharmaceutical and biomedical applications: a systematic mini-review. Pharmaceutics 14. https://doi.org/10.3390/pharmaceutics14091799

Petchwattana N, Naknaen P, Cha-Aim K, Sanetuntikul J (2021) Application of antimicrobial plates in food packaging as an alternative way for food waste minimization. Int J Sustain Eng 14(4):600–608

Piacentini E, Russo B, Bazzarelli F, Giorno L (2022) Membrane nanoprecipitation: from basics to technology development. J Membr Sci 654:120564. https://doi.org/10.1016/j.memsci.2022.120564

Pires JRA, Souza VGL, Fernando AL (2018) Chitosan/montmorillonite bionanocomposites incorporated with rosemary and ginger essential oil as packaging for fresh poultry meat. Food Packag Shelf Life 17:142–149. https://doi.org/10.1016/j.fpsl.2018.06.011

Pisoschi AM, Pop A, Georgescu C et al (2018) An overview of natural antimicrobials role in food. Eur J Med Chem 143:922–935. https://doi.org/10.1016/j.ejmech.2017.11.095

Pola CC, Medeiros EAA, Pereira OL et al (2016) Cellulose acetate active films incorporated with oregano (*Origanum vulgare*) essential oil and organophilic montmorillonite clay control the growth of phytopathogenic fungi. Food Packag Shelf Life 9:69–70. https://doi.org/10.1016/j.fpsl.2016.07.001

Prakash B, Kujur A, Yadav A et al (2018a) Nanoencapsulation: an efficient technology to boost the antimicrobial potential of plant essential oils in food system. Food Control 89:1–11

Prakash A, Baskaran R, Paramasivam N, Vadivel V (2018b) Essential oil based nanoemulsions to improve the microbial quality of minimally processed fruits and vegetables: a review. Food Res Int 111:509–523. https://doi.org/10.1016/j.foodres.2018.05.066

Prakash B, Kiran S (2016) Essential oils: a traditionally realized natural resource for food preservation. Curr Sci 110:1890–1892

Prakash B, Mishra PK, Kedia A, Dubey NK (2014) Antifungal, antiaflatoxin and antioxidant potential of chemically characterized *Boswellia carterii* Birdw essential oil and its in vivo practical applicability in preservation of *Piper nigrum* L. fruits. LWT - Food Sci Technol 56:240–247

Pudlarz A, Szemraj J (2018) Nanoparticles as carriers of proteins, peptides and other therapeutic molecules. Open life sciences 13(1):285–298

Qin Z, Zou Y, Zhang Y, Wang P, Zhang H (2022) Electrospun pullulan nanofiber loading *Zanthoxylum bungeanum* essential oil/β-cyclodextrin inclusion complexes for active packaging. Int J Biol Macromol 210:465–474. https://doi.org/10.1016/j.ijbiomac.2022.04.155

Qiu H, Caffrey M (2000) The phase diagram of the monoolein/water system: metastability and equilibrium aspects. Biomaterials 21(3):223–234. https://doi.org/10.1016/S0142-9612(99)00126-X

Rafiqah SA, Khalina A, Harmaen AS et al (2021) A review on properties and application of bio-based poly (butylene succinate). Polymers 13(9):1436. pp. 1-28

Ramos M, Valdés A, Beltrán A, Garrigós MC (2016) Gelatin-based films and coatings for food packaging applications. Coatings 6(4):41–61

Rather AH, Wani TU, Khan RS et al (2021) Prospects of polymeric nanofibers loaded with essential oils for biomedical and food-packaging applications. Int J Mol Sci 22. https://doi.org/10.3390/ijms22084017

Ravishankar S, Jaroni D, Zhu L et al (2012) Inactivation of *listeria monocytogenes* on ham and Bologna using pectin-based apple, carrot, and hibiscus edible films containing carvacrol and cinnamaldehyde. J Food Sci 77:M377–M382

Razavi MS, Golmohammadi A, Nematollahzadeh A et al (2022) Cinnamon essential oil encapsulated into a fish gelatin-bacterial cellulose nanocrystals complex and active films thereof. Food Biophys 17(1):38–46

Rehman A, Jafari SM, Aadil RM et al (2020) Development of active food packaging via incorporation of biopolymeric nanocarriers containing essential oils. Trends in Food Sci Technol 101:106–121

Rehman A, Ahmad T, Aadil RM et al (2019) Pectin polymers as wall materials for the nanoencapsulation of bioactive compounds. Trends Food Sci Technol 90:35–46. https://doi.org/10.1016/j.tifs.2019.05.015

Reis DR, Ambrosi A, Luccio Di M (2022) Encapsulated essential oils: a perspective in food preservation. Future Foods 5:100126. https://doi.org/10.1016/j.fufo.2022.100126

Reyes-Jurado F, Navarro-Cruz AR, Ochoa-Velasco CE et al (2019) Essential oils in vapor phase as alternative antimicrobials: a review. Crit Rev Food Sci Nutr 60(10):1641–1650

Rodrigues R, Patil S, Dhakane-Lad J et al (2021) Effect of green tea extract, ginger essential oil and nanofibrillated cellulose reinforcements in starch films on the keeping quality of strawberries. J Food Process Preserv 46:e16109. https://doi.org/10.1111/jfpp.16109

Romão S, Bettencourt A, Ribeiro IA (2022) Novel features of cellulose-based films as sustainable alternatives for food packaging. Polymers 14(22):4968–4995

Rowinski P, Bilewicz R, Stebe MJ, Rogalska E (2004) Electrodes modified with monoolein cubic phases hosting laccases for the catalytic reduction of dioxygen. Analytical Chem 76:283–291. https://doi.org/10.1021/ac034612

Roy S, Rhim J-W (2021) Carrageenan/agar-based functional film integrated with zinc sulfide nanoparticles and Pickering emulsion of tea tree essential oil for active packaging applications. Int J Biol Macromol 193(Part B):2038–2046

Rupert R, Rodrigues KF, Thien VY, Yong WTL (2022) Carrageenan from *Kappaphycus alvarezii* (*Rhodophyta solieriaceae*): metabolism, structure, production, and application. Front Plant Sci 13:859635

Sabo VA, Knezevic P (2019) Antimicrobial activity of *Eucalyptus camaldulensis* Dehn. Plant extracts and essential oils: a review. Ind Crop Prod 132:413–429. https://doi.org/10.1016/j.indcrop.2019.02.051

Safaya M, Rotliwala YC (2020) Nanoemulsions: a review on low energy formulation methods, characterization, applications and optimization technique. Mater Today Proc 27:454–459. https://doi.org/10.1016/j.matpr.2019.11.267

Said NS, Sarbon NM (2022) Physical and mechanical characteristics of Gelatin-based films as a potential food packaging material: a review. Membranes 12(5):442–468

Satyal P, Setzer WN (2020) Chemical compositions of commercial essential oils from Coriandrum sativum fruits and aerial parts. Nat Prod Commun 15(7). https://doi.org/10.1177/1934578X20933067

Sayadi M, Amiri S, Radil M (2022) Active packaging nanocomposite gelatin-based films as a carrier of nano TiO2 and cumin essential oil: the effect on quality parameters of fresh chicken. J Food Meas Charact 16:420–430. https://doi.org/10.1007/s11694-021-01169

Scaffaro R, Maio A, Gulino EF et al (2020) The effects of nanoclay on the mechanical properties, carvacrol release and degradation of a PLA/PBAT blend. Materials (Basel) 13. https://doi.org/10.3390/ma13040983

Senthilkumaran A, Babaei-Ghazvini A, Nickerson MT, Acharya B (2022) Comparison of protein content, availability, and different properties of plant protein sources with their application in packaging. Polymers 14(5):1065

Senthil-Nathan S (2020) A review of resistance mechanisms of synthetic insecticides and botanicals, phytochemicals, and essential oils as alternative larvicidal agents against mosquitoes. Front Physiol 10. https://doi.org/10.3389/fphys.2019.01591

Sepúlveda CT, Alemán A, Zapata J et al (2021) Characterization and storage stability of spray dried soy-rapessed lecithin/trehalose liposomes loaded with a tilapia viscera hydrolysate. Innov Food Sci Emerg Technol 71:102708. https://doi.org/10.1016/j.ifset.2021.102708

Shaikh S, Yaqoob M, Aggarwal P (2021) An overview of biodegradable packaging in food industry. Curr Res Food Sci 4:503–520

Sharma A, Sharma U (1997) Liposomes in drug delivery: progress and limitations. Int J Pharm 154:123–140. https://doi.org/10.1016/S0378-5173(97)00135-X

Sharma S, Mulrey L, Byrne M et al (2022) Encapsulation of essential oils in nanocarriers for active food packaging. Foods 11:1–21. https://doi.org/10.3390/foods11152337

Sharma C, Bhardwaj NK (2019) Bacterial nanocellulose: present status, biomedical applications and future perspectives. Mater Sci Eng C 104:1–18

Shi C, Zhou A, Fang D et al (2022) Oregano essential oil/β-cyclodextrin inclusion compound polylactic acid/polycaprolactone electrospun nanofibers for active food packaging. Chem Eng J445:136746. https://doi.org/10.1016/j.cej.2022.136746

Shukla A, Naik SN, Goud VV, Das C (2019) Supercritical CO_2 extraction and online fractionation of dry ginger for production of high-quality volatile oil and gingerols enriched oleoresin. Ind Crop Prod 130:352–362

Silva F, Caldera F, Trotta F et al (2019) Encapsulation of coriander essential oil in cyclodextrin nanosponges: a new strategy to promote its use in controlled-release active packaging. Innov Food Sci Emerg Technol 56:102177. https://doi.org/10.1016/j.ifset.2019.102177

Silva FAGS, Dourado F, Gama M, Poças F (2020) Nanocellulose bio-based composites for food packaging. Nano 10(10):2041–2070

Simionato I, Domingues FC, Nerín C, Silva F (2019) Encapsulation of cinnamon oil in cyclodextrin nanosponges and their potential use for antimicrobial food packaging. Food Chem Toxicol 132:110647. https://doi.org/10.1016/j.fct.2019.110647

Singh IR, Pulikkal AK (2022) Preparation, stability and biological activity of essential oil-based nano emulsions: a comprehensive review. OpenNano 8:100066. https://doi.org/10.1016/j.onano.2022.100066

Singh R, Lillard JW (2009) Nanoparticle-based targeted drug delivery. Exp Mol Pathol 86:215–223

Souza AG, Ferreira RR, Paula LC et al (2021) Starch-based films enriched with nanocellulose-stabilized Pickering emulsions containing different essential oils for possible applications in food packaging. Food Packag Shelf Life 27:100615

Steed W, Turner DR, Wallace K (2007) Core concepts in supramolecular chemistry and Nanochemistry. John Wiley and Sons

Sun X, Wang J, Zhang H et al (2021) Development of functional gelatin-based composite films incorporating oil-in-water lavender essential oil nano-emulsions: effects on physicochemical properties and cherry tomatoes preservation. LWT-Food Sci Technol 142:110987. https://doi.org/10.1016/j.lwt.2021.110987

Sun Y, Tang W, Pu C et al (2022) Improved stability of liposome-stabilized emulsions as a co-encapsulation delivery system for vitamin B_2, vitamin E and B-carotene. Food and Function 5. https://doi.org/10.1039/D1FO03617C

Surendhiran D, Roy VC, Park J-S, Chun B-S (2022) Fabrication of chitosan-based food packaging film impregnated with turmeric essential oil (TEO)-loaded magnetic-silica nanocomposites for surimi preservation. Int J Biol Macromol 203:650–660. https://doi.org/10.1016/j.ijbiomac.2022.01.178

Szoka F Jr, Papahadjopoulos D (1980) Comparative properties and methods of preparation of lipid vesicles (liposomes). Annu Rev Biophys Bioeng 9:467–508. https://doi.org/10.1146/annurev.bb.09.060180.002343

Tang Y, Zhou Y, Lan X et al (2019) Electrospun gelatin nanofibers encapsulated with peppermint and chamomile essential oils as potential edible packaging. J Agric Food Chem 67:2227–2234. https://doi.org/10.1021/acs.jafc.8b06226

Tarique J, Sapuan SM, Khalina A (2021) Effect of glycerol plasticizer loading on the physical, mechanical, thermal, and barrier properties of arrowroot (*Maranta arundinacea*) starch biopolymers. Scientifc Reports 11(1):13900

Tavares AG, Andrade J, Silva RRA et al (2021) Carvacrol-loaded liposome suspension: optimization, characterization and incorporation into poly(vinyl alcohol) films. Food Funct 12:6549–6557. https://doi.org/10.1039/D1FO00479D

Teixeira RF, Balbinot Filho CA, Borges CD (2022) Essential oils as natural antimicrobials for application in edible coatings for minimally processed apple and melon: a review on antimicrobial activity and characteristics of food models. Food Packag Shelf Life 31:100781

Teixeira-Costa BE, Andrade CT (2022) Natural polymers used in edible food packaging-history, function and application trends as a sustainable alternative to synthetic plastic. Polysaccharides 3(1):32–58

Thokchom SD, Gupta S, Kapoor R (2020) Arbuscular mycorrhiza augments essential oil composition and antioxidant properties of *Ocimum tenuiflorum* L. – a popular green tea additive. Ind crops. Prod 153:112418. https://doi.org/10.1016/j.indcrop.2020.112418

Tu XF, Hu F, Thakur K, Li XL et al (2018) Comparison of antibacterial effects and fumigant toxicity of essential oils extracted from different plants. Ind Crop Prod 124:192–200

U.S. Code of Federal Regulations. (2016). Title 21, title 21efood and drugs Part 182, Section (Accessed 10 December 2022) https://www.accessdata.fda.gov/

Van den Bogaart G, Mika JT, Krasnikov V, Poolman B (2007) The lipid dependence of melittin action investigated by dual-color fluorescence burst analysis. Biophys J 93(1):154–163. https://doi.org/10.1529/biophysj.107.106005

Van Den Hoven JM, Metselaar JM, Storm G et al (2012) Cyclodextrin as membrane protectant in spray-drying and freeze-drying of PEGylated liposomes. Int J Pharm 438(1–2):209–216. https://doi.org/10.1016/j.ijpharm.2012.08.046

Varoma S, Martín Á, Cicero MJ (2011) Liposomal incorporation of Lavadin essential oil by a thin-film hydration method and by particles from gas-saturated solutions. Ind Eng Chem Res 50:2088–2097. https://doi.org/10.1021/ie102016r

Vianna TC, Marinho CO, Júnior LM, Ibrahim SA, Vieira RP (2021) Essential oils as additives in active starch-based food packaging films: a review. Int J Biol Macromol 182:1803–1819

Viuda-Martos M, Mohamady MA, Fernández-López J et al (2011) *In vitro* antioxidant and antibacterial activities of essentials oils obtained from Egyptian aromatic plants. Food Control 22:1715–1722. https://doi.org/10.1016/j.foodcont.2011.04.003

Walia S, Kumar R (2021) Wild marigold (*Tagetes minuta* L.) biomass and essential oil composition modulated by weed management techniques. Ind crops. Prod 161:113183. https://doi.org/10.1016/j.indcrop.20

Wang J, Chen C, Xie J (2022) Loading oregano essential oil into microporous starch to develop starch/polyvinyl alcohol slow-release film towards sustainable active packaging for sea bass (*Lateolabrax japonicus*). Ind crops. Prod 188, Part B:115679

Weiss VM, Naolou T, Hause G et al (2012) (2012) poly(glycerol adipate)-fatty acid esters as versatile nanocarriers: from nanocubes over ellipsoids to nanospheres. J Control Release 158(1):156–164. https://doi.org/10.1016/j.jconrel.2011.09.077

Wüpper S, Lüersen K, Rimbach G (2021) Cyclodextrins, natural compounds, and plant bioactives-a nutritional perspective. Biomol Ther 11(3):401–422

Zanetti M, Carniel TK, Dalcanton F et al (2018) Use of encapsulated natural compounds as antimicrobial additives in food packaging: a brief review. Trends Food Sci Technol 81:51–60. https://doi.org/10.1016/j.tifs.2018.09.003

Zhang T, Wang H, Qi D et al (2022a) Multifunctional colorimetric cellulose acetate membrane incorporated with *Perilla frutescens* (L.) Britt. Anthocyanins and chamomile essential oil. Carbohydr Polym 278:118914. https://doi.org/10.1016/j.carbpol.2021.118914

Zhang MK, Zhang XH, Han GZ (2022b) Magnetic alginate/PVA hydrogel microspheres with selective adsorption performance for aromatic compounds. Sep Purif Technol 278:119547

Zheng H, Tang H, Yang C, Chen J, Wang L, Dong Q, Shi W, Li L, Liu Y (2022) Evaluation of the slow-release polylactic acid/polyhydroxyalkanoates active film containing oregano essential oil on the quality and flavor of chilled pufferfish (*Takifugu obscurus*) fillets. Food Chem 385:132693

Zhou Y, Miao X, Lan X, Luo J, Luo T, Zhong Z, Gao X, Mafang Z, Ji J, Wang H, Tang Y (2020) Angelica essential oil loaded electrospun Gelatin nanofibers for active food packaging application. Polymers 12:299–310

Zhu Z, Min T, Zhang X, Wen Y (2019) Microencapsulation of thymol in poly(lactide-co-glycolide) (PLGA): physical and antibacterial properties. Materials 12(7):1133

Zielińska A, Carreiró F, Oliveira AM, Neves A, Pires B, Venkatesh DN, Durazzo A, Lucarini M, Eder P, Silva AM, Santini A, Souto EB (2020) Polymeric nanoparticles: production, characterization, toxicology and ecotoxicology. Molecules 25(16):3731